Digital Signal Processing and the Microcontroller

Dale Grover and John R. Deller

with illustrations by Jonathan Roth

Prentice Hall PTR
Upper Saddle River, NJ 07458
http://www.phptr.com

ISBN 0-13-081348-6

Library of Congress Cataloging-in-Publication Data
Grover, Dale.
    Digital signal processing and the microcontroller / by Dale Grover
and Jack Deller
        p.   cm.
    Includes bibliographical references and index.
    ISBN: 0-13-081348-6
    1. Digital control systems. 2. Signal processing—Digital techniques. I. Deller, John R.
    II. Title.
    TJ223.M53G76 1998
    621.382'2—dc21   98-19115
                                                                                                                           CIP

Editorial/production supervision: *Nicholas Radhuber*
Manufacturing manager: *Alan Fischer*
Acquisitions editor: *Bernard Goodwin*
Editorial assistant: *Diane Spina*
Marketing manager: *Miles Williams*
Cover design: *Scott Weiss*
Cover illustration: *Jonathan Roth*
Cover design director: *Jerry Votta*

© 1999 by Motorola University Press
Published by Prentice Hall PTR
Prentice-Hall, Inc.
A Simon & Schuster Company
Upper Saddle River, New Jersey 07458

Prentice Hall books are widely used by corporations and government agencies for training, marketing, and resale. The publisher offers discounts on this book when ordered in bulk quantities. For more information, contact:
    Phone: 800-382-3419, Fax: 201-236-7141
    E-mail: corpsales@prenhall.com
    or write:
    Corporate Sales Department
    Prentice Hall PTR
    1 Lake Street
    Upper Saddle River, NJ 07458

All product names mentioned herein are the trademarks of their respective owners.
All rights reserved. No part of this book may be reproduced, in any form or by any means, without permission in writing from the publisher.

The authors and publisher of this book and CD-ROM have used their best efforts in preparing this book and CD-ROM. These efforts include research, implementation, and testing of the algorithms, programs, and design information to determine their effectiveness. The authors and publisher make no warranty of any kind, expressed or implied, with regard to the programs, algorithms, data, or design information contained in this book and CD-ROM. The authors and publisher shall not be liable in any event for incidental or consequential damages in connection with, or arising out of, the furnishing, performance, or use of these programs, algorithms, data, or design information.

Printed in the United States of America
10 9 8 7 6 5 4 3 2 1

ISBN 0-13-081348-6

Prentice-Hall International (UK) Limited, *London*
Prentice-Hall of Australia Pty. Limited, *Sydney*
Prentice-Hall Canada Inc., *Toronto*
Prentice-Hall Hispanoamericana, S.A., *Mexico*
Prentice-Hall of India Private Limited, *New Delhi*
Prentice-Hall of Japan, Inc., *Tokyo*
Simon & Schuster Asia Pte. Ltd., *Singapore*
Editora Prentice-Hall do Brasil, Ltda., *Rio de Janeiro*

To my father, Lawrence Grover, who patiently let me experiment with, and frequently destroy, his stock of LEDs, ICs, and op-amps.

—*D.G.*

To Antonia Jiang Deller. A loving family and a future filled with unimaginable possibilities are yours.

—*J.D.*

Table of Contents

Preface xiii
The Purpose of This Book xiii
Topics We'll Cover xiv
Who You Are xiv
Reading This Book xvi
Availability of Software xvii

Acknowledgments xix

List of Figures xxi

Chapter 1: The Big Picture 1
1.1 Overview 1
1.2 What is DSP? What Can It Do? 1
    1.2.1 A Definition of Digital Signal Processing 2
    1.2.2 What Can DSP Do With Signals? 5
    1.2.3 Applications of DSP 7
1.3 DSP Versus Analog Electronics—Why Bother? 7
    1.3.1 Programmable 9
    1.3.2 Fixed Performance—No *Component Drift* 9
    1.3.3 Economic 10
    1.3.4 Functions That Cannot Be Done With Analog Electronics 10
    1.3.5 High-Volume, Widely Available Parts 10
    1.3.6 DSP Tasks Can Share Hardware With Other Tasks 10
    1.3.7 Design Tools Are Available 10
1.4 DSP and Microcontrollers 11

1.5 Limitations of DSP 12
1.6 Why DSP Might Look Difficult 15
1.7 Summary 16
1.8 Resources 16

Chapter 2: Analog Signals and Systems 19

2.1 Overview 19
2.2 Sources of Analog Signals 20
2.3 Describing Signals and Systems in Time and Frequency 21
 2.3.1 Analog Signals 21
 2.3.2 Describing Analog Signals in the Time Domain 21
 2.3.3 Describing Signals in the Frequency Domain 24
 2.3.4 Describing Systems 46
2.4 Nonelephant Biology and Linear, Time-Invariant Systems 57
 2.4.1 The Amazing Properties of Linear Systems 58
 2.4.2 Real Systems and Nonlinearity 60
2.5 Summary 61
2.6 Resources 62

Chapter 3: Analog Filters 63

3.1 Overview 63
3.2 Purpose of Filters 64
3.3 Examples 64
3.4 Ideal Versus Real Filters 67
 3.4.1 The "Brick-Wall" Magnitude Response 67
 3.4.2 Distortion 67
3.5 Filter Specification 68
 3.5.1 Types 68
 3.5.2 Specifying Magnitude Response 74
 3.5.3 Specifying Phase Response 81
 3.5.4 Transient Response 82
3.6 Analog Filter Implementation 82
 3.6.1 Passive 83
 3.6.2 Active 85
3.7 Poles and Zeros 85
3.8 Filter Order 91
3.9 Summary 91
3.10 Resources 92

Chapter 4: Discrete-Time Signals and Systems 93

4.1 Overview 93
4.2 Sources of Discrete-Time and Digital Signals 94
 4.2.1 Analog, Discrete-Time, and Digital Signals 94
 4.2.2 Sampling 95
4.3 Describing Discrete-Time Signals and Systems 106

	4.3.1	Discrete-Time Signals in the Time Domain	106
	4.3.2	Discrete-Time Signals in the Frequency Domain	106
	4.3.3	Describing Discrete-Time Systems	108
4.4	Quantizing—Continuous to Discrete Amplitude Values		118
	4.4.1	Linear Quantization	119
	4.4.2	Nonlinear Quantization	120
	4.4.3	Quantization and Noise	122
4.5	Analog-to-Digital and Digital-to-Analog Conversion		124
	4.5.1	Analog-to-Digital Conversion	124
	4.5.2	Digital-to-Analog Conversion	141
4.6	Digital Filters, an Overview		147
	4.6.1	Digital Filter Design Process	148
	4.6.2	Selecting a Filter Type: FIR Versus IIR	150
	4.6.3	Diagram, Notations	153
4.7	Summary		153
4.8	Resources		155

Chapter 5: FIR Filters—Digital Filters Without Feedback ... 157

5.1	FIR Overview		157
5.2	Intuitive Convolution—How FIR Filters Work		161
5.3	Design Process Overview		164
	5.3.1	Filter Design Tools and Methods	165
5.4	Generating Coefficients		166
	5.4.1	Parks-McClellan or Optimal Equiripple	167
	5.4.2	Windowing	172
	5.4.3	Frequency Sampling	183
	5.4.4	*Ad-Hoc*	195
	5.4.5	Fast Convolution	197
5.5	Lowpass-to-Highpass Conversion		199
5.6	Structures for FIR Filters		199
	5.6.1	Direct and Linear-Phase Structures	199
	5.6.2	Cascade Structures	201
5.7	Summary		203
5.8	Resources		204

Chapter 6: IIR Filters—Digital Filters with Feedback ... 207

6.1	Overview		207
6.2	Design Process Overview		209
	6.2.1	Implementation	210
6.3	Direct Design Methods		210
	6.3.1	*Ad Hoc*— "Manual" Placement of Poles and Zeros	210
	6.3.2	Time-Domain Methods	211
	6.3.3	Frequency-Domain Methods	211
6.4	Indirect Design Methods		212

  6.4.1 Analog Filter Prototypes 213
  6.4.2 Mapping from s to z 240
6.5 Highpass, Bandpass, and Bandstop Conversions 250
6.6 IIR Filter Structures 252
  6.6.1 The Cascade Structure 253
  6.6.2 The Parallel Structure 257
6.7 Summary 258
6.8 Resources 261

Chapter 7: Microcontroller Implementation of Filters 263

7.1 Overview 263
7.2 Architecture Issues 264
  7.2.1 Single-Chip Programmable Digital Processors 266
  7.2.2 Operations 269
7.3 Programming Issues 277
  7.3.1 Languages 277
  7.3.2 Data Representations 287
  7.3.3 Optimizing for Speed 291
  7.3.4 Optimizing for Size 297
  7.3.5 Floobydust on Programming 298
7.4 Finite Word-Length Effects 301
  7.4.1 Coefficient Quantization 301
  7.4.2 Limit Cycles 303
7.5 FIR Filter Implementation 303
7.6 IIR Filter Implementation 315
  7.6.1 An Example IIR Implementation 321
7.7 Summary 329
7.8 Resources 330

Chapter 8: Frequency Analysis 333

8.1 Overview 333
8.2 What Do You Want? 334
  8.2.1 A Philosophical Question 334
  8.2.2 Using Individual Filters 335
  8.2.3 Looking for a Small Set of Frequencies 336
8.3 The Discrete Fourier Transform and Fast Fourier Transform 337
  8.3.1 A Complete Frequency-Domain Description 337
8.4 Using the DFT 337
  8.4.1 Input Data 337
  8.4.2 Why Use Windows? 340
  8.4.3 Interpretation of Output 349
  8.4.4 Increasing Frequency Resolution 352
8.5 Implementing the DFT 353
  8.5.1 Overview 353

    8.5.2 From the DFT 353
    8.5.3 ... to the FFT 356
    8.5.4 The Goertzel Algorithm.. 362
8.6 Implementation of the FFT on the 68HC16.. 362
    8.6.1 Design Issues 373
    8.6.2 Support Routines 375
    8.6.3 Additional Comments on the 68HC16 Code 377
8.7 The Inverse DFT/FFT 377
8.8 Time-Frequency Analysis... 377
    8.8.1 Multiple FFTs 378
    8.8.2 Wavelets 378
8.9 Miscellaneous Topics.. 378
    8.9.1 Fast Convolution 379
    8.9.2 Other Transforms... 379
8.10 Summary... 381
8.11 Resources 382

Chapter 9: Correlation 383

9.1 Overview... 384
9.2 Crosscorrelation 384
    9.2.1 Definition... 384
    9.2.2 Correlation and Convolution 388
    9.2.3 Applications... 388
    9.2.4 Implementations 396
9.3 Autocorrelation 396
    9.3.1 Definition... 396
    9.3.2 Applications... 397
    9.3.3 Implementation 401
9.4 Pseudo-Noise (PN) Signals... 401
    9.4.1 Characteristics of PN Sequences 403
    9.4.2 Software Implementation of Feedback Shift Register Generators 404
9.5 Signal Averaging 409
    9.5.1 Stimulus/Response... 409
9.6 Summary... 410
    9.6.1 Resources... 412

Chapter 10: Changing Sampling Rates.. 413

10.1 Overview... 413
10.2 Applications 414
    10.2.1 Matching Fixed Sampling Rates.. 414
    10.2.2 Reducing Input/Output Hardware Filters 414
    10.2.3 Reducing Computation 415
10.3 Decimation.. 415
    10.3.1 Efficient Decimation.. 415

10.3.2 Buying More Time With Buffers and Decimating in Stages 417
10.4 Interpolation 418
    10.4.1 Interpolate/Filter Structure... 418
    10.4.2 Efficient Interpolation... 418
    10.4.3 Interpolation Example 420
    10.4.4 Interpolation Results... 424
10.5 Rational Interpolation/Decimation.. 425
    10.5.1 Original Structure 425
    10.5.2 Combining *D* and *I* Filters 425
10.6 Summary 427
10.7 Resources 428

Chapter 11: Synthesizing Signals 429

11.1 Overview.. 429
11.2 Random Numbers 430
    11.2.1 Why Do We Want Random Numbers? 430
    11.2.2 What Do We Want?... 430
    11.2.3 Generating Pseudo-Random Numbers 431
    11.2.4 Nonuniform Amplitudes and Frequencies....................................... 437
11.3 Functions.. 439
    11.3.1 Polynomials 439
    11.3.2 Sine, Cosine, and Some Others... 439
    11.3.3 Arbitrary Waveforms... 448
11.4 Summary 448
11.5 Resources 449

Chapter 12: Parting Words... 451

12.1 Signals and Linear Systems 451
12.2 Math 453
    12.2.1 Transforms... 453
    12.2.2 Statistics and Stochastic Processes 454
12.3 Processing Multidimensional Signals 454
12.4 Processing Music 455
12.5 Processing Speech.. 455
12.6 Numerical Methods.. 457
12.7 Embedded Systems 457
12.8 Summary 458

Appendix 1: Useful Mathematics 461

Appendix 2: Useful Electronics.. 469

Appendix 3: 68HC16 Sample FIR Program..................................... 477

Appendix 4: 68HC16 Sample IIR Program...................................... 491

Appendix 5: 68HC16 Sample Interpolation Program..................... 505

Glossary 517

Index 527

Preface

The Purpose of This Book

This is a practical book about digital signal processing (DSP) and using microcontrollers for DSP.

DSP is powerful and fascinating—at least that's what we, Dale and Jack, think. One example that stands out for Dale is a DSP-based voice synthesizer for people with disabilities that he worked on a few years ago. The first time it spoke—a tentative "heh-low"—was just short of miraculous. And not because this was unexpected or had never been done before. But because here was a voice that was a manifestation of pure mathematics, a flow of numbers from random number generators and digital filters. Something deep and very interesting was happening in that tiny DSP chip.

Although Dale has a somewhat practical bias, Jack, on the other hand, is the "egghead" of this writing team. His wife, Joan, is fond of telling their friends that, in spite of being an electrical engineering professor, he is utterly useless when it comes to doing anything practical with electricity or electronics around the house. Although he did recently manage to fix their microwave oven, there is some truth to Joan's contention for two reasons. First, as we will discover, the field of signal processing (SP), although usually considered a branch of electrical engineering, is really one involved with discrete mathematics (mostly adds and multiplies), and some SP engineers can spend an entire (satisfying and productive) career having precious little contact with resistors, capacitors, inductors, transformers, transistors, and all of those classical EE things.[1] Secondly, Jack does SP research in fields that tend to be more algorithm- and theory-oriented—speech recognition, biomedical signal processing, system identification (whatever

---

1. Dale says this sounds distinctly unsatisfying!

that means) and adaptive filtering. Once Jack finishes his work on a problem, it takes someone with other SP skills[2] to come in and consider the hardware and implementation details — types of chip sets, parallel processing solutions, interfacing, sampling, and, yes, all the electronics involved in the practical solutions.

In this book, we hope to reach a compromise between theory and practice. Our goal is that you will be able to use digital filtering and other DSP techniques in your future microcontroller applications. However, good engineering requires a firm grasp of the underlying theory, especially in the field of DSP. Without this knowledge, you're skating on thin ice! We're going to take a decidedly less mathematical, more intuitive approach to the theory of DSP, one that we hope won't elicit too many yawns or sudden "nap attacks." On the way, we'll try to give you the *big picture* of how DSP is profoundly changing the face of electronics and, in fact, the world. This is probably not the only DSP book you'll ever need, but we hope you will find it useful as a gentle introduction to DSP.

We take a practical approach that avoids a lot of mathematics, but this still leaves us with a *language* problem. DSP is a branch of electrical engineering; DSP texts assume you know the language and concepts of electrical engineering. (If you've looked through other DSP texts, you know what we mean!) This is also the language that SP specifications are written in. So sometimes we'll need to take a side trip to explain some concept that may not seem crucial to DSP itself.

If you don't have a background in electrical engineering, this book is also for you. You might want to concentrate on the overall picture initially, then come back to pick up the details.

Topics We'll Cover

Our focus in this book is on digital filtering and the use of the "fast Fourier transform" (FFT), an algorithm that lets us look at the frequency content of signals. By the end of this book you should feel comfortable specifying, designing, and implementing digital filters on microcontrollers (and other microprocessors). You'll also know how to use the FFT, both as a stand-alone tool and as a shortcut for digital filtering. These are common and useful types of DSP; they are also the basis for more complicated processing.

The microcontroller we'll use as an example in this book is the 16-bit Motorola M68HC16, but the techniques can be applied to even modest 8-bit microcontrollers or, of course, full-blown digital signal processors. As in most of engineering practice, good use of DSP will involve making trade-offs, especially between cost and performance.

Who You Are

The audience for this book includes:
- Engineers who may not have a background in signal processing

2. "Wizards," Jack calls them.

- Technicians
- Students in electrical engineering and computer science
- Programmers, especially of embedded systems

In general, this book is for anyone interested in applying DSP techniques to microcontroller-based products.[3]

We expect that you have at least some familiarity with programming microprocessors at the assembly language level. As far as mathematics goes, most of the math that really counts (no pun intended) will be algebra. One thing we can't avoid is complex numbers[4]; if you aren't familiar with the subject, most of what you need to know is in Appendix 1 (Useful Mathematics).

You'll find an annotated bibliography of useful resources at the end of each chapter. This book can't tell you everything, and you may need to pursue some subjects in greater detail. A more rigorously mathematical approach to DSP reveals many connections and relationships among the topics covered here. Often, we'll just have to wave our hands and hope you'll take our word about these issues, but most "hard-core" DSP textbooks will take you through the proofs and give you a solid basis for a much deeper understanding. We've also listed resources like magazines, journals, and conference proceedings, as new techniques that improve performance or perform new functions are emerging every day.

A First Look at the Motorola M68HC16 Microcontroller

The Motorola M68HC16 is a 16-bit microcontroller with special registers and instructions to support efficient DSP. Upwardly code-compatible with the 8-bit M68HC11 family, the chip is designed around modules that share an internal, high-speed bus. The variation we'll use in this book, the M68HC16Z1, has the following modules:

- 1K byte high-speed RAM,
- 10-bit analog-to-digital converter (>100,000 samples/sec),
- high-speed serial connect to peripherals,
- timer/counter module.

The M68HC16Z1, running at 16.78 MHz, can execute up to 1.4 million 16-bit by 16-bit multiply/accumulate operations per second—an operation central to many DSP algorithms. (Recent releases in the M68HC16 family run even faster.)

---

3. However, Jack says this book is definitely *not* for students trying to learn the proper rules of grammar and sentence construction!

4. Actually, we could avoid complex numbers, but the cure would be worse than the disease! If the name bothers you, don't worry.

Reading This Book

We've tried to write this book in a rather informal tone, as though you had asked a colleague to bring you up to speed on DSP. We've always found that just reading about a topic is rarely enough, and we're sure you'll get much more out of this book if you take the time to work out examples for yourself as you're reading along.

At the start of each chapter, you'll find a brief note pointing out previous chapters and sections that would be helpful, any mathematics you might need, and so on. Don't let that stop you from reading a chapter! These notes are to let you know where you can go for background.

The figure below shows the organization of the book and how chapters relate to each other.

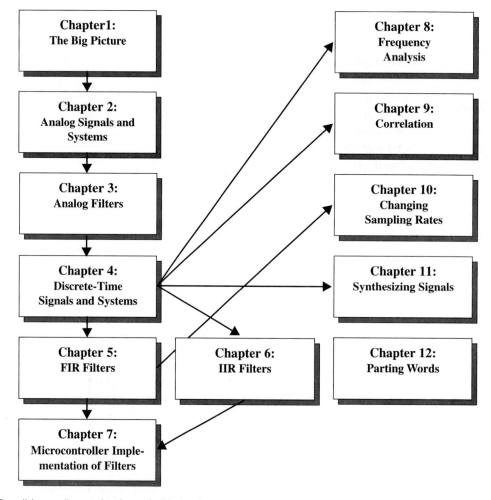

Possible reading paths through this book.

Availability of Software

We might as well tell you DSP's dirty little secret right now. A lot of DSP falls into two nice categories. First, there are algorithms that you generally never need to modify in order to use in your program, somewhat like a sorting routine. These are usually optimized for the particular processor you're using, and often you can just plug them in and go. The FFT is an example of such an algorithm.

The second category of algorithms is those that are very straightforward and even trivial—they may not take more than one to three lines of assembly language! The core of most digital filters is a prime example. The complexity here is in picking exactly the right coefficients.[5] You can usually calculate these coefficients by hand, but most people use one of the many available programs to generate "optimized" coefficients automatically.

There are a number of digital filter design packages available both commercially and in the public domain. A selection of microcontroller and DSP software is available from Motorola at *http://www.motorola.com*. It's not crucial you have this software as you read the book, but it will be useful in the chapters on digital filters. It might also be helpful to have manuals on the CPU16 processor core (the CPU of the M68HC16 family) and/or the specific microcontroller version you're using (e.g., M68HC16Z1). Consult the list of resources in Chapter 7 (Microcontroller Implementation of Filters) for specific titles. Documentation can be ordered online from Motorola or obtained through a Motorola sales office.

The CD-ROM included with this book contains example source code and manuals (in Acrobat® format) for the M68HC16. (See the CD-ROM for a detailed description of contents.)

---

5. Coefficients are constants involved in a calculation.

Acknowledgments

The concept for this book originated in the Marketing Department of the now-defunct Motorola Advanced Microcontroller Division. It began as a part of the M68HC16 product introduction. The M68HC16Z1 was the first MCU to incorporate DSP support, and many control system designers were unfamiliar with DSP concepts. After surveying the available texts, the marketing team felt that there was a real need for an informal introduction to DSP concepts and decided to put one together. This was a monumental undertaking, on an aggressive schedule, by a group of very busy people.

The team, led by Arie Brish and Judy Racino, mapped out a comprehensive, practical treatment of control-oriented DSP. Mark McQuilken and James LeBlanc were the primary technical contributors. Robert Chretien, Mark Glenewinkel, Amy Goldman, Charles Melear, Tanya Tussing, and James Waite also made significant contributions. The project was edited by Wendy Heath.

Large quantities of midnight oil were burned, but there simply wasn't time to complete the project as planned. To meet the deadline, key sections on control-oriented DSP were edited into a smaller booklet, the first incarnation of *Digital Signal Processing and the Microcontroller*, which was included in the box with M68HC16Z1 evaluation systems.

One artifact of the rush to make the HC16 product announcement must be addressed: Very much to his chagrin, the booklet mistakenly identified then-graduate student James LeBlanc as a Cornell University faculty member. The mistake proved to be an accurate prediction, however—he has subsequently become an Assistant Professor in the Klipsch School of Electrical and Computer Engineering at New Mexico State University.

The booklet was well received, but Motorola remained committed to completing the project as planned. Rob Roberson, documentation manager of the AMCU Division, took on the task. Talent was scarce, funds were short, corporate divisions came and went, but Rob persisted (Rob wrote this part).

Meanwhile, an M68HC16Z1 evaluation kit had come into Dale's hands. The DSP booklet in the box was just the kind of basic introduction to DSP concepts Dale wanted to provide his customers, so he called Motorola to find out about distribution rights. Dale (eventually) connected with Rob, who validated his marketing credentials by persuading Dale to consider completing the book.[1]

Dale, in turn, proved that, although perhaps a bit "reality-impaired," he at least had enough sense to recruit Jack to the cause. Now, as to Jack's sense in agreeing...

Dale and Jack found Jonathan Roth, an excellent cartoonist whose illustrations grace the pages to follow. In addition to illustrating books and doodling on placemats, he does two regular cartoon strips, *Living in Sincerity* and *Luke Warm*, which appear in various papers, magazines, and birthday cards to his grandmother.

Throughout writing this book, Rob was always there—unsticking paperwork, encouraging us, and keeping the project going. (We're writing this part.) We hope he's happy with the results.

Special thanks are due the following people whose thoughtful review at various stages helped us immeasurably: Mike Catherwood, Ra'id Khalil, Lawrence Grover, Fred Harrington, Charlie Melear, Bernie Fehr, and John Smallwood. Thanks are also due Martin King, Debbi Schaubman, David Leighton, Jim Senneker, and the folks at Tegic Communications.

One of the most pleasurable parts of writing this book was talking to some of the designers of the MC68HC16: Jim Nash, project leader; Mike Catherwood, the architect of the DSP section; and Kirk Livingston, micro-architect. Any profound insights into the 68HC16 architecture that happen to crop up in this book represent their influence. (Mike Catherwood's document on the internal architecture of the 68HC16 was both very helpful and too much fun.) John Bodner, technical marketer, Charlie Melear, applications engineer, and Derrick Klotz, field applications engineer, also kindly spent long phone calls discussing both the 68HC16 architecture and its programming.

Nick Radhuber and Bernard Goodwin at Prentice Hall and Susan Zilkowski at Motorola University Press turned what appeared to us as a simple job into a time-consuming, complex set of interconnected tasks, and made it look simple to us.

The opportunities for errors and omissions in a book of this kind are many, and we've no doubt made more than a few. On the other hand, we can't claim credit for the insights and powerful tools we hope to show you in this book; these are the products of extraordinary minds—mathematicians, scientists, and engineers—and even so, in many cases our simplified treatment will not do proper justice to the elegance or power of their work.

---

1. Though we have tried to keep the intuitive, practical approach of the original, we decided to write this from scratch. So you can blame McQuilken, et al., for the idea that a book like this could be written, but you'll have to blame Dale and Jack for writing one.

List of Figures

	Possible reading paths through this book.	xvi
1-1	Two examples of transducers	3
1-2	*Sampling* and *quantizing* turn an analog signal into a digital signal	4
2-1	North Inlet Estuary Meteorological Station, 1991	22
2-2	Human voice, "oh".	22
2-3	Waveform for DTMF key "8"	23
2-4	Frequency-domain representation of DTMF key "8"	24
2-5	Air temperature, North Inlet Estuary Meteorological Station, 1991	26
2-6	Finding the magnitudes of frequencies with a narrow filter	27
2-7	An example signal to be decomposed	28
2-8	Frequency components of the signal in Figure 2-7	29
2-9	(a) and (b) Decomposing a squarewave into its first four components. (c) Building a squarewave from the first four and (d) fifty components.	30
2-10	Frequency components of 1 Hz square wave.	31
2-11	A signal formed by adding two sinusoids.	31
2-12	Changing the phase of the 10-Hz sinusoid component from Figure 2-11.	32
2-13	Sum of first 50 frequency components of a squarewave, with random phase	32
2-14	Representing magnitude and phase with complex numbers.	33
2-15	Phase shift as a displacement in time.	33
2-16	Radians and degrees around the circle.	34
2-17	A signal with DC bias.	37
2-18	Bandpass and baseband signals.	38
2-19	Comparing linear and logarithmic frequency scales.	39

2-20 A magnitude spectrum plotted with linear magnitude. 40
2-21 A magnitude spectrum plotted with logarithmic magnitude. 41
2-22 Plotting phase with and without wrapping. 46
2-23 The impulse function. 47
2-24 Impulse response of fourth-order elliptical filter. 48
2-25 Labeling a system in the time domain. 49
2-26 The step function. 50
2-27 (a) Input and (b) output of the example system at 60 Hz. 51
2-28 (a) Input and (b) output of the example system at 190 Hz. 52
2-29 The (a) magnitude and (b) phase response of the system. 53
2-30 Example waveforms with complex frequency. 54
2-31 Signals with complex frequency, placed on the s-plane. 57
2-32 The scaling property. 59
2-33 The superposition property. 59
2-34 A nonlinear transducer. 61
3-1   Adding 60-Hz interference to a signal. 66
3-2   Filtering to remove 60-Hz noise. 66
3-3   Ideal vs. real lowpass filter magnitude response. 67
3-4   Example lowpass filter magnitude response. 69
3-5   Example highpass filter magnitude response. 70
3-6   Example bandpass filter magnitude response. 70
3-7   Example narrow bandpass filter magnitude response. 71
3-8   Example bandstop filter magnitude response. 71
3-9   Example arbitrary passband filter magnitude response. 72
3-10 Example all-pass filter magnitude response 73
3-11 Example all-pass filter phase response. 73
3-12 Example comb filter magnitude response. 74
3-13 A partial magnitude spectrum of a squarewave. 75
3-14 Lowpass filter parameters— "analog filter" style. 75
3-15 Lowpass filter parameters— "digital filter" style. 76
3-16 Highpass filter parameters— "digital filter" style. 76
3-17 Narrow bandpass filter parameters. 79
3-18 Magnitude and phase spectra for a linear-phase filter. 81
3-19 Physical and schematic representation of an RC filter. 83
3-20 Example RC filter magnitude response. 84
3-21 A voltage divider. 84
3-22 A pole/zero plot in the s-plane. 86
3-23 $|H(s)|$ for a one-pole, one-zero system. 87
3-24 $|H(s)|$ for the example filter. 88
3-25 $|H(s)|$ evaluated along s=0. 89
3-26 Example poles and zeros in the s-plane. 90
4-1   The relationship between analog, digital, and discrete-time
       signals. 94
4-2   Sampling an analog signal to produce a discrete-time signal. 96

List of Figures

4-3   Sampling a 200 Hz signal at 1 kHz. 97
4-4   Sampling a 490 Hz signal at 1 kHz. 97
4-5   Sampling a 600 Hz signal at 1 kHz produces aliasing. 98
4-6   "Real-world" frequency and aliased frequency. 98
4-7   Regions of reversed frequency spectrum in a sampled signal. 100
4-8   Magnitude spectrum for blood flow signal example. 101
4-9   Magnitude spectrum of sampled blood flow signal. 101
4-10 Purposeful undersampling to recover the radio station signal. 103
4-11 Magnitude spectrum of undersampled signal. 104
4-12 The magnitude spectrum of a DT signal converted to a
       CT signal. 107
4-13 Magnitude response of a simple DT filter. 109
4-14 Locating frequency and exponential growth/decay in the z-plane. 110
4-15 The z-plane. 112
4-16 $|H(z)|$ for a simple system. 113
4-17 Magnitude response for the system of Figure 4-16. 116
4-18 The transfer characteristic of an ideal linear converter. 119
4-19 μ-Law compression. 121
4-20 The analog-to-digital conversion process. 126
4-21 Example analog signal, noise spectrum. 129
4-22 Spectrum of DT signal and noise showing aliasing. 130
4-23 Two-bit flash converter and encoder. 134
4-24 Transfer characteristic for 3-bit signed converter/encoder. 136
4-25 Interfacing an ADC using a (synchronous) serial port. 137
4-26 Resolution vs. speed for ADCs. 139
4-27 Operation of a successive approximation ADC. 140
4-28 The digital-to-analog process. 141
4-29 (a) Magnitude spectrum of DT signal. (b) Magnitude response
       of zero-order hold (ZOH). (c) Magnitude spectrum of output of
       ZOH. 144
4-30 Expanded view of Figure 4-29c. 145
4-31 (a) Magnitude spectrum of DT signal. (b) Magnitude response of
       zero-order hold (ZOH). (c) Magnitude spectrum of output of
       ZOH. 146
4-32 Expanded view of Figure 4-31c. 147
4-33 Direct form structure of an IIR filter. 148
4-34 Direct form structure of ann FIR filter. 149
4-35 Common notation for IIR filter structure. 153
5-1   A very simple FIR filter. 159
5-2   A rearrangement of the simple filter of Figure 5-1. 159
5-3   Magnitude response of the simple filter of Figure 5-2. 160
5-4   Magnitude response of a filter using "better" coefficients. 161
5-5   Relationship between the impulse response and coefficients of an
       FIR filter. 162

5-6 Convolution as the sum of impulse responses. 163
5-7 Magnitude response of filter designed using Parks-McClellan. 171
5-8 Ideal impulse response of a lowpass filter. 174
5-9 Truncated ideal lowpass filter impulse response. 174
5-10 Magnitude response using coefficients of Figure 5-9. 175
5-11 Magnitude response using twice as many coefficients as in
      Figure 5-10. 175
5-12 Smoothing the truncated impulse response. 176
5-13 Magnitude response using windowed coefficients. 177
5-14 Shifting (windowed) coefficients to produce a causal filter. 178
5-15 The time domain and magnitude response of some common
      windows. 180
5-16 A plot of $h(n)$ (noncausal!) for a filter designed using
      windowing. 183
5-17 Magnitude response for an example filter designed using
      windowing. 184
5-18 Zeros of a simple comb filter ($M=8$). 187
5-19 Magnitude response of a comb filter ($M=8$). 188
5-20 Magnitude response of a simple coefficient frequency sampling
      lowpass filter. 189
5-21 Structure of a simple coefficient frequency sampling filter. 190
5-22 Canceling a zero of a comb filter with a pole. 191
5-23 Magnitude response of example simple coefficient
      frequency-sampling filter. 194
5-24 Placing zeros in the z-plane by hand. 196
5-25 Zero placement of simple bandstop FIR filter. 196
5-26 Magnitude response of simple bandstop FIR filter. 197
5-27 Magnitude response of lowpass and related highpass filters. 200
5-28 Direct form of FIR filter. 201
5-29 Linear phase structure for FIR filter (M odd). 202
5-30 Linear phase structure for FIR filter (M even). 202
5-31 A cascade of second-order FIR filters and detail of one section. 203
6-1 Direct I form of IIR filter. 208
6-2 Regions of stable poles in the s-plane and the z-plane. 213
6-3 The "big picture" of IIR design methods. 214
6-4 Magnitude response of the Butterworth analog filter. 216
6-5 Poles of a sixth-order Butterworth. 218
6-6 Butterworth filter parameters. 218
6-7 Magnitude response of Butterworth filter example. 222
6-8 Poles of example Butterworth filter. 223
6-9 Magnitude response of Chebychev I filter, $N=2, 4, 6$. 224
6-10 Pole locations for sixth-order Chebychev I filter. 226
6-11 Magnitude response of Chebychev I example filter. 228
6-12 Pole locations for example Chebychev I filter. 229

6-13 Magnitude response for Chebychev II, $N=2, 4, 6$. 230
6-14 Pole and zero locations for example sixth-order Chebychev II filter. 232
6-15 Magnitude response of Chebychev II example filter. 234
6-16 Pole and zero locations for example sixth-order Chebychev II filter. 234
6-17 Magnitude response for elliptic filter, $N=2, 4, 6$. 236
6-18 Magnitude response for example elliptic filter. 239
6-19 Pole and zero locations for example elliptic filter. 239
6-20 Backward-difference mapping between s- and z-planes. 241
6-21 Bilinear z-transform mapping from s- to z-planes. 242
6-22 Relationship between frequencies in the s- and z-planes with the BZT. 242
6-23 Magnitude response of example digital Butterworth filter. 249
6-24 Cascade or series structure for IIR filter. 253
6-25 Direct form I structure for IIR filter. 253
6-26 An IIR structure based on Equation 6-58. 254
6-27 Canonic or direct form II structure for IIR filter. 254
6-28 Pole-zero plot for digital Butterworth filter example. 258
6-29 Parallel structure for IIR filters. 259
7-1 Bit position values for 16-bit unsigned, signed, and 1.15 fractional formats. 290
7-2 Multiplying two numbers in 1.15 format. 290
7-3 A generic "tool chain," from source to executable object code. 300
7-4 CPU16 (e.g., 68HC16) register model. 305
7-5 FIR response using simulated quantized coefficients and math. 309
7-6 Magnitude response of actual and ideal FIR filter. 315
7-7 Canonic or direct form II structure of a second-order section for an IIR filter. 315
7-8 Modified structures for first- and second-order sections. 317
7-9 Final structure of first- and second-order sections. 321
7-10 Ideal and actual IIR filter magnitude response for example program. 328
8-1 A sequence of samples. What is the spectrum of this signal? 334
8-2 An ideal magnitude spectrum for the signal in Figure 8-1. 335
8-3 Using a bank of filters to analyze frequency content. 336
8-4 The DFT's view of the signal in Figure 8-1. 340
8-5 The relationship between window shape and a windowed signal's spectrum. 342
8-6 Common windows and their time- and frequency-domain representations. 343
8-7 Main lobe width for the Blackman window ($N=64, 128, 256$). 344
8-8 The effects of four different windows on the same data. 345
8-9 Details from Figure 8-8 showing resolution of nearby frequencies. 345

8-10 Varying the length of a Blackman window. 346
8-11 Detail of Figure 8-10. 346
8-12 Main lobe width and distinguishing nearby frequencies. 348
8-13 Symmetry of magnitude and phase of DFT output. 350
8-14 DTFT spectrum (*solid line*) and DFT sample (*dots*). 351
8-15 Relationship between zero padding and window length. 354
8-16 Structure of an eight-point DIT FFT. 358
8-17 The FFT "butterfly". 358
9-1 Crosscorrelation (numeric example). 385
9-2 Crosscorrelation (graphic example, same data as in Figure 9-1). 386
9-3 Sending a signal over a noisy channel. 389
9-4 Crosscorrelation used to detect a signal in additive noise. 390
9-5 Application of crosscorrelation to determine distance. 392
9-6 Example of crosscorrelation used to find time delay. 393
9-7 Determining the system function of an unknown system. 394
9-8 Finding $h(n)$ for an unknown system using the least-squared error method. 395
9-9 Autocorrelation of "oh". 398
9-10 A linear predictor system. 400
9-11 Calculating the "error" in a linear predictor. 400
9-12 Structure of the feedback shift register (FSR). 401
9-13 Forms of shift instructions. 405
9-14 A periodic waveform and with noise added. 410
9-15 Results of averaging over 5 and 100 periods. 411
10-1 The decimation process. 417
10-2 The interpolation process (naive approach). 419
10-3 Spectra of signals during the interpolation process. 420
10-4 Waveforms from interpolation example 10-12. 426
10-5 Rational interpolation. 427
10-6 Combining anti-imaging and antialiasing filters in rational interpolation. 427
11-1 An example of a "normal" distribution. (This is actually an approximation—see text.) 437
A1-1 The real number line. 462
A1-2 The complex number plane. 463
A1-3 Rectangular and polar forms of complex numbers. 463
A2-1 Series and parallel connections. 471
A2-2 Voltage divider. 472
A2-3 RC filter. 474

CHAPTER 1

The Big Picture

This chapter presents a general overview of DSP. No math is used, but we assume some familiarity with microprocessor terminology.

1.1 Overview

We begin by describing DSP and what makes it so useful. DSP is not just about digital filters; it has a pivotal and evolving role in electronics today. We also need to look at some of the limitations of DSP, especially the limitations important to microcontroller-based designs.

Chapter 2 (Analog Signals and Systems) will look at some of the fundamentals of signal processing (SP). Digital filters, frequency analysis, and other DSP areas will then follow. But first, the big questions.

1.2 What is DSP? What Can It Do?

In the preface, we alluded to the fact that SP is a very broad discipline—a branch of applied mathematics, really—that is often considered a part of "electrical engineering," but has had an impact on many disciplines and application areas. Once we reduce the problem to a computer solution in the form of discrete operations, the same operations and equations can frequently be used to solve problems in many areas. Solutions from SP, for example, have been applied in mechanical problems (modeling of automotive dynamics), acoustic engineering (speech processing, noise cancellation), chemical and nuclear engineering (control of substrate deposition), civil engineering (intelligent transportation systems), biomedical engineering (medical imaging, augmentative communication systems for persons with disabilities), and even economics ("econometric models" to predict the effects of tax policies). To illustrate just how far the SP field is from what once was considered electrical engineering, Jack likes to tell this story about his first job interview:

> I was only 26 years old, about to become a new Ph.D., and interviewing in academia for one of the first times. In the obligatory lecture on my signal processing thesis work, I had carefully and nervously laid out the digital model of speech production. The room was warm and dark, and this senior professor had immediately dozed off in the back. His light snoring was disconcerting to me, but the other members of the faculty had apparently become accustomed to this scenario. Suddenly, somehow the word *acoustic* had penetrated his deep slumber, and he woke with a start to ask, "What does that have to do with a violin?" Then, "And what in the world does it have to do with electrical engineering?" I had prepared for many possible questions about my research, but this one I had not anticipated.

1.2.1 A Definition of Digital Signal Processing

When we use a digital process to modify a digital representation of a signal, we are doing *Digital Signal Processing (DSP)*.

1.2.1.1   Digital Process

A *digital process* can be realized by any device that performs operations digitally. This can be a simple logic circuit with a handful of AND and OR gates, or a highly specialized chip integrating millions of transistors. Sometimes the hardware performs its operations without need for a controlling program, thus limiting that device's operation to a specific function such as filtering. (This "hard-wired" solution might be especially cheap and/or fast, however.) The type of digital process that we'll talk about in this book is usually an algorithm (program) running on a microprocessor—for example, a microcontroller. Some examples of hardware for digital processing include:

- "Generic" digital signal processors (e.g., Motorola MC56K DSP family)—microprocessor-like hardware optimized specifically for DSP, often single-chip.
- "Generic" microprocessors (e.g., Motorola PowerPC)—general purpose processors, may have specific support for DSP.
- **Microcontrollers** (e.g., 68HC16, 68HC11)—general purpose controllers, may have specific support for DSP operations and analog interfacing.
- Programmable and nonprogrammable logic (e.g., FPGAs, custom gate arrays)—general-purpose logic hardware that is programmed to perform DSP.

We'll get very specific later about what makes various DSP operations efficient on a given microprocessor. What's important at this point is that DSP can be performed on a wide variety of hardware and, in fact, in a wide variety of computer languages—not just at the native (machine or assembly language) level. It's when we start adding constraints like how fast the signal is changing or how much noise we can tolerate that we start narrowing down the type of hardware or language that can be used.[1] For example, a spreadsheet program running on your home PC will do a fine job of DSP, though not fast enough for processing audio from your stereo.

1.2.1.2 A First Look at Signals

A signal is a physical quantity of interest that is usually a function of time.[2] The voltage in a telephone line (electrical potential vs. time) sound waves produced by speaking (pressure vs. time), even the value of a stock (price vs. time), are all examples of signals. In our context of real-time SP using microcontrollers, most signals originate as measurements in the real world—measurements of sound (e.g., voice, music), temperature, light, strain, and so on (see Figure 1-1).

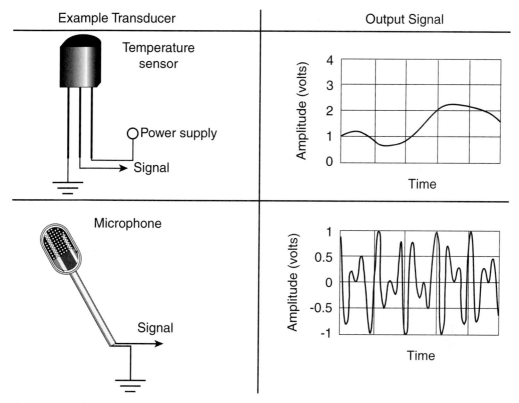

Figure 1-1 Two examples of transducers.

---

1. Strictly speaking, it's not the language we're concerned about, it's the performance of the code that's generated from the program source (which could be C, Pascal, Fortran, BASIC, etc.). More on this in Chapter 7 (Microcontroller Implementation of Filters).

2. Signals can also be a function of space, e.g., as with a photograph (light intensity vs. x and y position).

Because the microcontroller usually cannot directly measure phenomena such as temperature or the amount of light, *transducers* are used to translate the signal into more useful form. Many transducers produce a continuously varying electrical output (usually voltage). The opposite function is also required if we hope to create a signal such as sound, as when a speaker takes an electrical signal and converts this to sound pressure.

Many of the signals that we're interested in processing are *analog* signals, signals that are present at every moment in time and can be any amplitude within a certain amplitude range. However, digital processes cannot directly handle analog signals. The processes *sampling* and *quantization* produce a stream of numbers that represents, approximately, the original analog signal. See Figure 1-2. This stream of numbers is a *digital* signal. It's not an exact copy as we don't know what the amplitude of the signal is between samples, and the amplitude at a given moment is limited by what range of numbers we use to represent the amplitude of the signal. These are the two main issues to be decided—how fast we make measurements (leaving less to guess between samples) and with what resolution or range of numbers we represent the amplitude. Each of these decisions has a different effect on the resulting sampled signal. We'll explore these effects in Chapter 4 (Discrete-Time Signals and Systems).

It is *mathematically* easier to deal with an intermediate type of signal, the *discrete-time, continuous-amplitude* signal (usually shortened to *discrete-time* signal), so you'll often see the phrase "discrete-time signal processing" when you might be expecting "digital signal processing." As it turns out, generally DSP designs are initially based on discrete-time methods. Once the results are acceptable, then you can look at the effects of quantizing the amplitudes and coefficients of the system. These effects include noise and stability. We look at discrete-time signals and systems specifically in Chapter 4 (Discrete-Time Signals and Systems), but it will be at the core of everything we do in the digital domain.

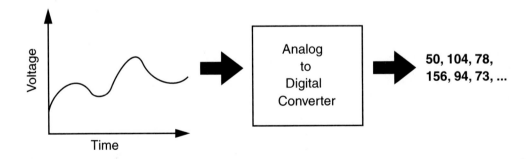

Figure 1-2 *Sampling* and *quantizing* turn an analog signal into a digital signal.

In some cases, a system creates a digital signal internally—for example, in speech synthesis or when generating tones for telephone dialing. Random-number generators, trigonometric functions, table-lookups, and other algorithms allow us to generate a wide variety of signals. If we change these tables or other parameters dynamically, we can end up with quite sophisticated output signals. This is how the speech synthesizer mentioned in the preface worked—a stream of numbers was generated by a random-number generator and a periodic waveform function, then filtered by more than a dozen different filters. The parameters of all these blocks changed dynamically. The basics of generating, or synthesizing, signals are covered in Chapter 11 (Synthesizing Signals).

1.2.1.3 Multidimensional Signals

The examples above are *one-dimensional* signals. DSP can also be applied to signals of two or more dimensions. For example, image processing uses DSP in two dimensions. This can take a lot of processing power if we require real-time speed, but if we can take a bit more time to, say, process a static image for a robot vision system, microcontrollers may be up to the task.

Many times algorithms for more than one dimension are close relatives of the one-dimension techniques discussed in this book. So, although these interesting topics are outside the scope of this book, a sound understanding of DSP in one dimension will be useful in those fields.

1.2.2 What Can DSP Do with Signals?

So far we've discussed the *means* for doing the processing and the entity that we are processing (a digital signal), but the interesting question remains, What can we do with this signal? The same things we can do with analog electronics? Better things?

The short answer is that, *subject to certain limitations,* we can do nearly everything in DSP that can be done in analog electronics. But the real power of DSP lies in operations that *can't* be done in analog electronics. It's almost like the difference between a mechanical adding machine and a personal computer. DSP is pretty powerful stuff!

Analog electronics use the physics of various components to process signals. Scientists found that placing two conductive plates parallel to each other produced a useful electrical function.[3] Same for a coil of wire (also known as an *inductor*). The fact that we have mathematical equations to describe the operation of these physical devices is very different from saying that for any arbitrary equation we can construct a physical circuit. In most cases we can't do that. The mathematics are only an approximate *model* of the physical circuit's behavior, and there are only certain classes of mathematical expressions we can realize in physical circuits.

This is in contrast to DSP. There is not a "physics" of numbers; if you like, *the equations are the reality.* There is no gravity, momentum, mass, or other factor imposing constraints on

---

3. This is a capacitor, and one useful function it has is to produce a current that is proportional to the derivative of voltage with respect to time. (Well, electrical engineers find this useful!)

what we do mathematically. If you can describe your "fractally inverted kryptonite transflibber" mathematically, you can process signals with it.[4] The universe of SP that can be described mathematically is vastly greater than what is possible using analog electronics.[5] On a macro level, the computing machine takes the place of circuitry.

The major DSP topics we'll discuss in this book are:

- Filtering: We often construct filters to modify the frequency content of a signal, like a bass or treble control on an audio amplifier. Filtering is a popular area of DSP and we'll devote several chapters to the different types of digital filters. DSP filters can essentially duplicate analog filters and can even do types of filtering that aren't possible with analog electronics.

- Spectral analysis: Sometimes we'd like to know what frequencies are present in a signal. For example, the display on some graphic equalizers shows us the frequencies present in music in five or so bands of frequencies. This topic is intimately linked with the idea of transformations—mathematical tools that allow us to move between two descriptions of a signal. Later on, we'll call these the *frequency domain* and *time domain* representations of a signal. Chapter 2 (Analog Signals and Systems) will discuss the frequency domain itself, and in later chapters we'll see that using the frequency domain representation of a signal can make processing easier.

- Synthesis: DSP systems can generate everything from simple tones to realistic human speech. Synthesis tools include trigonometric functions, look-up tables, random-number generators, digital oscillators, and so on. Chapter 11 (Synthesizing Signals) gives an overview of this topic.

- Correlation: You could view this as a special case of filtering, where the filter will pass only a specific signal[6]. We also use correlation to identify periodicity in a signal by comparing earlier sections of a signal with the current section, an operation called *autocorrelation*. One example of this is in speech processing, where autocorrelation is sometimes used to determine the pitch of the human voice. Correlation is described in Chapter 9 (Correlation).

These building blocks combine to form other functions. For example, speech compression for efficient transmission or storage can involve analysis, filtering, and correlation. And there are

---

4. *Please* be *very* careful using it, though.

5. Not to burden the point, but not everything that can be expressed mathematically can actually be realized in DSP systems—for example, some systems we might want require knowing future values of a signal. For another example, information theory quantifies the fact that there is an upper bound on the amount of information that can be sent through a communications channel—no amount of signal processing can get past this limit, but DSP is often used to get close to the theoretical maximum data rates.

6. Also known as a matched filter.

other operations like modulation that involve the operations above and other building blocks, so you should view this list as only a beginning point.

1.2.3 Applications of DSP

Whether DSP is done on generic microprocessors or on specialized hardware, the costs are constantly decreasing while performance is increasing. This is pulling DSP into applications where it is replacing analog electronics and creating new applications that wouldn't be possible without DSP. Table 1-1 lists just a few applications and the type of processing they use.

There is a strong parallel between the impact of microprocessors and microcontrollers on digital electronics in the 1970s and 1980s, and the impact of DSP on analog electronics now. In both cases a general-purpose chip replaces circuits made up of dozens, sometimes hundreds, of parts. For example, an earlier digital circuit might be made up of dozens of tiny integrated circuits (ICs), each performing a specific function and connected to perform a specific, "hardwired" task. With the advent of microprocessors, a program could specify the function and, because the same microprocessor could be used in hundreds of different designs, costs were low.

From Table 1-1 you can see the same general trend. In some SP, resistors, op-amps, capacitors, and so on are being replaced by a microcontroller or a specialized DSP chip running DSP software. We now have the equivalent of thousands, if not millions, of discrete components available in compact packages. Again, the same type of chip is being used in hundreds of different products, reducing the costs. Just as with microprocessors, products can include additional functions or expand into completely new product areas (e.g., medical imaging).

Keep these trends in mind as you read. Just a few years ago, engineers and students confronted microprocessors for the first time. DSP is another revolutionary technology—with certain limitations, of course.

1.3 DSP Versus Analog Electronics—Why Bother?

Analog techniques are well established and deeply ingrained in the engineering literature. Sometimes it's difficult to give up something that works (and that we have worked hard to understand) for new technologies or methods that promise certain benefits. In the early days of SP, many analog SP engineers decried DSP as a fad that would pass when their (often younger) colleagues got tired of fighting with slow and error-prone computers. Jack once had a colleague who taught DSP only because "it is good for the students to learn the mathematics, but the field will never amount to anything."[7] Technology nay-sayers are sometimes correct, but in the case of DSP, they were clearly wrong. Those who have refused to learn *digital* signal processing over the last few decades have, with some exceptions perhaps, been left behind in a bygone era. This will become only more and more true in the future.

---

7. Then there was Professor "Tubes" Jones, who based his 1970s electronics courses entirely on vacuum tubes because he considered the emerging integrated circuit to be of little future value.

	Application	Filtering	Analysis	Transformation	Synthesis	Correlation	(Linear) Control
DSP enhances existing products	Automobile Engine Control	X			X	X	X
	Automobile Active Suspension	X			X		X
	Answering Machines	X	X	X	X	X	
	Portable Phones	X	X	X	X	X	
	Cellular Phones	X	X	X	X	X	
	Television	X			X		
	Radio	X			X		
	Hard Disk Drive Electronics	X					X
	Electronic Music	X			X		
	Speech Synthesis	X	X		X		
DSP creates new products	High-Speed Modems	X	X	X	X	X	
	Speech and Image Recognition	X	X	X		X	
	Medical Imaging Equipment	X	X	X		X	X
	Active Noise Cancellation (ANC)	X	X	X	X	X	
	CD (Music), CD-ROM	X					X
	Digital Audio Tape (DAT)	X					X

Table 1-1 Examples of applications using DSP.

Bold claims? Just what are these convincing reasons for adopting DSP over traditional analog designs? Below we've listed the major benefits.

1.3.1 Programmable

We can split programmability of a DSP system into two categories. First, a standard part can be programmed for DSP operations at a number of different stages:

- During the chip manufacture
- Prior to product assembly
- After assembly

Each of these stages is associated with a different memory technology. For example, ROM (read-only memory) is programmable only by the manufacturer, while EEPROM (electrically erasable, programmable read-only memory) can be programmed even after assembly of the product. Generally, parts that allow programming later in the production sequence use more expensive memory, but for many applications the flexibility to change the program (rather than throw away the chip) more than offsets the price difference. For example, modems and PC motherboards often use EEPROM that allows the user to update the algorithms (*DSP algorithms* in the case of the modem!) by downloading new code to the chip.

In the second category, when the entire DSP program or selected parameters can be changed by the part itself, the door is opened to even more powerful functions. A prime example is a high-speed modem that must adapt to dynamically changing conditions of the telephone channel, including echoes of unknown timing. An entire field of research deals with these *adaptive filters*.

1.3.2 Fixed Performance—No *Component Drift*

The memory and logic of a processor do not deteriorate with age.[8] For example, the filter performance of a DSP algorithm will not change during the life of the product or in different environments (say, a change in temperature). The same cannot be said of analog designs, in which initial tolerances for capacitors may be in the 10–20% range and conditions such as temperature result in pronounced changes in values. Even in an analog design that is reasonably immune to drifting, it may be necessary to "tune" each manufactured device individually to meet tight performance tolerances.

Caution: While the behavior of a DSP algorithm won't change over time, that behavior won't always be what you expect if you haven't been careful with details. If you use only 8 or even 16 bits for filter coefficients, for example, the resulting filter performance might be very different from the theoretical expectation, especially for high-performance filters. This is due to

---

8. Of course, we usually have some interfacing to the outside, analog world. Analog-to-digital converters (see Chapter 4, Discrete-Time Signals and Systems) and other circuitry can change slightly with time, temperature, and so on.

the difference between *discrete-time* (infinite resolution) and *digital* (finite resolution) signals and processing. (More on this later, we assure you!)

Going a bit further, as digital signal processing algorithms are purely mathematical, it's possible to run the same algorithm on a number of different platforms and compare the outputs (this is what Dale did to get the voice synthesizer running). If the same precision is used for mathematical operations, the results of the same algorithm written in C and running on a desktop computer should match the output (down to the bit!) of the same algorithm coded in assembly language running on a microcontroller.

1.3.3 Economic

As the performance of an analog filter is enhanced—say, how well it differentiates between signals to be passed and those to be stopped—analog circuit complexity increases. Such filters require more and more op-amps, board space, power, and so on. With digital filters, once we've paid for the DSP chip we may be able to implement both additional individual functions and better instances of those functions on the same hardware.

1.3.4 Functions That Cannot Be Done with Analog Electronics

There are some DSP operations that are impossible to implement using analog electronics. DSP gives us flexibility by using arbitrary mathematical operations to process signals. (In a moment we'll see that analog electronics still has a speed advantage, though.)

1.3.5 High-Volume, Widely Available Parts

Whether we do DSP on a standard microprocessor or on a special DSP chip, these are usually high-volume parts. Further, manufacturers can take advantage of technologies developed for other ICs (like high-density memory chips) and apply them to DSP chips and microcontrollers. Digital ICs tend to be easier to manufacture than analog ICs because noise is "swallowed up" in digital logic, but imperfections in analog circuits add up. Costs for DSP chips continue to drop while performance spirals upwards—exactly the kind of trend engineers love to see!

1.3.6 DSP Tasks Can Share Hardware with Other Tasks

The same hardware that does the DSP in an application might be able to perform other product functions. In fact, most DSP chips have most of the same elements as general-purpose microprocessors and can run general-purpose programs fairly well. If the processing load is relatively small for the SP task, other tasks, such as user interface or control, can run concurrently.

We can turn this around and take a microcontroller already doing some other job and have it also do DSP. In a moment, we'll look at the microcontroller resources that are useful for DSP.

1.3.7 Design Tools Are Available

In DSP much of the work—such as formula derivations and algorithms—is already available.

More often than not, you'll identify an appropriate approach, implement a straightforward algorithm, and supply a set of coefficients that—contrary to what you would infer from most DSP texts[9]—you'll probably generate using a "canned" computer program.

Development systems exist for automating virtually the entire process, allowing you to enter block diagrams specifying the desired function and generating C or assembly language code customized to your hardware. Most of these tools target DSP-specific chips, but support is available for some microcontrollers, including the M68HC16. However, *all* DSP design requires thought,[10] and even with the most advanced tools you'll still need to understand the basics to create an economic and efficient design. Conversely, DSP tools in the hands of someone who does not understand basic DSP are a ticket to many unpleasant surprises—some ultimately coming from supervisors and managers. This book is here to assure you only pleasant experiences.

1.4 DSP and Microcontrollers

The relatively simple processing required by DSP tasks can often be done on modest microprocessors. Microcontrollers combine microprocessor units with internal program memory and other peripherals, such as:

- Data memory
- Digital I/O ports
- Analog I/O ports
- Timers and counters
- High-speed serial I/O

In many cases, all the hardware we need for DSP is on a single chip! Obviously, if we already have a microcontroller in a product, the additional cost to add DSP could be minimal.

Let's briefly look at the M68HC16Z1, a specific part in the Motorola M68HC16 family. Table 1-2 summarizes useful M68HC16Z1 resources.

---

9. "The problem with most of the books in this field is that their authors have forgotten that (digital) filters exist not for the purposes of mathematical manipulation but to filter out some signals and let others through. You would never in a million years guess that from casual reading in the field." Foster [1981, p. 98].

10. Oh no! Not *thinking* again!

Memory	20-bit address bus allows access of 1 megabyte of program/data memory (external hardware can decode signals to provide separate 1-MB program and data spaces) 1-K bytes of internal, no-wait state (i.e., *fast*) RAM; can be used for program, data, or coefficient storage
Analog inputs	Eight analog inputs 10-bit analog-to-digital converter (8.58 μs conversion time) capable of automatic conversion
Timer/counters	Two 16-bit counter/timers
Serial I/O	High-speed serial interface (connect to peripherals such as external analog-to-digital or digital-to-analog converters, or to other microcontrollers)
Digital I/O	Extensive digital I/O; many pins can be used for general purpose I/O if their special function is not necessary
Data width	Supports 8- and 16-bit data movement 16 x 16-bit multiplies (integer and fractional), 32-bit by 16-bit division, and special repeating 16-bit x 16-bit multiply with accumulate

Table 1-2 M68HC16Z1 resources.

The catch is that most microcontrollers are foolishly[11] designed for control applications (see sidebar below), not SP. Some microcontrollers, such as the M68HC16, do have special enhancements for DSP, but most microcontrollers have been optimized for digital control operations such as I/O and bit manipulation, not for number crunching. This makes DSP on microcontrollers harder; for the same performance, we'll need to work a bit more than with DSP-specific chips. When we can use them, however, microcontroller solutions may be a fraction of the cost of a fancier DSP-specific solution. In the future, we can expect to see both enhanced general-purpose mathematical operation and DSP-specific functionality in microcontrollers.

1.5 Limitations of DSP

We've hinted at some of the limitations of DSP. Here's a partial list, emphasizing points that are important for microcontrollers:

- Component cost: There is a certain minimum investment in the DSP chip and supporting hardware (like analog-to-digital converters and memory). Just the *socket* for a microcontroller is more expensive than a simple analog filter!
- Speed: *This is a major issue.* Being programmable generally makes a technology 10 to 100+ times slower than the corresponding nonprogrammable ("hard-wired") technology. For example, you can generate a logical AND function using a microcontroller, but a sim-

---

11. Our humble opinion.

ple AND gate will do the job a hundred times faster. DSP on a microcontroller, especially those without DSP support, absolutely requires good design. Even with an optimal design, it may be difficult to process large bandwidths (e.g., high-quality audio) if the filter has stringent requirements. Even DSP-specific chips have upper limits far lower than traditional analog electronics. (However, see Section 2.3.3.3 where we discuss the crucial issue of bandpass signals—you may not need as much bandwidth as you think!)

- Processing: Basic DSP operations are very simple, but they do need to be done quickly. Hardware multiplication is almost a necessity, and the ability to multiply and accumulate in one instruction cycle is even better.
- Memory: In general, higher-order (higher-performance) filters require more memory for both data and coefficients. Operations like frequency analysis also need storage for hundreds or thousands of samples. Both data and coefficients may need to be 16 bits wide or wider. These may not be huge requirements for most microprocessor-based systems, but it could be a tight squeeze if you are trying to fit everything into on-chip memory in a microcontroller! The good thing is that the actual DSP *program* (not including coefficients) is usually very compact.
- Precision: Most microcontroller implementations of DSP use *fixed-point* math with limited precision—perhaps 8 or 16 bits for data and coefficients. For certain types of filters, this lack of precision can lead to *stability* problems or, less seriously, differences between what you designed and what you actually get.
- Multiple technologies: We still have to interface with the real, analog world. This means converting to and from digital data, which means additional choices and additional places

---

Oh, you mean *control*!

A brief aside on what we mean by *control* in this book. In digital electronics, control is taken to mean the digital sequencing of other circuits or devices that a microprocessor or microcontroller might do. This includes turning on lights, moving a stepper motor, and other tasks that are discrete in nature. However, using DSP, we can create *compensators*—SP units that control analog devices (like motorized assemblies) that have inertia, friction, and other characteristics. In this context, *control* is a type of SP specific to control engineering. Control system theory is outside the scope of this book, but many of the topics we cover, especially transform theory, spectra, filtering, and implementation—will be useful if you need to pursue the subject.

Most folks will be from one side of this fence or the other; we thought we should make you aware of the possible confusion.

for problems to occur. Using DSP in your design doesn't exempt you from careful analog design—it requires it, *plus* careful *digital* design. (The hardware side of DSP designs falls into a category often called *mixed signal processing*, as the designer must account for both digital and analog signals in the design.)

1.6 Why DSP Might Look Difficult

We found DSP a difficult subject at first[12] and expect that other people have too. We'll close this chapter with our impression of why this is so.

- Electrical engineering (EE) background: DSP uses the language of electrical engineering and, in particular, many concepts from analog electronics, signals, and linear systems. For someone with a background in microprocessors, computer science, or even digital electronics, the terms *poles*, *zeroes*, and so on comprise a foreign (if not amusing) language. We'll cover the appropriate concepts as they come up. We'll also provide quick definitions in the Glossary.

- Tedious design process: Finding good coefficients for a digital filter is tedious to do by hand. In some cases, there are no fixed formulas for coefficients, and an experimental approach is required. Fortunately, there are plenty of programs available to help in the design process, allowing you to concentrate on the bigger picture. In this book, we'll show you the formulas for the filter coefficients, but we emphasize that it is far more important to know what other choices you must make earlier in the design process, and so use these tools effectively.

- Mathematical basis: The fact that DSP can be exhaustively analyzed and derived means, of course, that in most textbooks it usually is.[13] You will definitely need to investigate DSP more fully for some applications, and this will eventually involve some gory mathematics. This book attempts to present a subset of useful DSP techniques and a framework for applying them as part of good engineering, while keeping the mathematics as friendly as we can.

- Diverse elements: Finally, DSP isn't a monolithic area of study. It draws from a number of disciplines and is also shaped by the fields to which it is applied. Just a few of these are: analog electronics, microprocessor architecture, digital electronics, programming, mathematics (numerical techniques, advanced analysis), control theory, communications, and probability theory. You don't need to be an expert in every aspect of DSP, but because each area has contributed concepts and terminology, you will almost certainly find that your horizons widen as you get deeper into DSP.

---

12. If you had told Jack in 1975 that he would someday be writing books on DSP...

13. An engineering editor for a major book publisher once gave Jack some good advice when he said, "Many authors forget that the purpose of a textbook is to educate naive students, not impress expert colleagues."

1.7 Summary

DSP occurs when a digital process (e.g., an algorithm implemented on a microprocessor) is used to modify the digital representation of a signal. These modifications include filtering, frequency analysis, transformations between the time domain and frequency domain, synthesis, and correlation. Among the advantages of using DSP are programmability, consistency over time, low cost, unique function, ability to share hardware, and the availability of design tools. Microcontrollers combine many of the functions needed for DSP on one chip and may already be used in a product, though DSP-specific hardware may be necessary for demanding applications. Finally, DSP has real limitations, due to initial costs, speed, memory, and the need for familiarity with both analog and digital electronics.

Resources

Foster, Caxton, *Real Time Programming—Neglected Topics*, Reading, MA: Addison-Wesley, 1981.

In later chapters, we'll be listing resources to each chapter. Below are some resources of general interest.

IEEE Signal Processing Magazine (ISSN 1053-5888).
A bimonthly magazine published by the IEEE (Institute of Electrical and Electronics Engineers) Signal Processing Society. Jack is the past editor-in-chief. What a shameless plug!

Personal Engineering and Instrumentation News (ISSN 0748-0016)
This covers a range of topics around desktop computing in engineering with a surprisingly consistent emphasis on digital signal processing. Also has coverage of mathematics and filter design software. The ads are also a good source of information.

Electrical Engineering Times (ISSN 0192-1541)
A weekly newspaper covering the entire electronics industry, from cutting-edge research to worldwide politics. In between are articles on microprocessor architecture, DSP chips, and so on.

Summary

CHAPTER 2

Analog Signals and Systems

This chapter uses basic trigonometry, logarithms, and some complex numbers. (If you're comfortable with trigonometry, complex numbers won't be hard to pick up.) If you've already studied continuous signals and systems, feel free to skim. On the other hand, some of the concepts may be very strange if you've never seen them before and they might take some time to digest. Don't get bogged down here—it's important stuff, but it's not the Good Stuff of later chapters. On the other hand, the Good Stuff will not be accessible without some grasp of this material.

2.1 Overview

Our goal in this chapter is to foster an understanding of signals and systems in what we might call their "natural state," continuous time. Most of these topics carry over into digital signals and systems and provide the background for Chapter 3 (Analog Filters). Yes, this is supposed to be a book about *digital* signal processing, but the language of filter design is quite similar whether we're talking digital or analog. And as we mentioned before, one form of digital filter (the "IIR") often begins life as an analog filter, so this subject really won't be out of our way.

This chapter is a bit more theoretical in parts than most of the chapters to follow, but our goal is quite clear. To understand SP we need good ways of describing both signals and systems in the time and frequency domains. We begin this chapter by considering these two complementary views of a signal. What does a frequency-domain description of a signal look like? How do we produce one? And finally, why would we go to this trouble?

The *system* of this chapter's title is something that modifies an input signal and produces an output signal—a filter is one example. We can combine the frequency domain descriptions of signals and systems to predict the output of a system in response to a particular input, which is pretty powerful.[1]

If there's anything special about the topics in this chapter, it's the remarkable shortcuts that have been found to analyze signals and systems. If you've studied differential equations, you know the tedious math these shortcuts help us to avoid! But beyond the time savings, tools like the *s*-plane or frequency response curves help us visualize what's going on with complicated systems—things that just can't be done by staring at an equation. With these tools and concepts under your belt, analog *and* digital systems will be much easier and, we hope, enjoyable.

2.2 Sources of Analog Signals

Most analog signals in DSP tasks are first measured by transducers, devices that translate some measurable phenomenon into another form. Usually we're talking about translating one form of energy, such as heat, light, or sound, into another form of energy, usually electrical. For our purposes, we need a signal in terms of a voltage since that's what (most) analog-to-digital converters need. It's often necessary to go through one or two steps to create this voltage signal. For example, a pressure sensor might first translate air pressure to strain on a diaphragm, translate this strain into a change in resistance, translate resistance to a current, and finally translate the current to a voltage. Table 2-1 lists just a few types of common transducers and the types of signals they produce, though you can often buy transducers that include the additional circuitry to produce a voltage output.

Phenomenon	Example transducers	Output
Temperature	Thermistor	Resistance
Light intensity	Phototransistor	Current
Air pressure	Monolithic pressure sensor	Voltage
Current	Resistor	Voltage (across resistor)
Mechanical strain	Strain gauge	Resistance

Table 2-1 Common transducers and their output.

What's important is that the voltage signal we obtain is the product of a number of preprocessing stages. Each stage has imperfections, and the overall performance of a DSP system is highly dependent on the quality of the input signal which has passed through each imperfect stage.

---

1. Later on, we'll qualify this statement, but it's true in the sense that we'll most often care about.

As an example, noise can be introduced at almost any stage from the transducer to the analog-to-digital converter. A strain gauge—intended, of course, to measure strain—might also respond to a change in temperature; or electrodes monitoring brain waves (electroencephalogram) may pick up electromyographic (muscle) activity due to eye blinks. Cables connecting various measurement devices act like antennas, adding radio signals and interference from nearby motors to the desired signal. Nor are signals safe when they finally reach the analog-to-digital converter; noise from power supplies or from voltage references of the converter can corrupt the signal. This is all before the digital filter sees the signal; there are similar concerns on the other side, converting the digital signal into analog again.

We won't dwell on general SP in this book, but here are two brief observations. First, even a fancy digital filter won't be able to compensate for the poor quality of an input signal. This relates to the second observation, which is that the later in the development process you address a problem, the more expensive the solution is. Some of the references at the end of this chapter may be helpful in engineering good *systems*, one part of which may be a good digital filter. In the end it may save you quite a bit of work.

Enough of where signals come from. We now turn to methods for describing signals using graphs and equations.

2.3 Describing Signals and Systems in Time and Frequency

2.3.1 Analog Signals

Analog signals are continuous in both time and amplitude values—that is, the signal is well defined at every point in time from when it starts to when it ends, and can hold any amplitude value between its minimum and maximum. It is possible to take an analog signal and make it discrete in either time or amplitude, or both; the resulting signal is no longer analog, and a different set of tools for describing the signal is necessary. This is the topic of Chapter 4 (Discrete-Time Signals and Systems).

2.3.2 Describing Analog Signals in the Time Domain

Figure 2-1 is a plot of the air temperature measured at the North Inlet Estuary Meteorological Station on Crab Haul Creek near Georgetown, South Carolina, from January 1, 1991 to December 31, 1991.[2] Because each value of time has exactly one value of temperature associated with it, we can say this temperature is a function[3] of time. For example, we can use the notation $d(t)$ to stand for the temperature at some time t. $d(t)$ would then be a description of the signal as a function of time. (The choice of d was arbitrary, but t is rather universal for time.) Now we can

2. As if it weren't obvious.
3. We're not saying that we could necessarily find some concise *mathematical description*, like $d(t)=sin(t*20)$, for this plot. To say function d is a function of time is only to say that for any value of t within reason, there is exactly one value of the function d.

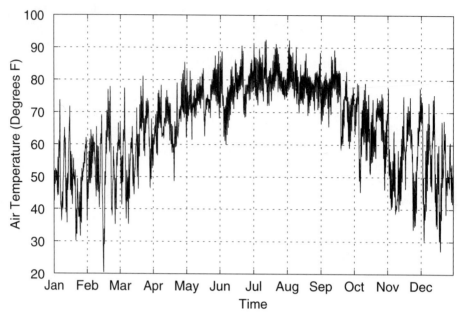

Figure 2-1 North Inlet Estuary Meteorological Station, 1991.

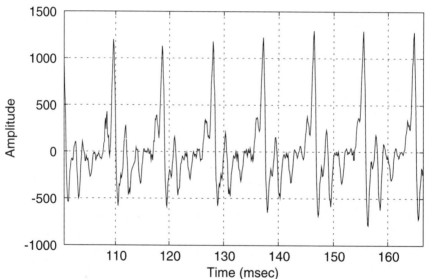

Figure 2-2 Human voice, "oh".

ask for the value of d, i.e., temperature, when $t=3$ A.M., March 5, 1993, which appears to be 48°F.

Figure 2-2 shows a segment of speech (a person saying "ohhhh"[4]). Again, we can describe the signal with a function that takes time as an argument and returns the value (sound pressure,

Describing Signals and Systems in Time and Frequency

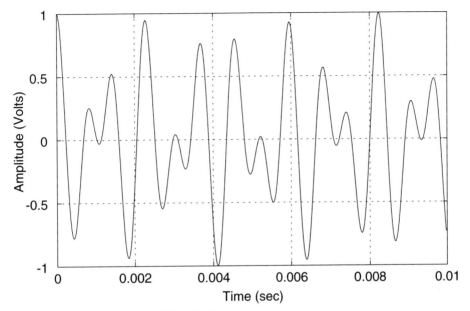

Figure 2-3 Waveform for DTMF key "8."

translated to voltage) of the signal at that time. If we use *g(t)* to be this particular function, *g(120 milliseconds)= –250* millivolts.

Finally, Figure 2-3 shows an ideal example of the tone generated for the number "8" on a Touch-Tone® phone (also known as Dual Tone Multiple Frequency, or DTMF, tone). If we call this signal *e*(t), then *e*(2 milliseconds)=–0.4 volts.

The DTMF tone is entirely described by *e(t)*. We could put a tiny person in a box with a variable power supply and, armed with this description of the signal, the "miniperson" could reproduce this signal completely. With twelve such descriptions, they could dial any number you wanted (though this isn't the best way to generate these tones, as we'll see). Direct digital waveform synthesizers (fancy function generators) use this technique to reproduce arbitrary waveforms, though they employ microprocessors and software to avoid the liability issues of having tiny people in the boxes. Solid state implementations are also somewhat lighter and have lower maintenance.

By the way, just so we're on the same track, recall that *e(t)* is for any *t*, not just every second, or every microsecond, but every instant.

If this time-domain representation is complete, what need is there for a frequency-domain representation?

4. Perhaps the initial sound a person would make on falling off of the North Inlet Estuary Meteorological Station into the North Inlet Estuary on March 5, when it was 48°F.

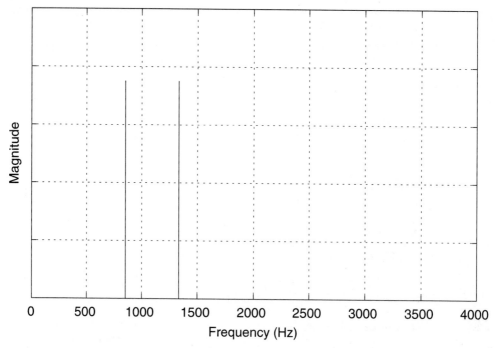

Figure 2-4 Magnitude spectrum for the DTMF signal for the "8" key.

2.3.3 Describing signals in the frequency domain

There is nothing "missing" from the description of a signal in the time domain; the frequency-domain description is an alternative view of the signal. Just why this different view is so useful is something we'll talk about in a moment. Right now look at Figure 2-4, which shows the same DTMF signal as Figure 2-3, but this time in the frequency domain. According to the plot, the only frequencies present in the DTMF signal are at about 850 Hz and 1300 Hz (the precise values are 852 Hz and 1336 Hz). Both frequencies have about the same magnitude—that is, there's the same amount of power at each frequency.

Since at each point on the graph there is a particular value—the magnitude or "amount of the signal" at that frequency—we can write this as a function, this time a function of frequency, like $e(f)$. Note that this is the same function name, since we're talking about the same signal. Of course, for the DTMF signal, this is a pretty boring function, as it is zero everywhere except at $e(852)$ and $e(1336)$. Figure 2-5 shows the magnitudes of frequencies present in the estuary data from Figure 2-1. While this isn't nearly as simple as the DTMF frequency-domain description, you can clearly see a peak at a frequency of one day (also two, three, four, and so on).

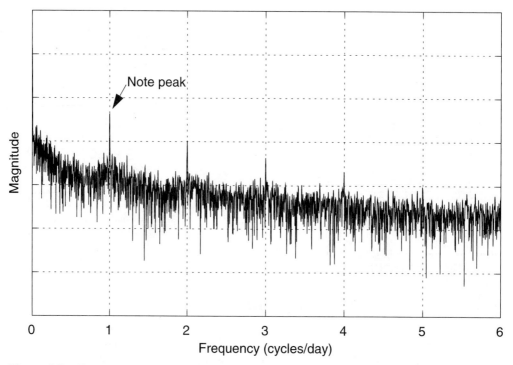

Figure 2-5 Air temperature, North Inlet Estuary Meterological Station, 1991.

What does it mean to say that a signal has some magnitude at some frequency—or, alternatively, that a signal is composed of different *frequency components*? This is a question we'll deal with rigorously in Chapter 8 (Frequency Analysis) when we talk about the FFT. For now, we've got two different suggestions for how to look at this.

(By the way, we're assuming that the signal has "constant dynamics" while we do the analysis. This means that the signal has the same frequency components, no matter where we look in time. Such a signal is called *stationary* or *time invariant*. Not all signals fall into this category. Human speech is always changing, and it's only for a fairly short duration (say, 5–20 milliseconds) that speech is relatively stationary.

As a first approach, let's assume we have a very nice filter that will allow only a narrow range of frequencies to pass, and that we can vary this filter such that the center of this narrow range can be at any frequency (see Figure 2-6). You could think of this as tuning in a radio station—only one station at a time comes through (ideally!); the rest are suppressed. Now "tune" this filter from its lowest to highest frequency, carefully noting how much of the signal is let through at each frequency—kind of like mapping what radio stations are coming over the air and how loud or faint each is. In Figure 2-6, the (AC) voltmeter is used to measure the "amount" of signal coming through the filter. In fact, there are some types of *spectrum analyzers* based on the

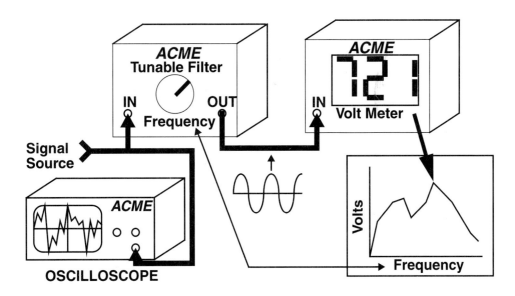

Figure 2-6 Finding the magnitude of frequencies with a narrow filter.

general principle of sweeping a "narrow" but tunable filter across the frequencies of interest. Conversely, instead of one filter that is swept across a number of frequencies, you can have many filters that are fixed in frequency and rather close spaced. Many graphic equalizers employ this technique for displaying the energy in frequency bands. The downside to this is that each band requires a separate filter. For example, a stereo, 10-band-per-channel display requires 20 filters.

A second method is to ask instead, What simple signals could we add together to recreate the original signal? If the signals we add are sine waves, which have only a single frequency, we could find the right combination of signals to add (the right frequencies and amplitudes) and could get a description of the frequencies present. Figure 2-7 shows a complicated signal while Figure 2-8 shows the sine waves that can be added together to produce that signal. Figure 2-9 shows how even a squarewave can be decomposed mathematically[5] into sine waves. Plotting the frequency and magnitude of each of the component sine waves gives the frequency-domain description. The catch is that for some waves, an infinite number of sine waves is required—or, put another way, the spectrum of such a signal is infinitely wide. A partial plot of the spectrum of a squarewave is given in Figure 2-10.

We'll revisit these topics in Chapter 8 (Frequency Analysis), which is about signal analysis. For now, what's important is that a signal can be described in terms of what frequencies (and how much of each) are present. But is this really enough to describe a signal?

5. Using the "Fourier series."

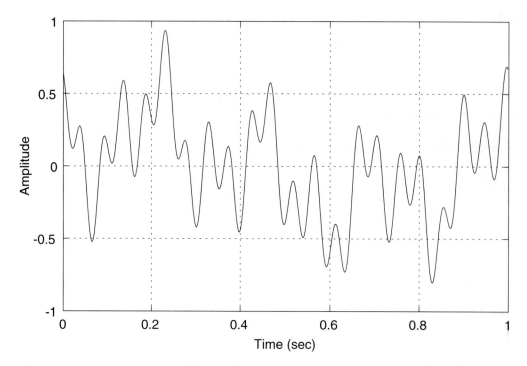

Figure 2-7 An example signal to be decomposed.

2.3.3.1 The "Complete" Picture—Magnitude and Phase

It turns out that we will need more than just the magnitudes of the frequencies in a signal to fully represent that signal. We're going to need the *phase* of the different frequency components as well.

Recall that one approach to finding the frequency components of a signal is to ask what sine waves are necessary to add together to get the original signal. Just noting the amplitude of the signal may seem sufficient, but as Figure 2-11 and Figure 2-12 show, changing the phases (time displacements or delays) of the sine waves will change the signal in the time domain. What happens if we take the frequency components of a squarewave and add them up with random phase? Did you expect the waveform in Figure 2-13? This waveform has the same *magnitude* frequency spectrum as the waveform in Figure 2-9d—but a different *phase* spectrum. To completely describe a signal in the frequency domain, we need two plots—one for magnitude and one for phase.[6]

---

6. We humans don't seem to pay much attention to phase as far as our hearing goes. All kinds of phase distortions are introduced in recording, playback, and communication processes, but the voice and music still sounds okay. This is a lucky thing—phase can be tricky to preserve, especially for an important class of filters called *IIR filters*.

Describing Signals and Systems in Time and Frequency 29

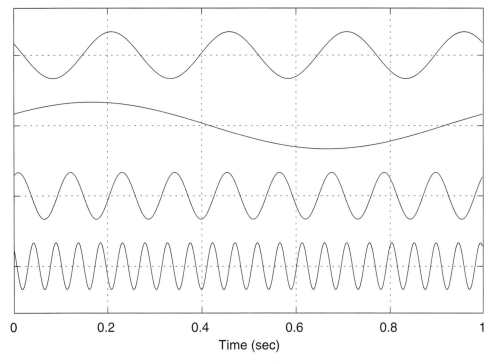

Figure 2-8 Frequency components of the signal in Figure 2-7.

But wait a second! Now we have *two* plots for the frequency domain. Does this mean we need twice as much information to represent a signal in the frequency domain as we do in the time domain?

The answer is no. The information in the signal is *split* between the magnitude and the phase. It may seem like most of the information is in the magnitude part—that's where we started this discussion, after all. And the truth is that for many filtering applications, the magnitude is the primary concern. But the phase characteristics of both signals and systems are crucial for many applications, including video and systems that control movement (control systems). Get the wrong phase here and your cruise control could have you barrelling down the road at 90 mph.[7]

Bundling Up It turns out that it is very convenient to bundle both magnitude and phase into a single value. Later in this chapter, this will let us construct some powerful descriptions of systems using just a single equation, instead of separate equations for magnitude and phase. (It might not make sense at the moment, but we're also going to describe systems (e.g., filters) in terms of magnitude and phase. Stay tuned.) *Complex numbers* are the key. Figure 2-14 shows

---

7. Which in the case of Jack's wife, Joan, would be creeping along.

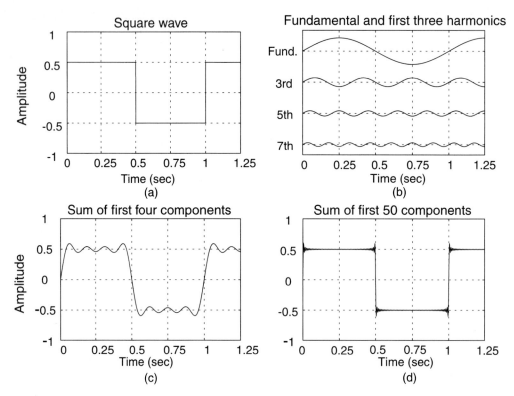

Figure 2-9 (a) and (b) Decomposing a squarewave into its first four components. (c) Building a squarewave from the first four and (d) fifty components.

how a single complex number can carry both phase and magnitude—magnitude carried in the absolute magnitude of the number, and phase in the angular measure of the number.

Appendix 1 is a brief summary of complex arithmetic if you need a refresher. Later on, we'll use complex numbers to "bundle up" some other related quantities to give some real time savings and a very useful way of "mapping" filters. You might ask whether this means that DSP routines require complex math. Fortunately for microcontrollers, most filter *implementations* will use only *real* (that is, noncomplex) mathematics. However, most of the design work for DSP, especially filters, uses complex algebra. Because there are so many calculations (and thus, opportunities to make mistakes), most folks greatly prefer using software to perform filter design. This is just automating the math, though. You'll still need a good grasp of complex math to avoid problems.

Phase Is Relative—to What? Our graphs in the time domain have had a nice starting point labeled "$t=0$." We can likewise choose either the sine or cosine wave to represent the reference phase (often the cosine is used). But in the real world the starting time is arbitrary; so, too, is any idea of absolute phase. Instead, phase is given relative to some understood reference;

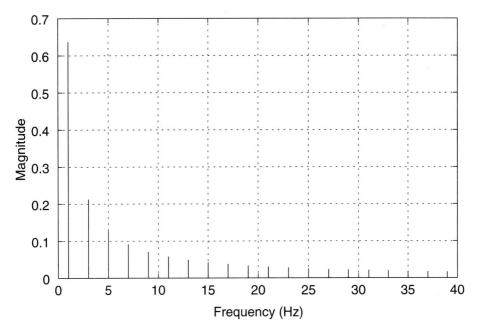

Figure 2-10 Frequency components of 1-Hz squarewave.

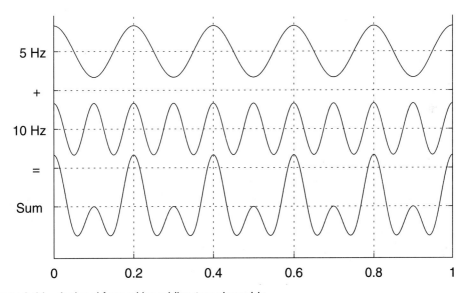

Figure 2-11 A signal formed by adding two sinusoids.

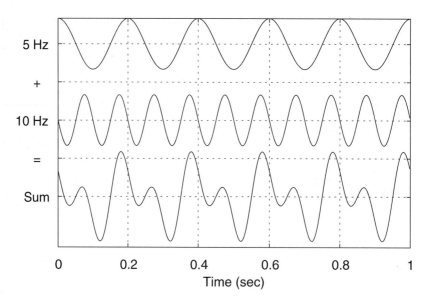

Figure 2-12 Changing the phase of the 10-Hz sinusoid component from Figure 2-11.

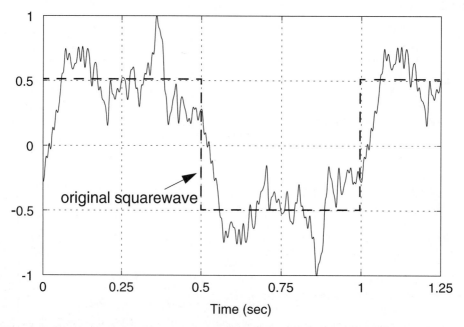

Figure 2-13 Sum of first 50 frequency components of a squarewave, with random phase.

Describing Signals and Systems in Time and Frequency

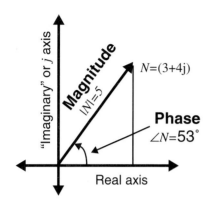

magnitude: $|N| = \sqrt{Re(N)^2 + Im(N)^2} = 5$

phase: $\angle N = \operatorname{atan}\left(\dfrac{Im(N)}{Re(N)}\right) = 53.13°, 0.927$ rad

($Re(x)$ is the real component of the complex value x; $Im(x)$ is the imaginary component.)

Figure 2-14 Representing magnitude and phase with complex numbers.

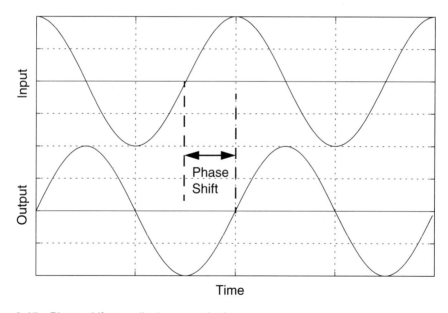

Figure 2-15 Phase shift as a displacement in time.

for example, the phase characteristics of a system are given relative to the phase of the input (see Figure 2-15). If you know the phase of both input, θ_S, and output, θ_R, with respect to some arbitrary point in time, the input-to-output phase is $\theta = \theta_S - \theta_R$.

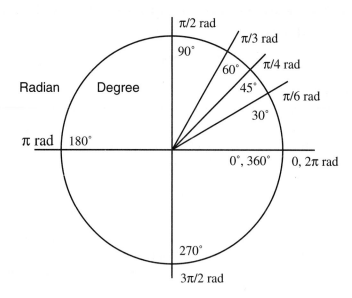

Figure 2-16 Radians and degrees around the circle.

Degrees and Radians It is mathematically easier to use the angular measure of *radians* instead of degrees. Figure 2-16 shows the relationship between degrees and radians on a circle. Because there are exactly 2π radians in a circle, we can use the following equations to convert between degrees and radians (r is in radians, d in degrees): $d = r \cdot 180/\pi$ and $r = d \cdot \pi/180$. Some common angles in degrees and radians are in Table 2-2. It's sometimes convenient to work out the math keeping π as a symbol, so we've shown two columns for radians, one in terms of multiples of π, the other in decimal form.

The use of radians also extends to frequency; in addition to Hertz (cycles per second), we can speak of radians per second (rad/s).[8] In the DSP literature, the uppercase omega, Ω, is often used as the variable for radian frequency, while F is reserved for frequency in Hertz. The conversions are $F = \Omega/(2\pi)$ and $\Omega = 2\pi F$. (See the sidebar on notation on the opposite page.)

---

8. Someone once suggested that rad/sec be given the designation "Avis." Fortunately, this was not adopted as a standard. (Dale didn't get this at first. "Hertz" is also the name of the number one rental car company in the U.S., and "Avis" is number two—"We try harder.")

A Note About Notation—A "Frequent" Problem with ω, Ω, f, and F

Here's the problem. Engineers conventionally use the *lowercase f* and ω (Greek lowercase letter omega) to stand for frequency. But in DSP, we have two very different frequencies, one associated with continuous time and another with discrete time. Because discrete time is a relative latecomer to the field, it makes some sense that frequency in discrete time should get the less desirable *uppercase F* and Ω. Which, of course, means that every group of textbook authors does whatever they feel like.

Well, it's not that bad. What actually happens is that most textbooks have an emphasis on either continuous time or discrete time, and they use the lowercase symbols for that system. Although we will work a lot with continuous-time signals and systems in this book, we finally decided to stick with the convention that Oppenheim and Schafer established (Oppenheim and Schafer [1989]), which is:

4 legs good, 2 legs bad

No, wait—that's Orwell's *Animal Farm*. Make that

Continuous-time frequency: F **(Hz) and** Ω **(rad/sec)**

Discrete-time frequency: f **(dimensionless) and** ω **(rad)**

Not all texts on digital signal processing will have this convention, so you'll need to find out (if it isn't clear) what each author has done.

Using radians for measuring angles or frequency may seem unwieldy, but most of the equations in electrical engineering use radians. Adding to the confusion, of course, engineers will often use degrees and Hertz to express the initial parameters and final results, converting back and forth as necessary. We'll follow in that great tradition, mostly because 120π rad/s doesn't have the same zip to it that 60 Hz does.

Period is the reciprocal of frequency; $T = 1/F$, where T is in seconds and F is in Hertz. Using radians/second, $T = 2\pi/\Omega$.

Degrees	Radians	
360	2π	6.2832
270	3π/2	4.7124
180	π	3.1416
90	π/2	1.5708
60	π/3	1.0472
57.2958...	1	1
45	π/4	0.7854
30	π/6	0.5236
1	π/180	0.0175

Table 2-2 Common angles and their radian equivalent.

Two other units having to deal with frequency are relative measures: *octaves* and *decades*. An octave is a doubling in frequency; for example, one octave above 2 kHz is 4 kHz. A decade is times ten; a decade above 2 kHz is 20 kHz.

Time Delay and Phase Changes in phase are the result of time delays, but as phase is a measure that is based on the frequency of the signal, phase and time delay are not equivalent. The relationship is:

$$t_d = -\frac{\theta}{\Omega} \qquad (2.1)$$

where t_d is the time delay in seconds, θ is the phase shift in radians, and Ω is the frequency (in radians/sec). Note that a negative phase shift corresponds to a delay in time, while a positive phase shift corresponds to an advance in time. Practically, we're dealing with time delays (vs. time advances) in the case of filters (hence, negative phase shift), but in a few pages we'll see that plots will often "wrap" the phase such that it appears there is a positive phase shift.

When we get to the subject of filters and phase we'll discuss some special types of phase relationships—specifically, the case of constant time delay (linear phase).

2.3.3.2 DC, or Frequency 0

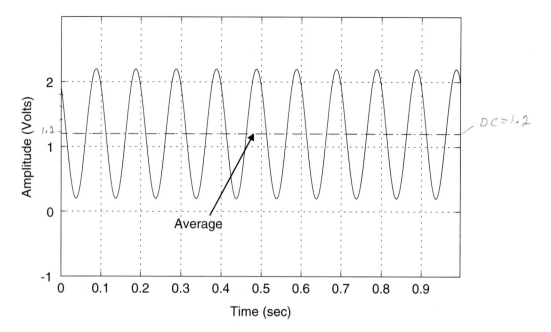

Figure 2-17 A signal with DC bias.

You might have noticed that the frequency axes of our plots begin at 0. What is a frequency of 0? Chapter 8 (Frequency Analysis), where we talk about the Fourier transform, will provide a mathematical reason, but basically a frequency of 0 corresponds to the *bias* or constant offset that has been added to a signal—it's just the average of the signal over all time. See Figure 2-17 for an example of a signal with a bias of 1.2 volts. This is also the *DC* (direct-current) value of the signal, in contrast to the *AC* (alternating-current) value.[9] In many cases the DC value is of little interest—we may know it or it may hold no information for us. In other situations, our signal may be so slowly changing that we can effectively treat it as DC. In either case, although there is something a bit special about a frequency of 0, we can also treat it mathematically as just another frequency, constructing filters that pass or stop this frequency.[10]

2.3.3.3 Baseband, Bandpass, and Bandwidth

Now seems as good a time as any to mention three frequency-domain terms. (We'll cover more in Chapter 3 (Analog Filters).) Figure 2-18 shows the frequency domain representation of

---

9. It may not be clear just what an AC value means—the peak-to-peak voltage? the amplitude? and what about signals with many frequencies present? A useful measure is the voltage that, in DC terms, would deliver the same power as the AC signal. This is known as the *RMS value* (from "root mean square").

10. Almost, that is. The Fourier transform's "lack of symmetry" at DC causes a few nuances in frequency domain analysis of DC signals.

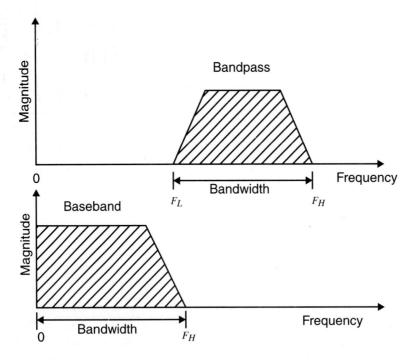

Figure 2-18 Bandpass and baseband signals.

some signals and the *bandwidth* (BW), or the frequency span, of each signal. Not every signal necessarily starts and ends perfectly in frequency, so we often use the point where the signal has one-half the peak power. (As we'll see later, this corresponds to $1/\sqrt{2}$, or roughly 0.7 of the largest value in the spectrum when talking about voltage or current.) The ideal signals in Figure 2-18 are shown as having a definite bandwidth.

Baseband and *bandpass* are terms from communications theory. Baseband refers to signals whose frequency range starts at 0 Hz, while bandpass signals don't include 0 Hz. Figure 2-18 shows two examples. What's important to see is that the highest frequency present in a signal (F_H) is not necessarily the same as the BW of a signal. Don't forget this! Later, when we talk about sampling, we'll see that this can sometimes let us save a lot of processing for bandpass signals.

A second use of the term *bandpass* occurs in filter theory, where it describes a filter that passes only a certain contiguous range of frequencies that does not include zero (DC).

2.3.3.4 Plotting the Frequency Domain

You would think plotting in the frequency domain wouldn't be that complicated, but it turns out that we often don't use the nice, *linear* (that is, equally spaced) scales that we almost always use for the time domain. Here's the scoop.

Describing Signals and Systems in Time and Frequency 39

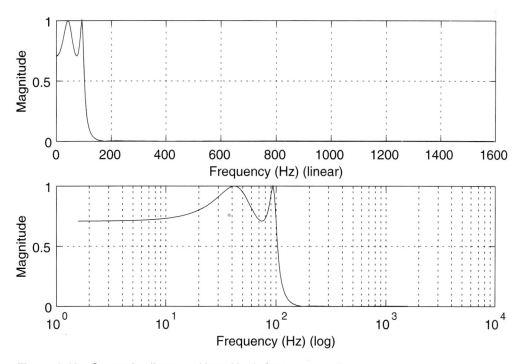

Figure 2-19 Comparing linear and logarithmic frequency scales.

Plotting Frequency Plots of magnitude and phase are functions of frequency, and, in theory, there's nothing that would prevent us from plotting them with a simple linear frequency scale. But most of the time we want both to cover a large frequency range and preserve a lot of detail at the lower frequencies. See Figure 2-19 for an example comparing two plots of what we'll later call the *frequency response* of a system, one with a linear and one with a logarithmic scale. As you can see, the logarithmic scale gives us a lot of detail at the lower frequencies and, like a spherical rearview mirror ("Objects are closer than they appear"), condenses peripheral frequencies into a small summary.

This is usually a very appropriate view for continuous systems, by the way, as it matches well with the underlying physics of systems. When we discuss filters later, you'll see that there just isn't much detail of interest out in the "boonies," so condensing them doesn't hurt us at all.

Another reason for using a logarithmic scale is that certain types of plots (exponential functions) emerge as straight lines when plotted on a log-log scale.[11]

Plotting Magnitude and Power

Magnitude In the frequency domain, we almost always use a logarithmic scale for the magnitude, because we'll be very concerned about very small values of magnitude. For example, let's

---

11. A plot with both the x and y axes expressed logarithmically. Read on for just such a case.

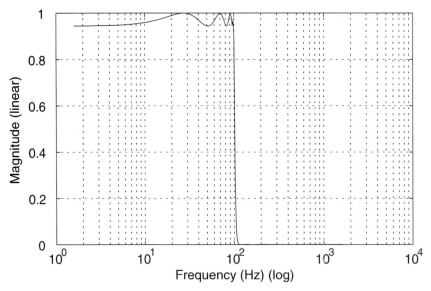

Figure 2-20 A magnitude spectrum plotted with linear magnitude.

say a signal was passed through some filter, and we now wish to plot its frequency spectrum. (In this case, the signal contained roughly the same amount of energy at every frequency in the range we're plotting, and was passed through a pretty good filter). Did the filter attenuate the higher frequencies to 0.001, our design goal, or could it be as high as 0.01? Figure 2-20 shows the magnitude spectrum using a linear plot.

Hard to tell, isn't it? Now see Figure 2-21, which has the same data but is plotted using a logarithmic scale. In exactly the same way as we compressed frequency, the logarithmic scale for magnitude shows us lots of detail at the low end. Now it's clear that our filter works fine—for high frequencies, the magnitude is never above 0.001 of the input magnitude.

By the way, *magnitude* is defined as the absolute value of the amplitude of the signal, usually indicated with bars, like so: $|x(F)|$. Logarithms of negative numbers would not make any sense, and the phase captures the information lost in taking the absolute value.

Yes, you're right. For continuous systems, we'll use logarithmic scales on both the magnitude and frequency axes.

Voltage vs. Power We mentioned that most signals you'll see are of voltages, but for practical purposes, we often want to talk about the *power* in that signal. If we convert that signal to sound, for example, our ears would respond to the power of the signal fed to a loudspeaker, not the voltage. The relationship is $power \propto voltage^2$, and so you'll often see plots with the magnitude squared: e.g., $|x(F)|^2$.

dB and "Relatives" Linear systems don't change the frequencies of frequency components that pass through them—that is, you'll never put a 10-kHz signal in one side of a linear

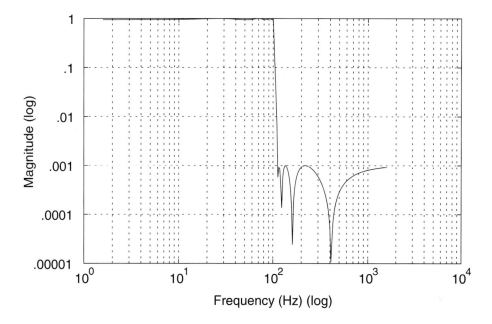

Figure 2-21 A magnitude spectrum plotted with logarithmic magnitude.

system and get 20 kHz out the other. On the flip side, they potentially change the amplitude and phase like crazy—that's their job. It would be convenient to be able to talk about the amount the signal is amplified as it passes through a system, and in a logarithmic way so that very small quantities aren't lost and very large factors don't overwhelm us. That is, we'd like a relative, logarithmic measure. The *decibel* (dB) is just that unit of measure.

The *decibel* is defined as 1/10th of a Bel, the Bel itself being a unit no one ever uses, so don't feel compelled to commit that fact to memory.[12] (And yet, because we say this, years from now you'll know this trivial fact, but forget your anniversary. Go figure.) The decibel is a logarithmic measure of the ratio of the *power* of two signals, and is defined by:

$$\text{relative magnitude in dB} = 10 \cdot \log\left(\frac{p_a}{p_b}\right) \qquad (2.2)$$

where p_a and p_b are measures of the *power* of the two signals (usually in watts).

---

12. The Bel is named after Alexander Graham Bell.

Because we can often assume roughly the same impedance (resistance) for both signals, using the relation $power = \frac{voltage^2}{resistance}$ we find that we can also calculate decibels as:

$$\text{relative magnitude in dB} = 20 \cdot \log\left(\frac{V_a}{V_b}\right) \quad (2.3)$$

where v_a and v_b are *voltages*. This is usually the relation given, but now you know why there's a 20 in the equation instead of a 10 ("deci").

By the way, there's the tendency to feel much more comfortable with voltage measurements than with current or power, because we can directly measure voltage easily (even see very fast voltage waveforms using an oscilloscope), and power and current are not as easily measured. Eventually, you'll need to feel relatively comfortable with each—the "voltage-only" view is unnecessarily constraining.

2.3.3.5 Using Decibels for Absolute Levels

We said that the decibel is a relative measure only, signal *a* versus signal *b*. What if *b* were a constant we all agreed on? We could then say some signal measured 12 dB, or that some noise was 115 dB, as long as we all agreed on the common value of *b* for both cases. We would retain the advantage of having a logarithmic scale. Different fields have done exactly this. The most familiar instance is in sound, where the threshold of hearing (corresponding to a sound intensity of 10^{-12} W/m^2) is used as 0 dB (i.e., as "signal *b*").[13] Telephone signals use a reference of 1 volt peak-to-peak into 600 ohms as their 0 dB. Properly speaking, it is necessary to somehow indicate the reference used (often tacking them onto the end of the dB, as in dBm, where *m* stands for milliwatt), but within an industry you'll often find the reference dropped—you may need to infer whether the measurement is relative or based on some reference. Common reference levels include dBm (1 milliwatt [mW] of power), dBW (1 watt of power), and dBV (1 volt into a specific impedance).

It's often the case that we use some *component* of a signal as the reference for other components. For example, we can pick the power of the signal at some frequency as the 0-dB reference. In a similar way, when describing a system, we might choose the gain[14] (actual or ideal) in the passband as the reference; the gain at other frequencies will be relative to the gain of frequencies passed. (We'll talk about describing systems in just a few pages.)

2.3.3.6 Attenuation and Gain, Negative dB

If *a* is smaller than *b*, the decibel measure is negative. Thus, a gain of -6 dB, if we're talking voltage signals, is a gain of about $10^{(-6/20)}=0.5$. We can alternatively say this is an *attenuation* of 6

---

13. Note that many sound measurements are concerned with how human ears respond to sound; hence, most sound measurements are in fact weighted by a frequency-dependent function to better describe how our ears respond, rather than just describe how much energy there is.

14. *Gain* can be read as *amplification*.

Describing Signals and Systems in Time and Frequency 43

dB (no negative sign). Note that because we're always comparing the magnitude ("absolute value") of a signal, which is always positive, it doesn't make sense to talk about a negative linear gain. In this sense, linear gain will always be positive; the question is whether it is smaller or larger than 1.[15]

2.3.3.7   0 dB and 0 Gain

You might have noticed that 0 dB is a gain of exactly 1. This is also called *unity gain*. You might also have noticed we can't talk about the decibel equivalent of a gain of 0—this would be "negative infinity" in terms of dB. This usually isn't a problem, as the filters and systems we'll create aren't perfect—there's always a little response creeping through.

2.3.3.8   -3 dB, the Half-Power Point

One figure that comes up time and time again is the value -3 dB. Plug this into the equation for decibels and you'll get approximately 0.707 (also known as $1/(\sqrt{2})$) as the linear ratio of two magnitudes. What is the ratio of the *power* of these two signals? Squaring both magnitudes is equivalent to squaring their ratio, so the ratio in terms of power is 1/2. So -3 dB means the output of a system has half the power of the input signal or that, at this frequency, a signal has half as much power as at its peak power frequency.[16] The -3 dB point is sometimes called the *half-power point*.

The -3 dB point is a common reference for SP—often a bandwidth is measured using this value. But note that this is still a relatively strong level for signals like audio; in most cases it's too much attenuation for signals you really want to pass and not enough attenuation for signals you want to stop. When we talk about real filters, we'll almost always have more stringent requirements.

See Table 2-3 for some examples of the relationships between decibels, power, and voltage.

---

15. How do we express a gain of -1 in decibels? The negative sign gets folded into the phase, and the gain of 1 (that is, the absolute value or |-1|) is just 0 dB.

16. The peak power is used as the reference in this case.

Decibels	Voltage ratio	Power ratio
∞ dB	∞	∞
120 dB	10^6	10^{12}
100 dB	10^5	10^{10}
80 dB	10^4	10^8
60 dB	10^3	10^6
40 dB	10^2	10^4
20 dB	10	10^2
10 dB	3.16...	10
6 dB	≈ 2	≈ 4
3 dB	$1.414... = \sqrt{2}$	2
0 dB	1	1
-3 dB	$0.70795... = 1/\sqrt{2}$	0.5
-6 dB	≈ 0.5	≈ 0.25
-10 dB	0.316...	0.1
-20 dB	0.1	0.01
-40 dB	10^{-2}	10^{-4}
-60 dB	10^{-3}	10^{-6}
-80 dB	10^{-4}	10^{-8}
-100 dB	10^{-5}	10^{-10}
-120 dB	10^{-6}	10^{-12}
$-\infty$ dB	0	0

Table 2-3 Voltage, power, and decibels for common values.

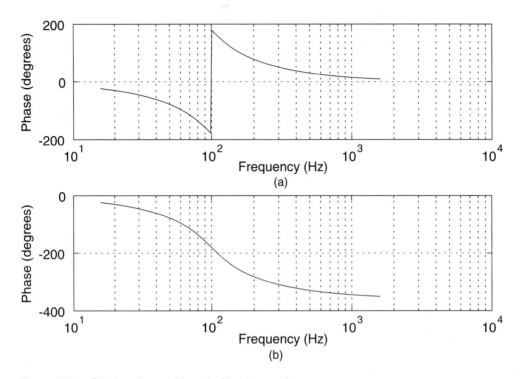

Figure 2-22 Plotting phase with and without wrapping.

Plotting Phase What about phase? Phase is usually happy with just a linear plot, and it's not uncommon to plot the phase in degrees rather than radians. The one anchovy on the pizza, as it were, is that usually just the range of +180 degrees to -180 degrees (+π to -π radians) is shown, and if the phase goes beyond this, it is just wrapped around to the other side. (See Figure 2-22.) Just be aware of this. Like we said earlier, for better or worse, a lot of filtering just ignores the phase, either because it's not important in the application or because it's understood that only certain types of filters that inherently don't adversely affect the phase will be used. Control systems, on the other hand, will bite you hard if you even think about ignoring phase.

2.3.4 Describing Systems

Filters are just one example of a *system*, something that takes an input signal, makes some modifications, and produces an output signal. Generally, a system has one or more inputs and one or more outputs, but most of the systems we'll talk about in this book are simple single-input, single-output systems. The input and output signals of a system don't have to be electrical—for example, it's useful to describe the human vocal tract in terms of a system, with signals consisting of acoustic energy. The same analysis can be done for economies, biological systems, mechanical systems, and so on, using appropriate inputs and outputs. We'll concentrate on elec-

Describing Signals and Systems in Time and Frequency 47

trical systems in this book, but it's not difficult to include electromechanical, hydraulic, or other types of systems in the analysis, which is done in texts on the related field of control theory.

Note that in some cases we're describing an existing, perhaps unchangeable, system, and in other cases we'd like to design a system to match some arbitrary description (as in filter design).

Just like signals, there are two different domains in which to describe systems: the time domain and the frequency domain. The problem is that systems are not signals—a system *modifies* signals. What we *can* do is describe what a system does to signals—its *response*.

2.3.4.1 Time-Domain Response

There are two common and mathematically useful ways to characterize the dynamics of a system in the time domain. The more important way is to observe its response (output) to an *impulse* input. The other, which is of less use in filter design, is to observe the response to a *step* input.

The Impulse Response *h(t)*   An impulse is, to a first approximation, a spike (see Figure 2-23). Both before and after the impulse, the signal is at 0, the impulse width is very small, and its amplitude is very large. In fact, mathematically, the width is infinitesimally small and the height infinite; however, if you integrate the signal (that is, find the area under the pulse), you get a total area of 1...but we really don't need to get this detailed. The important aspect of the impulse is that it excites a system equally at all frequencies, the same way you can take the same hammer and hit several different bells, each producing a different sound. The impulse is the "calibrated hammer" we'll use to hit systems, and see what "notes" come out. The impulse delivers all of its energy instantly, without interfering with the dynamics of the system. Figure 2-24 shows the impulse response associated with a simple system.

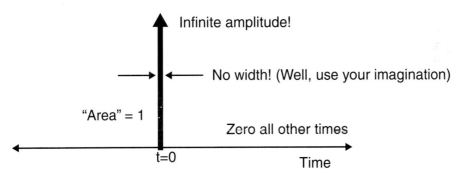

Figure 2-23 The impulse function.

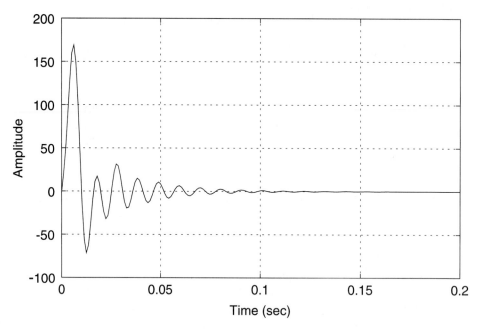

Figure 2-24 Impulse response of fourth-order elliptical filter.

By the way, just because we got bored and didn't plot the impulse response past 0.2 seconds doesn't mean the impulse is necessarily 0. In this case, we happen to know it's not, just very close. Because we always use a linear scale for signals in the time domain, it's difficult to see that there are still ripples long after the linear plot shows a straight line. The whole issue of whether the impulse response is finite or infinite in length (even if it is as close to zero as anyone could hope to calculate) is central to some things we'll do later with filters. (Not to spoil it, but we even split digital filters into two groups, those with infinite impulse responses and those with finite impulse responses.)

There are other "hammers" we could use to hit systems—the step signal below is one—but the impulse is particularly nice because, in theory, we can take the time-domain response (output of the system), convert this to the frequency domain, and have the frequency response function of the system, which we describe below. So the impulse response is a special type of response, and we'll hear a lot more about it when we design filters.

The notation $h(t)$ is commonly used for the impulse response of a system, and we often label a system with its impulse response. The notation for the impulse signal itself is $\delta(t)$. (See Figure 2-25.)

So where do you get an impulse generator? For our purposes, it is purely a mathematical tool, and the only systems we'll be hitting with it are mathematical ones. This doesn't mean that $h(t)$ is unimportant—it plays a central role in the design of digital filters. It's just not something

Describing Signals and Systems in Time and Frequency 49

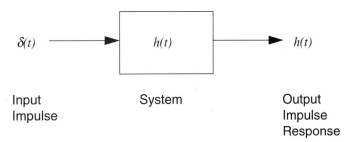

Figure 2-25 Labeling a system in the time domain.

we encounter in the physical world because in its pure form, it takes an infinite amount of energy to produce.

By the way, architects do use "impulse generators" on occasion. To assess the acoustics in auditoriums, they sometimes fire a gun![17] Of course, we have been to certain concerts where some minor gunplay might have led to an appreciable esthetic improvement, but the idea with the architect is to analyze the resulting "impulse response," rather than, say, to wing an off-key singer. Certainly there is some synergy between these two approaches, however. Killing two birds with—well, never mind.

Step Another signal used to characterize systems is the *step* (Figure 2-26). This signal is primarily used in control-system design; it's rarely used in digital filter design. The infinitely sharp edge of the step would, like the impulse, require an infinite amount of energy, but it's possible to generate a signal that's a pretty good approximation. Like the impulse, the quick-rising edge excites the system at all frequencies[18]; the new DC level is a kind of "DC" excitation. The output is known as a *step response*. (Are you surprised?)

2.3.4.2 $H(j\Omega)$: The frequency response function

The step function has utility for control work, and the impulse has important design uses, but it's really the *frequency response function, $H(j\Omega)$*, that we'll find most useful. For a system,[19] $H(j\Omega)$ describes both the gain and phase shift that a signal of frequency Ω experiences when going through the system. When considered for all Ω's, $H(j\Omega)$ gives a function of frequency that relates gain and phase shifts for any Ω. The frequency Ω is expressed in radians/second. The annoying j is just along for the ride, and for the moment you can just ignore it. Later on, it makes life easier.[20] $H(j\Omega)$ is also called the *transfer function* of the system, and it's also written in

---

17. With an audience? Actually, this might be preferable, as the acoustics are affected by the presence of an audience and even the acoustic properties of what they're wearing.

18. In the step case, however, frequencies are excited unequally. This must be accounted for in interpreting the step response.

19. Linear, time-invariant (LTI) systems, the type we're discussing in this book.

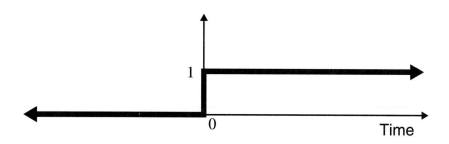

Figure 2-26 The step function.

some texts without the *j*, as just *H(Ω)*. Also, recall that ω and Ω are sometimes interchanged, depending on the text, so be prepared to see H(jω) too!

As an aside, *H(jΩ)*, as used in this book, is not the complete frequency-domain description of a system; it describes the behavior of the system only for repetitive input signals (signals that repeat periodically), and only once the system has had time to settle down. This is known as the *steady-state* condition. The factors ignored, such as glitches at start-up or when a signal changes quickly, are sometimes important, though they're outside the scope of this book. At the end of this chapter we've listed some texts that discuss these issues in detail.

H(jΩ) is complex-valued—that is, if you evaluate *H(jΩ)* at some frequency, the result is a complex number with both a magnitude $|H(j\Omega)|$, and phase $\angle H(j\Omega)$ (sometimes known as the argument or angle of the complex number). The magnitude is the ratio of the magnitude of the output to the magnitude of an input at that frequency, and the phase is the difference in phases between the output and input.

An example: If $H(j\Omega) = 142120/(142120+533j\Omega-\Omega^2)$, what is this system's (or filter's) output if the input is a sinusoid with a magnitude of 4 and a frequency 60 Hz? What about 190 Hz?

First, convert from Hertz to radians/second: 60 Hz=60·2π rad/sec, or roughly 377 rad/sec. This is Ω. Noting the *j* in the denominator, which means the calculations will have to be done using complex arithmetic, we end up with $H(j\Omega) = -0.000008 - 0.70729j$, or (in polar form) $H(j\Omega) = 0.70729 \angle -1.57$ (angle expressed in radians). Equivalently, we can write $|H(j\Omega)| = 0.70729$ and $\angle H(j\Omega) = -1.57$ rad. It's sometimes convenient to express the angle in terms of π; in this case, -1.57 rad = -π/2. Using our expression for t_d (Equation), we can calculate the phase delay as (positive) 0.004167 sec.

20. Notice we didn't say for *whom* it makes life easier.

Describing Signals and Systems in Time and Frequency

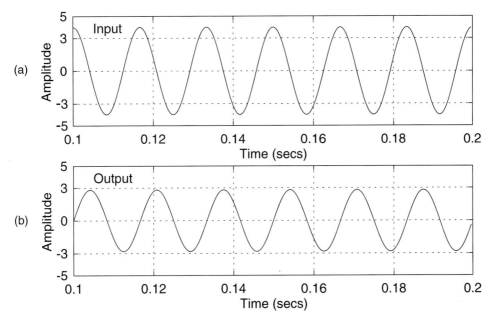

Figure 2-27 (a) Input and (b) output of the example system at 60 Hz.

Is there anything special about this output? Converting the magnitude ratio to decibels,[21] 0.70729 is very close to -3 dB (the "half-power" point). Likewise, converting from radians to degrees, we find the angle is -90°. So, for an input of magnitude 4 and frequency of 60 Hz, the output will have a magnitude of 4×0.70729 or 2.83, and the phase will differ by -90° from input to output, as shown in Figure 2-27.

Following the same procedure, we evaluate $H(j\Omega)$ at frequency 190 Hz (i.e., Ω=1194 rad/sec). The complex-valued result is $H(j\Omega) = -0.0889 - 0.0441j$, or, in the more useful magnitude/phase form, $H(j\Omega) = 0.0992 \angle -2.68$. Converting to decibels, the magnitude is close to -20 dB, while the phase difference is -154°. Using the same input magnitude as before, 4, the output magnitude is now 0.4 (that is, 1/10th the input). (See Figure 2-28.)

(In case you are wondering, we've started plots in Figure 2-27 and Figure 2-28 at 0.1 seconds to avoid the start-up behavior of the filters. Even in mathematical models, reality intrudes!)

$H(j\Omega)$ is a compact description of what a system (filter) does to a sinusoid at a particular frequency. The magnitude, $|H(j\Omega)|$, gives the scaling factor, and the angle, $\angle H(j\Omega)$, gives the change in phase for an input of frequency Ω (once the system reaches steady state). What is $H(j\Omega)$ of this particular example describing? Let's plot $H(j\Omega)$ for a number of values and see what the overall behavior is.

Figure 2-29 shows the preferred way of plotting $H(j\Omega)$ as two graphs,[22] each with the same horizontal axis, with a decibel scale for the magnitude and a logarithmic scale for the fre-

---

21. Unity gain (i.e., 1) is taken as 0 dB for these types of calculations.

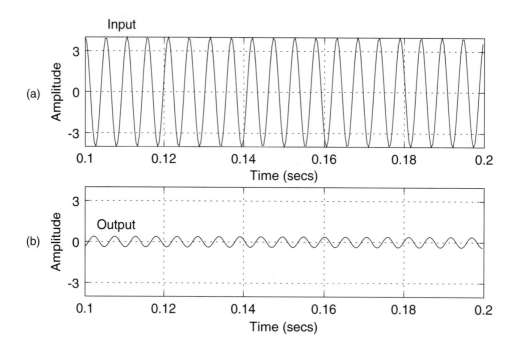

Figure 2-28 (a) Input and (b) output of the example system at 190 Hz.

quency—just as with signals. The magnitude and phase at 60 Hz and 190 Hz, from the examples above, are marked. We can see now that $H(j\Omega)$ describes a system that passes lower frequencies with no attenuation, but starts attenuating signals above 300 rad/sec or so. The phase goes through a transition as well, eventually approaching close to $-\pi$ rad, or $-180°$. This $H(j\Omega)$ describes a *low-pass* filter. The plot of magnitude versus frequency is known as the *magnitude response* or *magnitude spectrum*, while the plot of phase versus frequency is known as the *phase response* or *phase spectrum*.

2.3.4.3 $H(s)$: The System Function

$H(j\Omega)$ provides us with a very useful description of a system, and plots of magnitude response even show up on the back of cassette tape wrappers or home audio systems. However, most design work is done using a more general description, that of the *system function* or $H(s)$. The "only" difference between the two functions is that instead of using the simple frequency Ω, the *complex frequency s* is used. This difference will give us an extremely powerful tool for working with signals and systems. So, what is complex frequency?

22.   Some texts use the term *Bode plot* for these two graphs; other texts reserve this name for plots created by a set of rules that approximate the actual plot with straight lines.

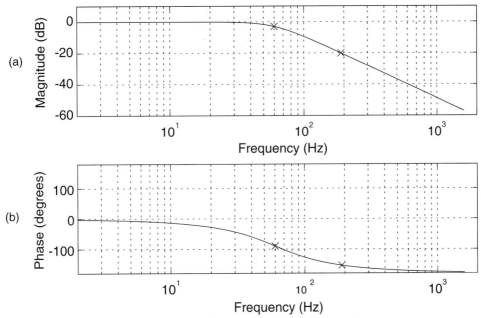

Figure 2-29 The (a) magnitude and (b) phase response of the system.

Signals with complex frequency are just ordinary sinusoids with exponential growth or decay of the amplitude. Instead of just describing the frequency alone, the complex frequency also describes this exponential growth. For example, we can write a time domain description of an "ordinary" sinusoid as $x(t) = A\cos(\Omega t + \theta)$. Here, A is just the magnitude of the signal[23] and θ (theta) is the phase. (The phase is not crucial to this discussion.) Of course, Ω is the ordinary frequency. To this we need only to multiply by the exponential term $e^{\sigma t}$ to get $x(t) = Ae^{\sigma t}\cos(\Omega t + \theta)$, an example of a signal with complex frequency. The σ (sigma) term affects how quickly and in what direction amplitude growth occurs. When $\sigma=0$, there's no change ($e^0 = 1$), when $\sigma < 0$, the signal decays, and for $\sigma > 0$, the signal grows over time.

Figure 2-30 shows a few examples of signals and the complex frequencies that are associated with them. Signal X has a constant amplitude ($\sigma=0$); signal Y is decaying, and, thus, has $\sigma<0$; and signal Z, which is exponentially growing, has $\sigma > 0$.

Just as magnitude and phase can be "bundled" into a complex number (recall $H(j\Omega)$), we likewise can bundle both frequency and the exponential factor into a complex number, which we'll call s. To make the math come out right, the real part is the exponential growth factor σ, while the imaginary part is just Ω, the ordinary frequency from before. Any value of s can be written as $s = \sigma + j\Omega$.

Are complex frequencies "real"? Well, you certainly can generate a sinusoid that exponentially decays (or increases, though not forever) and, thus, has a complex frequency associated

23. A is taken to be > 0; if not, you can make it so by adding $180°$ to the phase.

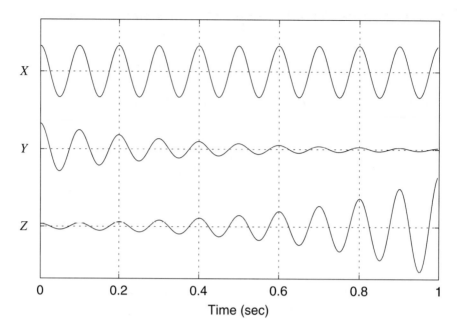

Figure 2-30 Example waveforms with complex frequency.

with it. But complex frequency is used as a mathematical tool to make working with systems a lot easier. In fact, whenever you see engineers using complex numbers, as they are wont to do, it's because it's easier than dealing with the individual components (e.g., magnitude/phase, frequency/exponential factor).

Complex frequency is used to represent signals and systems in the so-called *s-plane*. If we plot σ (the real part of *s*) along the horizontal axis (abscissa) and Ω (the imaginary part) along the vertical (ordinate), we can graphically describe systems by locating what we will later call the *poles* and *zeros* of the system. We'll do this mapping later, in the chapter on filters, but for now we can plot the complex frequency of a few signals in the *s*-plane, just to give you an idea of the relationship between frequency and exponential growth/decay as indicated in the plane (see Figure 2-31). For each of the twelve points indicated in the *s*-plane, we have drawn a (representative) signal associated with the complex frequency at that point. Needless to say, if someone asked you to plot the complex frequency associated with different signals on the s-plane, it is only necessary to indicate the point of the s-plane—our little waveforms are just for show.

Note that signals with complex frequencies in the left-half plane decay exponentially in time toward zero; signals with complex frequencies in the right-half plane, on the other hand, grow exponentially with time. Along the σ=0 (the vertical axis), the signal is a pure sinusoid—no change in magnitude over time.

Returning to the question of *H(s)* itself, by extending the description of a system, we end up with a far more complete description of the system than *H(jΩ)* provides. There are some aspects of *H(s)* we won't use here—such as the ability to account for the initial conditions of a

$H(\Omega)$, $H(j\Omega)$, and $H(s)$

Just so we don't feel too irresponsible, there's a bit more to the story of $H(\Omega)$, $H(j\Omega)$, and $H(s)$. In fact, these are all related to the *Laplace transform* (LT), which allows us to move between descriptions in the time domain (specifically, differential equations in time t) and descriptions in the s-domain (algebraic equations in complex frequency s). The great advantage is that derivatives and integrals in t turn into simple powers in the s domain—and believe us, polynomials are much easier to deal with than differential equations! $H(s)$ is, therefore, related by the LT to the differential equation that describes the system. (It's actually the LT of the impulse response of the system).

By the way, the mathematical definition of the LT is relatively simple-looking but can be difficult to evaluate for all but a handful of cases. Most folks just look up the appropriate equations in tables. Likewise, going from s to t gets even dicier for general cases, though for most real-world applications, inverse tables and some simple techniques are sufficient. We're going to leave you in peace on this subject, even though the LT is extremely powerful and useful in a variety of ways.

Because s of $H(s)$ is *defined* as $s=\sigma+j\Omega$, we can use the system function, when available, to give us the frequency response by just setting $\sigma=0$. This is why we chose to include the j in our system function—we've been covertly writing $H(\sigma+j\Omega)$ with $\sigma=0$, or $H(j\Omega)$.

And $H(\Omega)$? This is the *Fourier transform* (FT) of $h(t)$, the impulse response. The difference between the FT and the LT is that the FT is a function of simple frequencies only (not of complex frequencies); if you like, it is a limited case of the more general LT. The FT has an important role to play in steady-state analysis, but because we need the LT anyway, it makes sense to use just the LT notation all the time.

systems—but we'll make extensive use of the s-domain and $H(s)$ to design and describe filters. $H(s)$ still returns both a magnitude and a phase, just as $H(j\Omega)$ did, and if we set the exponential factor σ to 0, we'll end up with exactly the same plots as when we vary Ω in $H(j\Omega)$. In fact, we can use the relationship $s=\sigma+j\Omega$, reducing to $s=j\Omega$ when $\sigma=0$, to convert back and forth between the frequency response function $H(j\Omega)$ and the system function $H(s)$ as follows:

- To get $H(s)$ if you have $H(j\Omega)$, replace every Ω in $H(j\Omega)$ with s/j.
- To get $H(j\Omega)$ if you have $H(s)$, replace every s in $H(s)$ with $j\Omega$.

Example: A filter has a frequency response function $H(j\Omega) = \dfrac{1}{8+4j\Omega-\Omega^2}$. Find the system function.

Solution: Replace each Ω with s/j:

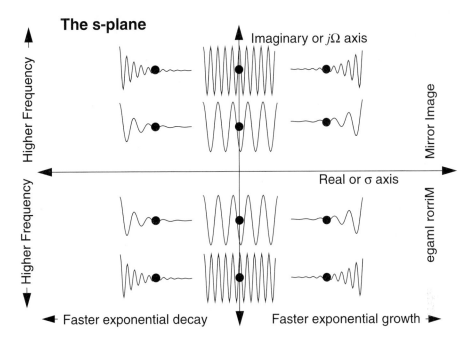

Figure 2-31 Signals with complex frequency, placed on the *s*-plane.

$$H(s) = H(j\Omega)|_{\Omega = s/j} = \frac{1}{8 + 4j \cdot \frac{s}{j} - \left(\frac{s}{j}\right)^2} = \frac{1}{s^2 + 4s + 8}$$

Reversing, we can find the frequency response function given the system function:

$$H(j\Omega) = H(s)|_{s = j\Omega} = \frac{1}{(j\Omega)^2 + 4j\Omega + 8} = \frac{1}{8 + 4j\Omega - \Omega^2}$$

That's it for now on the *s*-plane, system functions, and frequency response functions. In the next chapter we'll use these tools for something fun.

2.4 Nonelephant Biology and Linear, Time-Invariant Systems

There's an e-mail signature that goes like this: "Nonlinear physics is like nonelephant biology." We've been throwing around the phrase "linear system" as if it were something special, and we have some good news, bad news, and more good news for you about linear systems. It will explain a lot about what we can and can't do in analyzing systems. We also need to explain why we can get away with our "linear systems" blinders when the vast majority of the real world is nonlinear, as per the quote above.[24] But first, the first good news, the amazing properties of linear systems.

2.4.1 The Amazing Properties of Linear Systems

Almost all of the systems in this book are *linear systems*. We define linear systems as systems that have two very useful properties—scaling and superposition.

2.4.1.1 Scaling

The scaling property says that if a system produces $y(t)$ as an output when given an input of $x(t)$, then if $x(t)$ is scaled by some constant (say, k), the output will be similarly scaled ($k \cdot y(t)$). (See Figure 2-32.) Mathematically, you can verify that certain equations (which may be descriptions/models of analog systems or the actual system in the case of digital systems) have this scaling property, and others do not. Examples of equations that have the scaling property include

$y(t) = 3x(t)$

$y(t) = t \cdot x(t)$

$y(t) = dx(t)/dt$ (i.e., first derivative of $x(t)$)

The following equation does not have the scaling property:

$y(t) = x^2(t)$

The systems we'll create and analyze in this book will always obey the scaling property.

2.4.1.2 Superposition

Scaling seems pretty reasonable, but *superposition* will probably seem a bit strange at first. Superposition means that a linear system's effect on each frequency component of a signal is the same as if that component were the only signal passing through. The fact that other components (at different frequencies) are simultaneously present is irrelevant! Figure 2-33 shows an example. We'll take two different sine waves and pass them through the same system, and add the resulting outputs. Next, take the two original signals and add them together. Now pass this combined signal through the same system as before. The output of the system for the combined input is the same as the sum of the outputs for each signal individually!

Superposition may sound strange at first, but it makes a huge difference in how we approach SP. After all, we don't put just sine waves through our systems—we have extremely complicated signals to process. As long as our systems are linear, superposition means that the magnitude and phase response of our systems at a particular frequency are independent of what other signals are present. For example, the system amplifies or attenuates a 60-Hz frequency component the same way whether, say, a 1-kHz frequency component is also present. Now our earlier obsession with the frequency response function $H(j\Omega)$ might make more sense—it isn't just good for describing what happens to sinusoidal signals in isolation, but for sinusoids as components of arbitrarily complicated signals.

---

24. It must be said that, as important as they are, some engineers need nonlinear systems like elephants need bicycles.

Nonelephant Biology and Linear, Time-Invariant Systems

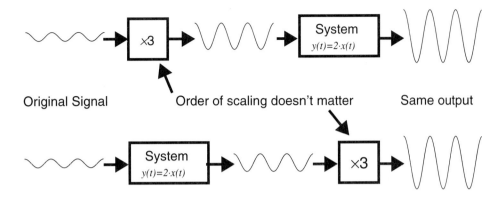

Figure 2-32 The scaling property.

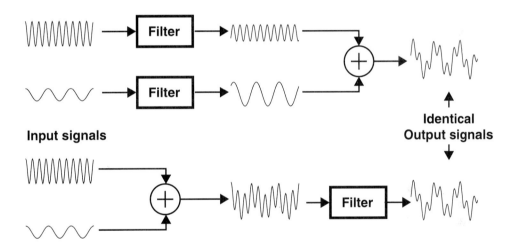

Figure 2-33 The superposition property.

2.4.1.3 Time-Invariance

Time-invariance is a separate property that a system may have. Linear systems don't have to be time-invariant, nor are time-invariant systems linear. However, most of the systems that we discuss in this book will be both linear and time-invariant. What does this mean?

Time-invariance is the property that the output of a system does not depend on *when* you present an input signal. You'll get the "same" output for a given input now or five seconds from now. Again, as we did for linear systems above, we can write the mathematical descriptions of some systems that are time-invariant and some that are *time-variant*. Time-invariant systems include $y(t)=3x(t)$, $y(t)=x^2(t)$ (even though it isn't linear!), and $y(t)=4$. Time-variant systems include equations that have t as a factor, such as $y(t)=t \cdot x(t)$. Unless we make a programming error, all of our systems will be time invariant.

2.4.2 Real Systems and Nonlinearity

The bad news about linearity is that there's practically nothing in the world that is truly linear. This was the point behind the quote about nonlinear physics. Look at any system, electrical or physical, and you'll always find limits to its linearity. For example, an audio amplifier can ouput signals only up to a certain amplitude, say 60 volts peak to peak; ask for anything more and the signal will become distorted ("clipped," in this case)—that is, nonlinear. Below this limit the amplifier is really only approximately linear—the real-life inductors, op-amps, and the like are contributing to nonlinearity.[25]

A common measure of nonlinearity is *harmonic distortion*. Harmonic distortion can be thought of as a measure of the distortion (the energy in other "manufactured" frequencies) in the output if the input is a perfect sine wave (i.e., one frequency). Certainly we can try to ensure the DSP systems we design are as close to linear as possible, but other elements in our system, such as mechanical arms, transducers, and so on, are to some extent nonlinear. So what's the good news?

The good news comes with some cautions, but basically it is this: Much of the time we can either find a reasonably linear region of a system to use, or can linearize[26] a nonlinear region. Because our tools assume a linear system, we'll try to make things look linear (or pretend they already are). For example, Figure 2-34 shows the resistance of a thermistor over a range of temperatures. (If you like, the thermistor has temperature as an input and resistance as an output. A transducer is a special case of a system.) As long as you don't need a wide range of temperature sensing, many sections of the response might be linear enough to use just as is. For a wider range, some linearization (e.g., a look-up table in software) might be necessary.

This really isn't any different from writing software—there are limits to how big or small arguments can be (even floating-point representations have a specific, though usually large, range of magnitudes they can represent). Any real-life linear system, to work well with other systems, has explicit or implicit limits to its actual linearity; stay within the acceptable input range, and the output will be reasonably close to linear.

The bottom line? It's a fundamentally nonlinear world, but the most prevalent tools in DSP assume linear (and time-invariant!) systems. We often get by with these tools since many systems are "linear enough" (or can be made so), but just because we have a flashlight doesn't mean

---

25. Capacitors can really depart from the ideal. Not only do they have leakage, variations with temperature, and the like—even some acoustic pickup—but some types exhibit a "vampire effect" (okay, most folks call it the *dielectric absorption effect*). Apply voltage to a capacitor (charging it up), discharge it rapidly down to zero, then watch with amazement as the capacitor "rises from the dead" and seems to charge itself up to maybe 1/10th the original charge. That sounds like an entirely new category of nonlinearity! (Bring a sharp wooden stake.)

26. A function, perhaps a mathematical equation or a look-up table, can take nonlinear values and map them to a linear range; this operation is called *linearization*.

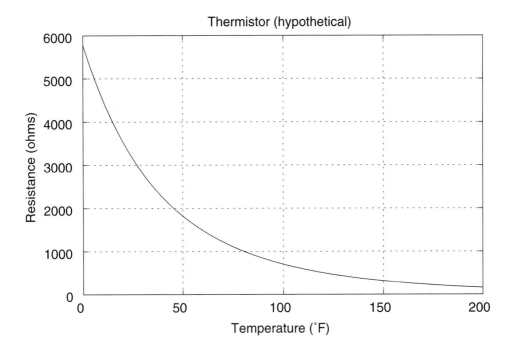

Figure 2-34 A nonlinear transducer.

it's daylight. In engineering, the "proof of the pudding" is often whether the design works well for the intended purpose. Linear techniques often yield excellent pudding.

2.5 Summary

Analog signals are continuous in both time and amplitude value. An analog signal can be described completely in either the time domain or the frequency domain. Descriptions in the frequency domain include both the magnitude and phase of the frequency components present.

Signals are plotted and described using such conventions as logarithmic relative measures (dB) and logarithmic frequency scales.

Systems can likewise be described in both the time and frequency domains. Because they aren't the source of a signal, instead we describe the signals that are the response of the system. In the time domain, we use the impulse signal as an input, and record the impulse response. In the frequency domain we ask how each frequency is affected (magnitude and phase) by the system.

The frequency response function $H(j\Omega)$ compactly describes the behavior of a system in the frequency domain. A more complete description, the system function $H(s)$, extends the idea of frequency to include complex frequency. At the moment we really haven't seen the real power of the *s*-domain, but it will be central to future material.

Linearity and time-invariance are two very important properties of systems; though most systems in real life are nonlinear, often we can find (or make) enough linearity to use the tools this chapter has described. Concepts including the frequency response function, system function, and future topics like "convolution using the impulse response" all require linear systems. By extension, the systems (digital filters) we design in the chapters that follow will also be linear systems.

Resources

A few books on linear systems and signals include:

Carlson, G.E., *Signal and Linear System Analysis*, Boston: Houghton-Mifflin, 1992.
Chen, C.-T., *System and Signal Analysis*, New York: Holt, Rinehart, and Winston, 1989.
Gabel, R.A. and R.A. Roberts, *Signals and Linear Systems* (2nd edition), New York: Wiley, 1980.
Glisson, T.H., *Introduction to System Analysis*, New York: McGraw-Hill, 1985.
Kamen, E.W., *Introduction to Signals and Systems* (2nd edition), New York: Macmillan, 1990.
Lathi, B.P., *Linear Systems and Signals*, Berkeley: Cambridge Press, 1992.
McGillem, C.D. and G.R. Cooper, *Continuous and Discrete Signal and System Analysis* (3rd edition), New York: Holt, Rinehart, and Winston, 1991.
Oppenheim, A.V. and A.S. Willsky, *Signals and Systems*, Englewood Cliffs, New Jersey: Prentice-Hall, 1983.

General Electronics:

Horowitz, Paul and Winfield Hill, *The Art of Electronics* (2nd edition), New York: Cambridge University Press, 1989.
Highly recommended book on electronics in general. Includes basic electronics, op-amps, digital electronics, analog-to-digital and digital-to-analog conversion, low-noise design, transducers, and a host of other topics. Emphasis on proper *and practical* engineering.

Pease, Robert A., *Troubleshooting Analog Circuits*, Boston: Butterworth-Heinemann, 1991.
Pease is by all accounts quite a character, and the book is written in his inimitable style. Not just troubleshooting, but issues that should find their way into the analog design process. Pease designs integrated circuits for National Semiconductor. (A fun read, too.)

CHAPTER 3

Analog Filters

To understand the material in this chapter you should be comfortable with the frequency domain descriptions of signals and systems that are covered in Chapter 2 (Analog Signals and Systems). We will build on our discussion of the frequency response function $H(j\Omega)$ and the system function $H(s)$ from the previous chapter, and so will again resort to the occasional complex number. You don't need to know anything about analog filters, filtering, or how to change the oil in your car; however, your engine will usually last longer with frequent oil changes.

3.1 Overview

Now that we know how to describe analog signals and systems, we turn our attention to filters, by which we usually mean systems that modify the frequency content of signals. We're starting with *analog* filters for two reasons. First, the same basic specifications are used for both analog and digital filters, so we can immediately begin discussing filter specification even though we have not yet gotten to the issues of discrete time (that's the next chapter). Second, and just as important, one class of digital filters, the IIR filter, uses an analog filter as a design starting point, so we need to be able to specify—and design—analog filters eventually, even if we plan never to build a single one. (We'll leave the designing to a later chapter.)

In this chapter, we'll also discuss poles and zeros, which fall out of the system function $H(s)$. Poles and zeros are an absolute necessity for our work in future chapters—they capture crucial information about filters, and we'll use them extensively in both the continuous- and discrete-time domains.

By the end of this chapter we'll be ready to tackle discrete-time signals and systems, a discussion which, in turn, leads to the digital filters you've been waiting for.

3.2 Purpose of Filters

In this book, when we talk about *filters,* we're talking about linear, time-invariant systems with the following characteristics:

- magnitude response: selectively pass (perhaps amplify) or reject (attenuate) certain frequencies;
- phase response: change the phase of the signal by different amounts at different frequencies;
- stability: produces a finite output in response to any finite amplitude input;[1]
- rise time: output changes at a certain rate in response to an instantaneous change in input;
- settling time: output takes a certain amount of time to settle down to a new value in response to an instantaneous change in input;
- overshoot: output may temporarily exceed the "proper" output value for an instantaneous change in input.

The first two are steady-state characteristics—that is, characteristics of the filter after it has been processing a signal for a while. Steady-state characteristics can be found from $H(j\Omega)$.

Any filter we design should be stable—that is, we don't want the filter to get into a vicious feedback loop and try to output ever-increasing values. Among linear, time-invariant systems, this is only a problem when the system incorporates feedback, so we'll let this topic lie for a couple of chapters. Besides, no stable person is going to ask you for an unstable filter.[2]

The remaining characteristics are *transient* characteristics, which for our purposes are specified in the time domain. High-frequency designers do need to worry about issues like the slew rate (related to rise time) of op-amps and such, and of course the control-systems folks worry about everything. We're not going to go into detail about transient characteristics in this book. For filter design, these effects are usually of secondary importance compared with the steady-state characteristics, and they end up involving quite a bit more math. We'll have our hands full enough with steady-state issues!

3.3 Examples

Figure 3-1 shows a signal with some 60-Hz interference added, perhaps as the result of using unshielded cable. Note that we aren't showing the phase plots in the frequency domain, just the magnitude. Figure 3-2 shows how we might apply a filter to remove 60-Hz signal, while

---

1. This is actually only one definition of stability.
2. However, an oscillator is technically not stable—sometimes it is said to be *marginally stable*—and you may actually want to design such a system.

leaving the original largely unchanged. The filter is an example of a "notch" filter, and here selectivity attenuates signals in the region of 60 Hz, while passing other frequencies. This filter has a frequency response function

$$H(j\Omega) = \frac{0.891\Omega^4 - 0.0255\Omega^2 + 0.000182}{\Omega^4 - 0.0192j\Omega^3 - 0.0290\Omega^2 + 0.000275j\Omega + 0.000204}$$

and a system function[3]

$$H(s) = \frac{0.891s^4 + 0.0255s^2 + 0.000182}{s^4 + 0.0192s^3 + 0.0290s^2 + 0.000275s + 0.000204}$$

This happens to be a fourth-order "elliptic" (aka "Cauer") notch filter, though that really doesn't matter right now. We'll talk about elliptic filters in Chapter 6 (IIR Filters—Digital Filters with Feedback).

Figure 3-3 shows the magnitude response of an ideal *lowpass* filter, superimposed (*dotted line*) over the plot of a typical real-life filter. The ideal lowpass filter passes signals (or components of signals) up to a certain frequency and attenuates all higher frequencies. The real filter is not so absolute—notice the uneven magnitude for low frequencies and the gentle slope instead of the sharp cutoff of the ideal filter. This is called the *transition region*, a range of frequencies neither fully passed nor fully attenuated. Why is this real filter so wimpy? Is it possible to create an ideal filter? Read on.

3. Recall, you can replace $j\Omega$ with s to switch between the two representations.

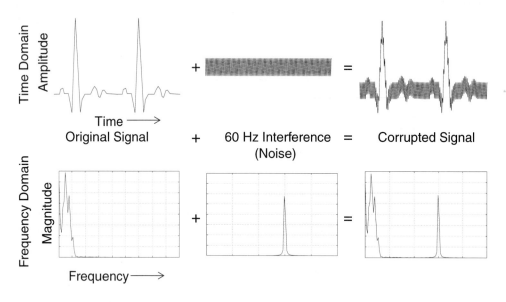

Figure 3-1 Adding 60-Hz interference to a signal.

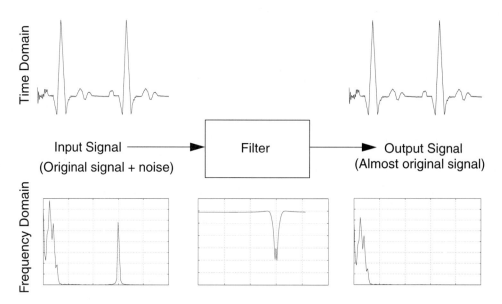

Figure 3-2 Filtering to remove 60-Hz noise.

Ideal Versus Real Filters

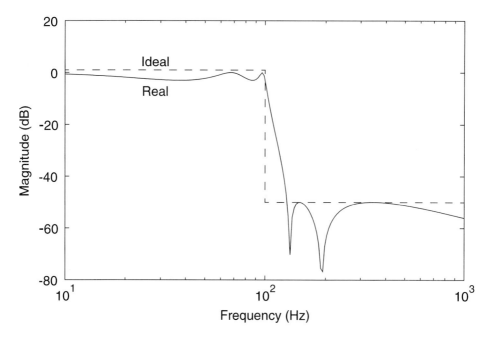

Figure 3-3 Ideal vs. real lowpass filter magnitude response.

3.4 Ideal Versus Real Filters

3.4.1 The "Brick-Wall" Magnitude Response

The ideal filter response in Figure 3-3 is sometimes called a *brick-wall* response, but the only place you'll see this response is in textbooks—the fact is, we can't make one in real life. Why? The full answer will have to wait for a few more chapters, but then we'll be able to show that to have a true brick-wall response would require a filter with an infinite number of coefficients *and* the ability to see into the future. Such a filter would be *noncausal*; that is, it would respond to inputs that had not happened yet. Obviously, we'll require all of our filters to be *causal*.[4] So the short answer is that with a finite number of coefficients and without strange psychic powers,[5] we're always going to have *some* slope to the filter response curve.

3.4.2 Distortion

Filters can produce three types of distortion:

  • Amplitude distortion: Different frequency components are amplified at different gains;

---

4. This is distinct from *casual* filters, which usually wear jeans and t-shirts.

5. We predict you will read this footnote. Here is a design project: Use the Internet to incorporate a 900-number "Psychic" line into a brick-wall filter design package. (You may keep any patent rights.)

- Nonlinear distortion: Caused by nonlinear systems (which, of course, we will avoid in our designs);
- Phase distortion: Different frequency components are delayed by different times.

We'll avoid nonlinear distortion by designing only linear systems (famous last words!). And the whole reason for designing filters is to distort the amplitude, right? But phase distortion is a problem for signals like video, where the shape of the signal must be preserved, which is the same as saying that each frequency component must experience the same time delay. (Recall what changing the phase relationships did to the squarewave earlier.) Uniform delay results in a *linear phase* response, which we'll talk more about below. Surprisingly, this is an ideal we can actually achieve with some filter designs.[6]

3.5 Filter Specification

3.5.1 Types

One way of organizing filters is by their frequency magnitude characteristics. Common categories are lowpass, highpass, bandpass, band-reject, arbitrary (or multi-passband), all-pass, and comb.[7] Below are the most common types of filters. In a few pages we'll look at a precise way of describing what we want in a filter. Right now we just want the general idea of each type. Note that each is named in terms of its magnitude response—phase gets the short end of the stick again.

3.5.1.1 Lowpass

Figure 3-4 shows the magnitude response of an example *lowpass* filter. Frequencies from zero to some specified frequency (the "cutoff frequency") are passed with relatively little change in magnitude, while higher frequencies are greatly attenuated. Lowpass filters are generally used when the signal of interest is in the range of DC (0 Hz) to some frequency, but where other high frequency signals (i.e., noise) are present. Note that there is a gain of about 1 (0 dB) at DC.[8] The lowpass is the most common filter type by a wide margin.

3.5.1.2 Highpass

Figure 3-5 shows a *highpass* filter magnitude response. In reality, of course, there will be limits on the upper frequency that can be passed—for analog circuits this might in the kHz, MHz, or even GHz range, while the sampling rate of our DSP hardware will set the upper frequency limit as we'll describe in the next chapter. Note that this filter type (ideally) has "infinite" attenuation at DC. When the cutoff frequency is placed relatively low (say 1–100 Hz), such an analog circuit

---

6. What? You thought there's some fine print or something? Really, it's true. Trust us.

7. Analog filters are also known by the design method or type of optimization associated with the filter. For example, a Butterworth lowpass filter is designed using a specific equation and has certain characteristics. We'll discuss these categories later in Chapter 6 (IIR Filters - Digital Filters with Feedback).

8. 0 dB relative to what? To the ideal passband gain.

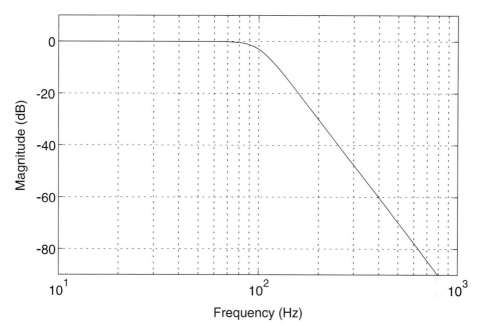

Figure 3-4 Example lowpass filter magnitude response.

might be used for "AC" coupling—that is, removing the DC from a signal and passing, in effect, everything else. (AC coupling is common in audio applications.) For highpass filters, the cutoff frequency is the lower limit of the passband.

3.5.1.3 Bandpass

Figure 3-6 shows a *bandpass* filter magnitude response. There are two cutoff frequencies, one on the low side and one on the high. You could get this response by "cascading" (follow one with the other) a lowpass and a highpass filter, but it's more common to design a single filter with this response. Figure 3-7 is also a bandpass filter, but its passband is much narrower. If the passband is fairly narrow (e.g., the ratio of the upper cutoff frequency to lower cutoff frequency is less than about two), it is more common to specify the center frequency of the passband and the bandwidth. A narrow bandpass filter may also be called a *resonator* or *peaking* filter.

3.5.1.4 Bandstop (Band-Reject, Notch)

The *bandstop* filter in Figure 3-8 is the opposite of the bandpass—this time a specific range of frequencies is attenuated. A common application for this type of filter is to remove a specific single frequency (rather than, say, a range of frequencies). Thus, 60-Hz band-reject filters are used to remove the 60 Hz "hum" that's often picked up from 120-volt AC house wiring. The example earlier in this chapter (Figure 3-2) illustrates this application.

3.5.1.5 Arbitrary or MultiPassband

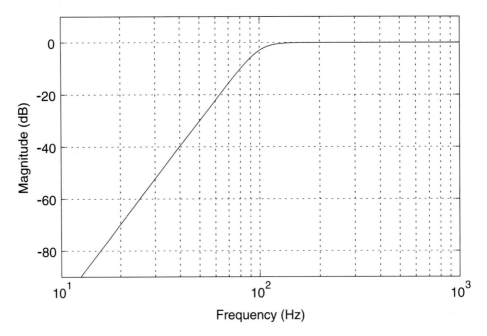

Figure 3-5 Example highpass filter magnitude response.

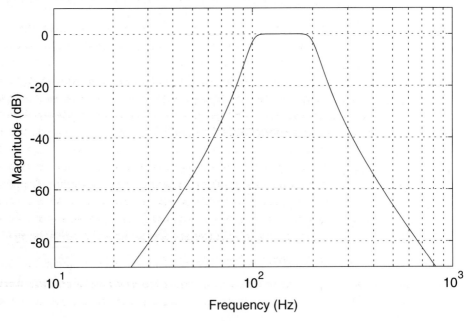

Figure 3-6 Example bandpass filter magnitude response.

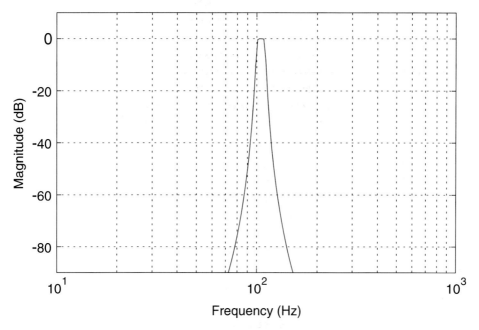

Figure 3-7 Example narrow bandpass filter magnitude response.

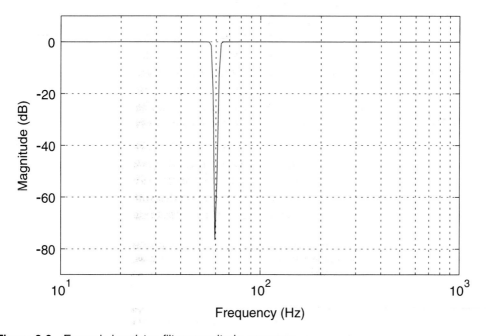

Figure 3-8 Example bandstop filter magnitude response.

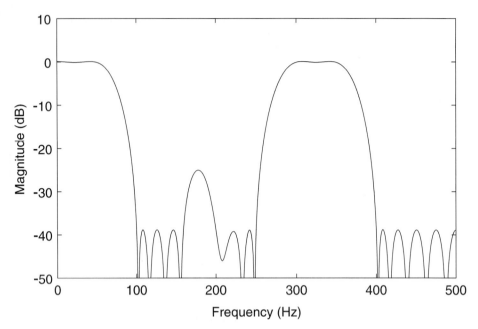

Figure 3-9 Example arbitrary passband filter magnitude response.

Using filter design programs, one may also specify multiple passbands as in Figure 3-9 or, alternatively, multiple reject bands. By the way, these plots are of actual filters, so the "humps" at low gains are not intended, just the best the filter design program could do under the constraints given.

3.5.1.6 All-Pass

We weren't straining ourselves plotting Figure 3-10! An *all-pass* filter doesn't look like it's doing anything at all—it passes all frequencies with uniform gain (again, up to a reasonable upper limit, as earlier). But it may be used to modify the phase of the signal, perhaps compensating for some prior operation (such as another filter). The phase response of a particular all-pass filter is shown in Figure 3-11. One application for the all-pass filter is to restore speech waveforms that have been distorted (i.e., due to nonlinear phase) by a microphone.

A filter whose output is simply a delayed version of the input is an all-pass filter, and one that has linear phase at that. In this case, none of the individual frequencies has its magnitude affected, while the delay produces a change in phase that is proportional to the frequency (a short delay produces a much larger phase shift for high-frequency signals than for low-frequency signals).

3.5.1.7 Comb

A special case of the multipass band filter is the *comb* filter, shown in Figure 3-12. In this case, there are specific relationships among the frequencies of attenuation, which are helpful in

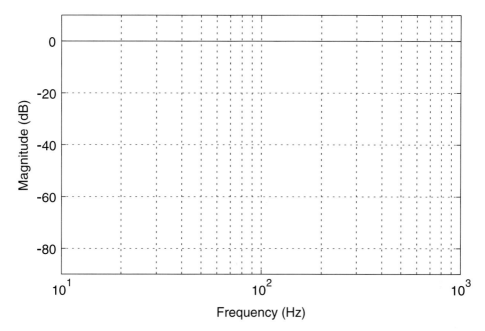

Figure 3-10 Example all-pass filter magnitude response.

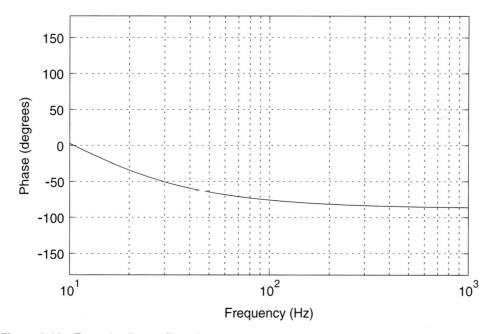

Figure 3-11 Example all-pass filter phase response.

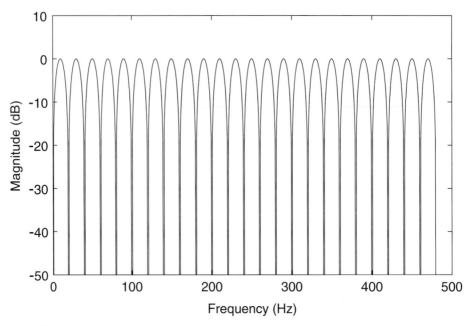

Figure 3-12 Example comb filter magnitude response.

removing both a specific unwanted frequency and its harmonics (multiples of that same frequency). The filter we plotted removes frequencies at roughly 20 Hz, 40 Hz, 60 Hz, and so on. Comb filters can be particularly simple to implement using digital filters (see Chapter 5 [FIR Filters—Digital Filters without Feedback]). Periodic signals other than sine waves will have harmonics, and comb filters are more efficient than having individual filters for each harmonic. For example, see Figure 3-13 for the spectrum of a squarewave at 20 Hz. (Note that a squarewave has only odd-numbered harmonics, so spectral lines appear at 20 Hz (the "fundamental"), 60 Hz (third harmonic), 100 Hz (fifth harmonic), etc.)

3.5.2 Specifying Magnitude Response

Figure 3-14 shows an example lowpass filter magnitude response with important characteristics marked.[9] And, well, Figure 3-15 shows an example lowpass filter magnitude response, too. The difference, you'll notice, is how the allowed variation in the passband is described—as peak-to-peak or just peak deviation. (We'll get into this in just a moment.) Figure 3-16 shows the same

---

9. And yes, we've kind of cheated with the decibel equivalents along the right edge—the plot itself is linear and not logarithmic—but you get the idea. Not to mention the fact that we're using F (frequency in Hz) when you know we'll have to convert to rad/sec eventually. However, most designs are specified in Hz to begin with.

Filter Specification

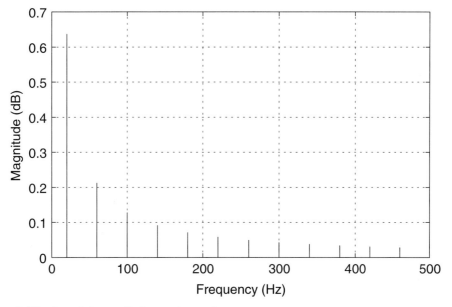

Figure 3-13 A partial magnitude spectrum of a squarewave.

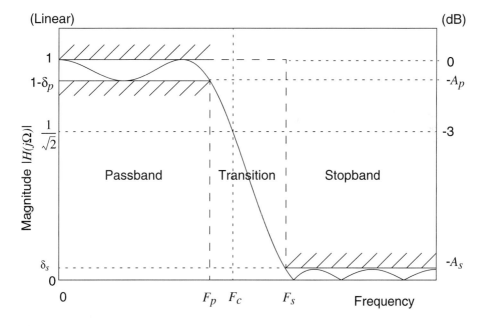

Figure 3-14 Lowpass filter parameters— "analog filter" style.

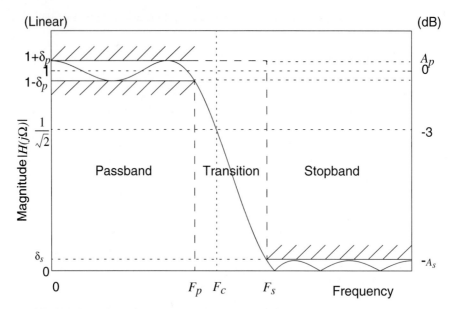

Figure 3-15 Lowpass filter parameters— "'digital filter" style.

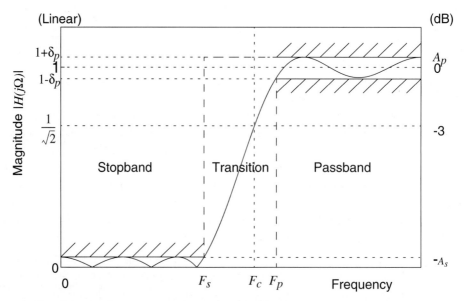

Figure 3-16 Highpass filter parameters— "'digital filter" style.

characteristics for a highpass filter (you can also use the style of Figure 3-14 for highpass filters). As we discuss each of these parameters in greater detail, we'll also see that a few other types of filters use additional or alternative parameters too, but the lowpass is a good example to start with.

Note that the gain of the magnitude frequency response is normalized (more or less) to 1 (0 dB) for a particular frequency, depending on the filter type; for lowpass filters the reference frequency is DC (0 Hz), while for a highpass filter it is at $F=\infty$ or at the highest frequency for the system (which for a digital system is *not* ∞!). By the way, this doesn't mean your filter can't have a gain of more than one; it's just a convention that most filter design methods assume. It's no problem to scale the output as you like. Once you have the overall magnitude response curve (in decibels), you can just add or subtract a constant decibel gain, which is equivalent to multiplying the whole curve by a constant linear gain.

3.5.2.1 Passband Cutoff or Edge Frequency

The *passband cutoff frequency* or *edge frequency* marks the edge of the passband; put another way, it is the highest frequency with a magnitude response that is within the passband ripple tolerance. Frequencies lower than the cutoff frequency are in the passband, and frequencies higher than this are in the transition band (and then the stopband).

F_C, the 3-dB point, is sometimes used as the implicit passband cutoff frequency; it also plays a role in some filter design methods (e.g., Butterworth, discussed in Chapter 6[IIR Filters—Digital Filters with Feedback]) even when both the passband and stopband cutoff frequencies are used. For most designs, F_C will be in the transition band, as shown in Figure 3-14.

3.5.2.2 Passband Ripple

Passband ripple is a measure of the allowed variation in magnitude response in the passband of the filter. Passband ripple is often specified in terms of δ_p, also known as the *passband deviation*. There are two different ways of defining ripple, which we can roughly call *analog* and *digital style* (our names). One method specifies the maximum deviation measured from a gain of one and is associated with analog filter design. (For passive analog filters, the maximum gain is less than or equal to one.) This is the method illustrated earlier in Figure 3-14. A second method measures the deviation from the ideal passband magnitude to the minima and maxima (vs. peak-to-peak of the first method.) The difference between the two methods is roughly a factor of two, so it's worth being clear on definition when a passband ripple is specified. The safest and most common way would be to interpret the deviation in the analog filter style (i.e., peak-to-peak).

In addition to worrying about measurement of passband ripple, the units given may be in one of three different forms: decibels (A_p), deviation (δ_p), or in terms of ε_p. (ε_p is rather specific to certain types of filter design, so we'll just leave this to Chapter 5 [FIR Filters—Digital Filters without Feedback].) The relationship between A_p and δ_p depends on the style; for the analog style, you can use:

$$A_p = -20 \log_{10}(1-d_p) \qquad (3.1)$$

and

$$\delta_p = 1 - 10^{-A_p/20} \qquad (3.2)$$

while for the digital filter style, you can use:

$$A_p = 20 \log_{10}(1+\delta_p) \qquad (3.3)$$

and

$$\delta_p = 10^{A_p/20} - 1 \qquad (3.4)$$

For example, a filter may be specified as having a magnitude response in the passband within +/- 0.01 of unity gain—that is, δ_p=0.01, and we're clearly using the "digital style." Using Equation 3.3, A_p is 20 $\log_{10}$(1.01)=0.086 dB.

3.5.2.3 Stopband Ripple/Minimum Attenuation

Stopband ripple (*minimum stopband attenuation*) describes the maximum gain (or minimum attenuation) you want for signals above the stopband cutoff frequency. Like passband ripple, it can be specified as either a linear or a decibel value. A positive value in decibels, like 40 dB, should be read as minimum stopband *attenuation*, while a negative value (e.g., -40 dB) would be maximum *gain*. In both cases, the linear gain for this example would be 0.01. In this book, we'll generally use attenuation.

3.5.2.4 Stopband Cut-Off or Edge Frequency

As with the passband cutoff frequency, all frequencies above this frequency will meet the stopband ripple tolerance. Stopband cutoff is usually abbreviated as F_s or F_{stop}. (Note that sometimes F_s is also used for the sampling frequency.)

3.5.2.5 Center Frequency

Figure 3-17 shows a narrow bandpass filter. The *center frequency* is the point of maximum gain (or maximum attenuation for a notch filter). Note that because we often use a log scale for frequency, the center frequency *appears* to be in the middle of the high and low cutoff frequencies for many types of bandpass filters. In fact, the center frequency is at the geometric mean, which is $\sqrt{F_{high} \cdot F_{low}}$.

3.5.2.6 Bandwidth

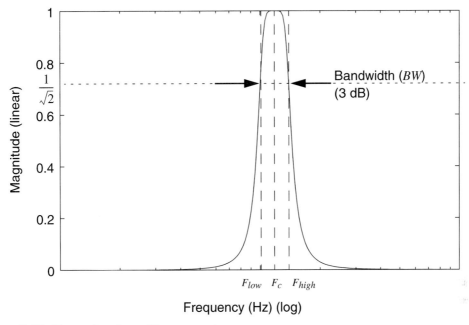

Figure 3-17 Narrow bandpass filter parameters.

The *bandwidth* is the difference between the two cutoff frequencies for a bandpass filter. If there is no maximum attenuation specified in the passband, the -3 dB points are often used. (Recall that -3 dB are the points of half-power). $BW=(F_{high}-F_{low})$.

3.5.2.7 "Q" or "Quality" Factor

Another way of describing a narrow peak or notch filter is in terms of Q, or the *quality factor*.

$$Q = \frac{F_{center}}{BW}, \text{ or } Q = \frac{\sqrt{F_{high} \cdot F_{low}}}{F_{high} - F_{low}} \qquad (3.5)$$

If expressed in terms of Ω, we need to include a 2π term in the first equation:

$$Q = \frac{\Omega_{center}}{2\pi BW}, \text{ or } Q = \frac{\sqrt{\Omega_{high} \cdot \Omega_{low}}}{\Omega_{high} - \Omega_{low}} \qquad (3.6)$$

There are many cases in which having a high Q factor is desirable—in radio receivers for example, to pick out a single carrier frequency to the exclusion of nearby channels. With analog circuits, maintaining a high Q can be difficult using components such as resistors, capacitors,

Filter Specification

and inductors. However, even digital designers are familiar with a very high-Q device that exhibits relatively little drift compared to these components—quartz crystals, the core of crystal oscillators that are all-pervasive in digital and microprocessor design. So somewhat ironically, any high-Q digital filters we might build eventually actually end up depending upon the high-Q analog characteristics of quartz crystals for a clock signal.

3.5.3 Specifying phase response

As noted above, we usually don't worry about the phase characteristics of a filter except in two cases: first, when we require linear phase; and second, when we are trying to compensate for phase distortion induced by some prior operation.

3.5.3.1 Linear phase

If every frequency experiences the same time delay, the result will be a linear change in phase. (For example, delaying 1 msec won't affect the phase of a 1 Hz signal much, but means a huge change to a 500-Hz signal. Signals in between will have some intermediate change in phase. Equation 2-1 describes mathematically the relationship between phase, time delay, and frequency.) Figure 3-18 shows a linear phase response for an example filter. Note that in the passband the phase wraps a few times (jumps of 2π), but in the stopband the phase has actual discontinuities in the phase, jumps of π that are due to the "zeros."[10] (For our purposes, we just

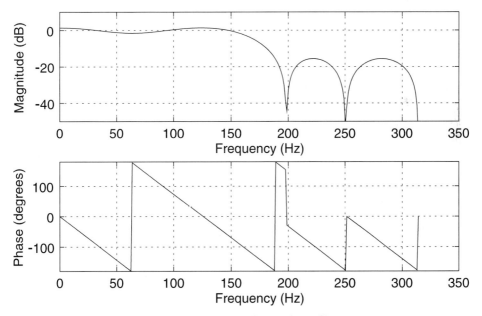

Figure 3-18 Magnitude and phase spectra for a linear-phase filter.

10. Zeros are discussed later in this chapter.

worry about linearity in the passband—stuff in the stopband is hopefully attenuated so much it doesn't matter what phase it has.)

Note that when we talk about "linear phase," we are always talking about the particular linear phase response that results in the same time delay for all frequencies. There are other phase responses that are linear, but not at the correct slope, and so they do not produce uniform delays at all frequencies. A particular case is that of *constant* phase. A constant phase response means frequencies will experience delays (inversely) proportional to their frequency—which will distort the waveshape. In contrast, filters with linear phase pass signals without phase distortion. This property is important in communications, music, data transmission, and other applications where the temporal relationships between different frequency components are important.

What about zero phase? Unfortunately, this would leave us zero time to perform any filtering, a quite serious design constraint, to say the least![11] Put another way, this implies a non-causal system, because the output is simultaneous with the input. If phase distortion is an issue, linear phase is more than sufficient and requires only that we tolerate an overall time delay through the system.

Under certain circumstances, linear phase digital filters can be designed—in fact, we just designed one for Figure 3-18 . These filters will have no feedback, as we'll see in a few chapters, and are a special case of what are known as FIR filters. We get an added bonus from such filters as there is symmetry to their coefficients. This symmetry can be exploited to reduce the storage and number of math operations to implement the filter. On the flip side, linear-phase filters usually require more operations than filters without linear phase, so linear phase is normally not something we specify unless there is a compelling reason to preserve waveshapes.

3.5.4 Transient Response

Fast enough for the user not to notice.[12]

3.6 Analog Filter Implementation

This isn't a book about analog filters per se, but it's useful to briefly discuss the common implementations of analog filters for a couple of reasons. First, as we'll discuss later, analog filtering is required in front of most analog-to-digital converters and again on the output of digital-to-analog converters. You may not be doing that design, but it will be helpful to know the general concepts. Second, the job of engineers is to solve problems, and for as much as we might like to use DSP techniques, sometimes the best solution is an analog filter.

---

11. That is, unless you have implemented the "900-number psychic line" filter design program suggested earlier.

12. Well, we told you that we weren't going to talk any more about it.

3.6.1 Passive Analog Filters

Analog filter implementations can be split into two categories, passive and active. *Passive* means there is no amplification in the circuit, while in an *active* filter an op-amp or transistors ("active" devices) provide gain. The most common filter of any type is the lowpass *RC* filter of Figure 3-19. Made up of a resistor (*R*) and capacitor (*C*), for a total cost of perhaps 10 cents, you're guaranteed to find some wherever there are analog signals. Even a single resistor or capacitor is often part of an RC filter, using the inherent capacitance or resistance in some other device to form the complete filter.

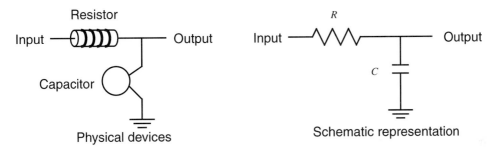

Figure 3-19 Physical and schematic representation of an RC filter.

Figure 3-20 shows the magnitude response of an RC filter consisting of a 16 K-ohm[13] resistor and a 0.1 μF capacitor. Right off you can see one shortcoming of the RC filter—it has a pretty slow transition from passband to stopband. You could cascade an additional stage to get a steeper transition, but it is rare to see more than a few stages because of the "loading" that each stage places on the preceding stage. (You need to take that loading into account, by the way, or your final filter won't be at all what you wanted.) Nonetheless, the RC filter response is sufficient for many noncritical filtering jobs.

How do we figure out $H(s)$ for an RC filter? Using the *s*-domain and a few simple equations, it's not that hard, and we'd feel quite remiss if we passed up this quick point. (Appendix 2 [Useful Electronics] discusses this in more detail.) Let's start with the *voltage divider* circuit of Figure 3-21, which is the same as our filter but with the capacitor replaced by a second resistor. The equation for the ratio of the voltage-out to voltage-in of such a circuit is $V_{out}/V_{in} = (R2/(R1+R2))$. Now, if we somehow could get a "resistance" for the capacitor, we could just plug it into this equation and be home free. However, the time-domain equation for a capacitor involves derivatives (or integrals, if you like—in any case, not very friendly) and so isn't immediately useful if we want to look at the behavior of the RC filter in the frequency domain. Instead, what we can do is look for the *impedance* (roughly speaking, the "resistance" a device exhibits as a

---

13. It's common to use Ω in place of ohm, the unit of electrical resistance. So the 16K-ohm resistor above can also be written as 16 KΩ. Fortunately, there's rarely any confusion between Ω as resistance and Ω as frequency.

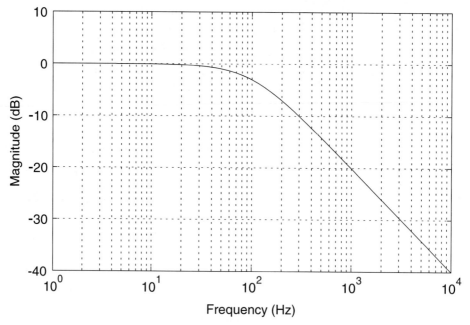

Figure 3-20 Example RC filter magnitude response.

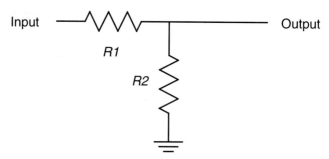

Figure 3-21 A Voltage divider.

function of frequency), and ideally we'd like this impedance as a function of s. The voltage-divider equation still applies for impedance just as for resistance, and we can end up with an $H(s)$ as the ratio of the input to output voltages (again, as a function of s).

Here's the s-domain impedances (Z) for common devices:

Resistor: $Z_R(s)=R$ (i.e., impedance not affected by the frequency of the signal)
Capacitor: $Z_C(s)=\dfrac{1}{Cs}$
Inductor: $Z_L(s)=Ls$

The new voltage divider equation using impedances is:

$$H(s) = \frac{V_{IN}(s)}{V_{OUT}(s)} = \frac{Z_C(s)}{Z_R(s) + Z_C(s)} = \frac{1}{1 + RCs}$$

Inductors and capacitors can be also used to build quite high-performance filters, but except at relatively high frequencies, filters using inductors are far less common than are filters made of *R*s and *C*s.

3.6.2 Active Analog Filters

The design of active filters is an art unto itself that involves much more than just adding amplification to "hop up" a passive filter. Most designs employ op-amps, resistors, and capacitors, though single-chip filters are also available. Generally speaking, filters built using op-amps will have two poles[14] per op-amp stage, and it is not unusual to have an eighth-order (i.e., eight-pole) filter on a chip. (The simple RC filter above has one pole.) However, analog filters with more than a dozen stages are not common, for reasons we mentioned in Chapter 1 (The Big Picture).

Analog filter implementation, both passive and active, are covered very well in Horowitz and Hill's *The Art of Electronics* [Horowitz and Hill, 1989].

3.7 Poles and Zeros

Two important related concepts are *poles* and *zeros*—you can't open any electronics text without being "poled" and "zeroed" to death. Poles and zeros figure into the method of mapping systems on the *s*-plane we mentioned earlier and drop out fairly easily from the system function *H(s)*.

For our purposes, *H(s)* is always of the form

$$H(s) = \frac{\text{some polynomial in s}}{\text{some other polynomial in s}}$$

such as

$$H(s) = \frac{0.0026458 s^4 + 0.44714 s^2 + 9.5249}{s^4 + 1.8 s^3 + 11.09 s^2 + 13.252 s + 13.47}$$

There are two interesting sets of values for *s* in this equation. First, there are values of *s* such that the numerator of *H(s)* is equal to 0. These values are the *zeros* of *H(s)* and are the *roots* of the numerator polynomial. In a similar fashion, *poles* are those values of *s* such that the denominator equals zero. The magnitude of *H(s)* is very large near these locations in the *s*-plane—in fact, it's infinite right at these locations!

Although *H(s)* is often expressed in terms of polynomials as above, the factored form is also useful, as it explicitly shows the poles and zeros:

14. If you don't know what a pole is yet, fear not! That's the next topic.

$$H(s) = A \cdot \frac{(s-z_1)(s-z_2)\ldots(s-z_{n_z})}{(s-p_1)(s-p_2)\ldots(s-p_{n_p})}$$

where there are n_z zeros, z_1 through z_{nz}, and n_p poles, p_1 through p_{np}. We'll always assume $n_z < n_p$, which will be the normal case for any filters we design. Note that we need to have a gain term A to make this equal to the original $H(s)$. (Leave A out and you'll have the same overall filter magnitude response shape, but the gain at every point will be off by this gain term.)

For example, the $H(s)$ above can be factored into poles and zeros to look like this:

$$H(s) = A \cdot \frac{(s-(12j)) \cdot (s-(-12j)) \cdot (s-(5j)) \cdot (s-(-5j))}{(s-(-0.2+3j)) \cdot (s-(-0.2-3j)) \cdot (s-(-0.7+1j)) \cdot (s-(-0.7-1j))}$$

where A is about 0.002646 to get the proper gain.

Put another way, the zeros of this $H(s)$ are at (0+12j), (0-12j), (0+5j), and (0-5j), while the poles are located at (-0.2+3j), (-0.2-3j), (-0.7+1j), and (-0.7-1j).

We can plot these poles and zeros in the s-plane, as shown in Figure 3-22. The usual notation is to use circles for the zeros and Xs for the poles.

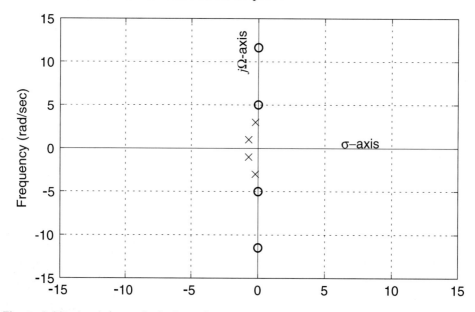

Figure 3-22 A pole/zero plot in the s-plane.

Poles and Zeros

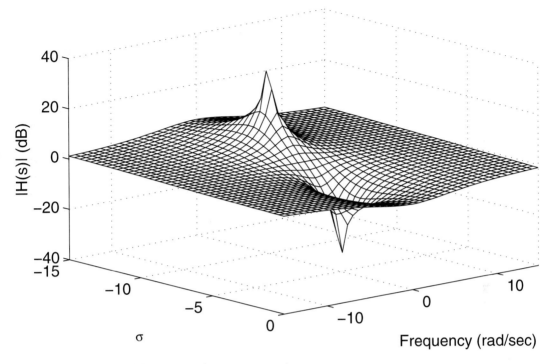

Figure 3-23 $|H(s)|$ for a one-pole, one-zero system.

Now comes the interesting part—plotting $|H(s)|$. This will end up being a three-dimensional plot, as s is a complex variable.[15] By the way, you'll never need to generate one of these plots, because the useful information can be shown on a two-dimensional plot, as we'll show. However, it shows very nicely how the s-plane, $|H(s)|$, and $|H(j\Omega)|$ are related. Because $|H(s)|$ will be bigger near poles, and smaller near the zeros, it's useful to imagine that $|H(s)|$ is a sheet of rubber being poked up near poles and pulled down near zeros. Let's start with a simple case, one pole and one zero. Figure 3-23[16] shows $|H(s)|$ for a pole at $s=-10$ and a zero at $s=-5$.

The next step is to plot the same figure for the poles and zeros of the example $H(s)$ above; this is done in Figure 3-24. Although $|H(s)|$ can be evaluated anywhere, we're showing only the region right around the poles and zeros, and have cut through the surface right along the line $s=j\Omega$ (i.e., $\sigma=0$). Recall that this is exactly the relationship we use to get $H(j\Omega)$ from $H(s)$, so we should expect to see the frequency response function along this cut—and we do! Starting at $\Omega=0$

15. If you've met the Laplace transform before, you know we're glossing over a few points here, including the fact that, mathematically, $H(s)$ doesn't exist for much of the 3-D plots we're going to show! The equation for $H(s)$ is only half the picture—the Laplace transform also gives us a statement telling us for what regions in the s-plane this $H(s)$ is valid. For the systems we'll use, the valid region always includes the $j\Omega$ axis, so don't worry about that. Just understand that we're taking some liberties with our 3D plot of $H(s)$.

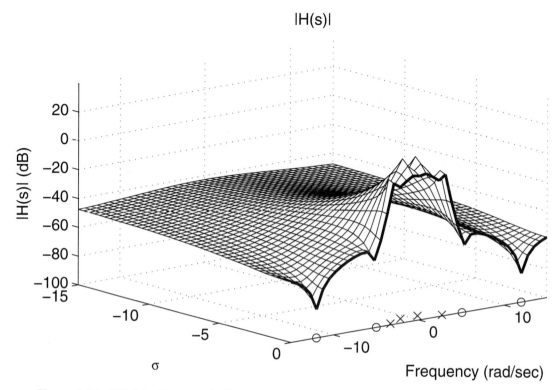

Figure 3-24 $|H(s)|$ for the example filter.

and going on up, we see exactly the frequency response function we get by converting $H(s)$ to $H(j\Omega)$ and plotting $|H(j\Omega)|$, which we've done in Figure 3-25.

In terms of magnitude, poles and zeros have the following effects:

- A pole at $s=\sigma+j\Omega_0$ will cause an increase in $|H(j\Omega)|$ near Ω_0.
- A zero at $s=\sigma+j\Omega_0$ will cause a decrease in $|H(j\Omega)|$ near Ω_0.
- The amount of effect a particular pole or zero has on $|H(j\Omega)|$ depends on how far from the $j\Omega$ axis the pole or zero is; that is, how large σ is.

16. Because the points we're plotting happen to miss the exact pole and zero locations, it isn't clear in this figure that poles correspond to infinite gain and zeros to infinite attenuation. (Well, sort of, as the Laplace transform doesn't exist at poles, neither does the concept of "gain" there. But, this is being needlessly picky.) Again, the problem of "connecting the dots" has obscured the underlying reality! This problem might occur in many plots you make (2-D or 3-D), as the points you (or a program) choose to evaluate will rarely fall directly on a pole or zero at other than a few special points (like $s=0$). And when you do evaluate points at a pole or zero, your mathematics or plotting package may balk at calculating or plotting $+\infty$ or $-\infty$. When we're faced with such problems, we sometimes limit extremely large or small values to reasonable values. This smaller range can be plotted without scaling the axis so much that we couldn't see the truly interesting information at locations other than at poles and zeros.

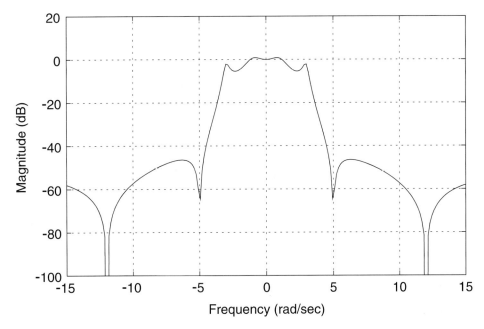

Figure 3-25 |H(s)| evaluated along σ=0.

- Stable filters may have zeros located anywhere in the *s*-plane. Zeros located on the *j*Ω axis (σ=0) completely block signals at that frequency.
- Stable filters may have only poles in the half of the *s*-plane with σ<0 (also called the *left-half plane*). Poles *on* the *j*Ω axis correspond to constant oscillation (*marginally stable*); poles with σ > 0 lead to exponentially increasing outputs and are associated with unstable filters.

Figure 3-26 shows some poles and zeros and their relations to stability in the *s*-plane.

If you accept for the moment that filters (or systems) can be created or evaluated in terms of poles and zeros, that we can mathematically represent these in the complex frequency plane (*s*-plane), and that the frequency response function is a slice of this figure along the *j*Ω axis, the question remains, how we figure out where to put these poles and zeros to get the frequency response function we want?

One option is just to put down (mathematically) some poles and zeros and move them around until you have a satisfactory |H(*j*Ω)|. Although this might work for some simple filters, in fact, a lot of people have analyzed the problem of filter design and have come up with design methods to locate poles and zeros to produce certain types of frequency response functions. This is what filter design is all about. These methods target a specific characteristic, such as how sharp the cutoff is between pass- and stopbands or how much ripple there is in the passband. Because some types of filters will often be translations of analog filters, we'll discuss such design types as Butterworth, Elliptical, and Chebychev filters in Chapter 6 (IIR Filters—Digital

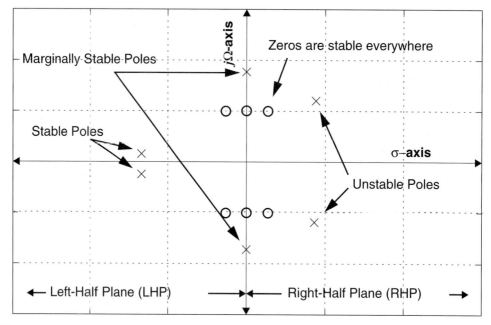

Figure 3-26 Example poles and zeros in the *s*-plane.

Filters with Feedback). The example we've just been using is called an *elliptical* filter. This filter trades off ripple in both bands for a steep transition between the passband and the stopband.

We asked earlier whether complex frequency (*s*) is "real." It's not. It is just a useful way of describing signals and systems. Are poles and zeros real? Can you go to the electronics store and buy a pole?[17] No, poles and zeros are also just useful abstractions. In physical circuits, the behavior we associate with poles and zeros emerges from the interaction of different circuit elements; in DSP systems, poles and zeros are a result of the coefficients and mathematical operations we perform on the digital signal. It's somewhat like asking whether degrees Fahrenheit are real. Although a "degree" is meaningful and very useful, it is not a physical entity in the air or water. Rather it is a manifestation of physicists' attempts to model much more complex and real physical processes. Same with poles and zeros.

By the way, perhaps it bothers you that the *s*-plane includes negative frequencies. This is a mathematical necessity of using complex frequency, though it may seem like just unnecessary computation. One justification is as follows: Using only real coefficients for the polynomials of $H(s)$ means that all of the poles (or zeros) will either be real (that is, they'll be along the σ axis, where $\Omega=0$) or will come in matched pairs, one with a positive Ω, one with a negative. (There is a theorem to this effect tthat we'll not get into here.) We call these matched pairs *complex conjugate pairs*.[18] You could say negative frequency gives the negative Ω complex conjugate a place

---

17. Obviously, you buy a pole at a tackle shop.

to go but, in fact, negative frequency is like i (or, as we've been saying, j) in that it provides us the necessary "elbow room" and symmetry to do the math we'd like to do.

What about the effects of poles and zeros on phase? Poles and zeros definitely affect phase! (Adding a zero at $s=-1$ and evaluating it at large and small values of Ω should convince you of this. You can add a zero at $s=-1$ by including a factor $(s+1)$ in the numerator.) The s-plane also provides a geometric means of finding the phase of $H(s)$ along $s=j\Omega$ (that is, $\angle H(\Omega)$) by measuring the angles from points on the $j\Omega$ axis to the poles and zeros. You're much more likely to use a math or filter design program to generate both the phase and magnitude plots, but this does show that the s-plane isn't just a gimmick—it provides a very useful and *complete* geometric interpretation of the system function $H(s)$ and the underlying poles and zeros.

3.8 Filter Order

Filter design is the science of producing systems that satisfy specific time- and frequency-domain characteristics, while minimizing the costs of those systems.

The *order* of a filter corresponds to the maximum number of poles or zeros in a filter, whether analog or digital. By adding poles and zeros to a filter, you can get a steeper transition, smoother passband, lower stopband gain, and/or optimize other design criteria. However, in an analog filter you also increase the number of op-amps, the circuit board area, the number of capacitors, and so on, when you increase the number of poles and zeros.

The number of poles and zeros also affects digital filters. As a general rule, a higher order filter requires more coefficients, more memory, more multiplications, and so on. Thus, the tradeoff is between how good the filter characteristics are and how much processing time or storage such a filter requires.

3.9 Summary

Filters are systems we use to modify signals. Usually, we hope to modify the frequency content in particular ways, including attenuating high frequencies, or passing only certain bands of frequencies. Though transient characteristics such as rise time or overshoot are often ignored, many applications also must meet certain transient criteria in addition to the specified magnitude and phase response characteristics that we refer to as steady-state characteristics.

Major types of filters include lowpass, highpass, bandpass, band-reject (notch), arbitrary (multipassband), all-pass, and comb filters. Two slightly different methods are used to specify the passband characteristics of filters (peak-to-peak vs. peak deviation) but, for the most part, the same general characteristics are specified for both analog and digital filters.

Phase is often a concern, and the ideal is a linear phase response, which corresponds to a constant delay for all frequencies of interest. This behavior can be achieved with a certain class of digital (but not analog) filters.

---

18. You might recall that the familiar "quadratic equation" for finding roots of a second-order polynomial can produce these complex conjugate pairs.

The RC filter is a particularly simple passive filter that finds widespread use, though it has rather unimpressive magnitude response characteristics. Using the idea of impedance and the voltage-divider equation, the system function for an RC filter is easily derived as $H(s)=1/(1+RCs)$

Poles and zeros represent the roots of the denominator and numerator of the system function $H(s)$, respectively, and are points of maximum gain and attenuation, respectively. The s-plane is a useful tool for visualizing the effects of poles and zeros on the magnitude and phase response of a system. Filter design can be viewed as the process of identifying the proper locations of poles and zeros to achieve a desired magnitude/phase response for a system. Likewise, identifying the poles and zeros associated with a system completely describes that system to within a constant gain factor. Although plots of the magnitude of $H(s)$ for all s are useful tools, our goal is always the values of $H(s)$ evaluated for $s=j\Omega$ (that is, $\sigma=0$, or the $j\Omega$ axis), which corresponds to $H(j\Omega)$.

Adding poles and zeros (if properly placed!) can increase the performance of a filter. However, additional poles and zeros require either additional components (for analog filters) or additional computation (for digital filters). Thus, filter design is the art of using a minimum filter order to meet filter design criteria.

Resources

Thede, Les, *Analog and Digital Filter Design Using C*, Upper Saddle River, New Jersey: Prentice Hall, 1996.

Volume Two of the *Maxim Engineering Journal* (mid-1990) contains an excellent article on switched capacitor filters and filters in general.

Many of the resources we've listed for Chapter 5 (FIR Filters--Digital Filters Without Feedback) and Chapter 6 (IIR Filters—Digital Filters with Feedback) discuss filters, particularly texts emphasizing digital filter design.

CHAPTER 4

Discrete-Time Signals and Systems

Because this chapter parallels Chapter 2 (Analog Signals and Systems), we'll be using many of the same concepts here. We're going to talk about the discrete-time counterpart to the s-plane, so be sure you feel comfortable with H(s), poles and zeros, and so on.

4.1 Overview

Discrete-time signals and systems are distinct from their counterparts in continuous time. We start this chapter by discussing how analog, discrete-time/continuous-value, and digital signals are related. This leads us right into a central issue in DSP—choosing the proper sampling rate. Just as with continuous-time signals, there's a frequency-domain representation of discrete-time signals, though it may not be what you would expect. We can also describe discrete-time *systems* in the time domain and frequency domain, for which we'll need the discrete-time equivalents of the impulse response and the system function (remember $H(s)$ and $H(j\Omega)$?).

As we'll see, discrete-time signals and systems are only "half-way" to digital signals and systems; when we also make signals discrete in value (quantized), we introduce other, somewhat less dramatic effects. This leads us to the topic of Analog-to-Digital Converters (ADCs), their pals the Digital-to-Analog Converters (DACs), and some supporting actors whose roles will become clear as we find what havoc sampling and quantization can cause.

This is all leading to the ability to take a description of a discrete-time system and express this as a very simple program that operates on sampled data. If that discrete-time system happens to be a filter, then the program will be a digital filter. This may sound like everything we need, but there's the small problem of how to specify an appropriate discrete-time description of a filter in the first place. To answer that problem, we need to look at the two major types of digital filters—*FIR* filters, described in Chapter 5 (FIR Filters—Digital Filters without Feedback), and *IIR* filters, described in Chapter 6 (IIR Filters—Digital Filters with Feedback). Although both are linear, time-invariant filters, there are some big differences between the two types; choosing between the two is one of the first tasks you face as a designer. We'll end this chapter with a general overview of digital filters, and the characteristics of the FIR and IIR types.

4.2 Sources of Discrete-Time and Digital Signals

Similarly to Chapter 2 (Analog Signals and Systems), we begin by looking at the sources of discrete-time signals. When discrete-time signals are produced by sampling analog signals, which is often the case, we have to account for some interesting effects. You may have heard of issues like the "Nyquist rate" or "aliasing."

4.2.1 Analog, Discrete-Time, and Digital Signals

Figure 4-1 shows the relationships among *analog*, *discrete-time/continuous-value*, and *digital signals*. Recall that analog signals are continuous in both time and amplitude value. Digital signals are at the other extreme—digital signals have an amplitude only at specific times, usually at regularly spaced intervals, and the amplitude is represented with a finite resolution. For example, if an 8-bit ADC is used, the magnitude is represented by one of only 256 different values. Most converters are at least 6-bit, with 8–16 bits or more both common and cheap, and you'll even find a few 24-bit converters and above. We'll have much more to say about ADCs later in this chapter.

		time	
		continuous "CT"	discrete "DT"
value	continuous	analog	discrete-time/continuous-value; sometimes called *discrete-time* or DT
	discrete	continuous-time/discrete-value (*rare*)	digital

Figure 4-1 The relationship between analog, digital, and discrete-time signals.

Discrete-time/continuous-value (we'll use the shorthand "DT"[1]) is one of the two possibilities "halfway" between analog and digital. The DT signal has a value only at specific—that is, *discrete*—times, but is continuous in amplitude. While we don't use DT signals in a digital processor, we'll almost always do our filter design work as if the signals and systems were DT, then, at the end of the design process, factor in the effects of also making the signals discrete in amplitude. Why is this?

When we produce a DT signal by sampling an analog signal (i.e., when we quantize the *time* dimension), we produce a signal that is not simply a "lower quality" version of the original, but a signal that can have fundamental differences from the original. For example, high frequencies in the original signal can appear to be low frequencies if the original signal is sampled too slowly. On the other hand, the question of amplitude resolution is more a matter of degree—of how much noise will be present in the signal, rather than major changes to the nature of the signal itself.

Further, attempting to account for quantization of signals in filter design work can often lead to an intractable mess. The payoff in "being careful" with amplitude quantization during filter design is usually not worth the mathematical pain of doing so, at least within the scope of this book. This is *not* to say that we can simply ignore quantization errors, but rather that these effects are dealt with using separate analyses or simulation.

We're shortly going to discuss the effects of both sampling and quantization and in a few chapters, another departure from the ideal, the quantization of the filter coefficients. It's just our lucky break that each of these issues is separate enough so that we aren't overwhelmed at one time with too many factors.

4.2.2 Sampling

One of the most critical decisions in DSP is choosing the rate at which an analog signal is sampled—this is the *sampling frequency* (or *sampling rate*), usually denoted by F_S (Hz) or Ω_S (rad/sec). The resulting signal, which is discontinuous in time but not in amplitude, is a DT signal. Figure 4-2 shows an analog signal and the DT signal that results from a particular choice of sampling rate. Sampling the same analog signal at different rates produces (sometimes radically) different DT signals, for reasons that will become clear in a moment.

The reciprocal of the sampling frequency is the *sampling period*, T_S (sec). This number is of critical importance to programmers, as it tells how much time (or how many machine cycles) we have to process between samples.

The choice for sampling rate is based on the frequency content of the signal to be sampled. The following observations apply:

1. Strictly speaking, digital signals are also DT—that is, discrete-time. But most of the time when folks talk about discrete-time, the understanding is that the signal is continuous-valued or at least that we're going to ignore the effects of quantization. Continuous-time—or CT—signals likewise include both discrete- and continuous-value CT signals, but we usually assume continuous-value (CT/discrete-value is rare).

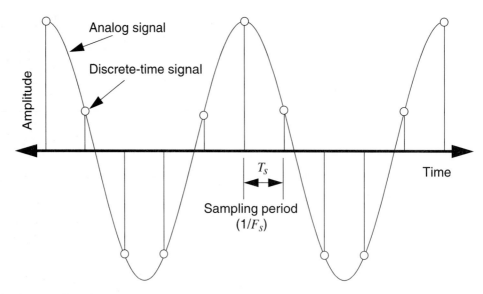

Figure 4-2 Sampling an analog signal to produce a discrete-time signal.

- Sampling a signal at sampling frequency F_s "folds" any frequencies higher than $F_s/2$ back into the frequency range of $0–F_s/2$. This is known as *aliasing*.[2] Put another way, signal components with frequencies higher than $F_s/2$ will produce the same samples as a signal with some frequency in the range of $0–F_s/2$ (i.e., there's no way to tell an aliased signal from the "real thing.") $F_s/2$ is known as the *Nyquist frequency* or *folding frequency*.
- The highest frequency present in a signal may be noise, not your intended signal. Noise gets folded back just as well as "legitimate" signals!
- Bandpass signals can be "folded down" intentionally by sampling more slowly than the highest frequency. Of course, no other signals should be in the way (i.e., at lower frequencies) for the bandpass signal to clobber.

Let's explore each of these observations with some examples.

Assume we will be sampling a signal at $F_s=1$ kHz. Figure 4-3 shows a 200-Hz signal with these sample times shown. Given the DT signal, you would have no problem in reconstructing the original signal reasonably closely.[3] Figure 4-4 shows a 490 Hz signal. You can see how there are fewer sample points per cycle, and we're on the verge of not having enough points to accurately recreate the original signal. Finally, Figure 4-5 shows a 600 Hz signal. "Connect the dots" (shown as dotted line) and, instead of 600 Hz, we appear to have a 400-Hz signal!

2. In everyday English, an *alias* is an alternative, often false, identity that a person adopts. In signal processing, it is as though the real-world signal masquerades as a signal of another frequency range.

3. Realize, though, that you could also "reconstruct" the signal with a sine wave at 800 Hz, or 1200 Hz, or 1800 Hz, and so on. Hmmm, could this be one of those foreshadowing footnotes?

Sources of Discrete-Time and Digital Signals

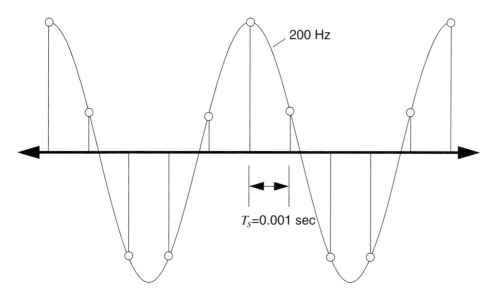

Figure 4-3 Sampling a 200-Hz signal at 1 kHz.

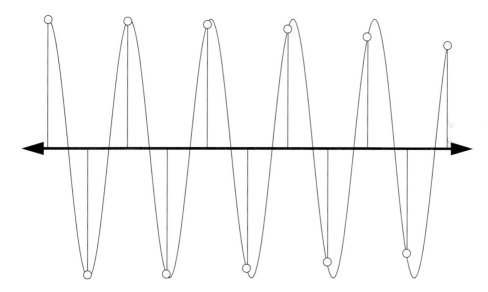

Figure 4-4 Sampling a 490-Hz signal at 1 kHz.

Signals above $F_s/2$ show up in the discrete time signal looking as if they were actually in the range of 0 to $F_s/2$. Figure 4-6 shows the cyclic nature of this folding or aliasing. Notice,

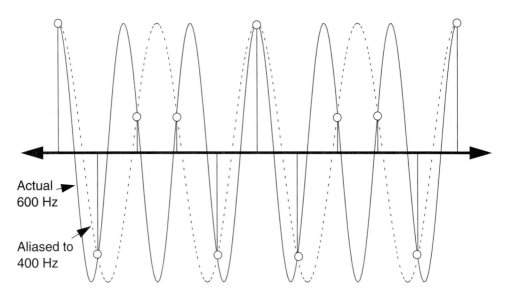

Figure 4-5 Sampling a 600-Hz signal at 1 kHz produces aliasing.

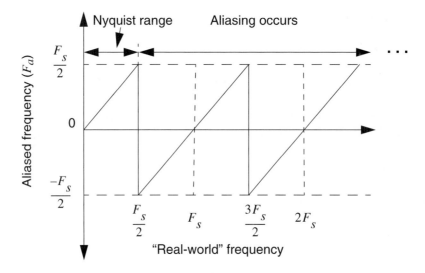

Figure 4-6 "Real-world" frequency and aliased frequency.

though, that some of the frequencies are aliased to "negative" frequencies (i.e., the range $-F_s/2$–0). What is "negative" frequency?

For our purposes, negative frequencies are mostly a mathematical accommodation, and you can take the absolute value of the aliased frequency ($|F_a|$) as its frequency. For example, you can just treat F_a=-50 Hz as if it were F_a=50 Hz. (What's really going on is that the negative sign represents phase information.)

How do we find the aliased frequency, given a "real-world" frequency? Just subtract F_s repeatedly from the frequency until you have a result in the range $-F_s/2$ to $F_s/2$. Then, if the result is negative, take the absolute value. For example, if F_s is 100 Hz—and thus $F_s/2$ is 50—a signal of frequency 30 Hz is not aliased, a signal of frequency 70 Hz is aliased to 30 Hz (70-100=-30), and a signal of frequency 110 Hz is aliased to 10 Hz (110-100). A signal of frequency 3157 Hz aliases to (just a second...) 43 Hz. (Subtract all the multiples of 100, leaving 57 Hz. 57-100=-43).

If you prefer, you can use the following equation to calculate F_a for a particular frequency component F

$$F_a = F - F_s \left\lfloor \frac{F + (F_s/2)}{F_s} \right\rfloor \qquad (4.1)$$

where $\lfloor x \rfloor$ means to round x down to an integer value. This is the same as the repeated subtraction method.

We can go in the opposite direction, too. If we have a signal with a known upper frequency of interest, we can double that frequency to get the *Nyquist rate*; that is, the sampling rate *above which* you *must* sample in order to avoid aliasing.

There's one more oddity about aliasing, and that's a strange effect where aliasing can reverse the spectrum of signals. In fact, this little twist is hiding in Figure 4-6 if you know where to look. What we mean by "reversing the spectrum" is that for a section of frequencies, higher frequencies in the real-world signal are aliased to lower frequencies in the DT signal. A tone that was increasing in frequency would, in this frequency range, alias to a tone that was decreasing!

Actually, you've seen this effect before in western movies. The sampling rate of the movie camera (say, 24 frames a second) aliases the motion of covered wagon wheel spokes, producing the occasional "high-speed" chase with slowly or even backward-rotating wagon wheels. If we could increase the speed of the actual wagon a bit, an apparently backward-turning wheel would slow down (reversed spectrum region), and an apparently forward-turning wheel would speed up ("nonreversed" spectrum region).

Figure 4-7 shows what's going on. "Real-world" frequencies A and B are aliased to frequencies A' and B' where A' < B', just as A < B. For our wagon wheel, this would be a "forward-turning" case. Not so in the region where C and D are—here D' is lower than C'. If we had a radio station with frequencies ranging from C to D and sampled this signal with F_s as shown, the spectrum of our DT signal would be reversed from what the radio station had transmitted. High-pitched sounds on the radio station would be low, and low-pitched sounds high—the resulting DT signal (if transformed back into a CT signal for our listening pleasure) would be unintelligible![4]

Why should we be concerned about what happens to the spectrum of aliased signals? You might already know that most of the time we're going to do our best to avoid aliasing to begin with, in which case we really don't care if the spectrum is reversed or not. But in some cases, folks actually alias signals on purpose. We'll come back to this in a moment.

Let's take an example involving a biomedical signal and see how aliasing and noise issues come into play. Say we're monitoring a blood flow in a human body; from experience we know that this signal has frequency components in the range of 0–20 Hz (most physiological signals are fairly low in frequency). However, our sensor also picks up some 60 Hz from nearby wiring, and even some signals from a nearby radio station. The situation may look a bit like Figure 4-8—your

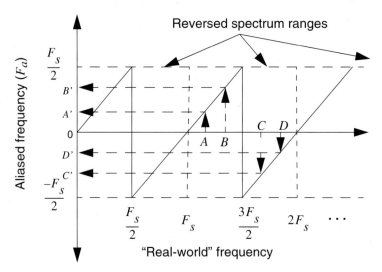

Figure 4-7 Regions of reversed frequency spectrum in a sampled signal.

---

4. Dale wrote a program using a slightly different method to reverse the spectrum for telephone-bandwidth speech (about 4 kHz). The output really was unintelligible, but repeating the process (flipping the spectrum twice) completely restored the speech. All of this mathematical strangeness has some validity to it!

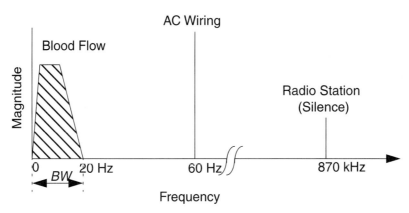

Figure 4-8 Magnitude spectrum for blood flow signal example.

desired signal is sitting at a fairly low range of frequencies, some 60-Hz noise from the wiring is mixed in, and much higher still is the radio station signal. To make it simple, we'll have the radio broadcast highlights from the regional mime convention (i.e., silence, which means we'll just see the carrier signal), and we'll put it at 870 kHz, near the low end of the AM broadcast range (540–1700 kHz). (In a moment, we'll let the station broadcast some music and see what we can do.)

Taking the highest frequency of interest in our *desired* signal (20 Hz) and doubling that, we have a Nyquist rate of 40 Hz. Any sampling rate above 40 Hz will avoid aliasing this desired signal, so let's pick a sampling rate of 50 Hz. The resulting spectrum of the sampled signal is shown in Figure 4-9.

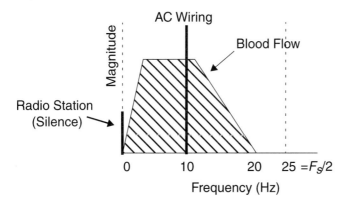

Figure 4-9 Magnitude spectrum of sampled blood flow signal.

As promised, the blood flow signal is captured with no problems. However, the 60-Hz noise from the wiring is now at 10 Hz (see Equation 4.1), and the radio station is sitting right at DC (0 Hz). And, as we've pointed out, aliased signals are indistinguishable from non-aliased signals of the same frequency—so, as far as we can tell, the blood flow data now includes a more-or-less constant 10 Hz hum (the 60 Hz) and, depending on how long the mime convention broadcast goes on, a constant DC bias (the aliased 870-KHz signal).[5] Obviously, we've got some problems! (Apart from the fact that we live near a radio station that would broadcast a mime convention.)

"Everything"—all the power—in the analog signal will show up in the DT version, regardless of its frequency. An *antialiasing filter*, prior to our sampling stage, can attenuate high frequencies so that there's little left to alias. The trade-off with this antialiasing filter is that, as we saw in Chapter 3 (Analog Filters), filters are limited in how sharply they distinguish between those frequencies to pass and those to stop. Furthermore, filters can affect the phase of the signal, which might be a problem for some signals. There is no single ideal antialiasing filter; sometimes a simple RC filter is sufficient, other times elaborate filters are necessary. We'll discuss antialiasing filters in a few pages, but it should be clear that they're an integral part of DSP design, where the name of the game is to maximize the power of the signal of interest and minimize the power of any unintended signals.

So far we've discussed the lower limit on a sampling rate, and it is probably obvious that the upper limit is a function of how fast an ADC we have and how fast we can execute any necessary code per sample. For example, let's use a 68HC16 microcontroller running at 16.78 MHz.[6] Each clock cycle is about 59.605 nsec. At a 10-kHz sampling rate ($T_s=1/(10$ kHz$)$, or 100 μsec), this is $(100 \times 10^{-6})/(59.605 \times 10^{-9})$ or about 1,677 clock cycles per sample. Looking ahead, the multiply-accumulate instruction of the 68HC16—which we'll depend on shortly—takes 12 clock cycles per iteration. Ignoring the inevitable overhead, this is close to 140 multiply-accumulates per sample. If the signal processing requires more than 140 multiplies, we're out of luck. For our blood flow example, sampling at the blazing-turtle rate of 50 Hz, we have 335,600 $((1/(50$ Hz$))/59.6$ nsec$)$ clock cycles, or roughly 28,000 multiplies!

We're not done with sampling, though. We promised to talk about "aliasing on purpose." Shannon[7] never said we couldn't sample a signal at a much lower frequency than the Nyquist rate—we just need to realize how the frequency content will be affected (this is known as the *Bandpass Sampling Theorem*). Let's say that, instead of wanting to recover the blood flow signal, we actually wanted the radio station. In place of just a single frequency, as we used in our

---

5. While this is only an example, we feel obliged to point out that it's actually unlikely all the timing here would be so perfect that the radio station would end up aliased right on 0 Hz. Just a tiny error in the sampling rate would be magnified enormously. About all we can say is that practically, the signal would show up somewhere in the range 0–25 Hz.

6. The clock can be derived from a low-cost external 32.768-kHz watch crystal; multiplied internally by 512, this yields 16.777 MHz (just in case you were wondering where this figure came from). More recent versions of the 68HC16 family can run at even higher speeds.

Sources of Discrete-Time and Digital Signals

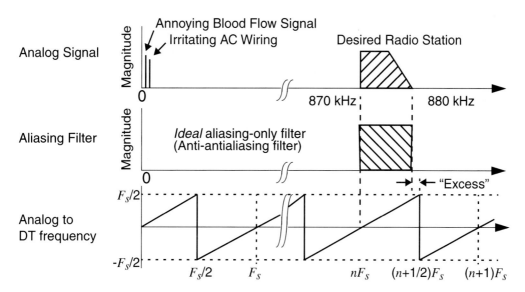

Figure 4-10 Purposeful undersampling to recover the radio station signal.

earlier example, we'll say the radio station actually has a 10-kHz bandwidth (the bandwidth of the audio signal we're hoping to recover) and ranges from 870 to 880 kHz.[8] (See Figure 4-10.)

First, we realize that, just as with the blood flow problem, we should filter out any other signals that are present, since regardless of their frequency, they'll end up in somewhere in our DT signal. In this case that means filtering out low frequencies (e.g., 0–20 Hz, 60 Hz) as well as any higher frequencies, so we want a bandpass filter with F_L = 870 kHz and F_H = 880 kHz as our "antialiasing" filter. (In this case it's almost an "anti-antialiasing filter," if you see what we mean.)

Reviewing Figure 4-7, we see that if we want to recover the audio signal from the station, we'll have to arrange the sampling rate so that $F_s/2$ is large enough to hold the signal without wrapping; or, put another way, $F_s/2 \geq BW$. We also need to make sure that a multiple of F_s falls right at 870 kHz (F_L), or the spectrum of the discrete-time signal may be partially or entirely

---

7. Claude Shannon (well known for his contributions to communication theory) is credited with *Shannon's Sampling Theorem*, which also seems to have been developed independently in the former Soviet Union by a gentleman named Kotel'nikov. As you've seen, a Mr. Nyquist got his name stuck on all of the rates and frequencies. In fact, everyone seems to have been building on the work of a 19th century mathematician named Whittaker, who obviously didn't have a very good public relations firm.

8. Actually, commercial AM is transmitted such that it takes twice the bandwidth, so we're still simplifying things. This type of AM wastes some power and bandwidth, but it made sense when vacuum tubes were the only game in town. However, amateur radio operators ("hams") use a version of AM that takes up only the bandwidth of the audio signal being transmitted—*single-sideband (SSB)*. So, if you like, pretend we're receiving SSB.

inverted. Subject to the condition that $2BW \leq F_L$ (otherwise this won't work out), F_s can be calculated as:

$$F_s = \frac{F_L}{\left\lfloor \dfrac{F_L}{2BW} \right\rfloor} \qquad (4.2)$$

In our case, F_s should be 20232.59... Hz, which is just a shade over the 20-KHz theoretical minimum (2 BW). (We'd get this minimum, for example, if F_L were 880 kHz.) The "excess" bandwidth goes toward making a multiple of F_s fall correctly with respect to the original signal so it aliases properly. Figure 4-11 shows the resulting spectrum.

Is this just a theoretical exercise? Can we really use a slow sampling rate to recover a signal like this? The answer is yes, and the technique of purposefully *undersampling* has a number of applications in communications. The trick is that the sampling must be of the signal at the right moment in time and of *only* that moment in time (versus, say, the value over a tiny but appreciable window of time). This is the job of a sample-and-hold amplifier (SHA), which we'll discuss later in this chapter.

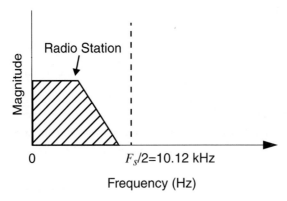

Figure 4-11 Magnitude spectrum of undersampled signal.

4.2.2.1 Generating a Sampling Rate

Now that we know how to calculate a reasonable sampling rate, how is a sampling rate produced?

In some cases, ADCs have an internally generated sampling rate (or can be configured so), usually derived from a higher-frequency external clock signal. In this configuration, the converter establishes the sampling rate, and the microcontroller either polls the ADC or is interrupted on completion of a conversion cycle.

Much more common, however, are arrangements in which the conversion is initiated by a command or signal from the microcontroller. For example, the conversion may be started when the microcontroller reads the results of the prior conversion, which means the microcontroller need only to issue one read per conversion. In other cases, a two-step process is necessary: First, start the conversion and, second, after a delay, read the results.

How can a microcontroller accurately issue these commands? Having the converter start a new conversion as soon as the previous conversion is completed might sound attractive, but a number of converters take a variable amount of time to complete a conversion. This would result in unacceptable variations in the sample rate. Software timing loops are possible and can provide very accurate sampling rates, but no other processing can take place during sampling without engaging in very painful programming.[9] As a separate issue, the resulting system is not easily upgraded in the future. This technique is not recommended.[10]

The usual approach is to derive the sampling rate using (internal) counters or timers that are based on the microcontroller's clock. In some cases, these signals are available externally so that they can be directly connected to the converter to initiate conversion cycles, while the converter signals the microcontroller to read the result using an interrupt. Initiating the conversion cycle from an interrupt routine is also possible, but the programmer must account for the *interrupt latency*,[11] which may vary too much to be acceptable. On-chip converters offer additional options, and external converters can be supplemented by additional hardware to meet almost any conversion need.

The built-in converter in the 68HC16Z1 can be configured to perform continuous conversions (within a limited range of sampling rates) or the program may initiate conversions at even slower rates. With a 16.78-MHz clock, the maximum conversion rate is 130 kHz (8-bit) or 116

---

9. Dale once foolishly agreed to program a microcontroller in just such a way. Everyone should do this at least once, carefully counting out cycles so each routine is exactly the same amount of time. Then swear never to do it again.

10. <Begin rant> Even though new architectures are introduced frequently, it's very common these days to see speed and architectural upgrades to *existing* microprocessors. Higher-speed parts execute all instructions faster, but architectural changes may also speed up specific instructions. Thus, it is no longer a subtle point that the time an instruction takes to execute is a *by-product*, not an inherent part of the instruction. For example, an ADD instruction's purpose is to add two operands and place the results somewhere, but *not* to take up a certain amount of time doing so. Timing loops in software depend on an aspect of the instruction set that won't necessarily be preserved as the architecture is improved over time, so are really not recommended. <End rant>

kHz (10-bit). The maximum bandwidth, therefore, is 65 kHz or 58 kHz, though we'll see, when we discuss ADCs in a few pages, that there's more than just sampling rate to worry about.

4.3 Describing Discrete-Time Signals and Systems

4.3.1 Discrete-Time Signals in the Time Domain

From the earlier table, recall that DT signals have a value only at specific times (i.e., they are discrete in time) but have a continuous range of values.[12] We've already seen from our discussion of sampling rates how we might go about plotting a DT signal using circles (or dots) at the end of line segments. You'll need to fight the temptation to "connect the dots" and turn the DT signal into a CT signal; it doesn't help that we "connected the dots" earlier, but that was for purely pedagogical purposes.

There are two different ways of denoting DT sequences. If $e(t)$ is the CT signal, $e(nT_s)$ is a DT representation where $n=0, 1, ...,$ and T_s is the sampling period. It is common practice to drop the T_s, which is equivalent to setting the sampling period to 1 sec. The DT sequence, therefore, is denoted $e(0), e(1), e(2), ...$

4.3.2 Discrete-Time Signals in the Frequency Domain

Having set T_s to unity (see paragraph above), all of our calculations for DT signals and systems can be carried out in terms of *normalized frequency*. The relationship between normalized frequency f (or ω) and *real-world*[13] *frequency* F (or Ω) is:

$$f = FT_s \text{ (dimensionless)} \qquad (4.3)$$

$$\omega = \Omega T_s \text{ (rad)} \qquad (4.4)$$

The spectrum of DT signals is restricted to a limited range (the Nyquist band) due to the aliasing effects. In terms of normalized frequency ω, this is the frequency range 0 to π, or in terms of f, from 0 to 1. Given this limited range, it is unnecessary to use a logarithmic scale for frequency; the question is merely whether you want to express the DT frequencies in terms of

---

11. Interrupt latency is a delay between the time an interrupt line is brought active and when the microprocessor actually begins executing the interrupt code. Some parts of this delay are constant (e.g., time in stacking certain registers), but other aspects of the delay vary according to the instructions being executed, when in the execution cycle an interrupt occurs, and so on.

12. Recall that, actually, *digital* signals are also discrete-time, but we usually take "discrete-time" to mean either discrete-time/continuous-value signals or digital signals where we're ignoring the effects of quantization.

13. *Normalized* is fairly standard terminology, but the label *real-world* for F is less so.

normalized frequency (ω=0 to π) or convert this to real-world frequency by means of Equation 4.4.

(Note that some filter design programs (notably, MATLAB) may define a normalized frequency that ranges from 0 to 2, which places the Nyquist frequency at 1. It's a little easier to work with these numbers, but don't get confused moving between the two conventions.)

Even though the time-domain DT signal is, by definition, discrete, the magnitude and phase frequency-domain representations are continuous in both f (or ω) and in the "value" at each f or ω. Later on, we'll see how to generate a discrete version of the frequency-domain representation—the DFT/FFT.

It is tempting to leave you with these half-truths and press onward, but our outstanding honesty and the fact that we'll get in a pickle later on force us to share a few more facts. First, we really can't say the spectrum is limited to a specific range of frequencies (say, ω=0 to π), because the math really doesn't have a built-in bias as to any particular range (e.g., the range ω=2π to 3π works just fine mathematically). The implication is that the spectrum of a DT signal, in fact, repeats infinitely, as in Figure 4-12. And, believe it or not, we'll actually see this effect when we convert our DT signal back into a CT signal, so the point really is worth bringing up.

The second issue has to do with negative frequencies. We need to extend our frequency scale not only out to positive infinity, but into the negative range as well, at least theoretically. Just as with the s-plane, there is some symmetry in the spectra such that we don't need to plot the negative frequency—this is because our signals are real. So why even mention it? It's an issue you'll come up against if you eventually read other books or even use some types of filter design software. Someday, you may even want to process complex signals. The thing to notice is that there is an even symmetry around ω=0 for magnitude ($|e(\omega)|=|e(-\omega)|$), and an odd symmetry for phase ($\angle e(\omega)=-\angle e(-\omega)$) (but again, this is only for real signals!). We don't lose any information just plotting ω from 0 to π (or F from 0 to $F_s/2$).

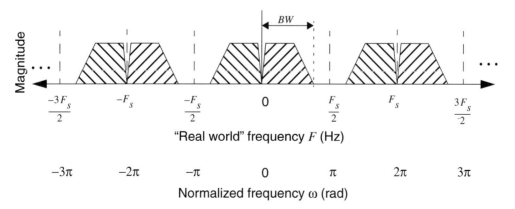

Figure 4-12 The magnitude spectrum of a DT signal converted to a CT signal.

4.3.3 Describing Discrete-Time Systems

4.3.3.1 The Discrete-Time Impulse Response

In Chapter 2 (Analog Signals and Systems), we discussed the (continuous-time) impulse, a useful signal (if only mathematically) that can be used to excite systems at all frequencies.[14] The DT counterpart of the impulse has a far less exotic definition: at sample $n=0$ (or the sample at time $0T_s$ if you like), the value is 1; all other samples have value 0. It is also true that this DT impulse (officially, the *Kronecker sequence*, as contrasted with the *Dirac impulse function* in the continuous-time domain) excites digital systems at all frequencies in the Nyquist bandwidth, but the two "impulses" are distinct and not at all interchangeable. A bit like faxing a pizza—you have to match up tools with material. In particular, the DT impulse is *not* just a sampled version of the CT impulse—if you recall, the CT impulse has *infinite* amplitude for a zero argument.

4.3.3.2 $H(e^{j\omega})$: The Steady-State Frequency Response

Exactly as with CT systems, DT systems can be characterized in the frequency domain in terms of their steady-state response—the magnitude and phase changes they impart to signals passing through them. Aside from only worrying about a limited frequency range (from 0 to π in radians), there isn't a lot to say—the notation we'll use is $H(e^{j\omega})$, the magnitude $|H(e^{j\omega})|$ is the gain a signal of frequency ω will experience and the angle $\angle H(e^{j\omega})$ is the phase shift. (And, as before, you'll see other notations for the DT frequency response function, including just $H(\omega)$, $H(\Omega)$, etc. The notation $H(e^{j\omega})$ makes things a bit more consistent later on.)

As an example, consider the following DT frequency response function:

$$H(e^{j\omega}) = \frac{0.3711}{1 - 1.5680e^{-j\omega} + 0.9391e^{-2j\omega}} \tag{4.5}$$

This happens to be the frequency response of a "digital resonator," a kind of bandpass filter that finds a lot of use in certain types of voice synthesizers. This digital resonator has a center frequency of 1 kHz and a 50-Hz bandwidth when a sampling rate of 10 kHz is used.[15] Let's just calculate ("by hand") two points, then ask a math program to plot the rest.

One easy point to calculate is at $\omega=0$; $|H(e^{j0})|$ is just 1, or 0 dB, and the phase, $\angle H(e^{j0})$, is 0. (This is easy because e^0 is just 1.)

Plugging in $\omega=0.628$ (i.e., $F=1000$ Hz, the center frequency), we find that $H(e^{j\omega})=12.3437-16.6123j$. The magnitude is 20.69 (26.3 dB with respect to the magnitude at 0 Hz), and the angle is -0.932 rad. Figure 4-13 shows the magnitude response from $F=0$ to $F_s/2$.

4.3.3.3 $H(z)$: The System Function

---

[14]. We noted, too, the architect's close approximation to the impulse, which would also be quite "exciting." And honestly, isn't that about the only thing you'll really remember from this book?

[15]. [Klatt, 1980]

Describing Discrete-Time Signals and Systems

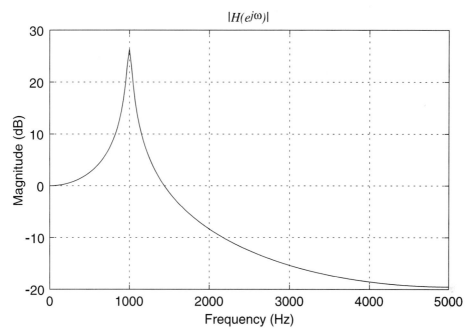

Figure 4-13 Magnitude response of a simple DT filter.

A logical question to ask is whether there's a counterpart in the DT domain to the system function $H(s)$. Fortunately,[16] the answer is yes—the DT system function $H(z)$. And though there are a few differences in the mapping and interpretation, we even have a *z-plane* to plot poles and zeros in, analogous to the *s*-plane earlier.

Although absolutely everything in this book is of crucial importance and your life will be immeasurably poorer for skipping any of it,[17] this $H(z)$ stuff is exceptionally useful. We'll use $H(z)$ to express digital filters, and it easily translates to *difference equations*, the expressions you'll actually write code to perform. If things have been a bit theoretical so far, we're getting close to the payoff.

Recall that the complex variable s is defined as $s=\sigma+j\Omega$.. σ is the (exponential) growth or decay factor, and Ω is the frequency of a CT signal. Recall, too, that we associated a complex frequency with this variable (the signal described by $e^{\sigma}cos(\Omega t+\theta)$ has such a complex frequency). The complex variable z is defined in a similar way and also is associated with a complex frequency, but is usually defined in terms of polar coordinates rather than rectangular coordinates. That is, instead of associating the real axis with exponential growth/decay and the

---

16. Jack notes that many students don't find this fortunate at all!
17. Including any footnotes you decide not to read!

imaginary axis with frequency, in the z-plane we associate angle with normalized frequency ω and radial distance with growth/decay. Real-world frequency is:

$$\Omega = \omega T_s = \frac{\omega}{F_s} \qquad (4.6)$$

$$F = \frac{\omega}{2\pi} T_s = \frac{\omega}{2\pi F_s} \qquad (4.7)$$

See Figure 4-14 below.

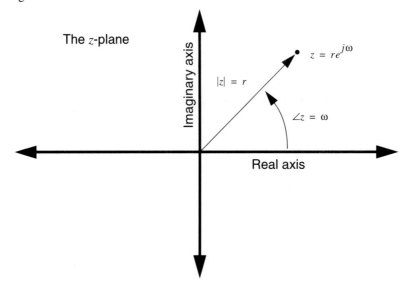

Figure 4-14 Locating frequency and exponential growth/decay in the z-plane.

z is defined as:

$$z = re^{j\omega} \qquad (4.8)$$

where r, its magnitude, is the (linear) growth/decay factor and ω, its angle (in radians), is normalized frequency.[18] r<1 is associated with decay, r>1 with growth, and r=1 no change. Alter-

18. By the way, a subtle point is that since e^{jx} defines points on a circle, the absolute value (or magnitude) of e^{jx} is always 1. So $|z|=|re^{j\omega}|=r$. If all this "e to imaginary power" business is making you unhappy, don't worry too much. We're using it mostly for its compactness, not to do amazing and tricky mathematical feats later on.

Describing Discrete-Time Signals and Systems

natively, you can write $r = e^{\sigma T_s}$, where σ is the exponential growth/decay we used in the s-plane. The definition in Equation 4.8 is based on Euler's identity, which relates the exponential function e^{jx} and the sine and cosine functions as follows:

$$e^{jx} = \cos(x) + j\sin(x) \qquad (4.9)$$

So, in fact we could write the definition of z as:

$$z = r[\cos(\omega) + j\sin(\omega)] \qquad (4.10)$$

although no one ever does. In fact, we're going to stick with the exponential notation as it's compact and more widely used.

This notation is entirely consistent with our earlier observation that the spectrum of a DT signal (or system) is cyclic, repeating every 2π (and symmetric around $\omega=0$). Getting rid of the sampling period T_s, if only temporarily, cleans up the math considerably.

With z defined as above, we can define a DT system function, $H(z)$, which relates the output $(Y(z))$ to the input $(X(z))$ of the system. As with $H(s)$, $H(z)$ can be expressed in terms of the ratio of two polynomials:

$$H(z) = \frac{Y(z)}{X(z)} = \frac{\text{some polynomial in z}}{\text{some other polynomial in z}} \qquad (4.11)$$

(usually in terms of z in negative powers, rather than z in positive powers), such as:

$$H(z) = \frac{1 - z^{-2}}{1 - 0.25z^{-2}} \qquad (4.12)$$

or in terms of poles and zeros (i.e., factored), such as:

$$H(z) = \frac{(z-1)(z+1)}{(z+0.5j)(z-0.5j)} \qquad (4.13)$$

Our interpretation of the locations of poles and zeros in the z-plane is just a bit different from that in the s-plane.

The z-Plane Figure 4-15 shows the z-plane. All of the elements of the s-plane are here—regions associated with stable and unstable poles, a line along which the frequency response is evaluated, and separate coordinates for exponential growth and frequency. However, instead of splitting the plane into a left-half and right-half plane, a circle of radius one—the *unit circle*—is the crucial dividing line now. Poles inside the unit circle are associated with stable systems, while poles outside the unit circle are associated with unstable systems. Poles right "on the line" (in this case, the unit circle; in the *s*-plane it was the *j*Ω axis) are associated with marginally stable systems.

Growth or decay is now a matter of the distance (r) from the origin to the pole. The unit circle lies at radius 1, the equivalent of the *s*-plane's *j*Ω axis where σ=0 (e^0=1). Frequency is now reflected in the angle of the pole.

The *z*-plane captures the cyclic nature of frequency in the DT domain. Increasing the frequency past π rolls around to -π and then back to 0 (and so on *ad infinitum*[19]).

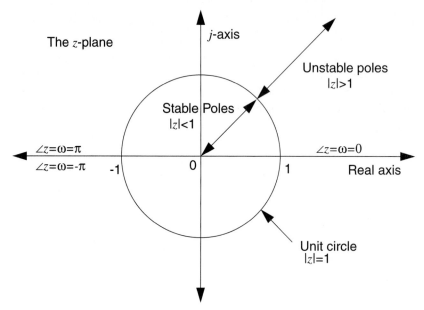

Figure 4-15 The *z*-plane.

---

19. Or, if you prefer, *ad nauseam*.

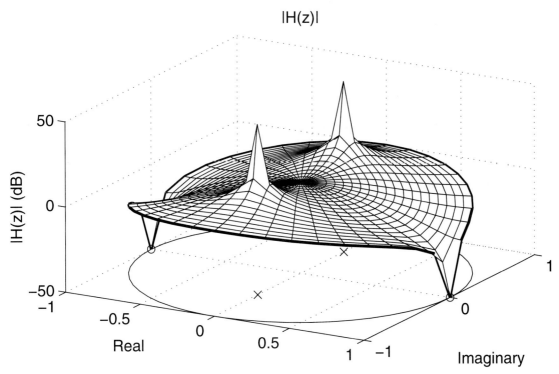

Figure 4-16 |H(z)| for a simple system.

Plotting |H(z)| We can plot |H(z)| in much the same way as we plotted |H(s)| as a three-dimensional plot, with peaks centered at poles and sinkholes at zeros.[20] Figure 4-16 shows |H(z)| for a system with poles at $z=+/-0.5j$ and zeros at $z=1$ and -1. (By the way, we could also do this with phase if we wanted.) This 3-D plot is rarely used; instead we plot the magnitude and phase along the unit circle, straightening it out into a linear plot. Figure 4-17 shows the plot we're more likely to see of the same system as in Figure 4-16. Here, r has been set to 1, and ω varies from $-\pi$ to π. (Sometimes you'll see these plots from 0 to 2π). You can see the symmetry around 0 Hz lets us drop the $-\pi$ to 0 section, which is what we'll usually do. If your system doesn't have real coefficients (i.e., poles and zeros in complex-conjugate pairs), you'll have to plot both, but none of the systems we talk about in this book lack that symmetry.

---

20. Just as with our 3-D plot in the s-plane, take this with a grain of salt. Mathematically, the z-transform doesn't exist in some of the regions we're showing, for the same reasons as in the s-plane. The z-transform gives us both an expression in terms of z and a statement of the regions where it this transform is valid. We're ignoring the latter in this discussion, though for our systems you can safely assume the region will always include the unit circle, and, thus, we are justified in evaluating the z-transform along the unit circle.

How $H(z)$ and $H(e^{j\omega})$ Are Related As you probably guessed, we're again using a particular notation for the frequency-response function, $H(e^{j\omega})$, so that it can be easily related to the z-transform, H(z). This is similar to what we did before with $H(j\Omega)$ and $H(s)$. The relationship[21] is $z = re^{j\omega}$, and by setting $r=1$ (i.e., evaluating the $H(z)$ along the unit circle, where radius r=1), we get $z = e^{j\omega}$.

$H(z)$, the system function, is the *z-transform* (ZT) of the DT impulse response (as well as the ratio $Y(z)/X(z)$). $H(e^{j\omega})$, the frequency-response function, is therefore found by evaluating the ZT around the unit circle (i.e., $r=1$). The frequency response function $H(\omega)$, the *discrete-time Fourier transform* (DTFT) of the DT impulse response, gives us the same information as found in $H(e^{j\omega})$, though the DTFT is a function only of "simple" frequency ω, not complex frequency z. So we have the same kind of situation as in Chapter 2 (Analog Filters); in the CT domain, we found that the Fourier transform (which led to $H(\Omega)$) is a special case of the more general Laplace transform. Here, the DTFT is a special case of the more general ZT. As long as the extra baggage of "ej" doesn't bother you, it's a bit less complicated to just use the ZT notation for both the system function and frequency-response function.

$H(z)$ and Difference Equations We're ready to make a very important connection between H(z) and the actual algorithms used to implement digital filters. If you've been skimming a bit, you might want to slow down for this section.

Recall from Chapter 2 (Analog Filters) that we can determine $H(s)$ using the Laplace transform. Specifically, $H(s)$ is the Laplace transform of the system's impulse response $h(t)$. The nice thing about using the Laplace transform is that we avoid work with differential equations; $H(s)$ is a polynomial, which is far, far easier to use than a differential equation, even if we have to deal with complex numbers.[22]

We have a similar situation with $H(z)$, the DT system function. $H(z)$ is the z-transform of the DT impulse response $h(n)$. However, DT systems are not described in terms of differential equations; instead, because of their discrete nature, they're described using *difference equations*. General difference equations express the current output of a system as a linear combination of current input samples, past input samples, and past output samples.

What's important for our work is that these equations can be translated very easily into computer programs. Let's first look at the form of difference equations, and then see how we can translate between $H(z)$ and difference equations.

21. By the way, you can also use $z = e^{sT_s}$ and the equations $s = \sigma + j\Omega$, and $\omega = \Omega T_s$ to get this relationship.

22. This is true until you have to go back to the time domain. If we're dealing only with steady-state responses, however, we avoid this headache.

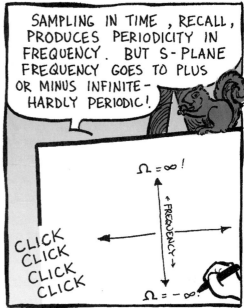
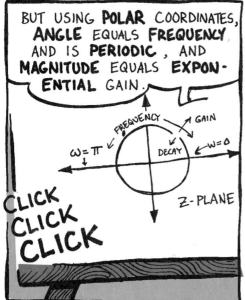

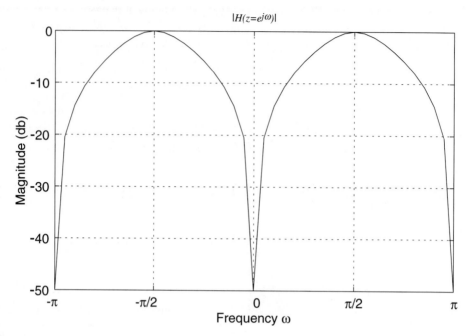

Figure 4-17 Magnitude response for the system of Figure 4-16.

Difference equations have the form:

$$y(n) = b_0 x(n) + b_1 x(n-1) + b_2 x(n-2) + \ldots + b_{n_z} x(n-n_z) \\ - a_1 y(n-1) - a_2 y(n-2) - \ldots - a_{n_p} y(n-n_p) \tag{4.14}$$

where

- $x(0), x(1), x(2), \ldots$ is the sequence of input samples
- $y(0), y(1), y(2), \ldots$ is sequence of output samples
- n is the sample index, which has values $0, 1, 2, \ldots$
- $x(n\text{-}k)$ is the kth input sample prior to the "current" sample—e.g., $x(n\text{-}1)$ is the prior sample. The same is true of $y(n\text{-}k)$.
- $b_0, b_1, \ldots, b_{n_z}$ and $a_0, a_1, \ldots, a_{n_p}$ are constant coefficients ("a_0" is usually set to 1; if not, it appears as a coefficient in front of $y(n)$). Each coefficient may be positive, negative, or zero.
- n_z is the number of zeros in the system; n_p the number of poles.

We can think of prior samples as "current samples" that have been delayed. That is, $x(n\text{-}2)$ was the "current" value two sample times ago, and has been delayed a total of two sample times.

Describing Discrete-Time Signals and Systems

Difference equations express the current output $y(n)$ in terms of the weighted sum of past inputs $(b_0 x(n) + b_1 x(n-1) + b_2 x(n-2) + \ldots + b_{n_z} x(n-n_z))$ plus the weighted sum of past output values $(-a_1 y(n-1) - a_2 y(n-2) - \ldots - a_{n_p} y(n-n_p))$. (We'll see in a moment why there's a negative sign in front of the "a" coefficients.) Note that for values of n less than n_z or n_p, as occur when we're just starting out, we get negative index values. Usually we just say that the values of x and y are 0 for samples $n<0$.

The form of difference equations is pretty simple. Happily, the mapping between difference equations and descriptions in the z-domain is equally straightforward. Delays in the difference equation map to multiplication by powers of z^{-1} in the z-domain. For example, $y(n)$, which has no delay, is mapped to $Y(z)$, but $y(n-1)$, with a delay of 1, is mapped to $z^{-1}Y(z)$. In general, $y(n-k)$ is mapped into $z^{-k}Y(z)$. Let's take a simple difference equation[23]:

$$y(n) = 3x(n) + 5x(n-1) + 7x(n-2) - 4y(n-1) - 6y(n-2) \qquad (4.15)$$

and see how the ZT represents it. Replacing each delay by an appropriate power of z^{-1} and $y(n)$ and $x(n)$ with $Y(z)$ and $X(z)$, respectively, we have the following:

$$Y(z) = 3X(z) + 5z^{-1}X(z) + 7z^{-2}X(z) - 4z^{-1}Y(z) - 6z^{-2}Y(z) \qquad (4.16)$$

Recall that $H(z)$ relates the output to the input, and is therefore the ratio of $Y(z)/X(z)$. Using simple algebra, we can get this ratio by moving all of the terms involving $Y(z)$ to the left side of the equal sign, terms involving $X(z)$ to the right, and factoring out the common $X(z)$ and $Y(z)$, to get

$$Y(z)[1 + 4z^{-1} + 6z^{-2}] = X(z)[3 + 5z^{-1} + 7z^{-2}] \qquad (4.17)$$

By a simple rearrangement, we have $H(z)$:

$$H(z) = \frac{Y(z)}{X(z)} = \frac{3 + 5z^{-1} + 7z^{-2}}{1 + 4z^{-1} + 6z^{-2}} \qquad (4.18)$$

It's just as easy going the other way, taking an expression for $H(z)$ and finding the corresponding difference equation. In fact, most folks don't bother with the algebra—they just read

---

23. No, this isn't anything special. We just made it up.

off the coefficients directly. Taking the ZT of Equation 4.14 and rearranging as we did above, $H(z)$ is

$$H(z) = \frac{Y(z)}{X(z)} = \frac{b_0 + b_1 z^{-1} + b_2 z^{-2} + \ldots + b_{n_z} z^{-n_z}}{1 + a_1 z^{-1} + a_2 z^{-2} + \ldots + a_{n_p} z^{-n_p}} \qquad (4.19)$$

Note that the coefficients $a_0, a_1, \ldots, a_{n_p}$ have positive signs in the system function (although a_i itself may be negative), a consequence of the negative signs in the difference equation.[24]

Although we usually don't deal with *differential* equations directly, difference equations are the heart of digital filters, so the relation between $H(z)$ and the corresponding difference equation is one of the most important things we study in this chapter.

Let's step back for a moment and see where we are:
- There are standard ways of characterizing filter behavior. We concentrated on frequency-domain characteristics (see Chapter 3 [Analog Filters]).
- We'll see in the following chapters that there are reasonably straightforward ways of translating these filter specifications into system functions, either in terms of $H(z)$ directly, or in terms of $H(s)$ and then transforming this into $H(z)$. $H(z)$ defines a DT system that implements the desired filter behavior.
- It is easy to transform the system function $H(z)$ into a difference equation.
- Difference equations are easy to implement in software (i.e., x and y are just arrays, n is an index). (See Chapter 7 [Microcontroller Implementation of Filters]) for a few hints to speed things up.)

The long detour through system functions might make a bit more sense now. These concepts let us take filter specifications and, assuming we can turn those into a DT system function, $H(z)$, generate actual microcontroller code.

We're going to back up a little to address the other way in which we make signals discrete, the quantization of value, then look at the analog-to-digital and digital-to-analog conversion processes in more detail. Following this digression we'll wrap up this chapter with an overview of the main classes of digital filters.

4.4 Quantizing—Continuous to Discrete Amplitude Values

DT signals exist only at specific, usually equally spaced, points in time. However, at these times, DT signals are continuous in value. Digital processors, be they microcontrollers or floating-point DSP chips, process data as values encoded in a finite number of bits, so the continuous-valued

---

24. (A word of caution—there are some texts that use a where we've used b and vice versa, and at least one text that places negatives in front of $a_0, a_1, \ldots, a_{n_p}$ in the system equation, which means there will be no negative signs in front of the a coefficients in the difference equation.)

signal must be *quantized*. This quantization is performed during the process of analog-to-digital conversion, though care must also be taken to preserve this resolution during subsequent processing. As we noted earlier, quantization leads to effects that are fundamentally different than those related to sampling rate; we usually lump quantization effects under the category of noise or harmonic distortion.

In this section, we're going to discuss the theoretical aspects of quantization—linear and nonlinear quantization, and the calculation of the noise introduced by quantization. In the following section, we'll look at the analog-to-digital process as a whole, which combines sampling (continuous to discrete in time) with quantization (continuous to discrete in value).

4.4.1 Linear Quantization

We'll begin with linear quantization, the most common type. Figure 4-18 shows the transfer characteristic of an ideal linear converter. The input range is divided into 2^b equal steps for a b-bit quantizer[25] and each step is associated with a unique numeric (binary) coding. (We've used binary integers here, but in a few pages, we'll explore other encodings that are used.) The acceptable range of input values is specified as the *full-scale range* (*FSR*) for signals that are

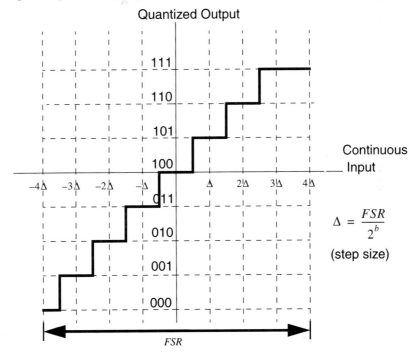

Figure 4-18 The transfer characteristic of an ideal linear converter.

---

25. In some texts b is defined as one less than the number of bits used by the quantizer. With that definition, each occurrence of b in the equations to follow should be replaced by $(b+1)$.

bipolar (signals with amplitudes above and below zero), while for *unipolar* signals (all amplitudes above zero) the term is just *full scale* (*FS*). Each step Δ has a size of *FSR/2^b* (or *FS/2^b*). We can talk, therefore, about the value represented by the *least-significant-bit* (LSB) of a converter, in terms of a voltage (*FSR/2^b*), scale factor (*1/2^b*), or percentage (*100/2^b*).

The *dynamic range* of a converter is the ratio of full-scale range to the step size; expressed in terms of decibels, it is just 20·log(2^b). (See Table 4-2.)

One thing you probably notice right away in Figure 4-18 is the offset to the decision points (the input values at which the corresponding output changes). The converter shifts all of the decision points toward the negative end by half a step. This offset makes the quantizer less sensitive to noise in low-energy signals (i.e., near 0), though at the expense of making the lowest step half the normal width and the highest step one-and-a-half steps wide. We usually don't even notice this offset, especially when using 10-bit or higher converters.

All converters depart from the ideal in several ways. Just a few of the many converter errors are:

- nonlinearity: Not all steps may be the same size
- offset error: The transition for the first step may not occur at exactly 1/2 step
- scale factor: If each step is the same size, the sum of all steps may still not be equal to FSR

Reading manufacturers' data sheets on ADCs can be a real education, and some caution is required to make sure that the performance you require will be met, especially for higher-resolution ADCs (say, 14-bit and above). For example, you may get only 14 or 15 bits of effective resolution from a 16-bit ADC, due to the errors above. But before getting too bent out of shape, it's worth considering that the step size for a 16-bit converter with FSR of 5 volts is 76 microvolts (0.000076 volts)—that's a very small voltage! Just delivering a signal without adding noise above 76 microvolts could be a challenge, so we shouldn't be too harsh on the shortcomings of real-world ADCs. Extraordinary engineering goes into the consumer electronics (e.g., CD players, PC sound cards) that boast 16-bit or higher resolution.

4.4.2 Nonlinear quantization

Nonlinear quantizers represent an alternative to linear quantizers. Before we go further, it's important to note that we really can't do standard SP directly on signals that have been quantized using a nonlinear scale. Instead, nonlinear quantization is useful in communication systems where it allows the useful information in a signal to be compressed. Audio signals, for example, can be quantized in a logarithmic fashion, coding small signals with high resolution and large signals with less resolution. If a logarithmically quantized encoded signal is decompressed using an inverse mapping, the resulting signal sounds fine to our ears, because our ears respond to audio logarithmically. The information that has been lost is information we wouldn't pay attention to.

Telephony makes extensive use of nonlinear quantizers in the form of companding CODECs,[26] devices that combine nonlinear ADCs with matching nonlinear DACs, usually along with other audio interfacing circuitry and a serial communications channel. (The resulting serial, digital data is a form of *pulse-coded modulation* (PCM), which occurs whenever a continuous signal is converted to a digital signal at a set rate and communicated via a serial data stream.)

Telephone-grade audio requires 12 or more bits of resolution for low-level signals. There are two standard logarithmic mappings that take this original >12 bits of resolution and produce a compressed 8-bit value. The U.S. and Japan use a standard known as µ-*law*, while Europe uses a slightly different *A-law* mapping. Figure 4-19 shows the µ-law encoding curve. If we tell you that the bandwidth of the telephone system is around 4 kHz,[27] you can guess the sampling rate used is twice that, or 8 kHz. A telephone conversation, then, can be sent at 8K samples/sec *8 bits/sample or 64K bits/sec with these mappings. Without the nonlinear mapping, the same conversation would take 50% more bits/sec (around 96K bits/sec)—and wouldn't sound appreciably better!

There are two reasons we mention nonlinear quantization and, in particular, telephony-grade companding CODECs: first, such devices often form excellent low-cost peripherals for DSP systems (you may even be able to hook a microphone and speaker directly to the CODEC); and, second, DSP is often used to process PCM data of this type. Our first reason may be getting

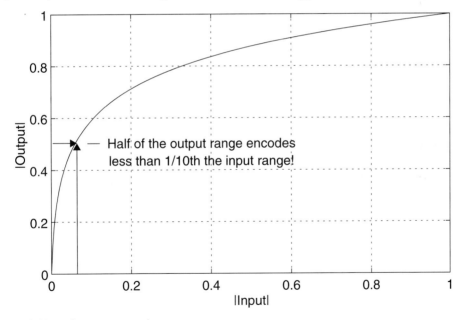

Figure 4-19 µ-Law compression.

26. *CODEC* comes from COder/DECoder; the term is now used to include ADC/DAC combinations that may or may not include any (nonlinear) compression.
27. The bandwidth used for voice transmission is actually less than this, around 3.3 kHz.

somewhat dated—the explosion in audio processing cards for PCs, along with other applications that require audio interfaces for DSP chips, has produced a flurry of noncompanding CODECs and "audio interface converters" that are just as handy, and linear to boot. However, the current telephone infrastructure will probably continue to employ "companding" as described for some time to come.

Table 4-1 below shows the standard mapping from a 12-bit value to a μ-law. (Note that the coded 8-bit word is inverted before being sent—this increases the density of ones in the resulting bit-stream, which is helpful in detecting and correcting errors in transmission.) Although a look-up table approach to conversion is possible, you can see that some well-placed shifts and masking could be used to make the translation. The problem, of course, is that any conversion takes away precious time from our other DSP tasks.

Sign	12-bit Signed Input (magnitude-sign format) Magnitude											Sign	8-bit μ-Law Output Segment			Position in Segment			
	11	10	9	8	7	6	5	4	3	2	1	0		$\bar{7}$	$\bar{6}$	$\bar{5}$	$\bar{4}$	$\bar{3}$	$\bar{2}$	$\bar{1}$	$\bar{0}$
S	0	0	0	0	0	0	0	Q_3	Q_2	Q_1	Q_0	S	0	0	0	Q_3	Q_2	Q_1	Q_0
S	0	0	0	0	0	0	1	Q_3	Q_2	Q_1	Q_0	S	0	0	1	Q_3	Q_2	Q_1	Q_0
S	0	0	0	0	0	1	Q_3	Q_2	Q_1	Q_0	X	S	0	1	0	Q_3	Q_2	Q_1	Q_0
S	0	0	0	0	1	Q_3	Q_2	Q_1	Q_0	X	X	S	0	1	1	Q_3	Q_2	Q_1	Q_0
S	0	0	0	1	Q_3	Q_2	Q_1	Q_0	X	X	X	S	1	0	0	Q_3	Q_2	Q_1	Q_0
S	0	0	1	Q_3	Q_2	Q_1	Q_0	X	X	X	X	S	1	0	1	Q_3	Q_2	Q_1	Q_0
S	0	1	Q_3	Q_2	Q_1	Q_0	X	X	X	X	X	S	1	1	0	Q_3	Q_2	Q_1	Q_0
S	1	Q_3	Q_2	Q_1	Q_0	X	X	X	X	X	X	S	1	1	1	Q_3	Q_2	Q_1	Q_0

Table 4-1 The standard 12-bit to 8-bit μ-Law mapping.

4.4.3 Quantization and Noise

Even when a quantizer is perfect, we're still replacing a continuous value with a "rounded-off" (quantized) value. In most cases, we can treat this effect as equivalent to adding noise to the signal—that is, we can model the difference between the continuous and quantized values as an added noise signal, just as we might model a noise signal encountered at any other stage. For a quantizer with many quantization levels, this effect is minimal; predictably, as we represent the signal with fewer and fewer levels, the "noise" increases in strength (i.e., power).

To talk about noise levels, it's necessary to have an objective measure. As with other measurements we've described, it's useful to talk about the relative power levels—in this case, the relative power of the desired signal (P_S) versus the power of the noise (P_N, all "undesired" components added to the "desired" signal). The *signal-to-noise ratio (SNR)* is such a measure and is expressed in dB as:

$$SNR_{dB} = 10 \cdot \log\left(\frac{P_S}{P_N}\right) \qquad (4.20)$$

This begs the question of how to calculate the power in a signal. For a DT signal of a length N, you can use the following equation:

$$P_x = \frac{1}{N} \sum_{n=0}^{N-1} x^2(n) \qquad (4.21)$$

That is, sum the squares of the amplitudes of the samples and divide by the number of samples.

We can use the following equation to find the *signal-to-quantization-noise ratio (SQNR)* introduced by b-bits quantization:

$$SQNR_{dB} = 6.02b + 1.76 \qquad (4.22)$$

The SQNR is a specific case of an SNR; in some cases where b is small, the SQNR may dominate the total system SNR, but SQNR is just one contributor, and other noise sources are usually present—as we've said, noise can creep in everywhere from the transducer back to the converter.

Table 4-2 summarizes the SQNR for a range of values of b, along with some other quantities that we'll use later on. To put the SQNR value in perspective, high-quality audio requires a (total, system-wide) SNR of around 90–96 dB, which means we must use a minimum of 16 bits for quantization. Seismology SP may use 24 bits of resolution; the SQNR is then 146 dB! (That's a power ratio of $4 \cdot 10^{14}$ to 1!)

Bits (b)	Quantization levels (2^b)	LSB in percent ($100/2^b$)	Resolution for 0–5V range ($5/2^b$)	Dynamic range ($20 \cdot \log(2^b)$)	Signal to Quantization Noise Ratio (SQNR) (Equation)	Min. stopband attenuation (Equation 4.24)	Time constants to settle to LSB (Equation 4.23)
(bits)	(levels)	(%)	(V)	(dB)	(dB)	(dB)	(τ)
6	64	1.56%	78.1 mV	36.1	37.9	43.9	4.9
8	256	0.391%	19.5 mV	48.1	49.9	55.9	6.2
10	1,024	0.0977%	4.88 mV	60.2	62.0	68.0	7.6
12	4,096	0.0244%	1.22 mV	72.2	74.0	80.0	9.0
14	16,384	0.00610%	305 µV	84.3	86.0	92.1	10.4
16	65,536	0.00153%	76.3 µV	96.3	98.1	104.1	11.8
18	262,144	0.000381%	19.1 µV	108.4	110.1	116.2	13.2
20	1,048,576	0.0000954%	4.77 µV	120.4	122.2	128.2	14.6
22	4,194,304	0.0000238%	1.19 µV	132.5	134.2	140.2	16.0
24	16,777,216	0.00000596%	298 nV	144.5	146.2	152.3	17.3

Table 4-2 Specifications for ideal b-bit conversion.

4.5 Analog-to-Digital and Digital-to-Analog Conversion

We've discussed in somewhat theoretical terms the two main operations necessary to convert an analog signal into a digital signal—sampling and quantization—and it's no leap of imagination to see that roughly the opposite functions probably are involved for digital-to-analog conversion. Let's now look at the practical aspects of both analog-to-digital and digital-to-analog conversion.

4.5.1 Analog-to-Digital Conversion

4.5.1.1 Functions

Analog-to-digital conversion actually involves a number of different operations and considerations. It's important to know how each piece fits together, even though some manufacturers have now integrated most or all of these functions on a single chip. Figure 4-20 shows how these operations are related. You'll note, by the way, that we haven't drawn a box around these and said, "Here's your ADC." Quantization and encoding are the only operations that are *always* found in ADC devices—other processes such as gain, antialiasing, and sample and hold may or may not be integrated, or you may need to augment on-chip resources with external devices.

Gain, Offset, Impedance Matching Real transducers usually produce signals that are a far cry from what we actually can convert. For example, a signal may have small amplitude swing, a constant (and undesired) DC bias or offset, or insufficient "drive" for a particular ADC. Often, we can address all three issues at the same time with a simple op-amp circuit; texts such as Horowitz and Hill [1989] cover the necessary design details. Among other things, the bandwidth, input and output range, noise, input and output impedance, and current consumption of the op-amp figure into the design. Most of the time, it is easy to condition a signal without introducing too much distortion or noise.

We should say a little bit about the concept of impedance[28] with respect to the input and output of devices. When we describe a signal by noting its voltage (amplitude) at a given time, we haven't described how much current is available "behind" that signal. If we ask for current from the signal, to what extent does that change the amplitude of the signal? Low-impedance-outputs are outputs that act like a perfect voltage source (i.e., they can supply an infinite amount of current) in series with a fairly small impedance; we can ask them to supply a relatively large amount of current and expect very little change in the amplitude of the signal. On the other hand, a high-impedance output acts like it has a fairly large impedance in series with its output. Ask for current from this signal and you can expect to see a big change in amplitude.

Why should we worry about this? The flip-side to output impedance is input impedance, the impedance that a signal "sees" looking into the device. A low input impedance asks any input signal to supply a lot of current—obviously a bad thing to ask of a signal coming from a high-impedance source! A high-impedance input, on the other hand, asks very little current; either a low- or medium-impedance source should work fine. All of the devices we use, op-amps, filter ICs, and especially ADCs, have a finite input impedance; depending on the characteristics of the source and input, you can end up with unintentional filters in your SP circuit.

Often, the impedance of an ADC will be specified in terms of its DC resistance and a capacitance. Usually, the resistance isn't much of a problem, but the capacitance will form a lowpass RC filter with the effective resistance of the signal. You can analyze this effect in the frequency domain (along the lines of the antialiasing calculations we discuss next) or in the time domain. Table 4-2 shows the amount of time, in terms of the time constant $\tau = RC$, required for

---

28. Recall that impedance is a kind of generalized resistance; it includes the frequency-varying "resistance" of capacitors and inductors, not just the usual frequency-independent resistance of resistors. Appendix 2 has additional discussion on impedance.

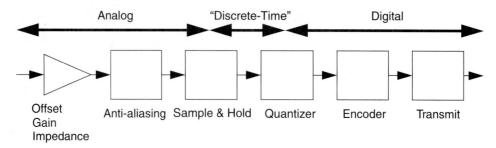

Figure 4-20 The analog-to-digital conversion process.

the output of an RC filter to drop to 1/2 LSB of a *b*-bit converter. Use the following equation for other values of *b*:

$$\text{time constants to } 1/2 \text{ LSB} = (b+1)\ln(2) = (b+1)(0.693) \qquad (4.23)$$

If we wished to sample a signal with an impedance of 50 ohms, for example, and the 68HC16Z1 ADC has a capacitance of 15 pF[29], our minimum sampling period due to this RC effect would be 7.6τ. Plugging in numbers, we find that τ=750 picosec (50×15×10^{-12}), and 7.6τ= 5.7 nsec, or a sampling rate of 175 MHz, not a limiting factor. If the impedance of our signal were, instead, 100 K ohms, the minimum sampling period would not only increase to 11.4 μsec (88 kHz), but there would be a significant error due to the 20K input impedance of the ADC. This error won't go away regardless of how slowly we sample. For the ADC on the 68HC16Z1, the input capacitance isn't much of a problem (input resistance could be) but, in general, if you are tempted to run a high-impedance sensor output directly to your ADC, you are well advised to check for this limitation.

Antialiasing We saw earlier that producing a DT signal by sampling can introduce a side effect called *aliasing,* in which frequency components of a signal higher than half the sampling rate are "aliased" into lower frequencies. These higher-frequency components may be valid signals from your sensor or noise from any number of sources—radio stations, fluorescent lights, etc. For example, you might be processing speech, which contains frequency components well above the 3-kHz bandwidth you plan on processing. If sampled at F_s=6 kHz, any frequency components above 3 kHz will be aliased back down on top of the signal in the range of 0–3 kHz. Even when your signal is *bandlimited* (all energy is within a specific frequency range), there's often noise with a wide frequency range. All of that noise will be aliased on top of your signal spectrum of interest!

By the way, when we talk about noise, there are really two flavors. First is the noise that occupies the same frequency range as the signal of interest. This *in-band* noise is problematic, as

---

29. pF is picofarad, or 10^{-12}F.

we can't use a simple highpass, lowpass, or bandpass filter to reduce the noise.[30] Instead, shielding and other techniques are used to keep the noise from getting into the Nyquist band in the first place. Second, we have noise at frequencies outside the spectral range of interest. This *out-of-band* noise *can* be filtered, though that doesn't mean we don't want to keep as much of it as we can out of our system, using shielding and so on. Now, both in-band and out-of-band noise may be coming from the same source, but the point is that there may be a tremendous amount of energy in the out-of-band noise, all of which will be aliased on top of our signal spectrum of interest.

The solution to these two issues is to filter the analog signal prior to sampling (i.e., to use an antialiasing filter). Frequencies outside the range of interest are attenuated to a level below what the ADC will see, thus, there will be no aliasing of either "extra" bandwidth of the signal or of out-of-band noise. (There's nothing we can do at this stage about in-band noise.) The minimum attenuation we need from the antialiasing filter is related to the resolution of the ADC, though different authors will give you different relations. A simple criterion is that the minimum attenuation should be less than the step size of the converter, essentially the dynamic range figure we list in Table 4-2. However a more conservative expression (based on the SQNR, or noise introduced by quantization) is:

$$A_{min} = 20 \cdot \log(\sqrt{3/2} \times 2^{b+1}) \qquad (4.24)$$

This works out to be 7.8 dB more than the dynamic range. For example, a 10-bit converter requires a minimum attenuation of 68.0 dB. (Compare this with the 60.2 dB attenuation we get using the step size.) This minimum attenuation is the stopband attenuation of the antialiasing filter.

As we saw in Chapter 3 (Analog Filters), we need a few more parameters in order to define a filter (in this case, an antialiasing filter). Assuming for the moment that the antialiasing filter is a lowpass filter (it may not always be!), we need to know the passband edge frequency, the stopband edge frequency, the passband ripple, and the filter order. (The stopband attenuation is, roughly, the minimum attenuation we found above.) Let's start with the easy ones.

The range of frequencies of interest defines the passband edge frequency of the antialiasing filter. Likewise, we can examine the overall system goals to see how much passband deviation (ripple) we can tolerate. (Note that digital processing, including digital filters, may add more ripple, so this isn't our whole "ripple budget.") Both of these figures are independent of the sampling rate.

It looks like we have only the stopband edge frequency and the filter order left to specify but, in fact, we also need to make adjustments to the stopband attenuation. Here are the issues we need to keep in mind in specifying these three parameters:

---

30. There are some filters that we could use—for example, a matched filter, which we discuss in a later chapter.

- The frequency range in which we want to avoid aliasing may be only a subset of the DT spectrum. For example, we may sample a speech signal at 20 kHz, giving us a 10-kHz bandwidth, but later pass this DT signal through a sharp digital filter that cuts off everything above 6 kHz. Therefore, we could tolerate aliasing in the range 6 kHz–10 kHz.
- The "minimum attenuation" figure actually should be called the minimum *signal* attenuation figure, as it's not the attenuation of the antialiasing filter that's important, it's the magnitude of the *signal* (above the stopband edge frequency) that we want to be below a certain amount. Thus, a signal that already has a lower magnitude at higher frequencies will not require as much attenuation to avoid aliasing. On the other hand, a signal that increases in magnitude outside the frequency range of interest may require a filter with a larger stopband attenuation.

Using these facts, we can summarize the parameters of an antialiasing filter (using "real-world" frequency!) as:

Passband ripple (A_p): Cannot exceed the overall (system) passband ripple tolerance. Common figures are 0.1–1 dB (peak-to-peak).

Passband edge frequency (F_p or Ω_p): The upper edge of the frequency range to be processed in the DT signal.

Stopband edge frequency (F_{stop}[31] or Ω_{stop}): The attenuation of the antialiasing filter at this frequency must be sufficient to reduce the magnitude of the analog signal to less than the ADC's minimum attenuation. Because aliasing is symmetric around the Nyquist frequency ($F_s/2$), a little arithmetic shows that aliasing in the range 0 to F_p will be avoided if F_{stop} is $\leq F_s - F_p$.

Stopband attenuation (A_s): Not the same as the minimum attenuation required by the ADC, since the magnitude of the analog signal at high frequencies (e.g., F_{stop} and above) could be more or less than that at lower frequencies (e.g., in the passband).

For an example, let's say that we want to process the signal whose spectrum is shown in Figure 4-21. The frequencies of interest are 0–5 kHz, and you can see that the magnitude of the signal already drops at high frequencies, perhaps due to some filtering that's already being done. Note that we also have some noise, down -30 dB with respect to the energetic part of the signal spectrum. Suppose that we select a 10-bit converter and require that no aliasing occur in the 0–5 kHz range. We will tolerate aliasing (of either signal and/or noise) above that frequency. This aliasing might be tolerable if we will be filtering the signal with a steep digital filter later on, for example. The maximum (peak-to-peak) passband ripple we'll accept for our antialiasing filter is 0.5 dB. Although it's not always the case that the sampling rate is fixed at this point, let's also assume that the sampling rate has been determined to be 20 kHz.

Using , we find the minimum attenuation to avoid aliasing is 68 dB. From our discussion of aliasing, we know that signals above $F_s/2$ will be aliased in a mirror-image, as in Figure 4-22.[32] Thus, this minimum attenuation must occur no higher than at a frequency as far above $F_s/2$ as F_p is below it; that is, $F_{stop} = F_s - F_p$. Because we know the sampling rate, this places F_{stop} at 20-5, or 15 kHz. We can now look at the magnitude of the analog signal at this frequency to see what

---

31. We'll use F_{stop} instead of the more compact F_s as we use the notation F_s for the sampling frequency.

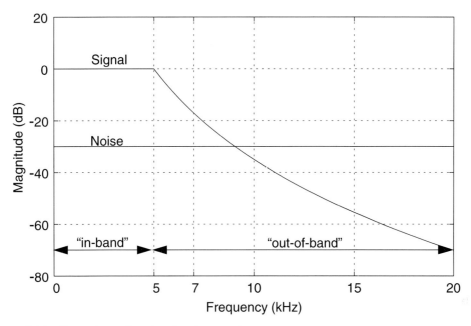

Figure 4-21 Example analog signal, noise spectrum.

attenuation we will need from the antialiasing filter. From Figure 4-21, it appears that the noise will dominate at 15 kHz (the signal is at about -55 dB, and noise is at -30 dB), so we'll need a minimum of 68-30, or 38 dB of stopband attenuation.

If we lower the sampling rate to 12 kHz, we would find that the antialiasing filter stopband attenuation would need to be 51 dB, as the magnitude of our signal at 7 kHz (12–5 kHz) is about 17 dB. The ratio of F_s to F_p is now much smaller than before (alternatively, the transition width is relatively smaller), which is an indication that the filter order will be higher than before. A lower sampling rate, however, increases the amount of intersample time we have to perform DSP on the signal.

We can now go to any analog filter design book or program and design antialiasing filters using these parameters.[33] It turns out that a fourth order "Chebychev I" filter will work fine for

---

32. Yes, there will be aliasing from higher frequencies; however, most physical systems tend to attenuate signals at higher frequencies, and most filters we'll use for antialiasing keep dropping off at a high rate (about $20N$ dB/decade, where N is the number of poles), so all we usually need to worry about is signals fairly close to $F_s/2$. Of course, if you have some really large signal present, even quite a ways above $F_s/2$ in frequency, you may need to be sure that you are also attenuating it below the minimum attenuation.

33. Actually, we provide enough information in Chapter 6 (IIR Filters—Digital Filters with Feedback) for you to determine the filter type (Chebychev, Butterworth, etc.) and the order of the filter required for most cases. But in that chapter we're interested only in the system function for analog filters as a way of getting to a digital filter; the information required to take a system function $H(s)$ and calculate the correct capacitors and resistors is outside the scope of this book.

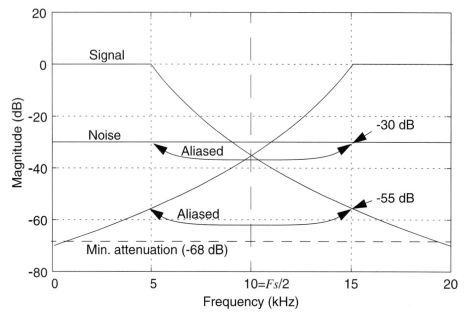

Figure 4-22 Spectrum of DT signal and noise showing aliasing.

the first filter, while a ninth order is required for the second. There are a number of other filters besides Chebychev that we could use—we'll talk about some of the characteristics of popular analog filters in Chapter 6 (IIR Filters—Digital Filters with Feedback). The Chebychev is desirable in that it has a small transition between passband and stopband.

To give you a rough idea of the material necessities, a fourth-order filter using discrete components will take two op-amps and eight or so resistors and capacitors. (Most analog active filter designs require one resistor and capacitor per pole and one op-amp for every two poles.) You can also buy a single chip filter with a fourth- to eighth-order filter on it. This would save some space. One chip manufacturer even has software that takes your design parameters and calculates the correct settings to make to the filter chip to get a specified response. (You "program" this chip with a few external resistors.)

We have a few more comments about antialiasing filters to make before moving on to the rest of the conversion process. First, like any other filter, antialiasing filters affect the phase of the signal. This might not matter for your application, but if you have any interest in the time-domain characteristics of your signal, you might need to augment your filter specifications with time-domain specifications such as settling time or overshoot. The filter type we chose above, the Chebychev, is not too hot in the time domain, which is the trade-off it makes for good frequency domain characteristics. To repeat ourselves, there is no brick-wall filter (perfect frequency domain response), nor a filter that has perfect time-domain response. All we have are some pretty good designs that come as close as we know how. Don't trust any company that tells you that their antialiasing filter is the only one you'll ever need![34]

Second, while most antialiasing filters are lowpass, if your frequency range of interest is not a baseband signal (i.e., 0 to some F_p) but, instead, is a bandpass signal (F_L to F_H), the antialiasing filter must be a bandpass filter. Even if your signal is the one that's being aliased (on purpose), don't let it be corrupted by a bunch of (low-frequency) noise.

Third, in spite of a need to meet requirements efficiently, we must allow for imperfections in analog filter performance and other analog elements (including transducer, wiring, connectors, etc.), and allow for changing environmental conditions. As we noted above, discrete implementations of filters have two poles per op-amp, so it isn't much cost to make all your filter orders even. Single-chip filters are also "chunky," offering four-pole filters as a unit. In some cases, then, you might just use an eight-pole when six is all you theoretically need. On the flipside, the SQNR of the converter limits the overall system SNR. If a four-pole filter takes you well below the minimum attenuation, an eight-pole filter just wastes resources.

Finally, eight poles is (currently!) a reasonable upper limit on economical filters you should consider; if your calculations show you need more than that (e.g., our nine-pole filter, above), you should probably look long and hard at raising your sampling rate, which will widen the transition from F_p to F_s, thus lowering the order of the filter you need. (Again, this "rule of thumb" will no doubt continue to change as better designs become available; on the other hand, don't mention our names if you decide to tell your boss that you need a twenty-second-order antialiasing filter.)

Sample and Hold   (You can safely skim this section—it deals with issues you normally don't worry too much about. Are we nice or what?)

The sampling operation that we've been discussing so far is a mathematical abstraction. In real life, we can't sample the instantaneous value of a signal, nor can we sample with exactly the same period from sample to sample. The closest we come is to create a rough copy of the value of a signal during a relatively brief amount of time, then hold this value (without too much *droop*, or change in voltage) while we take our time quantizing the value.

There are two terms for the devices used to create a temporary sample of the signal—*track and hold amplifiers* (THA) and *sample and hold amplifiers* (SHA). The distinction lies in whether we assume that the signal value changes appreciably during the time the device "looks at" the signal. A THA generates a signal that tracks (is a copy of) the input signal until we ask it to "hold" the value. Once the value is held, the quantizer can convert this static value at its leisure. An SHA expects to sample the input signal as briefly as it can, then hold this value. At high sampling rates, the two types merge, as the THA really isn't tracking very much versus how long it holds. We'll just assume the use of an SHA here.

---

34.   Unless you work for the company!

We worry about two issues with SHAs—*aperture time* and *aperture jitter*. Aperture jitter is the variation in sampling period from sample to sample, and aperture time is the length of time during which the SHA "looks at" the input signal.

The aperture time is crucial as it places a limit on the upper frequency we can convert—independent of the conversion (sampling) rate! To see why this is so, let's consider sampling a sine wave at a rate comfortably above the Nyquist rate. Rote application of the sampling theorem says we should be fine, but for an accurate sample to be taken we really can't have the signal change more than the resolution of the quantizer during the aperture time. If we use 1/2 bit of resolution as the most we allow a signal to vary during the aperture time, the following formula relates the maximum frequency F_{max}, b bits of resolution, and aperture time τ_a:

$$F_{max} = \frac{1}{2\pi 2^b \tau_a} \quad (4.25)$$

Or, if you prefer using Ω_{max} (and who doesn't!)

$$\Omega_{max} = \frac{1}{2^b \tau_a} \quad (4.26)$$

Let's see what a speech-processing application requires. Take the bandwidth to be 0–4 kHz, an 8-kHz sampling rate (let's ignore the transition of the antialiasing filter), and we'll use 13 bits of resolution. Now, we know the sampling period is 1/(8 kHz) or 125 μsec. What's the required aperture time? Plug in the numbers and you get τ_a=4.9 nsec!

Aperture jitter reduces the SNR of the signal. The equation relating the SNR due to aperture jitter for different full-scale sinewave inputs is:

$$SNR = 20 \cdot \log\left[\frac{1}{2\pi F t_a}\right] \quad (4.27)$$

where F is the frequency of the full-scale sine wave input and t_a is the (RMS value of the) aperture jitter. For example, for high-quality audio, we might want a full-scale sine wave input at 20 KHz, and an SNR due to aperture jitter of 96 dB. This is a (RMS) value of 126 picosec (that's $126 \cdot 10^{-12}$ seconds), which is pretty darn small.

You can make a THA or an SHA using op-amps and/or transistors, but for high-resolution conversion, it's better to buy it as a ready-to-go IC (you supply a capacitor and resistor, perhaps). All of the pesky issues like offsets (the DC difference between the input signal and output signal) and droop (the tendency of a "held" signal to change over time) are well optimized on a

Analog-to-Digital and Digital-to-Analog Conversion 133

commercial IC. Of course, your ADC may have an SHA or a THA integrated already (e.g., the 68HC16Z1's ADC). Even so, you may have some application that requires a faster SHA—undersampling, for example.

By the way, the "A" of SHA and THA means "amplifier," which generally means you'll have a nice low-impedance output and a fairly high impedance input. Just be sure to check the specifications to make sure that your input signal impedance is reasonable and that the input to the next stage will not present too big a load for the SHA's output (quite unlikely, but check).

Conversion (Quantization) Figure 4-23 shows one realization of a simple four-level quantizer. (The triangles are comparators—the output goes high if the + input is higher than the - input, low otherwise.) All that's really happening here is that each comparator is presented with a fixed reference from a voltage divider and continually compares the input against this reference. (Note that the resistor values are set to produce the 1/2 LSB offset in decision point.) The resulting output is a *thermometer code* for reasons you can see in Table 4-3. This particular converter is called a flash converter, since the conversion is done all at once by a massively parallel circuit.[35]

Input amplitude	Thermometer code	Encoder output	
Volts	C,B,A	R2,R1	decimal
2.5 to 4.0	111	11	3
1.5 to 2.5	011	10	2
0.5 to 1.5	001	01	1
0 to 0.5	000	00	0

Table 4-3 Thermometer codes and output of 2-bit converter/encoder.

Thermometer codes are not particularly convenient for transmitting information— a 12-bit flash converter with 4095 comparators requires 4095 bits (=wires or traces)! Instead, this form of converter is followed by an encoder that produces more compact codes, which we'll discuss in a second. A flash converter takes quite a lot of circuitry, and there are other types of converters that trade off conversion time for fewer components. For our purposes, however, we really don't care how the conversion is done, just how long it takes and any quirks it introduces. So we'll discuss the different converter forms later when we talk about different real-world ADCs.

There are two other conversion issues we need to discuss briefly—references, and single-ended versus differential (double-ended) inputs.

ADCs usually end up comparing the input signal against internally generated voltages. These voltages, whether multiple constant voltages (e.g., flash encoder above) or a voltage that

---

35. Oh, so three comparators isn't your idea of massive? Well, we agree, but an 8-bit flash converter has 255 comparators, and a 12-bit would have 4095—that is, you need 2^b-1 comparators for a b-bit converter.

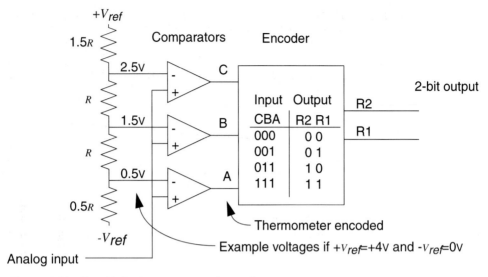

Figure 4-23 Two-bit flash converter and encoder.

changes over time (e.g., successive approximation converter, discussed below), usually are derived from one or more *reference* voltages. The stability and accuracy of the conversion depends on the quality (amount of noise, drift, etc.) of these references, while the full-scale range of conversion is set by the value of the reference.

Some ADCs generate reference voltages internally; in such cases, your design task is simpler, but you need to live with the accuracy the IC designer chooses and with the range that is, therefore, hard-wired into the converter. Other ADCs require an external reference (or two); in this case, you not only may provide a high-quality reference, but you likely have the option of customizing the range of the ADC to match the range of your signal, maximizing resolution.

There are many cases in which it is advantageous to convert the difference between two signals rather than a single signal or the two signals separately. Often a signal is sent via pair of wires, one carrying the desired signal and the other an inverted version. Any noise that might be picked up by the wires will ideally be picked up in equal amounts by each wire, so the original signal should be recoverable with minimal noise by just extracting the difference between the two signals and ignoring signals that are common to both. Differential (vs. single-ended) inputs to ADCs allow exactly this conversion of the difference between two inputs and are actually quite common. The name of the game is to achieve maximum SNR, and differential inputs are another tool for maximizing SNR.

Encode It's pretty obvious we want to end up converting the thermometer codes of the converter (or whatever codes are produced by other converters) to a compact binary form. But what form should that be? Just assigning integers 0 through 2^b-1 to the quantization levels from lowest to highest makes some sense, but many signals are bipolar and it would be convenient to

Analog-to-Digital and Digital-to-Analog Conversion

have a two's complement representation sometimes. Earlier, the table for converting linear to µ-law values seems to suggest a magnitude/sign format would be useful in that application. So what form of encoding is used by ADCs? All of the above, and none! Some ADCs may even offer two or more modes of encoding, as the 68HC16Z1 ADC does. Without a doubt, though, the only ADC that will fit on your board or will interface cleanly to your microprocessor will output in exactly the "wrong" code. This is not a big problem, as conversion from one format to another is usually very simple.

To convert from unsigned binary to two's complement (where the midpoint of the unsigned binary is the new zero) just invert the most-significant-bit (MSB). Do exactly the same to convert from two's complement to unsigned binary. Table 4-4, below, shows different code possibilities for a three-bit encoder. (Note the two different "zeros" possible with sign/magnitude!)

Example Unipolar Input (0 to +4)		Example Bipolar Input (-4 to +4)		Signed Encoding		Offset Binary (Unsigned)		Magnitude/Sign		
Range		Range		Bin	Dec	Bin	Dec	Sign	Mag.	Dec
3.25	4	2.5	4	011	3	111	7	0	11	3
2.75	3.25	1.5	2.5	010	2	110	6	0	10	2
2.25	2.75	0.5	1.5	001	1	101	5	0	01	1
1.75	2.25	-0.5	0.5	000	0	100	4	0	00	0
								1	00	"-0"
1.25	1.75	-1.5	-0.5	111	-1	011	3	1	01	-1
0.75	1.25	-2.5	-1.5	110	-2	010	2	1	10	-2
0.25	0.75	-3.5	-2.5	101	-3	001	1	1	11	-3
0	0.25	-4	-3.5	100	-4	000	0	1	(11)	(-3)

Table 4-4 Output encodings for a 3-bit unipolar or bipolar converter/encoder.

Figure 4-24 shows, for the bipolar case, the mapping between input value and output code for the -4 to +4 example of the table.

Transmit The final step in the analog-to-digital conversion is transmission of the encoded data. In some cases, this is trivial—the results are available directly on b pins, usually with another few signals indicating when the conversion is complete. These pins could be hooked directly to input pins on a microcontroller and the results read directly. Some ADCs can be configured to be accessed over a microprocessor bus, in which case they look like just another memory location, directly readable or writable. In cases where the bus size (often 8 bits for small

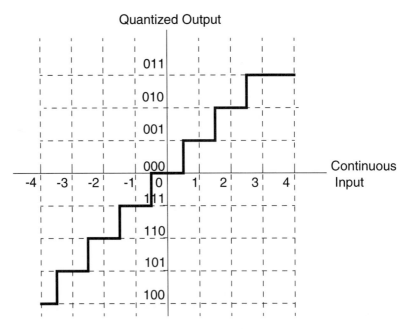

Figure 4-24 Transfer characteristic for 3-bit signed converter/encoder.

systems) does not match the conversion size (perhaps 10–16 bits), it is often possible to transmit the word in two reads (e.g., low byte, then high byte).

People who work with microcontrollers are particularly concerned with interfaces that minimize costs,[36] and this usually means serial interfaces. Instead of a parallel data transfer, the *b* bits are sent synchronously one at a time (that is, controlled by some clock). There are several standards for two- and three-wire synchronous data transfers, and some compatibility between families. This isn't exactly the place to go into hardware interfacing, but we think you'll agree that the interface shown in Figure 4-25 has a lot going for it—with just four wires, we connected a 10-bit, 11-input ADC. We could add at least three additional serial devices using the same data and clock lines by using a different chip select line for each device. The use of serial interfaces saves pins on both the microcontroller and the ADC device; multiple serial devices may reside on the same serial lines (though they may need their own *chip selects*); many different functions are available in serial interface, including ADCs, DACs, memory, sensors, etc.; and a smaller pin count means great reductions in the real-estate requirements for chips.

The substantial hardware benefits of serial interfaces do require either additional software or specific hardware support internal to the microcontroller. In the case of the 68HC16Z1, a special hardware module, the Queued Serial Module (QSM) manages the serial interface. The QSM

---

36. There are some who say that people who work with microprocessors are like those who play golf. They spend their time doing tedious, frustrating work, convinced that they are enjoying it.

Analog-to-Digital and Digital-to-Analog Conversion

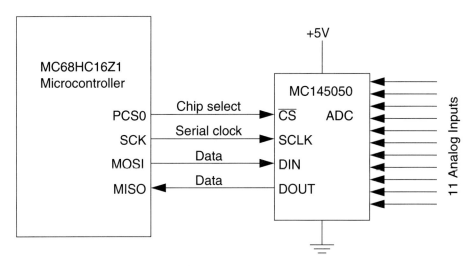

Figure 4-25 Interfacing an ADC using a (synchronous) serial port.

is rather powerful. More modest serial support is found on most microcontrollers, and a serial port[37] or two is practically a requirement on DSP chips. The alternative, a software-driven serial interface, is possible, but it is feasible only in cases of relatively slow sampling rates.

4.5.1.2 Types

ADCs come in a number of different types, some of which are pretty darn clever in cancelling out errors, reusing circuitry to minimize space, or even employing digital filtering techniques! We don't have the space here to delve deeply into the many available architectures, but we do want to present a basic overview. From Table 4-5 below, you can see that various architectures trade off complexity (complexity→size→cost), speed, and resolution. Figure 4-26 shows roughly where the different architectures fit on a scale of resolution versus speed. (There's overlap among several architectures; as of the mid-1990s, successive approximation is being encroached upon by delta-sigma from the North and multistage from the East.) Of course, any examples we give will look like the vacuum-tube era by the time you read this. Below, we briefly discuss the major characteristics of each ADC architecture.

By the way, ADCs are also available with built-in multiplexers, allowing you to convert several channels (often two, four, or eight) with a single chip. In rare cases you can sample all N channels at the same time (using multiple internal SHAs), so the converted values are all from the same point in time. Note that with multiplexed ADCs, the maximum sampling rate decreases as you convert more channels, but, it may be cheaper to have a single multiplexed high-speed

---

37. We have been careful to call these interfacing schemes *synchronous*. The common "serial port" on a PC (RS-232) is an *asynchronous* serial interface—that is, it is unaccompanied by a clock. So what kind of "serial port" is supported by your microcontroller? Perhaps both—though more recent microcontrollers are more likely to include support for synchronous than older microcontrollers.

Type	Needs SHA?	Cycles/conversion	Advantages	Disadvantages	Example (resolution, sampling rate)
Flash	no	1	fast!	expensive, may be power hog	6-bit 400 MHz 8-bit 300 MHz 10-bit 75 MHz
Half-flash and multistage	yes	≥2	fast, cheaper than full flash		8-bit 20 MHz 12-bit 2 MHz
Successive approximation (SAC)	yes!	b	cheap, fast	slower than flash	8-bit 500 kHz 12-bit 600 kHz 16-bit 400 kHz
Integrating	yes	varies	precise	slow	22-bit 20 Hz (!)
Voltage-to-frequency	(no)	N/A	precise	slow	N/A
Sigma-delta	no	(many)	mostly digital, linear, good resolution	complex digital design, more expensive than SAC	16-bit 100 kHz 18-bit 55 kHz 22-bit 4 kHz

Table 4-5 Characteristics of ADC architectures.

ADC than two slower speed ADCs. You can also (carefully!) add your own multiplexer in front of a single-channel ADC. Not only are multiplexers rather inexpensive, but certain ADC errors may be less problematic if all your measurements are made through the same ADC.

Flash Figure 4-23 shows a simple *flash* converter. Quantization takes place in one step, at the cost of one comparator per quantization level. Thus, this architecture is well suited for very high speed and low to medium resolution, although weekly advances continue to bring higher resolution (and lower prices!). Because the conversion is made at one moment, ideally, you can omit the sample and hold.

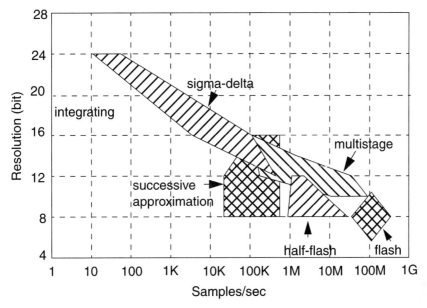

Figure 4-26 Resolution vs. speed for ADCs.

A variation of the flash, the *half-flash* or *subranging converter*, makes two passes at the signal—an initial "coarse" quantization, which is subtracted from the input signal, and a second "fine" quantization of what remains. Half-flash takes up less space on an IC and still offers relatively high speeds. However, now an SHA is needed to hold the signal value constant during both passes.

Multistage converters are related to half-flash converters but use three or more stages to complete a conversion. They generally have a higher resolution than half-flash, but at comparable speeds.

Flash converters are also known as *parallel encoding* (or *direct*) *converters*.

Successive Approximation A very popular architecture for medium-speed conversion, the *successive approximation converter* (SAC) uses a binary search method to determine the quantization level for the input. Instead of using 2^b-1 reference voltages and comparators, the reference voltage to a single comparator is generated by an internal DAC that then generates 2^b references—one at a time and in an optimal sequence.

Digital logic begins by setting the DAC to its midpoint (i.e., set the DAC's highest bit to 1), then looks at the comparator output. If the comparator output is high, the DAC value is less than the input value, and the highest bit is left set. If the input value is less than the DAC value, then this bit should not be set, so it's cleared. The converter then goes to the next lower bit of the DAC, sets it high, checks the comparator output, and either leaves the bit set or not. This process, which is essentially a binary search of the input voltage range, takes just b cycles, and the value of the DAC at the end of that time is the converted value of the input signal. Figure 4-27 shows the DAC and comparator output for an 8-bit successive approximation conversion for an input voltage of 3.653 V, which results in an unsigned output 10111011 (binary) or 187 (decimal).

SACs are relatively fast and don't take up as much IC space as flash converters. The longer conversion time requires that a sample and hold (or track and hold) be used.

Integration There are several methods that can be loosely grouped under the category of *integration*. The basic idea is to measure the amount of time it takes to charge up a capacitor to match the voltage of the input signal. Variations of this approach compensate for imperfections in the capacitor and charging circuit, and quite high precision is possible (e.g., 22 bits of resolution). However, these methods are slow, often in the tens of Hertz, so are useful only for high-resolution, slowly varying signals (e.g., medical).

Voltage-to-Frequency One useful conversion technique is to convert the signal voltage into a frequency. In a sense, this is related to integration, but the voltage-to-frequency (V-to-F) converter operates autonomously, continuously outputting a frequency that has a specific relationship to the input voltage. (For example, the frequency is proportional to the voltage, within a

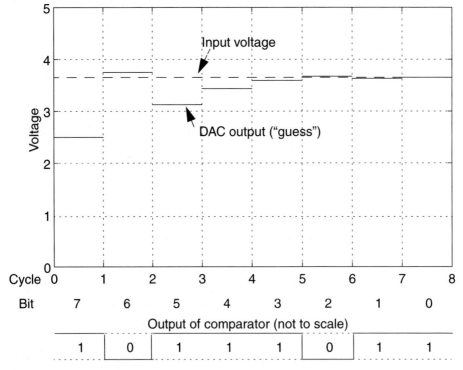

Figure 4-27 Operation of a successive approximation ADC.

specific range.) This can be very useful if we want to place the V-to-F converter remotely, then send the frequency-varying signal back to our microcontroller. On-chip timers can accurately time the period of this signal, giving an indication of the voltage at the converter.

V-to-F converters can resolve very small changes in signals, but cannot be used with rapidly changing inputs. The proliferation of serial-interface ADCs using successive approximation and other techniques may have eaten away at some of the applications for V-to-F converters, but you may find it's a good technique, especially if you have good timing resources on your microcontroller and can use the ability to transmit the frequency output instead. You might also find that your transducer (e.g., one that changes in resistance, capacitance, or inductance) could form part of an oscillator; this could lead to a very low-cost system by eliminating a lot of signal conditioning and a separate ADC stage.

Sigma-Delta Sigma-delta converters represent a modern converter design made possible by advances mostly in digital, rather than analog, IC manufacturing. A single-bit ADC is used in conjunction with a large amount of digital logic to produce relatively high-resolution readings at a medium conversion rate. (In fact, some of that digital logic is implementing a digital filter!) Sigma-delta converters have good linearity since they use only a single one-bit ADC internally and, being mostly digital, are easier to integrate with other digital devices on the same IC. Expect to see a lot more sigma-delta designs in the future.

4.5.2 Digital-to-Analog Conversion

4.5.2.1 Function

The flip-side to ADCs are DACs, digital-to-analog converters. We've already covered a lot of DAC concepts in talking about ADCs, and DACs are quite a bit simpler, so this won't be very long. We do have two theoretical issues to deal with: first, compensating for a distortion introduced by the DAC, and second, filtering to remove high-frequency components. Figure 4-28 shows the steps involved, including the decoding and converting steps that are analogous to those in analog-to-digital conversion.

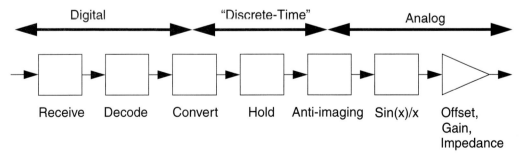

Figure 4-28 The digital-to-analog process.

DACs are also available with multiple outputs. In contrast to ADCs, DACs usually do not multiplex the outputs (though it's possible!); an eight-output DAC usually has eight separate DACs inside. Because DACs are relatively simple, they are available over a range of levels of integration. At one extreme, you can get DACs that embody only the convert stage (as DACs usually use a binary, rather than thermometer-type code internally); at the other end are devices that include audio-amplifiers and volume control. They are also generally much cheaper than ADCs of the same speed and resolution.

Receive Just as ADCs are available with a number of different interfaces, so, too, DACs are available with serial, parallel, and microprocessor-compatible interfaces. Serial interfaces are particularly popular in DACs that address high-quality audio applications such as CD players and computer audio.

Decode DACs may expect offset binary, two's complement, or even more exotic coding formats. Most are offset binary.

Convert In a sense, most DACs are "flash" DACs—the conversion takes place in a single operation. Unlike flash ADCs, there is no need for an intermediate thermometer-code representation as most techniques for producing b-bit resolution outputs using far fewer than 2^b elements. In general, for a given resolution and speed, DACs are easier to make, and are thus much cheaper than ADCs. (You'll recall that some ADCs even include DACs.)

DACs have several different forms. One basic distinction is the type of output a DAC provides—current or voltage output. Internally, most DACs produce a current that is proportional to the code they receive. This current is then, internally or externally, converted to a voltage over some range and, hopefully, with a fairly low impedance so as to drive other circuitry well. Here you are well advised to consult the manufacturer's databooks—they are chock full of application notes and sample circuits illustrating the factors that you'll need to take into account. We also note in passing that some DACs generate differential[38] current outputs, but most of the time, we try to get to a single-ended voltage signal just as quickly as we can.

Another distinction among DACs at the conversion stage is whether a DAC has an internal reference or whether one must be provided. Again, the trade-offs are much as they are with ADCs—live with what the chip designer gives you, or supply your own. In addition, you can also sometimes specify the output range by means of the reference. A special variety of DAC called the *multiplying* DAC is designed to accept a rapidly varying reference voltage. This DAC operates like a digitally controlled variable resistor ("volume control," if you like), attenuating the input (reference signal) according to the value written to the DAC.

Hold There are two interrelated issues that can be handled by the same mechanism, an SHA on the output of the DAC. The first issue involves maintaining the value of the last DAC output for the entire sampling period until the next sample is ready to be output. (Actually, you can get this effect by latching the DAC input value as well, which is much easier.) The second

---

38. If you like, akin to the differential inputs on some ADCs we mentioned earlier.

issue is that most DACs have an inherent problem in making smooth transitions in output, depending on the differences between the old and new input bit patterns. What's happening internally is usually some sort of switching of current through one set of resistors or another. But the binary switching scheme usually employed requires a number of these switches to be changed simultaneously. For example, the transition from 01111 to 10000 has all five switches of a 5-bit DAC switching at the same time. If any switch is slower (or faster), the DAC can momentarily output a giant spike or *glitch*.

Glitches vary widely in amplitude and are difficult to filter out; the preferred solution is to have an SHA on the output that samples only after all of the switches have settled to their proper values.

At this point, we need to say something about the spectrum of the "deglitched" output of the SHA in order to make sense of the next two stages. First, recall that the spectrum of a DT signal repeats an infinite number of times at regular intervals (these intervals are the sampling rate). We would actually see this repeating spectrum if the DAC could output a series of weighted (by the sample values) impulses. In this case, the spectrum would, for example, look like Figure 4-29a for a hypothetical signal with a bandwidth of nearly 5 kHz and a sampling rate of 10 kHz. The SHA, however, in extending those impulses to steps (each lasting T_s long), acts like a lowpass filter. Mathematically the magnitude frequency response of that filter is described by the $sin(x)/x$ function where $x=\pi F T_s$ or, alternatively, $x=\Omega T_s/2$. Figure 4-29b shows the magnitude response for a sampling rate of 10 kHz. Multiply Figure 4-29a by Figure 4-29b and the result is the spectrum shown in Figure 4-29c. The higher frequency images, while present, are attenuated by the lowpass filtering of the SHA. In fact, electrical engineers would call the SHA a *zero-order hold* in this application and, as you might guess, there are *first-order holds* and so on, which do a better job of attenuating those high frequencies. However, they're a lot more complicated than our simple SHA.

Figure 4-30 shows a closer view of the output of the SHA, this time plotted in dB. Not as pretty a picture, is it? On the one hand, the first image is coming on strong with as little as 4 dB of attenuation (recall we sometimes use -3 dB as a passband edge frequency!). On the other hand, we have as much as 4 dB of attenuation in the band we want, which is awful compared to the 0.5-dB ripple we might specify for our digital filtering.

These two problems, the presence of high-frequency images and the $sin(x)/x$ distortion of the desired output signal, must each be addressed for the output of the system to be useful. To solve the imaging problem, we'll use an *anti-imaging filter*, our next topic. The sin(x)/x correction is a bit more flexible—we'll have a couple of choices there involving either avoiding significant distortion (by raising the sampling rate) or introducing a compensating filter. We'll discuss this following anti-imaging.

Anti-Imaging Anti-imaging filters—also known as *reconstruction filters*—are for DACs what the antialiasing filters are for ADCs; in both cases, we are dealing with the effects of moving between DT and CT signals. Anti-imaging filters can often be less trouble than antialiasing filters for two reasons: first, the *sin(x)/x* effect of the zero-order hold aids somewhat by

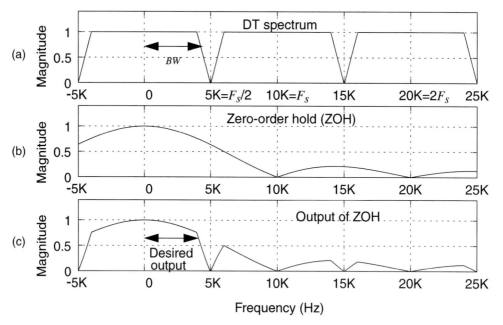

Figure 4-29 (a) Magnitude spectrum of DT signal. (b) Magnitude response of zero-order hold (ZOH). (c) Magnitude spectrum of output of ZOH.

attenuating the image frequencies; second, it is often easy (and cheap!) to increase the sampling rate of the digital signal just prior to the DAC, increasing the transition region so that lower-performance filters can be used for the anti-imaging. Let's look at this second issue in more detail.

Figure 4-30 shows the spectrum of the output of the SHA for a DT signal that covers practically the entire Nyquist range. Note the small transition width needed for the anti-imaging filter to avoid the first image. What if we instead were interested only in a relatively small percentage of the range 0 to $F_s/2$? The transition width could then be fairly large. (We made a similar argument earlier for antialiasing filters.) We can make this the case by increasing the overall sampling rate, though this would mean a faster ADC, more multiply-accumulates per second, and so on. An alternative, however, is to increase the sampling rate internally, just prior to the DAC.

We're going to discuss this topic in more detail later (see Chapter 10 [Changing Sampling Rates]), but, as a first pass, imagine that we generate additional samples "in between" the existing samples. Properly done, we'll have (approximately) the same frequency content, but with a higher sampling rate. Figure 4-31 repeats the same plots of Figure 4-29, but with a sampling rate of 40 kHz (vs. 10 kHz). Figure 4-32 shows a close-up of the resulting output spectrum. Note that now, for the same bandwidth of interest (0–3 kHz), we have much less passband distortion (perhaps so little that we can ignore it) and the first image is already down 19 dB—*before* the anti-image filter. This technique may reduce your anti-imaging filter needs considerably, but it must

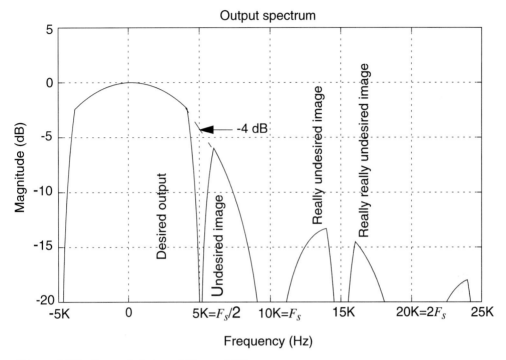

Figure 4-30 Expanded view of Figure 4-29c.

be balanced against the added complexity of increasing the sampling rate near the end of your DSP program.

How much attenuation (e.g., of the first image) is enough? A reasonable minimum is the SQNR of the digital system. Thus the magnitude of the first image should be down at least 62 dB for a 10-bit system (Table 4-2). One important point: sometimes you don't need to worry about images if the frequency response of the output transducer, the medium it is affecting, or the receiver (e.g., listener) has a limited bandwidth. For example, if the signal is controlling a large motor, the inherent physics of the system will effectively attenuate high frequencies. Too, human ears have a limited bandwidth (especially if you are, or were, fond of rock-and-roll concerts[39]). Knowing what your actual requirements are can save you some work.

Sin(x)/x Compensation Sin(x)/x compensation can be implemented, as indicated in Figure 4-28, as a compensating filter (x/sin(x)) on the output of the DAC. But it may have occurred to you that because this is just a filter, then maybe a digital filter (prior to the DAC)

---

39. When he was a teenager, Jack's mother agreed to pick up a record for him at the record store. She and the woman at the store searched in vain for the new rock hit "Light My Fire by the Door." "We couldn't find it," Jack's mom reported, "because we didn't know the artist." (For our readers not of that generation, Jack was hoping to receive "Light My Fire," a song by a group named *The Doors*.) This explains some of Jack's behavior some 30 years later.

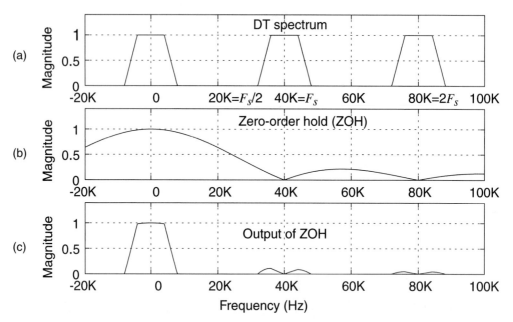

Figure 4-31 (a) Magnitude spectrum of DT signal. (b) Magnitude response of zero-order hold (ZOH). (c) Magnitude spectrum of output of ZOH.

could "precompensate" for the sin(x)/x effect. Both methods are possible, though it's certainly much nicer when the compensation is included as a filter in the DAC itself. This is often the case with very high-integration DACs (especially audio CODECs). A precompensating digital filter is the next best option—if you have the processing power for yet another stage, or can integrate the $x/\sin(x)$ response in an existing filter. As a last resort, a stand-alone analog filter can be used for the compensation.

Also recall that the technique of increasing the sampling rate to the DAC not only helps with anti-imaging, but reduces the sin(x)/x distortion as well. You may find you can reduce the sin(x)/x error enough to avoid the need for compensation. Between these two benefits, the additional complexity of increasing the DAC sampling rate might be well worth it.

Gain, Offset, Impedance Matching   Finally, our much-processed signal sees the light at the end of the tunnel.[40] The DAC and any SHA or filters we've added do not always provide the range, voltage offset, or output impedance ("drive") we need for the output signal. Although some DACs do offer hefty output drive, usually an op-amp or two can give you practically any gain or other conditioning needed. Audio signals destined for a speaker are usually happiest with an amplifier that's specifically designed to drive the low-impedance load of a speaker; there's quite a nice selection these days of such audio amplifiers.

---

40. Probably a train coming toward us, but let's press on.

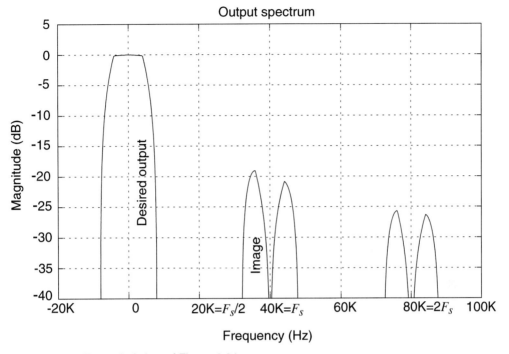

Figure 4-32 Expanded view of Figure 4-31c.

Well, that was a bit of a detour to check out ADCs and DACs, but we hope that we have provided not only an understanding of the processes, but also the notion that some of the abstract concepts like repeating spectra have practical significance. Of course, if you still don't believe in the images in the DAC output, that's your option. Please tell us if you plan on designing any products we might use, though!

4.6 Digital Filters, an Overview

Earlier, we found that using the z-transform, we could move between $H(z)$, a DT system function, and a difference equation relating the current output value to prior input and output values. The difference equation will be of the form:

$$y(n) = b_0 x(n) + b_1 x(n-1) + b_2 x(n-2) + \ldots + b_{n_z} x(n-n_z) \\ - a_1 y(n-1) - a_2 y(n-2) - \ldots - a_{n_p} y(n-n_p)$$ (4.28)

Figure 4-33 shows the flow of data graphically. This type of diagram is used (with a few modifications) for expressing most digital filter structures and represents the most general class

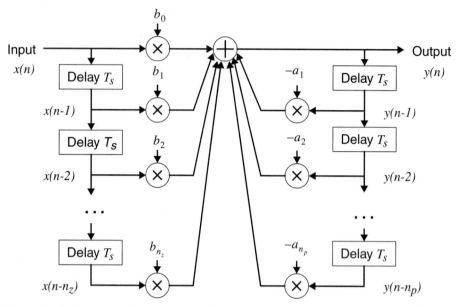

Figure 4-33 Direct form structure of an IIR filter.

of digital filters. Note that the structure is *not* a circuit but a simple depiction of the operations necessary to directly implement Equation 4.28.

When at least one of the a_i values in Equation 4.28 is nonzero, this class of digital filter is called an *infinite impulse response filter*, or *IIR filter* for short. It takes this name from its impulse response, which, in theory, is infinitely long due to the feedback of its output. IIR filters have both poles and zeros, and thus may be unstable if the poles are not properly located.

A second very useful class of digital filters is possible if we eliminate the feedback. This is equivalent to setting $a_1, a_2, \ldots a_{n_p}$ to zero and results in a system without any poles. Lacking feedback/poles, the impulse response of this filter is of finite duration and always stable. This class of filter is predictably called a *finite impulse response filter*, or *FIR filter* for short. Figure 4-34 shows the structure of an FIR filter, which is just the IIR structure without the feedback paths.

4.6.1 Digital Filter Design Process

The following are the main steps in designing digital filters:

• Specify the desired filter characteristics:

We discussed these characteristics in Chapter 3 (Analog Filters) and noted that these often emphasize the magnitude response of the filter rather than the phase response. However, you can specify time-domain characteristics instead, if your application requires.

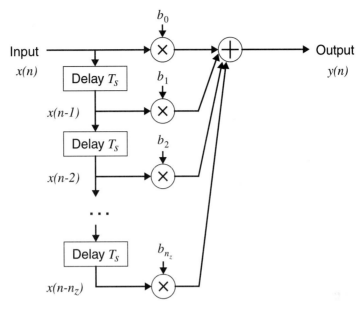

Figure 4-34 Direct form structure of an FIR filter.

This stage also defines the (minimum) sampling rate, which you'll use in calculating the filter coefficients. (This implies also knowing the bandwidth of your input signal.) The sampling rate and the speed of your processor also dictate how many machine cycles you have between samples to actually do the filtering.

• Select a filter type:

For our purposes, this amounts to deciding between the FIR and IIR filter types. We discuss the trade-offs involved below.

• Calculate the filter coefficients:

This is a tricky part of digital filter design. The most challenging task is to derive a system function $H(z)$ that meets the filter specifications. The process of finding $H(z)$ is quite different, depending on whether you want an FIR or an IIR filter, so we will devote two separate chapters to this task.

This is also the point where the *number* of coefficients, or order of the filter, becomes apparent—which is a direct indication of how much processing your filter will take. You can compare this figure with the computational and memory resources you have on your microcontroller to see if you're in the ballpark.

• Select a filter structure:

A number of issues, such as speed, memory space, sensitivity to the quantization of coefficients, and roundoff noise, are affected by the *structure* used to implement a digital filter. By *structure*, we mean the order and types of computations that our program will perform to implement the filter. From a purely mathematical point of view, two structures may produce the same result, but one structure might use half the number of multiplies! This issue is particularly important for embedded systems, since we're usually in short supply of instruction cycles, memory, and high precision math.

- Simulate effects of finite precision arithmetic:

This is an important step! It's almost always possible to simulate your digital filter using a PC, whether you write your own program in BASIC or use a high-end mathematics package. Check the actual frequency response. Run some data, real or simulated, through the filter. We'll discuss more strategies in Chapter 7 (Microcontroller Implementation of Filters).

- Write or generate the actual code:

The actual digital filter code can be as small as a few lines of assembly language, but it will require familiarity with your processor to obtain the best performance from it. Chapter 7 (Microcontroller Implementation of Filters) discusses microcontroller implementation.

- Verify with simulation:

Finally, you should verify your implementation against your design criteria.

You may need to go back to prior steps during the process of designing a filter—perhaps you need to relax some filter parameters, select a different filter order, use more bits for coefficients, and so on. Digital filter design isn't necessarily a one-pass process.

4.6.2 Selecting a Filter Type: FIR Versus IIR

You might think feedback would only minimally change the characteristics of a filter, but the difference between FIR and IIR filters is enormous. Choosing between the two types is your first job in filter design. We've summarized some of the major differences in Table 4-6.

Obviously, if phase is an issue for your design, you have just two choices—an FIR filter or, possibly, an IIR filter compensated by an FIR filter. Otherwise, you need to weigh a number of factors. For most embedded systems, speed is the critical factor, followed by memory space. Both factors favor the IIR filter. The available design tools play an important role, as some designs require special software to find optimum coefficients.

Although it may not be clear right now, the issue of feedback is intimately linked with whether a system has poles. FIR filters must have zeros; IIR filters might have zeros, but only

Digital Filters, an Overview

	FIR	**IIR**
Computation vs. performance	- Often more computation for the same magnitude response as IIR	+ Less computation for a given magnitude response
Phase	+ Can have exactly linear phase + Other phase responses possible	- Nonlinear phase leads to phase distortion of signal (distortion of "waveshape")
Stability	+ Guaranteed stability	- Must verify stability of final design; no guarantee of stability
Effect of limited number of bits for coefficients and math	+ Noise and errors are generally lower than for IIR	- More sensitive to quantization of coefficients and noise from rounding off calculations
Use of analog filter as model	- No direct conversion from an analog design to FIR (indirect methods exist)	+ Several methods for converting analog designs (this is the most common "pencil and paper" design method)
Arbitrary filter specifications	+ Possible even when no analog equivalent is possible	- (Much) more difficult to produce arbitrary designs
Implementation	+ Straightforward implementation; most DSP hardware supports directly and efficiently	- Large filters tend to be broken down into smaller stages; a bit more complicated than FIR - For optimum performance, careful design of filter stages is necessary
Design	- Optimum designs require computer programs (no closed-form solutions) + Some design methods are fairly simple	+ Well-known design processes, can be done manually

Table 4-6

IIR filters can have poles (feedback).[41] This will become clearer as we explore the DT system function $H(z)$ for both types of filters in the chapters to follow.

---

41. We might as well start apologizing now. Actually, we're going to show you how to play a little trick and use feedback to create a very efficient FIR filter. It will have a finite impulse response, but it'll also have a pole, which happens to be cancelled by a zero in the system. But this is a rare situation, so please forget you read this footnote.

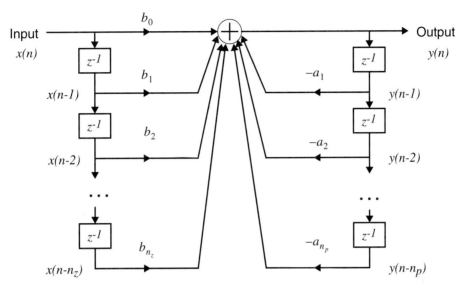

Figure 4-35 Common notation for IIR filter structure.

4.6.3 Diagram, Notations

Figure 4-35 shows a more compact notation for digital filter structures. Note that now the delay is represented by the z-transform of a delay (namely, z^{-1}), and multiplication indicated rather subtly by a coefficient placed near an arrow. (Yep, those arrows typically look just like the "directional flow" arrows. You have to look at the context.) You might compare this with the more expansive diagram of Figure 4-33. To re-emphasize, this is not a circuit! It's just a diagram showing the order and types of arithmetic operations and delays—the flow (and processing) of data, if you will, *not* the flow of electrons.

4.7 Summary

Discrete-time (DT) signals have values only at (regular) points in time but may be continuous in value. *Digital* signals are both discrete in time and also discrete in value (quantized). We often use the term DT for discrete-time/continuous-value or digital signals where we ignore the effects of quantization.

Sampling is the process of producing a DT signal from an analog signal. The sampling rate affects the frequency content of the resulting DT signal. Frequency components above the Nyquist frequency $F_s/2$ (one-half the sampling rate) are *aliased* down into the frequency range 0 to $F_s/2$. We also note that the spectrum in the range 0 to $F_s/2$ has a mirror-image in the range $-F_s/2$ to 0, and this spectrum is, in turn, repeated at regular intervals (spacing F_s) to positive and negative infinity.

It is possible to purposely sample an analog signal such that the desired frequency range is aliased. This *undersampling* requires careful choice of the sampling frequency such that the desired spectrum range is not made discontinuous or reversed.

When discussing DT signals and systems, normalized frequency or ω is used and is defined as $\omega=\Omega T$. Normally, we need only to examine the range $\omega=0$ to π; real signals and systems with real coefficients have symmetric spectra.

The frequency response function $H(e^{j\omega})$ provides the "same" information for DT systems as $H(j\Omega)$ did for analog systems. Likewise, the system function $H(z)$ furnishes a more complete description. The z-plane plays the same role as the s-plane does for CT systems; poles and zeros may be plotted and their effects determined graphically.

It is very simple to move between $H(z)$ of a system and the difference equation associated with that system. Difference equations relate the current output of a DT system with a linear combination of present and prior input values, and prior output values, and can be very easily programmed on a microprocessor. One can view the task of digital filter design as one of producing a system function $H(z)$ with certain desired characteristics, converting this to a difference equation, then implementing this difference equation on a digital processor.

Quantization is the process of representing a continuum of values with a discrete, finite set of values (in our case, the number of levels is always an integral power of two). Nonlinear quantization is useful for transmission and storage of certain types of signals (especially audio) but must be converted to a linear range for DSP to be performed. Linear quantization divides the input range into equal steps; these levels may be encoded in a number of different ways. Quantization produces an error that can be modeled as noise; it is then possible to talk about the signal-to-noise ratio resulting from quantization (SQNR). SQNR is strictly a function of the number of bits used to represent the signal level.

Analog-to-digital conversion encompasses both sampling and quantization but, practically, must be supported by additional functions. Signal conditioning in the form of gain, offset, and impedance changes makes up the first stage; antialiasing removes frequency components that otherwise would be aliased in the DT signal; and a sample-and-hold amplifier (SHA) is used to both sample the input signal during a brief interval and also to hold this value during the conversion stage. Quantization is followed by encoding; a transmit stage interfaces the ADC with other devices (e.g., microcontroller serial port).

Many types of ADC architectures exist, each of which is optimized for a certain set of parameters that might include resolution, conversion speed, cost of manufacture, and compatibility with digital IC manufacturing processes. Flash converters offer the maximum speed; successive approximation converters occupy a middle ground with moderate speed and low cost; and integrating and delta-sigma converters maximize resolution but offer slower conversion times.

Digital-to-analog converters (DACs) have fewer architectural variations and implement many complementary functions to ADCs. An anti-imaging filter is usually required to remove the high-frequency images present in the converted signal, while the SHA on the DAC output introduces a sin(x)/x distortion that must be compensated by an x/sin(x) filter. This x/sin(x) filter may be incorporated in the DSP already being done, or may be a physical filter. Increasing the output sampling rate can be beneficial in reducing both the imaging and sin(x)/x effects.

Digital filters fall into two main categories: FIR—filters with a finite duration impulse response; and IIR—filters with an infinite duration impulse response. The key difference between the two filters is the presence of feedback in the IIR filter. These two classes of filter are so fundamentally different that entirely different design methods must be used.

The main steps in digital filter design are as follows: Specify filter characteristics, select filter type (FIR or IIR), calculate filter coefficients, select filter structure, simulate effects of quantization, write actual code, and verify performance.

Resources

This chapter includes material from three different areas of study—discrete-time signals and systems, analog-to-digital conversion, and a brief bit about digital filters. Discrete-time signals and systems are often treated in the same books as continuous-time signals and systems—i.e., check out the "linear signals and systems" books of Chapter 2 (Analog Signals and Systems). Chapter 5 (FIR Filters—Digital Filters without Feedback) and Chapter 6 (IIR Filters—Digital Filters with Feedback) provide additional information and resources for digital filters. Which leaves the theoretical and practical aspects of analog-to-digital (and digital-to-analog) filters.

The theoretical aspects of ADCs and DACs are often discussed in discrete-time signal processing books (see the chapters to follow); however, for practical information, your absolute best bet is manufacturer literature, augmented by an electronics text such as *The Art of Electronics* (mentioned earlier).

Analog Devices, Burr-Brown, Maxim, Motorola, National Semiconductor, and Texas Instruments are just a few semiconductor manufacturers with data sheets and/or application notes that cover both linear ICs (such as op amps) and ADC/DACs. For example, Motorola has an application note AS-20 on laying out circuit boards for the best 68HC16 ADC performance. Application notes can be a great source of practical information on getting the best performance from both a specific product (e.g., a specific 16-bit serial DAC) and that type of product in general (e.g., DACs). If you happen to have an evaluation board, be sure to check out the schematic and sample code to see how the analog interfacing and operation is performed—engineers have paid particular attention to the design of these boards as they want the best possible performance. By the way, much of this information is now appearing on the World Wide Web, and is thus much easier to get than when information was available only in databooks.

CHAPTER 5

FIR Filters—Digital Filters Without Feedback

Finally, we get to some actual filters! In this chapter on FIR filters we won't use the s-domain much (that's later), but the z-domain will be central to the material in this chapter. You should be familiar with the filter terminology discussed in Chapter 3 (Analog Filters).

5.1 FIR Overview

At the end of Chapter 4 we briefly discussed the two major types of digital filters—finite impulse response (FIR) and infinite impulse response (IIR). Reiterating, the major benefits of the FIR filter are:

- can have linear phase
- guaranteed stability
- less sensitive to noise and errors from finite-precision math
- simple implementation

The FIR filter has some drawbacks. An FIR filter usually requires more computation than an IIR—at least, if there's an IIR with the same (magnitude) response. This stems from the fact that FIR filters, lacking feedback, have only zeros—no poles. (IIR filters can have both zeros and poles.) However, the FIR filter is a good match to a simple DSP architecture, and although the total number of computations required is an important factor, sometimes the low overhead of the simple flow of data can offset the additional computational requirements. Where's the crossover point? It will heavily depend on the particular architecture used. Some types of data moves may be very quick, some math operations very time-consuming. With the great variety of architectures present in microcontrollers, there isn't a fixed answer. (However, later we'll see that IIR filters require more work to turn a design into a useful program.)

It's tempting to think of FIR filters as black boxes, good only for filtering, but let us be philosophical for a moment and point out that they're excellent tools for solving a broad class of problems. For example, we can use them to model processes such as drug dispersion in the kidneys, or economic systems. A whole research field called *system identification* is central to the employment of dozens of professors, making our streets immeasurably safer at night.[1] *Adaptive filtering* (also occasionally done with IIR filters) now plays a crucial role in maximizing data flow across the phone system and in other communication channels, by dynamically adapting to the changing conditions of the channels. The subject we're discussing in this chapter has much broader applicability in engineering than conventional "filtering" operations.

Before overviewing the design process, let's take a look at a basic FIR filter, followed by a more complicated example. This first filter represents nothing more than an approach we've all used at some time or other, taking the average of the last N samples. Figure 5-1 shows one way we might express this in a diagram, first adding the current plus the last $N-1$ samples, then dividing by N (or multiplying by $1/N$) to get the average. Moving things around a bit, Figure 5-2 shows the same filter expressed in the form we'll use for other filters—each sample is multiplied by a coefficient, then added together to form the output. In this case, of course, all of the coefficients (which we've labeled $h(0), h(1), ..., h(N-1)$) are the same—$1/N$.

We'd expect the magnitude frequency response to show a lowpass filter response (intuitively, averaging should reduce the higher frequencies), and as Figure 5-3 shows, it does. If the purpose is to do lowpass filtering, this filter is not perfect, as you see. The response doesn't have a very sharp corner, and even when the response does go to a minimum, there are *lobes* ("humps" in the frequency response).[2] (However, if the purpose is to *average*, then this filter does a perfect job.) Can we do better?

The essence of good filter performance is picking "good" filter coefficients. Listing compares equal-valued coefficients for $N=8$ and a "better" set generated by the "Parks-McClellan program," which we discuss later in this chapter. Figure 5-4 shows the magnitude response using these slightly different coefficients. A point to recall, if you're not impressed by the change, is that the 6-dB difference in stopband attenuation represents a factor of two less signal power, and a factor four less voltage.

Our major job in this chapter is to figure out how to generate "good" coefficients. And if you recall our earlier admission—that in fact, we usually use programs to generate these coefficients—well, this chapter should be pretty brief. Something like a chapter on snakes in Iceland. Fortunately, we have several topics worth discussing.

1. System identification is also useful when you have a system that you have a rough idea of how to model, but don't know the specifics beforehand. The human vocal tract is an example—it has various noise sources and resonances, but we don't know beforehand what they are for a specific person at a specific time.

2. These lobes will be back shortly to haunt us. Haunting lobes can be terrifying!

FIR Overview

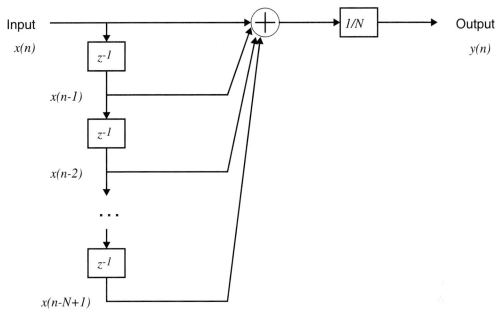

Figure 5-1 A very simple FIR filter.

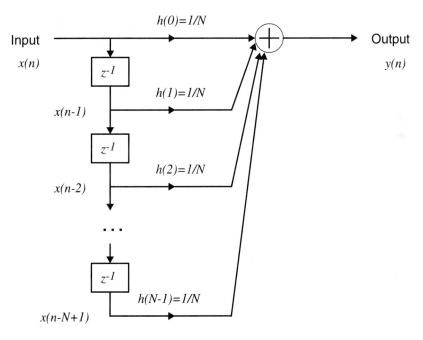

Figure 5-2 A rearrangement of the simple filter of Figure 5-1.

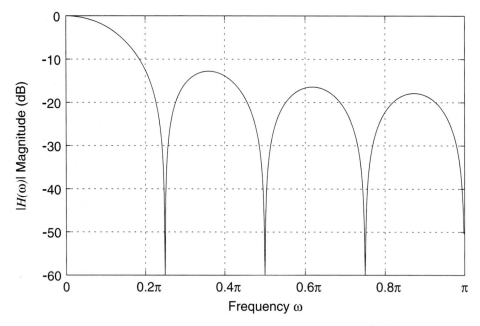

Figure 5-3 Magnitude response of the simple filter of Figure 5-1.

Equal coefficients	"Better" coefficients
h(0)=0.1250	h(0)=0.1032
h(1)=0.1250	h(1)=0.1056
h(2)=0.1250	h(2)=0.1368
h(3)=0.1250	h(3)=0.1544
h(4)=0.1250	h(4)=0.1544
h(5)=0.1250	h(5)=0.1368
h(6)=0.1250	h(6)=0.1056
h(7)=0.1250	h(7)=0.1032

Listing 5-1 Equal coefficients vs. coefficients found using the Parks-McClellan program.

First, with the background from Chapter 4 (Discrete-Time Signals and Systems), you're just a few diagrams away from understanding *how* FIR filters work; it's actually pretty neat, and we would feel remiss if we skipped over that. Second, there are several different design methods for FIR filters, but filter design programs might only use only one or two of the methods. Because each design method starts with different constraints or goals, it's useful to know what you're buying into when you choose some program—you might be getting far more (or less) than you want! Third, implementation choices can increase efficiency and decrease memory requirements, effects we need for embedded systems. Finally, in the course of discussing FIR filter design and implementation, we'll come across valuable concepts that relate to the FFT and other DSP areas.

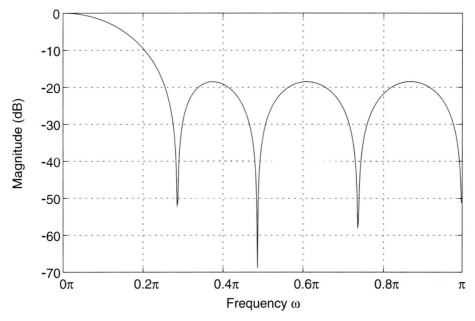

Figure 5-4 Magnitude response of a filter using "better" coefficients.

5.2 Intuitive Convolution—How FIR Filters Work

To see how FIR filters work, we need to believe in the following things:

- Assuming causality, there is a specific, one-to-one relationship between the impulse response of a digital system and the frequency response of that system.
- *Linear* systems—by definition—respond to individual frequency components of signals independently of which other frequencies are present. In addition, scaling (multiplying) the input by some factor scales (multiplies) the output by the same factor.
- *Time-invariant* systems respond the same to a signal, regardless of when it is presented—though we realize, of course, that the output will be delayed in time by the same amount as the input was delayed.
- FIR filters are linear, time-invariant systems.

For digital systems, the discrete Fourier transform (DFT) can be used to transform a digital system's impulse response into its (discrete or sampled) frequency response. But, as Figure 5-5 shows, the impulse response of an FIR filter is just the sequence of coefficient (filter multiplier) values, one right after another. (As the "1" of the impulse travels through the filter, only one coefficient at a time is output.) Very interesting!

What if we scaled the impulse we input to the system? Linearity applies to any signal, so we should expect the output to be a scaled version of the impulse response. Now, here's the leap to make—can't we look at the sequence of samples (the input signal) as just a sequence of scaled

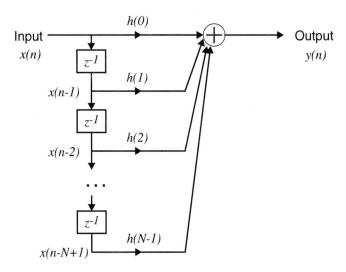

Impulse response (i.e., $x(n)$=1, 0, 0, 0, ... for n=0, 1, 2, ...)

	$x(n)$	$x(n-1)$	$x(n-2)$	...	$x(n-N-2)$	$x(n-N+1)$	$y(n)$
0	1	0	0	0	0	0	$h(0)$
1	0	1	0	0	0	0	$h(1)$
2	0	0	1	0	0	0	$h(2)$
...	0	0	0	...	0	0	$h(...)$
N-2	0	0	0	0	1	0	$h(N-2)$
N-1	0	0	0	0	0	1	$h(N-1)$
N	0	0	0	0	0	0	0
N+1	0	0	0	0	0	0	0

Figure 5-5 Relationship between the impulse response and coefficients of an FIR filter.

impulses, each of which produces a scaled and delayed impulse response? The total output of the filter, then, should be the impulse response (no delay) scaled by the first sample, added to the impulse response delayed by one sample period and scaled by the second sample, and so on. In Figure 5-6, we show how a DT signal (the bottom input shown) with three non-zero samples at n=0, 1, and 2 produces an output that is the same as adding the scaled, delayed impulse responses due to each individual sample. (The first sample has a value of 1, so the output due only to that sample is just the impulse response, unscaled and not delayed. You should verify that the output due only to the sample at n=1 is twice as large in amplitude, but delayed by one sample time.)

Intuitive Convolution—How FIR Filters Work

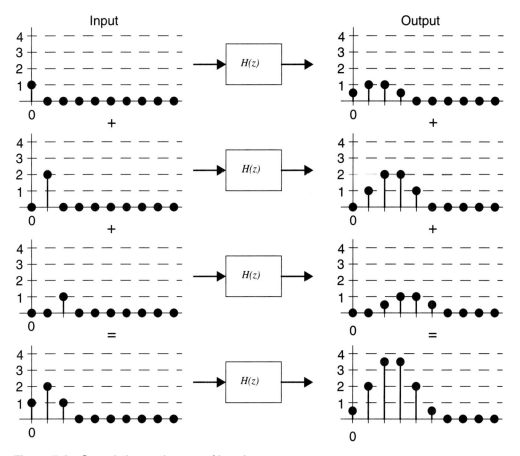

Figure 5-6 Convolution as the sum of impulse responses.

In words, the output of the FIR filter *at any given time* is the sum of the scaled and delayed impulse responses caused by the current and prior input samples. If $h(k)$ is the impulse response of the filter, and N the number of coefficients (also the length of the impulse response), we can write the following equation:

$$y(n) = \sum_{k=0}^{N-1} x(n-k)h(k)$$

Not only does this equation describe what is happening mathematically, but as Figure 5-5 shows, it's also a direct translation of the diagram we've been using for FIR filters. (It may also be the most feared and despised relation among electrical engineering students!)

One thing to notice is that as the *k* index increases, we go "back in time" with the samples. This idea of taking two sequences, reversing the order of one of them, multiplying each element with the corresponding element from the other sequence, then adding the products together, is given the name *convolution*. A fancy, if somewhat confusing name for the operation, but there you are. By the way, convolution is the act of *convolving* two sequences. Rather than announcing, "I forgot to *convolute* the signals, which means all our work this past year is trashed," one should instead say, "*Roger* forgot to *convolve* the signals, which means all our work this past year is trashed." Another tip for your edification.[3]

To summarize, the coefficients of an FIR filter are equivalent to the impulse response of that filter, and the output of the filter can be imagined to be the sum of a number of scaled and delayed impulse responses, where the scaling is taken from the input signal samples. Mathematically, and in real life if we like, we get these results by convolving the impulse response with the input signal.[4]

5.3 Design Process Overview

As mentioned before, designing a digital filter—and an FIR digital filter in particular—involves the following steps:
- Specify filter characteristics
- Select a filter type
- Calculate filter coefficients
- Select filter structure
- Determine effects of finite-precision coefficients and computation
- Write or generate actual code
- Verify using simulation

Chapter 3 (Analog Filters) discussed how filter characteristics are specified in terms of magnitude (low/band/highpass, stopband attenuation, etc.) and phase responses.[5] The next decision point, type of filter (i.e., IIR vs. FIR), is guided by the general characteristics of each filter type (recall Table 4-6), but may not be a cut-and-dried decision unless there are some characteristics (e.g., linear phase) that point to one filter type or the other.

From a theoretical viewpoint, the only challenge in digital filter design is in generating the appropriate filter coefficients. However, you'll note that we have a number of steps listed above that come after that, and, for microcontrollers, those may not be trivial. To begin with, our filters will often need to be executed in realtime. This places an upper limit on the number of computations a filter can require, since we have only a certain amount of time between samples.[6]

---

3. Jack finds himself surrounded by students who insist on "convoluting" sequences, which annoys the bejabbers out of him.

4. However, sometimes there are more efficient ways of implementing the filter than the straightforward convolution algorithm shown here. We'll talk about this later.

5. Again, a necessary caution that this isn't the whole story. These are just the most common parameters.

Another major concern is the limited precision with which we can represent samples and coefficients, and the limits on the precision of math operations. Recall that we usually don't worry about these effects during early stages of design, as they would quickly bog us down in complicated math. However, in extreme cases this could strongly influence the design process (we'll mention a particular filter design later that makes some trade-offs for purely integer arithmetic).

Oh, there's also the little issue of writing the actual program to implement the filter.

5.3.1 Filter Design Tools and Methods

The options in designing filters are a bit like the options you have in figuring your income tax.[7] At one extreme, there's the cup full of #2 pencils, a yellow legal pad, a trusty four-function calculator, and a pot of coffee. At the other extreme, there are software packages that interview, prompt for and collect data, show you tentative results on the screen, let you make modifications, print out the finished forms, and even file the form electronically.

Filter design offers similar options—and trade-offs. At an extreme, a designer can select a method for generating coefficients, estimate the number of coefficients required, and generate the coefficients using some reasonable equations—not exactly a #2 pencil job, but neither are taxes anymore. At the other extreme, software exists that will take your filter specification (or even system of filters and other processing) and generate ready-to-run software in your choice of popular computer languages, including assembly language.

The pencil-and-paper approach has some benefits, aside from the "because it builds character/understanding/familiarity" argument. As we'll see in this chapter (and the following, on IIR filters), several filter-design methods are possible for the same filter specification. However, no single method is optimum for every filter characteristic. It may be very important to be able to choose the filter-design method, rather than be restricted to choices someone else made—just as with tax software.[8]

The benefit of a "push-the-button-and-go" software package isn't necessarily the time saved in generating a single set of coefficients, but in the fact that having a quick way of generating a filter design allows you to make a number of iterations. For example, adjusting a passband cutoff frequency a little might shave off a few coefficients; you can explore many possibilities quickly if you can change a parameter and see the effect instantly. This is precisely the power we need, as microcontroller programmers don't have much processing bandwidth to waste. Just be sure you know what you're getting—obtaining less than optimal results quickly (e.g., using the "wrong" design method) can be of little value if you need the best solution.

6. Samples may also be processed in blocks, but the average number of cycles per sample still places an upper limit on the operation.

7. For those folks who have—or pay!—income taxes in their country.

8. Jack's wife Joan does their taxes using her cherished programs. However, Jack then checks the entire process by hand, taking special care to point out the idiocies inherent in the tax software. Used in this way, the tax software extends the tax-preparation time by a factor of four or five. However, as the cost of the software is tax-deductible, spending the extra time might be worthwhile in certain cases.

By the way, the differences in performance we're talking about can be quite large. This is not saving 2% here or 5% there; this is picking filters that take 50% less time and memory or choosing a filter that uses shifts and adds—no multiplies. These are often "make-or-break" issues in the microcontroller world.

So what's the middle ground between clicking a mouse ("click here to design filter") and wearing out your calculator buttons? What should your design environment look like? We strongly recommend you have a way of generating magnitude and phase frequency responses from filter coefficients and the ability to quickly change the filter design parameters. The flexibility to choose among different design methods is next on the list. And while we're starting our study with FIR filter design, your tools should support IIR filter design, too (see the next chapter).

References we've noted at the end of this chapter and others, especially the periodicals and Internet newsgroups, are a good source of up-to-date information on the tools available. Several powerful general-purpose mathematics packages now include, or have add-on, DSP toolkits. Substantial discounts are often available to students, making these excellent educational values. A number of filter-design programs are freely available on the Internet, and several books contain program listings for such programs. Further, it's not difficult to write your own programs for most design methods.

Useful design environments have the ability to:
- easily explore the effects of varying filter parameters,
- generate and export coefficients (in appropriate numerical format),
- choose design method and/or write your own,
- evaluate the designed filter (including graphically),
- evaluate the practical implementation of filter (e.g., effects of finite coefficients and amount of computation),
- easily, repeatedly, test theoretical output versus output from real system,
- generate code, either high-level or low-level.

5.4 Generating Coefficients

Finally, we're ready to generate the coefficients for a digital filter. As we saw above, these coefficients are actually identical to the (DT) impulse response of an FIR filter. And, though we've only been waving our hands so far, the so-far mysterious *DTFT* provides a means for transforming this impulse response to the frequency response. It seems reasonable, then, that one way of deriving filter coefficients, given the frequency response, is to use the (inverse) DTFT and generate the impulse response from the desired frequency response.

In fact, this method is quite viable, and we'll discuss just how to use it in a moment. Realistically, though, there's a far quicker method (assuming you have a computer), one that you'll use much more often. We're speaking of the vastly popular *Parks-McClellan* or *Optimum Equiripple Linear-Phase* FIR filter design method.

5.4.1 Parks-McClellan or Optimal Equiripple

If there is a "Quicksort"[9] of digital filter design, it's the program originally written by Parks and McClellan in the early 1970s for optimum equiripple linear-phase FIR filters. It is often the design method of first choice, gives excellent results, produces a variety of filter types (e.g., low-pass, bandpass, etc.), is widely and freely available, and is well tested. (There are other similar methods, but the availability, flexibility, and computational efficiency of the Parks-McClellan program combine to make it much more popular.)

The goal behind the Parks-McClellan program is to generate an FIR digital filter that has linear phase and is "optimal" in its the magnitude frequency response. This "optimal" is in terms of how much difference (error) there is between the actual and desired magnitude frequency response in the passband(s) and stopband(s). The program spreads out the error within a frequency band across the width of that band, which produces ripples of equal magnitude. The user can specify the relative importance of each band, so, for example, there would be less (but uniform) ripple in the passband than in the stopband. The program gets to do whatever it likes in the transition region—the more transition width allowed, the better job it can do with error in the passband and stopband. There isn't a closed-form solution known to this problem, so the algorithm is an iterative one. Once the program finds an acceptable solution, the impulse response (i.e., set of filter coefficients) is generated.

Many versions of the Parks-McClellan program are available in books and on-line, including several versions in C (the original is in FORTRAN). Equivalent routines are almost always available in filter design packages, sometimes under other names, such as *Remez*, *Remez-exchange*, and *optimal equiripple*. (The Parks-McClellan program uses an algorithm called the Remez-exchange, hence the aliases.)

We mentioned that the Parks-McClellan program generates linear phase filters. There are other ways of generating FIR filters with linear phase (e.g. windowing, frequency sampling), so the method is not unique in that respect. Further, in some cases, the linear-phase constraint might make your filter longer than you actually need if linear phase is not one of your requirements (IIR filters may be more appropriate). Finally, linear phase requires (or produces, if you want to think of it that way) filter coefficients that are symmetrical—you can take the first $N/2$ coefficients and reverse the order to get the second $N/2$ coefficients.[10] (We'll show you some coefficients in a moment.) With some clever rearrangement, filter implementations can take advantage of this symmetry to reduce both the amount of coefficient storage and the number of multiplies required. We'll explore this fact later in this chapter.

The Parks-McClellan program isn't limited to the simple filter parameters we discussed earlier. It can design filters with passbands that have a particular slope or even more complex

---

9. "Quicksort" is a particularly efficient algorithm for sorting numbers; most sorting algorithms take time proportional to n^2 if there are n elements to be sorted, while Quicksort is proportional to $(n)ln(n)$, which can be *much* smaller. Thus, it's generally the sort of first re-sort (sorry).

10. This is true if N is even. If N is odd, take the first $(N-1)/2$ coefficients, reverse, and tack them onto the first $(N+1)/2$ coefficients.

specifications. This may require that you modify the source code; however, some filter design packages using this algorithm let you specify arbitrary magnitude response using line segments (e.g., the magnitude response is given at the start and end frequency of each band, so is linearly interpolated for frequencies in between). Recall from Chapter 4 (Discrete-Time Signals and Systems) that we sometimes want to incorporate $\sin(x)/x$ compensation in the processing—if you already need an FIR filter, "adding in" the necessary $x/\sin(x)$ response could be an efficient strategy.

Let's design an optimal equiripple FIR filter using a version of the Parks-McClellan program. We'll use the following specifications:

Sample rate: 10 kHz

Passband: 0–2.5 kHz (i.e., lowpass), 0.5 dB maximum ripple

Stopband: 3.0–5 kHz, 50 dB minimum attenuation

We begin by estimating the number of coefficients required, N. A simple expression (but one that will impress your boss) that gives good results is:

$$\hat{N} = \frac{-10 \cdot \log(\delta_s \delta_p) - 13}{14.6 \Delta f} + 1 \tag{5.1}$$

where

$$\Delta f = \frac{\omega_s - \omega_p}{2\pi} = f_s - f_p \tag{5.2}$$

$\hat{N}$ should be rounded up to the nearest integer to give N. Note that if the number of coefficients is odd, there are no restrictions on the type of filter you can design with the Parks-McClellan program. However, choosing N even will always result in a zero in the spectrum at $\omega = \pi$; thus, an odd N is good only for lowpass and bandpass filters, not highpass or bandstop types.

For the present specifications $f_s = 0.3$, $f_p = 0.25$, Δf is 0.05, and we can use (from Chapter 3 [Analog Filters]):

$$\delta_p = 10^{A_p/20} - 1 \tag{5.3}$$

and

$$\delta_s = 10^{-A_s/20} \tag{5.4}$$

Generating Coefficients

to find $\delta_s=0.00316$, $\delta_p=0.0593$, and $N=34.25$. (Recall that this equation for δ_p is the "digital filter" version.) Rounding up, we'll try $N=35$ as the first estimate. (If we were designing a highpass filter, we would have to make sure N is odd. For a lowpass, it doesn't matter.)

The actual format for entering different parameters will depend on the version or implementation of the Parks-McClellan program you use. For one version (written in FORTRAN and compiled to run on PCs), the program expects the following parameters:

- *NFILT* (integer): the number of coefficients to generate. (If *NFILT* is even, you cannot have a highpass or bandstop filter response.)[11]
- *JTYPE* (integer): the type of filter to design: 1=multiple passband (including lowpass/highpass), 2=differentiator, 3=Hilbert transform. (You'll almost always use the multiple passband type.)
- *NBANDS* (integer): the number of bands—2 for lowpass/highpass (i.e., pass and stopband), 3 for bandpass/bandstop, >3 for multipassband.
- *LGRID* (integer): the density of the "grid" used for computation. Usually just set this to 0.
- *EDGES* (floating-point array): the lower and upper edges for each band in terms of f (i.e., ranging from 0–0.5, where 0.5 is the Nyquist frequency).
- *FX* (floating-point array): the "function array," which gives the ideal magnitude within each band. Ordinarily, use 1 for passbands, 0 for stopbands.
- *WTX* (floating-point array): the "weight array," which describes how much error can be tolerated in each band. (See below for a description of how to compute.)

For our example, we have all of the parameters except *WTX* (the weight array). The weight array determines how much ripple is permitted in each band relative to the other bands. A greater weight should be given to bands where it is important to have smaller error. Set $WTX_{sb}=1$ and $WTX_{pb}=\delta_s/\delta_p$. This gives us a *WTX* array (0.0533, 1).

Because we're using a stand-alone version of the Parks-McClellan program, we'll submit our parameters in a file, as shown in Listing 5-2. We've added comments to the listing to show you what parameters are where (comments are not allowed in an actual parameter file). You may find you'll have to watch the format of the numbers you enter—e.g., always including a decimal point for floating-point values—depending on the exact version you use, what language it was written in, and so on.

Input file:

```
35,1,2,0              NFILT, JTYPE, NBANDS, LGRID
0.0,0.25,0.3,0.5      EDGES (upper/lower edge freq/band)
1.0,0.0               FX (desired magnitude per band)
0.0533,1.0            WTX (weight given to minimize error)
0                     NFILT (0=halts program)
```

Listing 5-2   Example use of the Parks-McClellan program.

---

11. Although most programs derived from the original Parks-McClellan FORTRAN source use this definition, some filter design programs require N-1 (the filter order) instead of N (the number of filter coefficients).

Program output: :
```
************************************************************
                    FINITE IMPULSE RESPONSE (FIR)
                    LINEAR PHASE DIGITAL FILTER DESIGN
                       REMEZ EXCHANGE ALGORITHM

                          BANDPASS FILTER

                       FILTER LENGTH =  35

                ***** IMPULSE RESPONSE *****
         H( 1) =     361.922 = H( 35)
         H( 2) =     589.000 = H( 34)
         H( 3) =      52.556 = H( 33)
         H( 4) =    -538.095 = H( 32)
         H( 5) =     -58.657 = H( 31)
         H( 6) =     499.472 = H( 30)
         H( 7) =    -251.531 = H( 29)
         H( 8) =    -785.168 = H( 28)
         H( 9) =     381.999 = H( 27)
         H(10) =     812.822 = H( 26)
         H(11) =    -934.419 = H( 25)
         H(12) =   -1082.725 = H( 24)
         H(13) =    1547.666 = H( 23)
         H(14) =    1083.109 = H( 22)
         H(15) =   -3229.928 = H( 21)
         H(16) =   -1275.738 = H( 20)
         H(17) =   10268.660 = H( 19)
         H(18) =   17571.900 = H( 18)

                            BAND  1         BAND  2
     LOWER BAND EDGE       .0000000        .3000000
     UPPER BAND EDGE       .2500000        .5000000
     DESIRED VALUE        1.0000000        .0000000
     WEIGHTING             .0533000       1.0000000
     DEVIATION             .0568509        .0030302
     DEVIATION IN DB       .4802745     -50.3707100

   EXTREMAL FREQUENCIES—MAXIMA OF THE ERROR CURVE
         .0243056   .0572917   .0885417   .1197917   .1493055
         .1805554   .2100692   .2361107   .2500000   .3000000
         .3069444   .3225694   .3434026   .3677081   .3937497
         .4197912   .4458328   .4736105   .5000000

************************************************************
Stop - Program terminated.
```

Listing 5-2 (Continued) Example use of the Parks-McClellan program.

This program labels the output as the impulse response, which we know is identical to the set of filter coefficient values. Actually, we have a minor detail to worry about. This particular version of the program scales the coefficient values by 32768 prior to printing them out, so we

Generating Coefficients

need to take each of the 18 unique values and divide by 32768.[12] Some versions of the Parks-McClellan program don't do this scaling. Also note that, reflecting its FORTRAN roots, the indexing begins at 1 and ends at N; we usually work with indexing that begins at 0 and ends at N-1.[13]

As you can see, the coefficients are symmetric around the middle coefficient. This is because a necessary condition for linear phase is a symmetry of the impulse response.

We aren't quite done yet. Recall that N=35 was just an estimate. We need to check the actual frequency response to verify that N is large enough—or, if you're really close to using all your processing bandwidth, to see if there's enough slack to back N down some! Checking the output of the program, we see the maximum deviation in the passband is 0.48 dB and the stopband attenuation is 50.4—both meet the design parameters, so we have the coefficients.[14]

Once we've scaled the coefficients properly (dividing by 32768), we can plug these coefficient values into a filter program and plot the actual frequency response, as we've done in Figure 5-7. You can see the equal ripple in the stopband; in the passband, the ripples are also equal in magnitude.

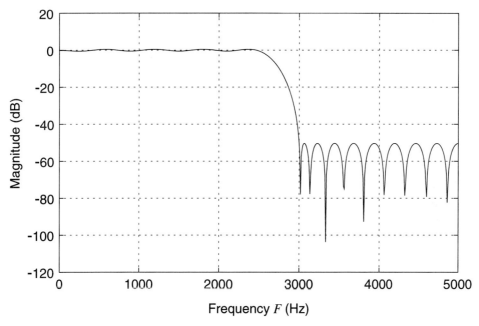

Figure 5-7 Magnitude response of filter designed using Parks-McClellan.

12. Why 32768? This is 2^{15}, and produces integers that represent 16-bit fixed-point coefficient values (1 bit is for the sign). We'll talk about numeric representations in a later chapter.

13. An artifact of the days when FORTRAN could not use zero to index an array. See what DSP students had to endure back in the "old days"?

14. It works the first time only in books.

5.4.2 Windowing

If the Parks-McClellan program is the Quicksort of coefficient generation, filter design by windowing is the "bubble sort." Straightforward, easy to understand, and occasionally the best choice. The basic idea is to take the ideal (continuous) frequency response and calculate its impulse response—which should be the filter coefficients, right? The problem is that the impulse response for a filter with any sharpness to its frequency response is infinitely long. This is bad news since we need a limited number of coefficients, by definition, for the filter to be FIR. *Windowing* is one way of getting around this problem, hence the name of the technique. But let's start at the beginning by calculating the impulse response, given an ideal filter response.

Without getting into details, there is a useful mathematical transform, the *inverse discrete-time Fourier transform* (IDTFT), that we can use for converting between the *continuous* frequency response (i.e., frequency-domain characteristics) and the *discrete*-time impulse response (i.e., time-domain characteristics). If you'd like to stare at the mathematical definition, here it is:

$$h(n) = \frac{1}{2\pi} \int_{-\pi}^{\pi} H(\omega) e^{j\omega n} d\omega \tag{5.5}$$

where $H(\omega)$ is the discrete-time Fourier transform (DTFT) and $h(n)$ is the impulse response. (If it's been too long since you had calculus, don't worry. We'll use this equation to get answers for common cases, then plug numbers into those answers rather than use this equation every time.)

Using Equation 5-5, it's relatively easy to compute $h(n)$ for the ideal lowpass, highpass, bandpass, and bandstop filter types, and we've listed the impulse responses in Table 5-1. Because we'll be dealing with both the ideal impulse response and the actual impulse response of our filter, we use the subscripted $h_d(n)$ for these "desired" (ideal) impulse responses.

There are two things to note about the desired impulse responses. First, the infinitely sharp cutoffs of the filters imply a noncausal response; thus, the impulse response extends into both positive and negative time. Our filters can deal only with $n \geq 0$, so we'll have to make this adjustment at some point. Second, due to the mathematics, we need to evaluate the cases of $n=0$ separately, and this equation is noted in a separate column in the table. The ideal impulse responses are doubly infinitely long as well.

Figure 5-8 is a view of the impulse response of an ideal lowpass filter ($\omega_c = \pi/4$ in this example). Because we require a finite number of coefficients, it's logical to ask what would happen if we truncated the impulse response after it gets fairly small.

Figure 5-9 shows the impulse response after truncating it to 21 points. We've plotted the corresponding frequency response in Figure 5-10. Note the "ringing" near the transition frequencies. Can we get rid of that "ringing" by taking more points?

It is logical to assume that if we get an ideal frequency response using an infinite number of samples and we get ringing when we use a small number, then increasing the number of sam-

Generating Coefficients

Type	$h_d(n)$, $n \neq 0$	$h_d(0)$	$h_d(n)$
Lowpass (0 to ω_c to π)	$\dfrac{\sin(n\omega_c)}{\pi n}$	$\dfrac{\omega_c}{\pi}$	
Highpass (ω_c to π)	$\dfrac{-\sin(n\omega_c)}{\pi n}$	$1 - \dfrac{\omega_c}{\pi}$	
Bandpass (ω_l to ω_u)	$\dfrac{\sin(n\omega_u)}{\pi n} - \dfrac{\sin(n\omega_l)}{\pi n}$	$\dfrac{\omega_u}{\pi} - \dfrac{\omega_l}{\pi}$	
Bandstop	$\dfrac{\sin(n\omega_l)}{\pi n} - \dfrac{\sin(n\omega_u)}{\pi n}$	$1 - \left(\dfrac{\omega_u}{\pi} - \dfrac{\omega_l}{\pi}\right)$	

Table 5-1 Ideal impulse responses for common filter types.

ples should reduce the ringing. Well, let's double the number of samples and plot the frequency response. (See Figure 5-11).

Comparing Figure 5-10 and Figure 5-11, although we see a sharper transition, there's no change in the peak magnitude of the ringing, roughly 1.09. This startling result is called *Gibb's*

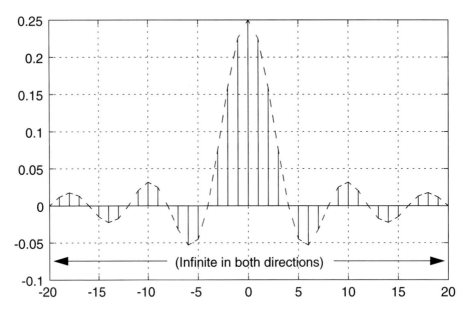

Figure 5-8 Ideal impulse response of a lowpass filter.

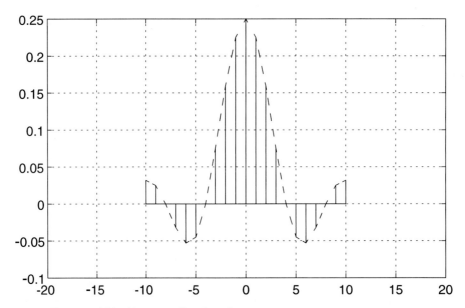

Figure 5-9 Truncated ideal lowpass filter impulse response.

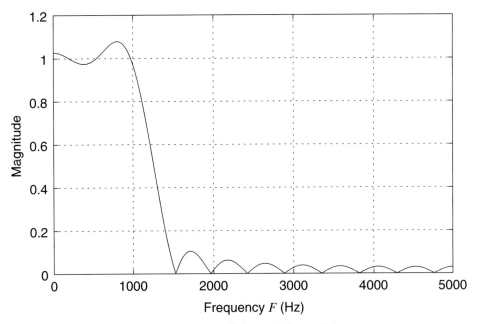

Figure 5-10 Magnitude response using coefficients of Figure 5-9.

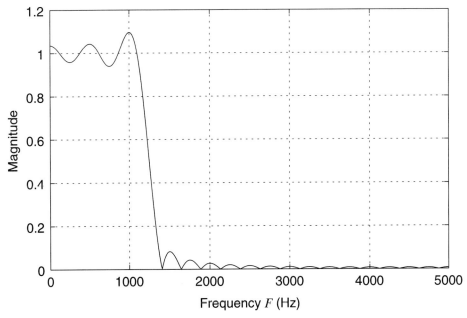

Figure 5-11 Magnitude response using twice as many coefficients as in Figure 5-10.

Phenomenon[15] and it's an unavoidable consequence of having a *discontinuity,* or abrupt change, in one domain (here, truncating in the time domain).

By the way, most folks didn't realize this would happen; they just assumed that as you increased the number of samples from the impulse response, the frequency response you got would just get better and better. In fact, when they built some machines (precomputer) to analyze and reproduce complex signals and saw this phenomenon, at first they thought there was something wrong with their machines! Without belaboring the point, this is one of those fundamental trade-offs built into the universe—sharp corners, in the impulse response or magnitude response, are associated with infinitely long representations in the other domain.

If the ringing problem is caused by discontinuities where the impulse response is truncated, why not "smooth out" the ends? In Figure 5-12, we've taken a special *window* function and smoothly attenuated our 21 coefficients to zero at both ends. We should expect this to affect the magnitude response somewhat, but check out the results (especially the ringing) in Figure 5-13. Not bad, eh? By windowing, we've traded off the ripples of Gibb's phenomenon for a wider transition band.

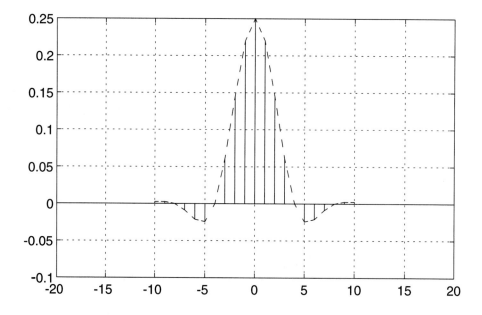

Figure 5-12 Smoothing the truncated impulse response.

15. Named after Barry Gibbs of the Bee-Gees. (See Chapter 12.)

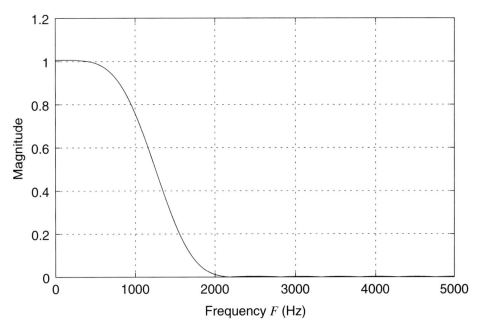

Figure 5-13 Magnitude response using windowed coefficients.

There are several popular windows from which to choose. Table 5-2 summarizes their key points, which we'll explain below in our example. Just as with the different analog filters like Chebychev, Butterworth, and so on, each of the windows is optimized for different characteristics. For example, in designing a lowpass filter, the Hamming window delivers low ripple throughout the spectrum and good stopband attenuation, at the cost of a wide transition band.

The steps for using windows are:
- Determine the window type (Hanning, Hamming, Kaiser, etc.) that will satisfy the stopband attenuation requirements.
- Determine the minimum size of the window (N) using the transition width. The resulting samples will range from n equals $-(N-1)/2$ to $(N-1)/2$ for odd values of N, and n equals $-N/2$ to $N/2$ for even values of N.
- Calculate the window coefficients $w_{win}(n)$ for all values of n.
- Generate the ideal impulse response $h_d(n)$ for the type of filter desired (e.g., lowpass) for all values of n.
- Multiply the window and the impulse response together, point by point (i.e., $h(n)=w_{win}(n) \cdot h_d(n)$) for all values of n.
- Make the filter causal by shifting the result by $(N-1)/2$ points so the indexing starts at 0. (Shift by $N/2$ if N is even.) See Figure 5-14. (Note that for $N=21$ in our example, that the last sample after shifting is at $n=20$.)

Window	Passband ripple (db)	Stopband attenuation (dB) A_m	First side-lobe (dB)	Transition width Δf (norm. Hz)	δ_m
Rectangular	0.7416	21	-13	0.9/N	0.0891
Kaiser, A=30, β=2.12	0.270	30	-19	1.5/N	0.0316
Hanning	0.0546	44	-31	3.1/N	0.00632
Kaiser, A=50, β=4.55	0.0274	50	-34	2.9/N	0.00316
Hamming	0.0194	53	-41	3.3/N	0.00224
Kaiser, A=70, β=6.76	0.00275	70	-49	4.3/N	0.000316
Blackman	0.0017	74	-57	5.5/N	0.000196
Kaiser, A=90, β=8.96	0.000275	90	-66	5.7/N	0.0000316

Table 5-2 Key properties of windows.

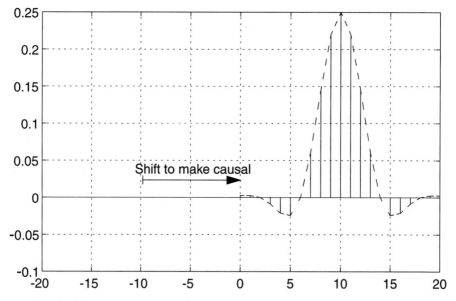

Figure 5-14 Shifting (windowed) coefficients to produce a causal filter.

We'll show an example of windowing in just a moment. First, let's examine the windows in more detail. Figure 5-15 shows the time domain and frequency domain characteristics of some popular windows. Below we briefly discuss each:

Rectangular: That is, effectively "no window" on the points retained. Offers the sharpest transition in the frequency domain, but at the expense of lessened attenuation in the stopbands. The stopband attenuation is only 21 dB, regardless of the window size. (All windows share the behavior that increasing the number of points decreases only the transition width, not the stopband attenuation. However, each window will have a different stopband attenuation, generally better than the rectangular window.)

$$w_{rect}(n) = 1 \tag{5.6}$$

(All of the window equations are for $-(N-1)/2 \leq n \leq (N-1)/2$. Outside of this range, windows have a value of 0. Replace $(N-1)/2$ with $N/2$ for even values of N.)

Hanning (other names: *von Hann, raised cosine, cosine bell*): Much wider transition (more than three times as wide as the rectangular window), but with a stopband attenuation of 30 dB.

$$w_{hanning}(n) = 0.5 + 0.5\cos\left(\frac{2\pi n}{N}\right) \tag{5.7}$$

Hamming: Yes, this is different from the "Hanning" window.[16] A bit wider transition than Hanning, but an additional 10 dB of stopband attenuation.

$$w_{hamming}(n) = 0.54 + 0.46\cos\left(\frac{2\pi n}{N}\right) \tag{5.8}$$

Blackman: Continuing the trade-off of transition width for stopband attenuation, the Blackman delivers 74 dB of stopband attenuation, but with a transition width six times that of the rectangular window.

$$w_{blackman}(n) = 0.42 + 0.5\cos\left(\frac{2\pi n}{N-1}\right) + 0.08\cos\left(\frac{4\pi n}{N-1}\right) \tag{5.9}$$

---

16. Apparently the similarity between Hann and Hamming was enough to encourage some folks to add on an "ing" to von Hann's name. Blackman and Kaiser were left out of the fun, though. "Blackmanning" and "Kaisering" windows, anyone?

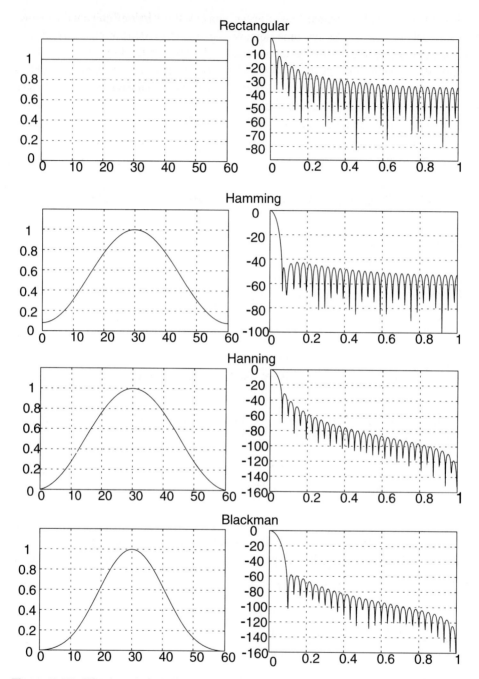

Figure 5-15 The time domain and magnitude response of some common windows.

Generating Coefficients

Kaiser (Kaiser-Bessel):[17] Not a single window, the "Kaiser window" is actually a family of windows generated from a common equation. Given a stopband attenuation, you can create a "custom" window; we generated parameters for four such windows for Table 5-2 (A_p=30, 50, 70, and 90 dB). This window is somewhat more complicated to produce than the others, but the ability to meet intermediate values of stopband attenuation or to exceed the Blackman window's attenuation can be valuable.

The Kaiser window begins with a calculation of β as follows:

β=0	(5.10)	$A \leq 21$ dB
β=0.5842(A-21)^0.4+0.07886(A-21)	(5.11)	$21 < A < 50$ dB
β=0.1102(A-8.7)	(5.12)	$50 \leq A$ dB

The equation for $w_{kaiser}(n)$ is:

$$w_{kaiser}(n) = \frac{I_0\left(\beta\left[1-\left(\frac{2n}{N-1}\right)^2\right]^{1/2}\right)}{I_0(\beta)} \quad (5.10)$$

where, believe it or not, $I_0(x)$ is the "zero-order modified Bessel function of the first kind." One way of calculating $I_0(x)$ is through the following equation:

$$I_0(x) = 1 + \sum_{k=1}^{L}\left(\frac{(x/2)^k}{k!}\right)^2 \quad (5.11)$$

L, the number of terms, is usually less than 25. If you do calculate a Kaiser window for other than the values we've provided in the table, you'll need to know the transition width, which you can get by using the following relationship between the stopband attenuation, transition width, and N, the number of window points:

$$N \geq \frac{A - 7.95}{14.36\Delta f} \quad (5.12)$$

---

17. By the way, Dr. James Kaiser is one of the pioneers of the modern DSP discipline and, in spite of his many distinguished accomplishments, is one of the nicest guys in the DSP field that you'll ever meet. What a gentleman. In fact, to this day, he never refers to the "Kaiser" window by that name.

The transition width, Δf, is defined as:

$$\Delta f = f_s - f_p = \frac{F_{stop} - F_p}{F_s} = \frac{\omega_s - \omega_p}{2\pi} \qquad (5.13)$$

As you can see, the equation for the Kaiser window is a bit complicated, but it does come very close to producing an optimal window for a given stopband attenuation. Remember that we use a window only once, to create the coefficients. We don't have to program these on microcontrollers or run them in real time.

Before showing you an example of a filter design using windowing, there are a few points about windowing that you should know. First, windowing produces filters with $\delta_p = \delta_s$. The practical implication is that you'll usually have far less passband ripple than you actually need, just to meet the stopband attenuation. Put another way, the minimum of your design parameters δ_p and δ_s sets the stopband attenuation and passband ripple of the final filter. Second (if done right!), windowing produces linear-phase filters—recall that the symmetric impulse response is a necessary condition for linear phase. Linear phase is a rather costly specification, and IIR filters have a lot going for them if you can live with their nonlinear phase response. Finally, the windows we've mentioned have fixed stopband attenuation (and/or fixed passband ripple, if you like); the only thing we can change is the transition width, which we decrease by increasing the size of the window (and hence, number of coefficients in the filter).

Let's take an example from filter specification to filter coefficients to see how windowing is used. We'll use the same filter design specifications as in the Parks-McClellan example, namely:

Sample rate: 10 kHz
Passband: 0 to 2.5 kHz (i.e. lowpass), 0.5 dB maximum ripple
Stopband: 3.0–5 kHz, 50 dB minimum attenuation

Again, we can immediately calculate δ_p and δ_s as 0.0593 and 0.00316 (Equation 5.3 and Equation 5.4). Clearly, our stopband deviation is the smaller, so we'll call that δ_m. Using Table 5-2, we look to see what windows will meet our maximum δ_m requirement. For $\delta_m = 0.00316$, the Kaiser ($\beta \geq 4.55$), Hamming, and Blackman all will do. To minimize the transition width (and minimize the number of calculations!), we choose the Hamming.

The next step is to calculate the size of the window (i.e., the number of coefficients of the filter). Our transition width, Δf, is (3000-2500)/10000, or 0.05. Using the relationship from the table for the Hamming window, we find $N = 3.3/\Delta f$, or 66. We're always going to use windows with N odd, so we'll use $N=67$ for our window.

What should ω_c be? It's not identical to ω_p, or to ω_s. But we'll need a good value to generate the ideal impulse response we plan on windowing. A reasonable approach is to place ω_c in the middle of the transition band; that is, $f_c = f_p + \Delta f/2$ (then translate this to ω_c). In this example, ω_c works out to be 1.73.

Generating Coefficients

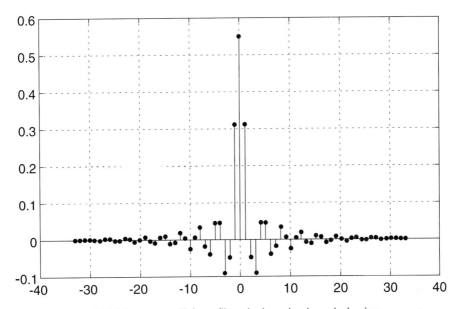

Figure 5-16 A plot of *h(n)* (noncausal!) for a filter designed using windowing.

The next step involves a lot of number crunching; Equation 5.8 must be evaluated from $n=-33$ to $n=33$ (a total of 67 points) for the windowing function $w_{hamming}(n)$, and the ideal impulse response $h_d(n)$ calculated for a lowpass filter from Table 5-1, using the same range of n. The actual impulse response, and our coefficients, are $h(n)=h_{hamming}(n) \cdot h_d(n)$. Figure 5-16 is a plot of $h(n)$, and Table 5-3 shows the numeric values. One last item before we plug this into the filter, though—we need to add $(N-1)/2$ to each index so that $h(n)$ will run from $n=0$ to $n=66$ (i.e., $N-1$). (This makes the filter causal, recall.) Figure 5-17 shows the magnitude response (you might want to compare it to the Parks-McClellan plot earlier.)

5.4.3 Frequency Sampling

The window design method takes an ideal magnitude response (and ideal phase response as well!) and uses the IDTFT to produce an ideal impulse response. This infinitely long impulse response is then creatively chopped off to give us a finite number of filter coefficients. The *frequency sampling* method takes a slightly different approach, with a different set of design goals. This difference in approach leads to an FIR filter design that can be implemented in a recursive form, and in some cases, using very simple integer coefficients. Hmmm, did that last phrase catch your attention? Simple integer coefficients sound like just the ticket for microcontrollers.

The frequency sampling design usually starts with an ideal magnitude frequency response, just as the window method did. However, instead of using the (inverse) DTFT to produce an impulse response, the ideal response is sampled at regular intervals, producing a sequence of discrete samples (in frequency, not time!). The discrete Fourier transform is used to produce a

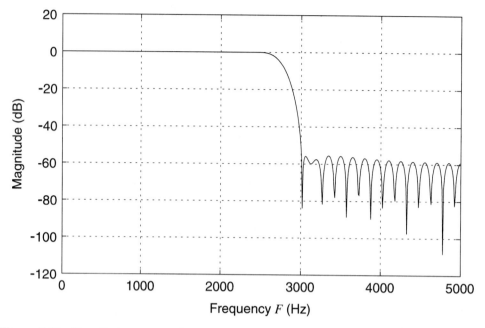

Figure 5-17 Magnitude response for an example filter designed using windowing.

discrete-time impulse response from these discrete samples in frequency. What's important to us, though, is that the resulting impulse response is finite in length—in fact, there's the same number of samples in the impulse response as magnitude and phase samples given as input. No need to worry about windowing!

Somewhere, someone is asking how this can be—didn't we start with an ideal magnitude response? What about this universal law that discontinuities in one domain lead to infinite lengths in the other? Here, we had the same ideal response but got a perfect, finite impulse length. The full answer will have to wait for Chapter 8 (Frequency Analysis) but briefly is similar to the issue of sampling in the time domain. Our sampling rate in either the frequency domain or time domain places limits on the amount of "detail" that we preserve in our sampling. As we increase the number of samples around a discontinuity in the frequency domain (e.g., the sharp transition between passband and stopband in an ideal filter), we see an increase in the length of the impulse response. However, as long as we use a finite number of samples, we'll never have a "sharp enough" transition to warrant an infinitely long impulse response.

The flip-side of the bargain, though, is that the actual (magnitude) frequency response of the filter will match our ideal only at those points that were sampled. In between, the response can be "whatever the DFT likes," which may be more than close enough or could be outside the limits you have. Bear in mind that if you've completely specified the frequency response for all of the frequencies of interest (e.g., you've sampled at the fundamental and harmonics for some signal), this may be exactly the response you need.

Generating Coefficients

h(n)=		=h(N-n-1)
h(0)	0.0004	h(66)
h(1)	-0.0008	h(65)
h(2)	-0.0001	h(64)
h(3)	0.0010	h(63)
h(4)	-0.0002	h(62)
h(5)	-0.0014	h(61)
h(6)	0.0008	h(60)
h(7)	0.0018	h(59)
h(8)	-0.0019	h(58)
h(9)	-0.0019	h(57)
h(10)	0.0034	h(56)
h(11)	0.0014	h(55)
h(12)	-0.0052	h(54)
h(13)	0.0000	h(53)
h(14)	0.0071	h(52)
h(15)	-0.0026	h(51)
h(16)	-0.0086	h(50)
h(17)	0.0066	h(49)
h(18)	0.0091	h(48)
h(19)	-0.0119	h(47)
h(20)	-0.0077	h(46)
h(21)	0.0184	h(45)
h(22)	0.0035	h(44)
h(23)	-0.0257	h(43)
h(24)	0.0047	h(42)
h(25)	0.0330	h(41)
h(26)	-0.0186	h(40)
h(27)	-0.0398	h(39)
h(28)	0.0427	h(38)
h(29)	0.0452	h(37)
h(30)	-0.0928	h(36)
h(31)	-0.0488	h(35)
h(32)	0.3137	h(34)
h(33)	0.5500	

Table 5-3 Coefficients for example filter designed using windowing. (Note symmetry!)

With the window method, we traded a wider transition width for lower ripple and/or stopband attenuation. The trade is much the same for the frequency sampling method. In this case, we explicitly incorporate a transition in our ideal frequency response and make sure that a sufficient number of samples is taken in that region. The question then becomes, How many samples during the transition are optimal (and where are they located)?

The answer gets a bit messy; there's no simple equation. Typically, a table is consulted that gives optimal values based on the desired stopband attenuation, number of samples in the transition band, and number of coefficients.

We've actually glossed over a few other points on just where the frequency response should be sampled. Do you start the sampling at 0? Do you end at the Nyquist limit? Depending on your answers, you'll end up with one of four different sampling methods, each of which has advantages and disadvantages in the final frequency response.

We're going to bail out and point you to the references at the end of this chapter. The general form of the frequency sampling method is a bit too messy to get into at the level of this book. That said, frequency sampling is very useful if you need specific frequency response characteristics at well-defined (and usually uniformly spaced) frequencies. It's also possible to sample the frequency response much more frequently than you want, then to apply the same types of windows that are used in the windowing method to reduce the number of coefficients. By a process called *warping*, it's possible to produce a non-uniformly spaced sampling—emphasizing the lower frequencies, for example.

We didn't come down the frequency-sampling path just to disappoint. Although the general form of the frequency sampling method is outside our scope, there's a special case that is useful for microcontroller implementations. Because it uses simple coefficients and is based on frequency sampling, we'll call this the *Simple Coefficient Frequency-Sampling Filter Design Method*.

5.4.3.1   Simple Coefficient Frequency-Sampling Filters

What makes the frequency-sampling method especially appealing for microcontroller applications is that it's possible to design an FIR filter that can be implemented in a recursive form (i.e., using feedback). Normally, FIR filters aren't recursive—they don't have any poles, and, in fact, some texts equate FIR filters with *nonrecursive* filters. But there is a recursive form of the FIR filter; not only can it save quite a bit of computation, but there's a certain type of recursive filter that uses *integer* coefficients to boot!

The idea behind this method is to use a *comb filter*, a simple FIR filter with zeros evenly spaced around the unit circle, and to selectively cancel one or two of those zeros with one or two matching poles. Cancelling the zeros provides a passband, while the remaining zeros provide attenuation for a stopband. Let's first examine the comb filter, then see how we can cancel selected zeros using integer arithmetic.

The system function of a simple comb filter is

$$H(z) = 1 - z^{-M} \tag{5.14}$$

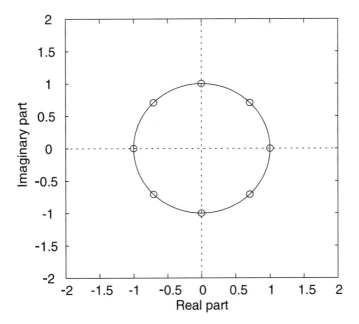

Figure 5-18 Zeros of a simple comb filter (M=8).

For example, Figure 5-18 shows the eight zeros of $H(z)$ for $M=8$, which are evenly spaced around the unit circle in the z-plane. This spacing occurs because the M roots of $H(z) = 1 - z^{-M}$ occur at $z_k = e^{j2\pi k/M}$ (for $k=0$ to M-1). If you remember Euler's identity, $e^{jx} = \cos(x) + j\sin(x)$ and plug in the various values of $2\pi k/M$ for $k=0$ to M-1, these values all fall neatly and exactly[18] on the unit circle as shown in Figure 5-18. The difference equation is just:

$$y(n) = \frac{1}{M}(x(n) - x(n-M))$$

For example, if $M=8$,

$$y(n) = \frac{1}{8}(x(n) - x(n-8))$$

Not bad, eh? The coefficients are all one, and the scaling factor in this case is an integral power of two so that the scaling is easily accomplished by three right shifts.

The magnitude response for this example is shown in Figure 5-19. The comb filter is a useful filter in some circumstances, as we discussed in an earlier chapter. But we can play a trick

18.   If you use infinite-precision math!

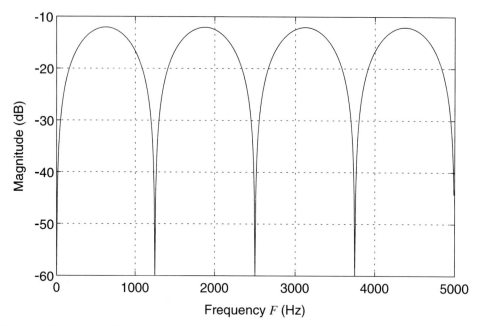

Figure 5-19 Magnitude response of a comb filter (M=8).

to make it into a lowpass, bandpass, or highpass filter by selectively cancelling out some of the zeros by putting poles right on top of them.

If this were on television, we'd be saying, "don't try this at home!" about now, especially if you are designing an elevator or a nuclear reactor, because normally you're asking for trouble when you put a pole on the unit circle. Recall that poles outside the unit circle are associated with unstable systems, and poles right on the unit circle are *marginally stable*, which is to say that although there's no exponential gain associated with them, there's no exponential decay either. They're right on the knife edge, and any rounding errors or imprecision in arithmetic will push them off—half the time into the land of now-would-be-a-good-time-to-revise-your-resume.[19]

There are two ways around the problem of placing poles—and zeros—on the unit circle. First, place the poles and zeros just *inside* the unit circle. This complicates the coefficients a bit but does give a margin of safety. Second, pick the locations of the to-be-cancelled zeros so that they have simple coordinates that can be expressed exactly with finite-precision representations. The only possibilities are +/-1 and +/-1j. Since +/-1j are complex conjugates, there are really only three different possibilities ($z=1$, $z=-1$, $z=+/-1j$). Cancelling the zero at $z=1$ produces a lowpass filter, cancelling the zeros at $z=+/-1j$ a bandpass filter, and cancelling the zero at $z=-1$ a highpass filter.

19. Oscillating elevators are not favored by most riders. As for oscillating nuclear reactors ...

Generating Coefficients

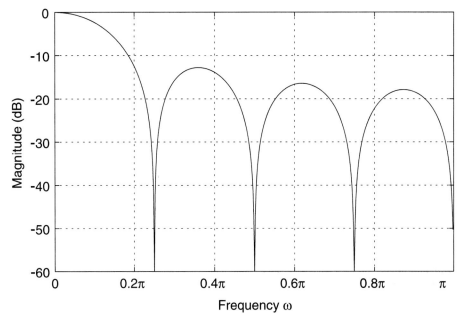

Figure 5-20 Magnitude response of a simple coefficient frequency sampling lowpass filter.

Option one, placing the zeros inside the unit circle, is certainly viable, and you may want to keep it in mind as a technique. Remember, this form reduces the number of nonzero coefficients, so the computation may still be reduced compared to FIR filters designed using other methods. However, the full treatment is outside the scope of this book. (For example, see Ifeachor and Jervis [1993].)

Option two is of much more interest in microcontroller-based DSP. No need to worry about mathematical errors, as we're dealing with simple integers. Properly chosen, in fact, we can avoid multiplication completely, perhaps using several shifts at most! Figure 5-20 shows the magnitude response of a filter that requires two additions, three shifts—and *no* multiplies.

H(z) for this filter is:

$$H(z) = \frac{1}{8} \cdot \frac{1-z^{-8}}{1-z^{-1}} \qquad (5.15)$$

and the difference equation (now recursive!) is:

$$y(n) = \frac{1}{8}[x(n) - x(n-8)] + y(n-1) \qquad (5.16)$$

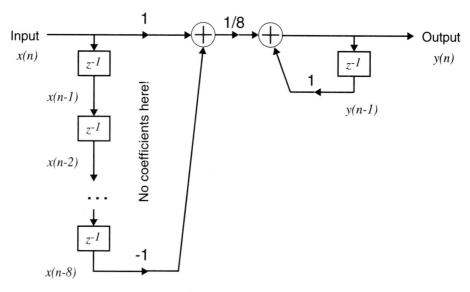

Figure 5-21 Structure of a simple coefficient frequency sampling filter.

Figure 5-21 shows the resulting structure. Note that the multiplication by 1/8 is equivalent to three right shifts (each right shift is a divide by 2).

Figure 5-22 shows what we're doing here. We've taken a comb filter (evenly spaced zeros on the unit circle, a *very* simple and compact difference equation) and put a pole (the denominator of $H(z)$) right on top of the zero at $z=1$. This effectively cancels the zero, leaving the rest of the zeros unaffected. The equation, in general, for this type of filter is:

$$H(z) = \frac{1}{M} \cdot \frac{1 - z^{-M}}{1 - z^{-1}} \qquad (5.17)$$

There are a few disadvantages to this type of filter:
- The passband edge frequency is entirely a function of M and the sampling frequency.
- The stopband attenuation is nothing to write home about, and is fixed by M.

Generating Coefficients

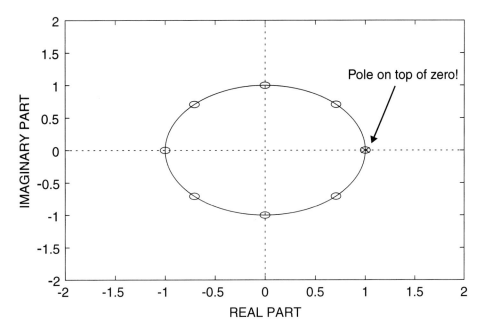

Figure 5-22 Cancelling a zero of a comb filter with a pole.

- Although there's minimal computation, storage is required for each delay stage. Due to overflow, filters with large values of M require care in performing the computations.

We can get a better understanding of the lowpass version of this filter by writing out the equivalent nonrecursive $H(z)$. We know this exists—we can just take the zeros that are not cancelled and produce a polynomial. If we do this for the general case, we get

$$H(z) = \frac{1}{M} \cdot \frac{1 + z^{-1} + z^{-2} + \ldots + z^{-M}}{1} \tag{5.18}$$

If this reminds you of the simple filter we started the chapter with, it should! The lowpass version of the simple coefficient frequency-sampling filter is just an equal-coefficient lowpass filter, but implemented in a tricky way![20] (You might want to compare the magnitude responses in Figure 5-3 and Figure 5-22.)

---

20. What's going on is that the sum isn't calculated from scratch for each new sample; instead, the oldest sample is subtracted and the newest added. $y(n-1)$ is the old sum.

Let's look briefly at the other filter types. For highpass filters, we require that M be even; to cancel out the zero at $z=-1$, $H(z)$ will be of the form

$$H(z) = \frac{1}{M} \cdot \frac{1 - z^{-M}}{1 + z^{-1}} \quad (5.19)$$

(i.e., the denominator now has a zero at $z=-1$). The equivalent nonrecursive equation is

$$H(z) = \frac{1}{M} \cdot \frac{1 - z^{-1} + z^{-2} - z^{-3} + \ldots \pm z^{-M}}{1} \quad (5.20)$$

A bandpass filter requires zeros at $+/- 1j$, so M must be a multiple of four. The denominator, since it has to have two poles to cancel the complex conjugate zeros, will have a second-order polynomial (i.e., z^{-2}):

$$H(z) = \frac{1}{M} \cdot \frac{1 - z^{-M}}{1 + z^{-2}} \quad (5.21)$$

and the equivalent nonrecursive $H(z)$ is

$$H(z) = \frac{2}{M} \cdot \frac{1 + 0z^{-1} - 1z^{-2} + 0z^{-3} + 1z^{-4} + 0z^{-5} - 1z^{-6} \ldots \pm z^{-M}}{1} \quad (5.22)$$

We mentioned the dismal stopband attenuation. A technique for increasing the stopband attenuation somewhat is to raise $H(z)$ to an integer power. For example, taking the lowpass filter and squaring it gives

$$H(z) = \left(\frac{1}{M} \cdot \frac{1 - z^{-M}}{1 - z^{-1}}\right)^2 = \frac{1}{M^2} \cdot \frac{1 - 2z^{-M} + z^{-2M}}{1 - 2z^{-1} + z^{-2}} \quad (5.23)$$

This requires a few more additions, but the factors of 2, top and bottom, are just shifts or adds, and M^2 is still a power of two (i.e., you can divide by shifting). Table 5-4 gives the (lowpass) stopband attenuation (i.e., first lobe attenuation), passband edge frequencies for 1 dB and 3 dB, and the stopband edge frequency for selected values of M and for $H(z)$ squared.

Let's design a filter with the following specifications: passband edge frequency 20 Hz, maximum passband attenuation 1 dB, and stopband attenuation a very lax 12 dB (i.e., a factor of 4).

M	First lobe/A_s (dB)	f_{pb} (-3 dB) (Hz)	f_{pb} (-1 dB) (Hz)	f_{sb} (Hz)
4	11.3	0.11	0.068	0.20
8	12.8	0.056	0.033	0.10
16	13.1	0.028	0.017	0.052
32	13.2	0.014	0.0085	0.026
64	13.3	0.007	0.0045	0.014
128	17.9	0.0035	0.0025	0.0075
4-squared	22.6	0.083	0.048	0.20
8-squared	25.6	0.041	0.024	0.10
16-squared	26.3	0.02	0.012	0.052
32-squared	26.5	0.01	0.006	0.026
64-squared	26.5	0.005	0.003	0.014
128-squared	26.6	0.0025	0.0015	0.007

Table 5-4 Characteristics of simple coefficient frequency sampling filters for selected values of M.

From Table 5-4 we see that any $M \geq 8$ will give the specified stopband attenuation; since higher values of M mean higher sampling rates, we'll choose $M=8$. The passband edge frequency (in normalized Hz) is 0.033; we can solve for F_s by using

$$0.033 \cdot F_s = 20 \qquad (5.24)$$

from which we find $F_s = 606$ Hz.

Something not shown in the table is the ratio of stopband frequency to passband edge frequency—this is close to 3 for $M=4$ through 128, and around 4 for the "squared" cases. This means the stopband edge frequency will be around 60 Hz, given a passband edge frequency of 20 Hz.

By the way, don't ignore the nice, evenly spaced zeros spaced at F_s/M intervals. If 60 Hz were a real problem, it would make a lot of sense to put the first noncancelled zero right at 60 Hz.

The difference equation for this filter is the same as given before, namely,

$$y(n) = \frac{1}{8}[x(n) - x(n-8)] + y(n-1) \qquad (5.25)$$

Figure 5-23 shows the magnitude frequency response.

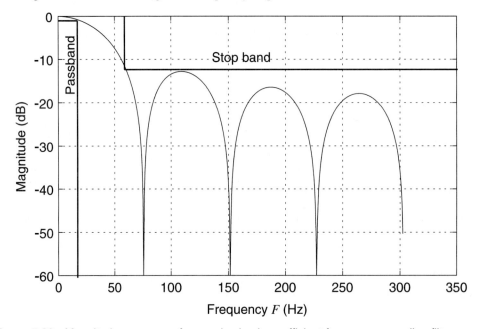

Figure 5-23 Magnitude response of example simple coefficient frequency-sampling filter.

To summarize, the frequency sampling method produces filters whose frequency response matches some ideal only at specific frequencies. Design tables are available for optimizing the filter by carefully specifying samples in the transition region. The recursive form allows for an efficient FIR filter by starting with a comb filter (evenly spaced zeros on the unit circle, or just inside) and selectively cancelling zeros with poles. Certain restricted forms of recursive filters can be constructed using integer arithmetic only, though, as we've seen, the magnitude response is not very good.

5.4.4 Ad-Hoc

When all of this careful use of equations and theory gets a bit too much, there's always the famous *ad-hoc* filter design method. Which is to say, put your zeros in the z-plane wherever you think they belong, see what filter response you get, then repeat the cycle until you're satisfied.

This is obviously a useful technique only if you have a quick way of determining the frequency response of a system, given its zero locations. But it's a quite viable design method for certain simple digital filter designs. Here's a quick review of the issues to keep in mind:
- The sampling rate determines the range of frequency covered by the 180 degrees of the upper half of the z-plane. (The lower half is "negative frequency.")
- Zeros can go anywhere in the z-plane without affecting stability. (Poles, which are normally associated only with IIR filters, must be inside the unit circle for stability.)
- The closer a zero is to the unit circle, the more it attenuates signals at that frequency (recall, angle in the z-plane corresponds to frequency). A zero right on the unit circle attenuates that frequency completely.
- To have real filter coefficients (which we assume), zeros must be placed in complex-conjugate pairs, unless the zero is right on the real axis (i.e., there is no imaginary component of the zero). Complex-conjugate pairs, recall, just differ in the sign of the imaginary component. For example, if you put a zero at $(0.5+0.5j)$, you also need one at $(0.5+0.5j)^*$, which is $(0.5-0.5j)$.
- For linear phase, you also need to place zeros in *reciprocal pairs*, meaning if z_1 is a zero, then you'll also need a zero at $1/z_1$. (Of course, to have real filter coefficients you'll also need z_1^* and $1/z_1^*$, the complex conjugates of those zeros.) Figure 5-24 shows some reciprocal and complex-conjugate pairs. Note that placing just one of the four zeros shown specifies the other three, if linear phase and real coefficients are desired.

As an example, we can design a simple bandstop filter, linear phase, with a zero at 1 kHz, and a sampling rate of 4 kHz. Figure 5-25 shows the pole-zero placement, and Figure 5-26 the frequency response.

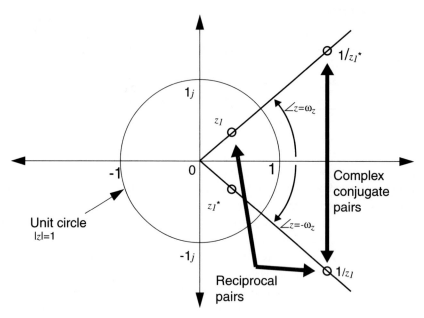

Figure 5-24 Placing zeros in the z-plane by hand.

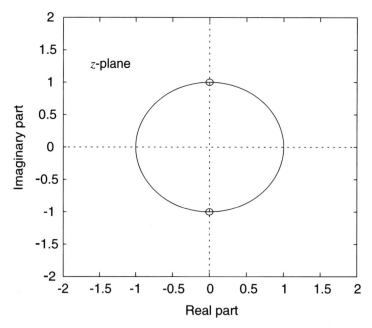

Figure 5-25 Zero placement of simple bandstop FIR filter.

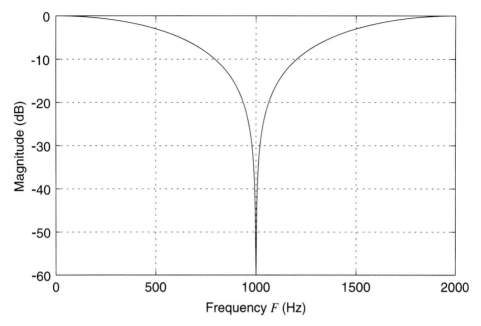

Figure 5-26 Magnitude response of simple bandstop FIR filter.

We can find $H(z)$ by just writing down the zero factors and multiplying them to get a polynomial in negative powers of z:

$$(z-j)(z+j) = z^2 + 1 \quad (5.26)$$

To make $H(z)$ causal, divide by the highest power of z (which is z^2) to get:

$$H(z) = 1 + z^{-2} \quad (5.27)$$

The difference equation is:

$$y(n) = x(n) + x(n-2) \quad (5.28)$$

5.4.5 Fast Convolution

Thus far, we've discussed filters that produce one output sample for each input sample. *Fast convolution* is based on the fact that multiplication in the frequency domain is equivalent to convolution in the time domain. A block of samples is transformed (using

discrete Fourier transform—see Chapter 8 [Frequency Analysis]) into the frequency domain; multiplying frequency components by either one or zero keeps or kills off those components, which are transformed back into the time domain using the inverse DFT.

This may sound like a roundabout way of doing filtering, especially given how simple it is to implement FIR filters, but for very large values of N it is actually computationally quicker to do the forward and inverse transforms and some simple multiplies. The exact crossover point where fast convolution is more efficient is highly dependent on the underlying processor architecture. We'll postpone more discussion on the DFT-based technique until a later chapter.

5.5 Lowpass-to-Highpass Conversion

While the Parks-McClellan (optimal equiripple), window, and frequency-sampling methods all support lowpass, highpass, and bandpass filter design, it is sometimes convenient to design a filter assuming a lowpass response, then convert the design to a highpass. Given the coefficients for a lowpass filter, $h_{lp}(n)$, the coefficients for a highpass filter, $h_{hp}(n)$ are $h_{hp}(n) = (-1)^n h_{lp}(n)$ (i.e., flip the sign of every other coefficient starting with the second). In the z-plane, the zeros of the filter are just flipped about the imaginary axis, so it is necessary only to negate the real parts of all zeros. The new cutoff frequency is, therefore, as far from $\omega=\pi/2$ as the original cutoff frequency was from $\omega=0$. (See Figure 5-27.) In fact, we can get the new frequency response by replacing the Ω in the original frequency response by $(\Omega_s/2)-\Omega$, where Ω_s is the sampling frequency: $H_{hp}(\Omega) = H_{lp}\left(\frac{\Omega_s}{2} - \Omega\right)$. In Chapter 6 (IIR Filters—Digital Filters with Feedback) we'll discuss some more general transformations from lowpass to highpass, bandpass, and bandstop.

5.6 Structures for FIR Filters

5.6.1 Direct and Linear-Phase Structures

Once the coefficients have been determined, the next step is to decide on the structure (form) of the filter. Recall that one way of representing the transfer function of an FIR filter is in the following general form:

$$H(z) = c\frac{(1-z_1 z^{-1})(1-z_2 z^{-1})(1-z_3 z^{-1})\ldots(1-z_M z^{-1})}{1}$$

where $z_1, z_2, \ldots, z_M$ are the zeros of the system, and c is some constant scaling factor. Because this is an FIR filter, there aren't any poles (i.e., roots of the denominator). Multiplied out, we get:

$$H(z) = \frac{b_0 + b_1 z^{-1} + b_2 z^{-2} + b_3 z^{-3} + \ldots + b_{M-1} z^{-(M-1)}}{1}$$

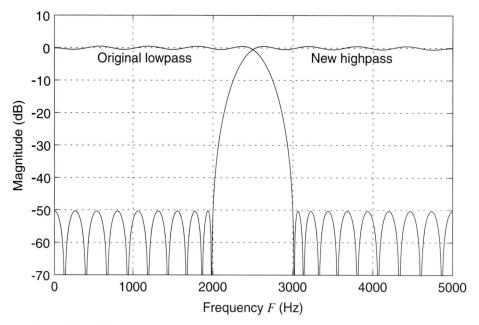

Figure 5-27 Magnitude response of lowpass and related highpass filters.

where the b_i values depend on the values of the zeros.

As we know now, b_i is just $h(i)$, the impulse response, and the negative powers of z correspond to delays. Equivalently, we can write this equation in a more compact form as:

$$H(z) = \sum_{k=0}^{M-1} b_k z^{-k}$$

Writing out the equivalent difference equation, we find

$$y(n) = b_0 x(n) + b_1 x(n-1) + b_2 x(n-2) + \ldots + b_{M-1} x(n-(M-1))$$

Figure 5-28 shows the *direct form* implementation of this difference equation, a familiar sight by now. However, recall that if the FIR filter is linear phase, the coefficients will be symmetric—i.e. $b_0 = b_{M-1}$, $b_1 = b_{M-2}$, etc. We can use this fact to create the *linear phase structures* shown in Figure 5-29 and Figure 5-30 (we need to account for both even and odd numbers of coefficients). The number of multiplies is cut in half by cleverly adding pairs of samples, then multiplying the sum by the coefficient they share. The trade-off is in some additional addressing complexity. Instead of progressing linearly through the sample array with a single pointer, we need two pointers, one starting at either end. However, if multiplies take much more time than

Structures for FIR Filters

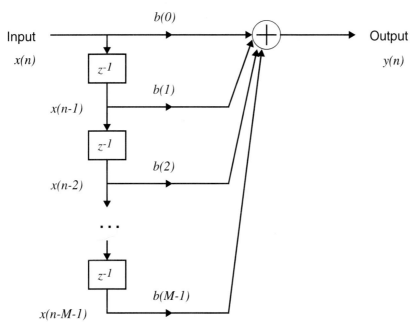

Figure 5-28 Direct form of FIR filter.

simple pointer increments and decrements (ordinarily the case), this form may save quite a bit of time.

5.6.2 Cascade Structures

Returning to the factored form of $H(z)$:

$$H(z) = c\frac{(1-z_1 z^{-1})(1-z_2 z^{-1})\ldots(1-z_{M-1}z^{-1})(1-z_M z^{-1})}{1}$$

we can break up $H(z)$ into a series of shorter equations:

$$H(z) = c \cdot \frac{(1-z_1 z^{-1})(1-z_2 z^{-1})}{1} \cdot \frac{(1-z_3 z^{-1})(1-z_4 z^{-1})}{1} \cdots \frac{(1-z_{M-1}z^{-1})(1-z_M z^{-1})}{1}$$

We actually know a bit more about these zeros $z_1, z_2, \ldots, z_M$. For example, they must occur in complex-conjugate pairs. If we group the zeros into these pairs and multiply out the expressions, we end up with a cascade of second-order equations like:

Chapter 5 • FIR Filters—Digital Filters Without Feedback

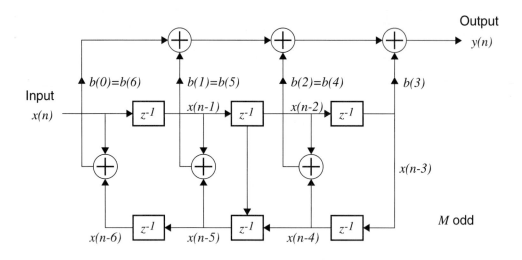

Figure 5-29 Linear phase structure for FIR filter (M odd).

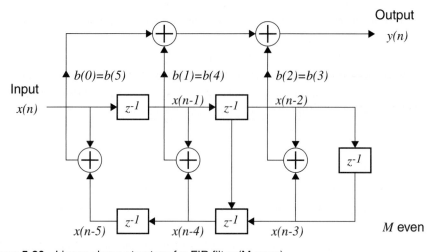

Figure 5-30 Linear phase structure for FIR filter (M even).

$$H(z) = c \cdot \frac{b_{10} + b_{11}z^{-1} + b_{12}z^{-2}}{1} \cdot \frac{b_{20} + b_{21}z^{-1} + b_{22}z^{-2}}{1} \cdot \ldots \cdot \frac{b_{K0} + b_{K1}z^{-1} + b_{K2}z^{-2}}{1}$$

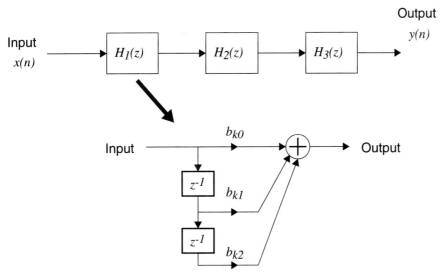

Figure 5-31 A cascade of second-order FIR filters and detail of one section.

Figure 5-31 shows a cascade of second-order FIR filters and the detail of a single second-order section. Continuing with the fact that we will have reciprocal zeros, we can also combine zeros into fourth-order sections. Note that in both the second and fourth order cases, we can adjust the gain factor c so that the initial coefficient for each stage is 1, eliminating a multiply.

Compared with the linear phase structures above, we have about the same number of multiplies, adds, and storage. However, splitting up the filter in this manner reduces roundoff errors, which may be critical for some applications. If roundoff is not a problem, the direct form (and/or linear phase) may involve less data movement and prove more efficient. (For processors with DSP support, the direct form may be the most efficient, even over the linear phase form, as there is no penalty for multiplication over addition.)

5.7 Summary

FIR filters have coefficients which, when viewed as a time sequence, are identical to the impulse response of the DT system. FIR filters convolve the impulse response with the input samples to produce an output.

The windowing technique takes the impulse response of an ideal filter, truncates the impulse response to a finite length, windows the coefficients to reduce the effects of truncation (at the cost of a wider transition), and shifts the coefficients to make the filter causal (no impulse response prior to $n=0$).

The Parks-McClellan algorithm uses sophisticated optimization algorithms to generate coefficients for equiripple filters. This technique requires the use of a computer (except for people with lots of time on their hands) and is generally the design method of choice.

Frequency sampling takes a sampled version of the frequency response of a desired filter, then uses the (inverse) discrete Fourier transform to produce a (finite) impulse response. The resulting filter has a frequency response that generally matches the ideal only at the sampled points.

A special case of the frequency-sampling technique produces an FIR filter with simple coefficient values. Although this form is extremely computationally efficient, the frequency response is rather poor, and there are limits on the types of filters possible. The lowpass simple coefficient frequency-sampling filter is identical to a computationally efficient moving average algorithm.

Zeros may be placed "by hand" in the *z*-plane to produce *ad-hoc* FIR filters; these are generally useful for only very simple designs but are quick and easy to produce.

Table 5-5 summarizes the advantages and disadvantages of the different filter design methods.

A number of mathematically equivalent structures exist for FIR filters; however, finite-word length effects or computational/programming efficiencies usually recommend one structure over others.

Resources

DSP Committee and IEEE ASSP Society, ed., *Programs for Digital Signal Processing*, New York: IEEE Press, 1979.

Ifeachor, Emmanuel C. and Jervis, Barrie W., *Digital Signal Processing: A Practical Approach*, New York: Addison-Wesley, 1993.

Orfanidis, Sophocles J., *Introduction to Signal Processing*, New Jersey: Prentice-Hall, 1996.

Lynn, P.A., "Frequency sampling filters with integer multipliers." In: *Introduction to Digital Filters* (Bogner, R.E. and Constantinides, A.G., eds.), New York: Wiley, 1975.

Proakis, John G., and Manolakis, Dimitris G., *Digital Signal Processing: Principles, Algorithms, and Applications*, 3rd Edition, New York: MacMillan, 1995.

Rorabaugh, C. Britton, *Digital Filter Designer's Handbook (Featuring C Routines)*, New York: McGraw-Hill, 1993.

Software:

Many versions of the Parks-McClellan program are available on the Internet, either in source code (FORTRAN or C) or compiled for PCs. Search for keywords like *FIR*, *DSP*, *Parks-McClellan*, *Remez*, and so on.

Many mathematics packages offer student versions. If you qualify, these are excellent educational resources. In some cases, filter design for both IIR and FIR filters are included, as is the ability to generate frequency-response plots.

Summary

Method	Advantages	Disadvantages
Parks-McClellan (Optimal equiripple)	Linear phase Symmetric coefficients Produces optimal filter Error distributed equally within a band (equiripple) Easy to use Program widely available in print and on-line Lowpass, bandpass, highpass, and multi-passband possible	Requires software
Windowing	Can design with simple equations Many windows available to optimize frequency response	Truncation of impulse response always changes resulting frequency response
Frequency sampling	Produces frequency response that exactly matches desired response at sampled frequencies Can be recursively implemented	Somewhat complicated design process Improving response requires use of special tables
Simple coefficient frequency sampling	Uses only simple integers as coefficients (can be powers of two!) Very few multiplies (i.e., most coefficients are zero, remaining can be powers of two) Easily designed	Wide transition bands, cutoffs are not steep Limitations on the types of responses possible
Ad hoc	Latin name, will impress some people, confuse others Quick, easy for some simple filters	Interactions between zeros become complex for high-order filters, limiting this method to simple filter designs
Fast convolution	May be computationally efficient compared to direct time-domain implementations for large N	Outputs calculated in blocks, not one at a time If more than one block is processed, requires special merging of blocks (i.e., more processing overhead)

Table 5-5 Advantages and disadvantages of FIR filter design methods.

CHAPTER 6

IIR Filters—Digital Filters with Feedback

This is one of the more mathematical chapters. Although we can avoid calculus completely, we'll be up to our ears in complex numbers and both the continuous-time system function H(s) and the discrete-time system function H(z) so you should be pretty familiar with both. All of this boils down to algebra and trigonometry, but because we're going to use a variety of Greek letters for miscellaneous values,[1] it may look a bit daunting. Don't let the letters intimidate you.

IIR filters don't have as much in common with FIR filters as you might think, so it's not crucial to have read Chapter 5 (FIR Filters—Digital Filters Without Feedback), but the FIR background may provide some useful perspective.

6.1 Overview

In the previous chapter, we discussed FIR filters, filters that have no feedback and can have only zeros.[2] In this chapter, we discuss the other major class of digital filters, *infinite impulse response* (IIR) filters, which can have both poles and zeros. Figure 6-1 shows the so-called direct

---

1. These Greek letters are used the same way by everyone—we aren't just making them up to cause you grief.
2. Any poles that appear in an FIR filter must be cancelled by identical zeros, in which case the difference equation and system function can always be "reduced" to reflect an all-zero system.

I form of an IIR filter. Writing out the difference equation explicitly shows the feedback of the output into the filter:

$$y(n) = b_0 x(n) + b_1 x(n-1) + b_2 x(n-2) + \ldots + b_{n_z} x(n-n_z) \\ - a_1 y(n-1) - a_2 y(n-2) - \ldots - a_{n_p} y(n-n_p) \quad (6.1)$$

The current value of y is based on a weighted sum of current and past values of the input x and of weighted values of the past n_p values of y. Here, n_p is the number of poles, and n_z is the number of zeros.

Adding feedback to a digital filter has far-from-subtle effects. For example, for a given magnitude response, it is not uncommon for an IIR filter to be 5–10 times "shorter" (fewer coefficients) than the equivalent FIR filter! On the other hand, IIR filters can never have absolutely linear phase (unlike FIR filters, where linear phase is a major selling point), and the design and implementation of IIR filters is much less straightforward than for FIR filters. Most implementations of IIR filters do not have the lengthy sum of products that is characteristic of FIR filters and which is often available as an optimized instruction on DSP-friendly microprocessors.

Thus, for the same number of multiplications and additions, an IIR filter will often take more processing time, due to additional data manipulation.

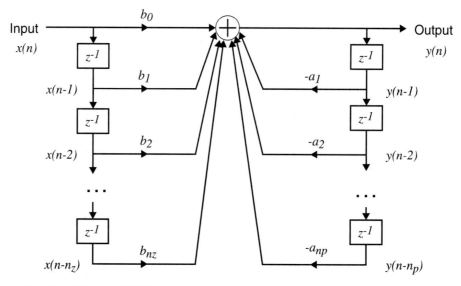

Figure 6-1 Direct I form of IIR filter.

Even with the drawbacks mentioned, the efficiency of IIR filters makes them indispensable in DSP. And the reality is that most IIR filter design is done using software that largely shields the engineer from the equations of this chapter. But let the engineer beware! Picking an IIR filter design method is a decision you should make carefully. Having at least a general understanding of the "equations" is essential to making that decision, and it will also give you a much greater appreciation of your software package!

6.2 Design Process Overview

The IIR filter design process ordinarily begins with the same information as that required for FIR filters, the desired behavior of the filter in the frequency domain, usually emphasizing the magnitude response rather than the phase response.[3] However, the presence of feedback means that the filter coefficients and the impulse response are no longer clearly related. This fact eliminates the techniques used for FIR filters as candidates for IIR filter design. Therefore, we need a new suite of filter design techniques for IIR filters, which we can divide into two categories, "direct" and "indirect," depending on whether the filter design is conducted entirely in the discrete-time (z) domain or whether an intermediate design is performed in the continuous-time (s) domain.

Direct design methods include our friend *ad hoc* (i.e., direct placement of poles and zeros in the z-plane) and some time- and frequency-domain techniques embodied in algorithms (programs) that can be thought of as the IIR *procedural* equivalent to Parks-McClellan. Placing poles and zeros "by hand" requires little information beyond that developed for the FIR case and is viable only for simple designs at that, so we will not dwell on it. Likewise, the exact nature of the algorithms used by the direct design IIR programs is probably not as interesting as knowing when one uses the programs, so our discussion will be on the practical side and rather brief.

Indirect design methods for IIR filters take the counterintuitive approach of first designing an analog filter with the desired characteristics, then mapping this design from the s-domain (CT) to the z-domain (DT). The reason this makes sense in practice is that research on analog filters has produced a number of highly optimized analog filter designs. Leveraging knowledge of analog filters sounds like a great idea; however, the mapping from the s-domain to the z-domain does not always preserve the quality of the s-domain filter. Briefly put, the problem lies in translating the infinite range of frequencies in the s-domain to the finite range of the z-domain. Several s-to-z mappings exist—for example, one wraps the $j\Omega$ axis of the s-plane around and around the unit circle of the z-plane, while another "warps" the frequency axis in a nonlinear fashion to compress the entire $j\Omega$ axis into one cycle in the z-plane. These compromises aren't without consequences, which we'll need to understand in order to avoid distorting the final filter performance.

---

3. Although we continue to give rather short shrift to the time-domain characteristics of filters—e.g., overshoot, settling time, etc.—there are many cases where these characteristics are the primary design goals. This subject is discussed in several of the books we list as references and is also a major topic in the field of control systems.

Contrary to the FIR case, most IIR filter design is done via indirect routes, rather than using direct-design methods. In part, this stems from the relative lack of concern we have about phase when dealing with IIR filters. (If we need linear-phase response, our choice *must* be FIR.) Because analog filters are generally optimum for a particular characteristic, and since we can compensate for the distortions introduced by the *s*-to-*z* mappings, the indirect methods are satisfactory for many designs that stress the magnitude response. It is the rarer cases that stress both magnitude and phase response, or perhaps an unusual magnitude response, that may benefit from the direct-design optimization algorithms. We'll get into this in a moment, after a word about implementation.

6.2.1 Implementation

The IIR filter's great strength is also its biggest weakness. Feedback makes IIR filters much more susceptible to the effects of finite word lengths, both for the flow of data (digital signals) and the representation of coefficients (which are, in turn, related to the poles and zeros). A number of mathematically equivalent computational structures are possible for IIR filters, but once finite word length effects are considered, certain structures become more practically favorable. (Compare with FIR filters, which are usually implemented in a straightforward way.) In addition, some design methods naturally lead to certain structures.

In some cases we can postpone the choice of structure until the end of the design process. However, the most desirable structures share a requirement that we know the poles of the system to be implemented. This is hardly an issue with most mathematical or filter design programs—if we have a polynomial (the denominator of the system function, for example), it is a simple matter to extract the roots, which are the poles of the system. But if the design is done "manually," there are plenty of places to make a mistake, and it makes sense to design in terms of poles and zeros (i.e., the factored form of the system function). (Contrast this with the Parks-McClellan program, where the zeros are not even provided!) We'll have more to say on this shortly.

6.3 Direct Design Methods

6.3.1 *Ad Hoc*— "Manual" Placement of Poles and Zeros

For simple filters, it is often possible to place poles and zeros in the *z*-plane directly. Although this approach is not useful for filters with complicated specifications, a relatively simple filter is sometimes required. The following observations apply:

- As one moves around the unit circle in the *z*-plane (through the Nyquist frequency range) to evaluate the magnitude or phase response of a DT system, the magnitude response is increased in the vicinity of a pole and decreased in the vicinity of a zero. The proximity to the unit circle of a pole or zero affects the magnitude response. The closer a pole or zero is to the unit circle, the more effect it has.

Direct Design Methods

- Zeros may appear anywhere in the z-plane; however, poles outside the unit circle are associated with unstable systems and must be avoided. Poles *on* the unit circle are associated with marginally stable systems and will cause oscillations.
- The frequency associated with a pole or zero in the z-plane is indicated by the angular position of the pole or zero. The frequency range $-\pi \leq \omega \leq \pi$ (normalized radian frequency) or $-1/2 \leq f \leq 1/2$ (normalized Hz) is represented by the angles between $-\pi$ and π radians (-180–180 degrees), moving clockwise from $-\pi$ rad.
- Poles and zeros either are real or occur in conjugate pairs. (This constraint must be met for the system functions to have purely real coefficients.)
- More than one pole may be located at the same spot in the z-plane; however, this always indicates that a better magnitude response is possible from a system with the same number of poles. Put another way, systems that are optimized for a particular magnitude response characteristic always have distinct poles.[4]
- Multiple zeros at the same location are common and often occur at $z=1$ or $z=-1$.

Even if you will not manually place poles and zeros, having a good engineering understanding of the effects of poles and zeros in the z-plane is extremely valuable.

6.3.2   Time-Domain Methods

There are several direct design methods for IIR filters that are based on matching the impulse response (i.e., time-domain characteristics) of a desired filter with an IIR filter of a particular order (say, N). Briefly, given a desired system function $H_d(z)$, we can calculate the associated impulse response. The problem then becomes how to determine the best coefficients of an Nth order IIR such that the error between the desired and actual impulse response is minimized in a particular sense (usually "least-square error"). Using a few tricks, this sticky problem becomes a set of N linear equations in N unknowns—a piece of cake for the mathematically inclined. The resulting IIR filter is, by definition, the best fit in terms of its impulse response—that is, in the time domain—but the frequency-domain response may be inadequate if N is too small. (See, e.g., Proakis and Manolakis [1992]).

6.3.3   Frequency-Domain Methods

There are also algorithms that produce IIR filters that match a frequency-domain description in some optimal sense. Much as Parks-McClellan does for FIR filters, one such algorithm uses an iterative approach to produce a filter with user-specified magnitude and phase characteristics. This iterative method is not without pitfalls—for example, it may mistake a *fairly good* answer for the *best* answer.

---

4. This leads to the observation that cascading multiple copies of the same filter won't give you the most efficient filter for a given number of poles. The cascaded system will have a better magnitude frequency response than just one filter, but you can always design a filter with the same number of poles and zeros that has a better frequency response.

Direct design methods are useful in situations in which the indirect methods don't work well. Such designs might include unusual magnitude responses or linear-phase response in certain frequency regions. However, in cases in which the phase is not of concern and the magnitude response is to be optimized in common ways,[5] indirect methods produce an equivalent or better (more efficient) filter. In addition, software support for direct design methods is not as common as that for indirect methods. For these reasons we'll concentrate on the indirect methods for the remainder of this chapter. At the end of this chapter, we've listed resources for direct design methods.

6.4 Indirect Design Methods

Indirect design methods produce very efficient IIR filters. They are appropriate when the desired filter emphasizes characteristics that have been addressed by optimal filter designs in the analog domain. These include features like the flatness of the passband or the steepness of the transition between passband and stopband. Because analog filters play a crucial role, we'll spend a few pages examining the most popular ones and provide the equations for generating their poles and zeros. (Now aren't you happy you know the s-plane?) But this is only half of what we need. The key to the indirect design methods is finding a useful mapping between the analog filters and digital filters, the s-to-z mapping.

Mapping an s-plane design to the z-plane is tricky business. Figure 6-2 shows the region of the s-plane representing stable pole locations (zeros may be located anywhere).[6] This region includes the entire left-half plane ($\sigma < 0$). In the same figure we see the diminutive area for stable poles in the z-plane—points inside the unit circle. The question is how to stuff an elephant into an airplane overhead storage compartment.[7] Although there are several approaches, the *bilinear z-transform* (BZT) is the most popular method, with a couple of runners-up. So, after our discussion of analog filters, we'll show how the various s-to-z transforms take optimized analog filters and give us the digital filters we need.

One comment before we get going. We're going to be using our old friend the lowpass filter as the basis for all of our methods. What about bandpass, bandstop, and highpass filters? Instead of having four different variations for every filter type, the usual procedure is to design a lowpass filter, then to use one of a set of transforms to convert the lowpass filter into bandpass, bandstop, and highpass filters. In fact, we need two sets of these transformations, since they depend on the domain (s or z) in which we are making the transition. Because the s-to-z transforms sometimes will limit the types of filters we can create (e.g., one type of transform is not

---

5. For example, some filters may need a passband with a magnitude response that slopes downward, or some other arbitrary magnitude response that isn't captured in the minimal set of filter parameters we've discussed so far, such as passband ripple or edge frequency.

6. Be aware we are omitting some subtleties in this discussion. All mappings involve the sampling rate inherent in a DT system; the periodicity this implies in the s-plane means we're really mapping a horizontal strip of the s-plane into the interior of the unit circle in the z-plane.

7. Or, at least under the seat of the person in front of you.

Indirect Design Methods

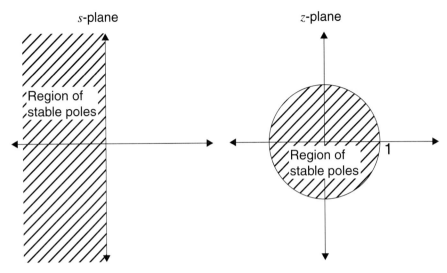

Figure 6-2 Regions of stable poles in the s-plane and the z-plane.

good for highpass filters), the domain in which we make the transformation is important. (We'll visit this point later.)

Figure 6-3 shows the big picture of IIR filter design and how indirect and direct methods are related. (We're leaving out the finer points like lowpass to highpass transforms. Within each box, we're also showing some of the different options for expressing the filter parameters—you need only one parameter per column.)

6.4.1 Analog Filter Prototypes

Table 6-1 summarizes the characteristics of the most popular analog filter *prototypes*, that is, analog filters used as models for the eventual digital filters. Pay special attention to characteristics marked *maximally*; these analog filters are the optimum filters with those characteristics. An optimization of one characteristic always implies a trade-off, and we've already drawn your attention to the abuse we heap on the phase response! But even in the magnitude response we often gladly accept a softer "knee" (sharpness of transition between passband and stopband) in exchange for less magnitude distortion (e.g., ripple) in the passband. And everyone is interested in approximating the "brick-wall" filter; this is where the Chebychev and Elliptic stand out.

A crucial parameter in analog design is the order of the filter. This directly affects the computational costs of a particular filter, and we may often reject a filter because of the order required to meet the design specifications. For example, both a Butterworth and a Chebychev filter can both be used to meet a sharp knee requirement, but the Chebychev might do this in half the order! Fortunately, there are estimates of the necessary order available for most filter types.

Because the equations for the Butterworth are a bit nicer, we'll start the discussion there. Then on to the Chebychevs, Elliptic, and Bessel. Fair warning, though—the gory details of the

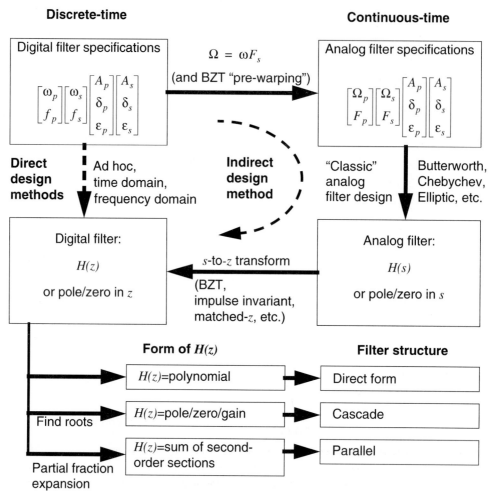

Figure 6-3 The "big picture" of IIR design methods.

Elliptic and Bessel filters are beyond what we can realistically include in this book, so for these we will have to retreat to being tour guides, at least as far as "do it yourself" filter design is concerned. (As the table shows, the Bessel will also turn out not to be as useful as we'd like.)

Oh—one more point. In reality, most folks will not be designing filters "by hand" or even by calculator! Even though we're going to give you the equations, please try to avoid doing filter design this way, if possible. Read our exciting narrative, try designing a simple filter, and get down the important characteristics of each type of filter. Then use mathematical or filter design software for your designs, especially if you can immediately generate frequency-response and/or pole/zero plots with which to check your designs. Get good tools, good insight, and use both!

Indirect Design Methods

Filter	Transition ("knee")	Passband	Stopband	Phase	Comments
Bessel	Knee? What knee?	Monotonic	Monotonic	Near-linear	s-to-z mappings distort phase. FIR usually more efficient for linear phase
Butterworth	Rounded	Maximally flat, monotonic	Monotonic	Nonlinear near cutoff	Easy to design by hand. Maple syrup is better on waffles
Chebychev I	Sharp	Ripples	Monotonic	Worse	Easy to design by hand
Chebychev II	Sharp	Monotonic	Ripples	Worse	Somewhat more complicated design than Chebychev I
Elliptic	Maximally sharp	Ripples	Ripples	Drunk fly on cross-country skis in tornado	Not viable for design by hand

Table 6-1 Characteristics of common analog filter prototypes.

6.4.1.1 Butterworth—Maximally Flat Passband

Figure 6-4 shows the lowpass Butterworth magnitude responses for some selected orders, all with F_c of 1 kHz and a passband ripple A_p of 1 dB. We've blown up the passband so you can see the behavior near the passband edge frequency. Below are the major characteristics of the lowpass Butterworth filter (you can "flip" some of these descriptions for highpass Butterworth):

- The magnitude response in the passband is *maximally flat*.
- The magnitude response overall (in both passband and stopband) is *monotonic*, which is to say the magnitude only decreases, or stays the same, as frequency increases.
- The Butterworth filter has only poles, no zeros.[8] These poles are evenly spaced along a half-circle in the left-half s-plane. (They won't stay in this shape when we map them to the z-plane, though.)

By the way, we're going to see a few circles and ellipses as we discuss analog filters. These circles are in the *s-plane*, and have nothing to do with the unit circle of the *z-plane*. The

---

8. Actually, we're lying just a bit. When we convert analog filter designs to digital, it will make sense to talk about analog filters having an equal number of poles and zeros. For filters that "have no zeros," like the Butterworth and Chebychev I, we'll say that the "missing zeros" are located at $s=\infty$. These "missing zeros" will pop back in the picture when we're in the z-domain.

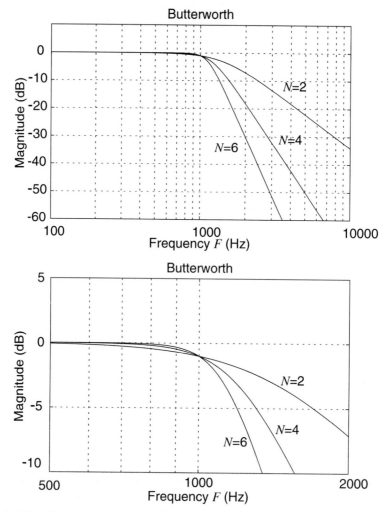

Figure 6-4 Magnitude response of the Butterworth analog filter.

unit circle in the z-plane is the dividing line between poles associated with stable and unstable systems; the circles and ellipses in the s-plane associated with analog filters are the lines along which poles are placed to produce desired magnitude responses.

Indirect Design Methods

It is customary to describe an analog filter in terms of its magnitude response function, so here it is for the Butterworth (note we're using magnitude-squared!):

$$|H(j\Omega)|^2 = \frac{1}{1 + \varepsilon_p^2 \left(\frac{\Omega}{\Omega_p}\right)^{2N}} \qquad (6.2)$$

where Ω_p is the passband edge frequency (real-world), N is the order (number of poles), and ε_p is a measure of the attenuation at Ω_p (we'll show you the exact relationship in a moment). Plot this function over a range of Ω and you have the magnitude (squared) frequency response. But this is just a description of the magnitude, and what we really want is the location of the poles in the s-plane.[9] The poles fall along a semicircle in the s-plane and are evenly spaced (about the real axis, since they're complex conjugate pairs); with a bit of math we can find the following equation for the poles:

$$s_i = \Omega_c \cos(\theta_i) + j\Omega_c \sin(\theta_i) \qquad (6.3)$$

where s_i is the ith pole, Ω_c is the -3 dB cutoff frequency of the filter, $i=1,2,...,N$, and N is the order of the filter. θ_i, the angle of the ith pole, will be useful in the Chebychev filters, as well as the Butterworth, and is defined as:

$$\theta_i = \frac{(2i + N - 1)\pi}{2N} \qquad (6.4)$$

It is clear from these equations that we need to calculate only N and Ω_c to be able to generate the locations of the poles. Figure 6-5 shows the arrangement of the poles in the s-plane for $N=6$—a semicircle of radius Ω_c with poles spaced evenly at π/N radians. Ω_c in the figure is about 1100 rad/sec, and the spacing with $N=6$ is $\pi/6$.

Determining the Order N Figure 6-6 shows the specifications for a lowpass Butterworth filter.[10] Recall from Chapter 3 (Analog Filters) that for IIR filters we take the maximum

---

9. Maybe you already see the problem—this equation: 1) didn't come from an actual $H(j\Omega)$; and 2) even if it did come from an actual $H(j\Omega)$, taking the magnitude squared means we can't just simply replace Ω with s/j to get $H(s)$.

10. Although it is common for texts to show Butterworth filter design based on Ω_c instead of the pair Ω_p and ε_p, this makes life difficult when you want to design to a particular passband attenuation other than 3 dB, which is most of the time. If you happen to be given Ω_c, you can back into Ω_p using Equation 6.5 and Equation 6.14, assuming you have an A_p in mind.

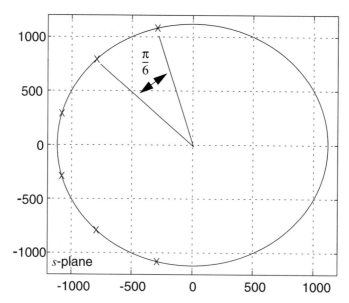

Figure 6-5 Poles of a sixth-order Butterworth. (Note that the circle shown is not the unit circle!)

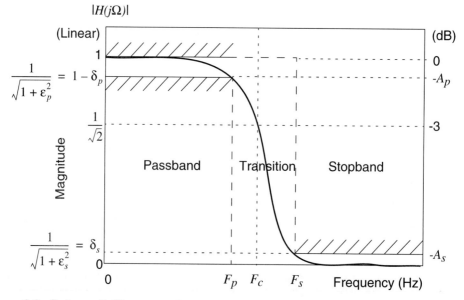

Figure 6-6 Butterworth filter parameters.

Indirect Design Methods

passband magnitude as 1 (0 dB) and measure the deviation down from there. Not that this distinction matters much for the Butterworth (it is monotonic in the passband), but for filters with ripple in the passband it's necessary.

Although it is convenient to have the -3 dB point (F_c or Ω_c) for Butterworth design purposes, other filter specifications use parameters sets such as $\{A_p, A_s, \Omega_p, \Omega_s\}$, $\{\delta_p, \delta_s, \Omega_p, \Omega_s\}$, or $\{\varepsilon_p, \varepsilon_s, \Omega_p, \Omega_s\}$. We'll usually be given A, δ, or, more rarely, ε, for the passband and stopband, along with Ω_p and Ω_s (or, more likely, F_p and F_s). In the first two cases we need to convert the passband and stopband magnitudes or deviations to ε_p and ε_s, which we can do with the following equations:

$$\varepsilon_p = \sqrt{10^{A_p/10} - 1} \tag{6.5}$$

$$\varepsilon_s = \sqrt{10^{A_s/10} - 1} \tag{6.6}$$

$$\varepsilon_p = \sqrt{\frac{1}{(1-\delta_p)^2} - 1} \tag{6.7}$$

$$\varepsilon_s = \sqrt{\frac{1}{\delta_s^2} - 1} \tag{6.8}$$

The inverse relationships are as follows:

$$A_p = 10 \cdot \log(1 + \varepsilon_p^2) \tag{6.9}$$

$$A_s = 10 \cdot \log(1 + \varepsilon_s^2) \tag{6.10}$$

$$1 - \delta_p = \frac{1}{\sqrt{1 + \varepsilon_p^2}} \tag{6.11}$$

$$\delta_s = \frac{1}{\sqrt{1+\varepsilon_s^2}} \qquad (6.12)$$

Once we have ε_p and ε_s, the formula for N is:

$$N = \frac{\log\left(\dfrac{\varepsilon_s}{\varepsilon_p}\right)}{\log\left(\dfrac{\Omega_s}{\Omega_p}\right)} \qquad (6.13)$$

N must be an integer, of course, and it is rounded *up*.[11] However, there are two considerations (that also apply to other filters). First, the fact that N is not "exact" means that the filter will meet one "end" (e.g., passband) of the design parameters exactly, but exceed (in a good way!) the parameters at the other "end" (e.g., stopband). For the Butterworth, the choice is yours—whether the "excess clearance" is in the passband (less passband deviation) or in the stopband (more stopband attenuation). It may not matter, but it might buy you a little extra room. The formula for Ω_c giving excess clearance in the stopband is:

$$\Omega_c = \frac{\Omega_p}{\varepsilon_p^{1/N}} \qquad (6.14)$$

and the formula that places the benefit in the passband is:

$$\Omega_c = \frac{\Omega_s}{\varepsilon_s^{1/N}} \qquad (6.15)$$

A second consideration concerns the filter implementation. Often, IIR filters are broken into *second-order sections* (SOS for short); we'll discuss SOSs below; they result from combining the complex-conjugate pole pairs together. It may make sense to make N even; otherwise there will be one section that is first-order and, although quicker to compute than an SOS, will nonetheless be a special case for computation. (This is mostly a space issue rather than one of code speed.) Not a killer issue; just keep it in mind.

11. Rounding down would mean the resulting filter would fail to meet one or more of the design specifications. Even a number like 5.1—or 5.01—should be rounded up, unless you are willing to fall short of the design specifications.

Indirect Design Methods

Once we have calculated the various constants, Equation 6.3 provides the pole locations $s_1, s_2, ..., s_N$, which we can use to create $H(s)$ as:

$$H(s) = \frac{G}{(s-s_1)(s-s_2)...(s-s_N)} \quad (6.16)$$

G is a gain term included so that the passband is normalized to 0 dB.

To illustrate, let's design a lowpass Butterworth filter with F_p=1000 Hz, F_s=2000 Hz, A_p=1 dB, and A_s=40 dB.

We begin by translating the filter parameters into the set $\{\Omega_p, \Omega_s, \varepsilon_p, \varepsilon_s\}$. Because $\Omega=2\pi F$, we find Ω_p=6283 rad/sec and Ω_s=12567 rad/sec. Equation 6.5 and Equation 6.6 can be used to find that ε_p and ε_s are 0.50885 and 99.95, respectively.

Next, using Equation 6.13, we find the minimum required order. The exact N is 7.62 in this case, which we'll round up to N=8.

Because we did not start out with a specified Ω_c, we can use Equation 6.14 to calculate this cutoff frequency; this version will have a little bit better stopband attenuation than we require and will meet the passband parameters exactly. We find Ω_c=6837 rad/sec.

Using Equation 6.4, we calculate the angles θ_i as {1.767, 2.160, 2.553, 2.945, 3.338, 4.123, 4.516} radians.

Because there are no (finite!) s-plane zeros in a Butterworth filter, the next step is to calculate the poles, using Equation 6.3, and the values for Ω_c and θ_i. We find the following poles: (-1333.8+6705.5j), (-3798.4+5684.6j), (-5684.6+3798.4j), (-6705.5+1333.8j), (-6705.5-1333.8j), (-5684.6-3798.4j), (-3798.4-5684.6j), and (-1333.8-6705.5j). (Note that the last four poles are complex conjugates of the first four.)

As we've hinted, we actually can stop here, as we'll be translating the poles individually to the z-domain, rather than the entire $H(s)$. However, just to see if we're on the right track, let's go ahead and find $H(s)$. Placing these poles into Equation 6.16 and multiplying everything out, we obtain the following expression:

$$H(s) = \frac{G}{a_0 s^8 + a_1 s^7 + a_2 s^6 + a_3 s^5 + a_4 s^4 + a_5 s^3 + a_6 s^2 + a_7 s^1 + a_8} \quad (6.17)$$

where the coefficients are as follows:

$a_0 = 1$

$a_1 = 3.50 \times 10^4$

$a_2 = 6.14 \times 10^8$

$a_3 = 6.98 \times 10^{12}$

$a_4 = 5.61 \times 10^{16}$

$a_5 = 3.26 \times 10^{20}$

$a_6 = 1.34 \times 10^{24}$

$a_7 = 3.58 \times 10^{27}$

$a_8 = 4.77 \times 10^{30}$

$G = 2.09 \times 10^{-31}$

Because we know the magnitude response of the Butterworth at $\Omega=0$ (thus, $s=0$) should be unity, G must be $1/a_8$.

Figure 6-7 shows the magnitude response and Figure 6-8 a plot of the poles for the filter we just designed. It looks like we met the design goals.

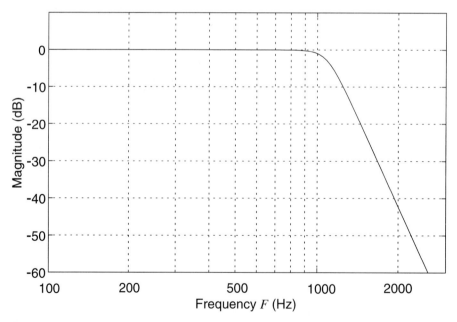

Figure 6-7 Magnitude response of Butterworth filter example.

Indirect Design Methods

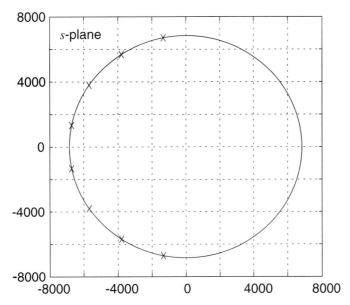

Figure 6-8 Poles of example Butterworth filter.

6.4.1.2 Chebychev I—Trading Ripple for a Sharp Transition

Figure 6-9 shows the lowpass Chebychev I magnitude responses for the same orders as the Butterworth filters in Figure 6-4. The Chebychev[12] filters, types I and II, give a sharper knee, but at the cost of ripple in the passband or the stopband, respectively. Chebychev I is a bit more popular, probably due to the fact that it contains only poles, no (finite) zeros. The Chebychev II filter puts the ripples in the stopband and provides a monotonic passband, which may be a useful feature. However, the Chebychev II requires (finite) zeros in addition to the same number of poles as the Chebychev I. We'll treat these two filters separately, though they share many features.

Chebychev I filters are characterized by:

---

12. By the way, you'll see *Chebychev* spelled a few different ways, including our favorite, Tschebyshceff.

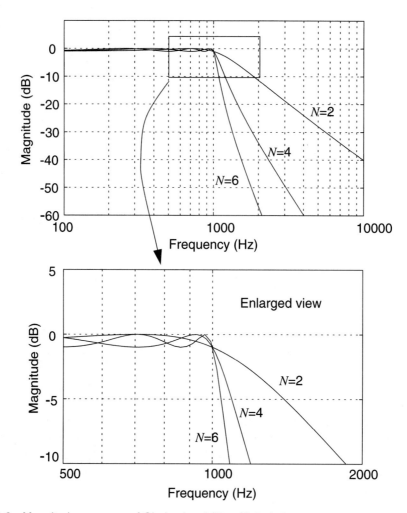

Figure 6-9 Magnitude response of Chebychev I filter, $N=2, 4, 6$.

- Ripples (equal-sized) in the passband,
- Monotonic stopband,
- Relatively sharp knee (transition between passband and stopband),
- All-pole realization ("no zeros"),
- Poles in the s-plane that lie on an ellipse. The angles of these poles are the same as those for the Butterworth filter,
- DC gain (at $\Omega=0$ or $F=0$) of either unity (for odd number of poles) or the maximum passband deviation (for even number of poles),
- Worse phase shift than Butterworth (nonlinear near passband edge frequency).

The magnitude-squared frequency response for the Chebychev I filter is

$$|H(\Omega)|^2 = \frac{1}{1 + \varepsilon_p^2 C_N^2\left(\frac{\Omega}{\Omega_p}\right)} \tag{6.18}$$

where ε_p and Ω_p are the same as for the Butterworth filter—ε_p is related to the passband deviation, and Ω_p is the passband edge frequency. $C_N(x)$ takes a bit of explaining—it's the Nth order Chebychev polynomial of the first kind. (We know you probably thought it was the *second* kind. We make the same mistake all the time.) It's actually a pretty neat function, producing equal-sized ripples between +1 and -1 for absolute values of x less than 1, then heading for the hills for values greater than 1. This gives us the equiripple effect in the passband, which is a way of minimizing the error over the passband and making it possible to take the sharp knee at the end of the passband. Here's one definition of the Nth order Chebychev polynomial of the first kind:[13]

$$C_N(x) = \begin{cases} \cos(N\cos^{-1}x) & |x| \leq 1 \\ \cosh(N\cosh^{-1}x) & x > 1 \end{cases} \tag{6.19}$$

The pole locations are given by the following equation:

$$s_i = \Omega_p \sinh(\alpha)\cos\theta_i + j\Omega_p \cosh(\alpha)\sin\theta_i \tag{6.20}$$

---

13. Just in case it isn't clear, sinh and cosh are hyperbolic sine and cosine, respectively. Most scientific calculators support such functions. While we're at it, the "-1" superscript stands for the inverse or "arc" function, not the reciprocal ($1/x$) of that function.

where α is defined as:

$$\alpha = \frac{1}{N}\sinh^{-1}\left(\frac{1}{\varepsilon_p}\right) \qquad (6.21)$$

and θ_i is, as for the Butterworth,

$$\theta_i = \frac{(2i + N - 1)\pi}{2N} \qquad (6.22)$$

These poles fall on an ellipse in the *s*-plane, with the same angles as the Butterworth filter. Figure 6-10 shows an example, with a circle drawn where the poles would fall for a Butterworth filter.

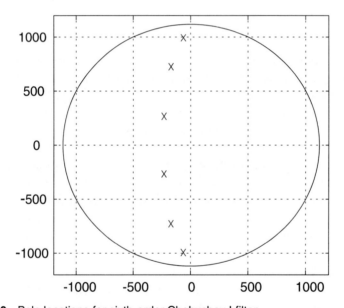

Figure 6-10 Pole locations for sixth-order Chebychev I filter.

Determining the Order N   Given ε_p, ε_s, Ω_p, and Ω_s, the minimum required order of a Chebychev I filter is:

$$N = \frac{\cosh^{-1}\left(\frac{\varepsilon_s}{\varepsilon_p}\right)}{\cosh^{-1}\left(\frac{\Omega_s}{\Omega_p}\right)} \qquad (6.23)$$

Because N must be an integer, it is rounded up. In this case, the "excess clearance" goes to the stopband (i.e., better stopband attenuation than specified).

In addition, recall that the DC gain is either at the maximum or the minimum of the passband ripple, depending on the number of poles. N odd gives unity gain (the maximum magnitude in the passband) and N even a gain of $1-\delta_p$ at DC. If the exact DC gain is important in your application, you may need to bump up N to be odd.

If you have been given design parameters A or δ instead of ε, you can still use Equation 6.5 through Equation 6.8 to obtain ε.

Let's take the same design parameters as for our earlier Butterworth and see what the Chebychev looks like: Design a lowpass Chebychev I filter with F_p=1000 Hz, F_s=2000 Hz, A_p=1 dB, and A_s=40 dB.

We first need to translate the filter parameters into the set $\{\Omega_p, \Omega_s, \varepsilon_p, \varepsilon_s\}$, just as in the Butterworth example. We can use those results, which are Ω_p=6283 rad/sec, Ω_s=12567 rad/sec, ε_p=0.50885, and ε_s=99.95.

N is found to be 4.54 by Equation 6-23, which we round up to 5. (Recall for the same parameters, the Butterworth was eighth-order!).

We next need to calculate the constant α using Equation 6.21; this is 0.2856.

Recall that the angles θ_i are the same for Butterworth and Chebychev, but because the order is different, we'll need to calculate a new set (N=5 vs. our prior N=8). The new set of θ_i is $\{1.885, 2.513, 3.142$ (i.e., π!), $3.770, 4.398\}$. Note that one of the poles falls right on the real axis—that is, at angle π. This happens whenever N is odd.

Because the Chebychev I filter has no zeros, we skip to calculating the poles using Equation 6.20. The poles are $(-562.1+6221.0j)$, $(-1471.6+3884.8j)$, $(-1818.9+0j)$, $(-1471.6-3884.8j)$, and $(-562.1-6221.0j)$.

The expression for $H(s)$ as the ratio of two polynomials[14] is:

$$H(s) = \frac{G}{a_0 s^5 + a_1 s^4 + a_2 s^3 + a_3 s^2 + a_4 s + a_5} \qquad (6.24)$$

---

14. The denominator is $(s-p_0)(s-p_1)(s-p_2)(s-p_3)(s-p_4)$, multiplied out.

where the coefficients are as follows:

$$a_0 = 1 \qquad (6.25)$$
$$a_1 = 5.89 \times 10^3 \qquad (6.26)$$
$$a_2 = 6.67 \times 10^7 \qquad (6.27)$$
$$a_3 = 2.42 \times 10^{11} \qquad (6.28)$$
$$a_4 = 9.05 \times 10^{14} \qquad (6.29)$$
$$a_5 = 1.20 \times 10^{18} \qquad (6.30)$$
$$G = 8.31 \times 10^{-19} \qquad (6.31)$$

Since the DC gain (found by evaluating $H(s)$ at $s=0$) is unity, G is $1/a_5$. If N were even, an additional factor would be necessary as well,

$$G \text{ for } N \text{ even} = \sqrt{\frac{1}{1+\varepsilon_p^2}} \qquad (6.25)$$

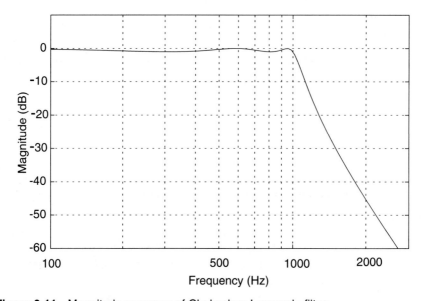

Figure 6-11 Magnitude response of Chebychev I example filter.

Indirect Design Methods

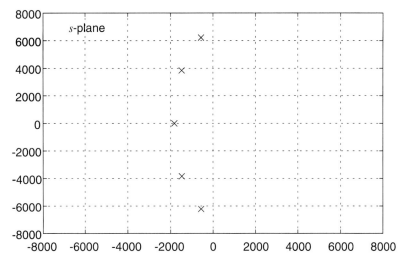

Figure 6-12 Pole locations for example Chebychev I filter.

Figure 6-11 shows the resulting magnitude response, and Figure 6-12 shows the arrangement of poles in the *s*-plane.

6.4.1.3 Chebychev II—Ripple in the Stopband

Figure 6-13 shows the lowpass Chebychev II magnitude responses for the same orders as we've plotted before. Here the ripple is in the stopband, versus in the passband for the Chebychev I filter, and the passband is monotonic. The sharpness of the knee hasn't changed, and we'll find that the order N is exactly the same as before (though we don't need to adjust its even/oddness to account for nonunity DC gain as for the Chebychev I). Chebychev II filters are characterized by:

- Ripples (equal-sized) in the stopband (i.e., "equiripple").
- Monotonic passband.
- Relatively sharp knee (transition between passband and stopband).
- Poles and (finite) *zeros*.
- Poles on an ellipse in the *s*-plane. The angles of these poles are the same as those for an equivalent-order Butterworth filter but are not at the same locations as for Chebychev I.
- Unity DC gain.
- Worse phase shift Butterworth (nonlinear near passband edge frequency).

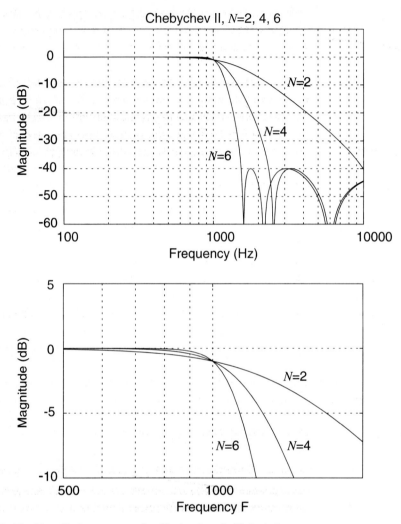

Figure 6-13 Magnitude response for Chebychev II, N=2, 4, 6.

The magnitude-squared response for the Chebychev II filter is:

$$|H(\Omega)|^2 = \frac{1}{1 + \varepsilon_p^2 \left[\dfrac{C_N^2\left(\dfrac{\Omega_s}{\Omega_p}\right)}{C_N^2\left(\dfrac{\Omega_s}{\Omega}\right)} \right]} \qquad (6.26)$$

Indirect Design Methods

The equation for the zeros is:

$$z_i = \frac{j\Omega_s}{\sin\theta_i} \qquad (6.27)$$

where $i=1, 2, \ldots N$. The poles are

$$s_i = \frac{\Omega_s}{\sinh(\alpha)\cos\theta_i + j\cosh(\alpha)\sin\theta_i} \qquad (6.28)$$

which are almost the reciprocals of the poles for an equivalent-order Chebychev I filter. We define α differently from the Chebychev I, as:

$$\alpha = \frac{1}{N}\sinh^{-1}(\varepsilon_s) \qquad (6.29)$$

though good old θ_i is, as always,

$$\theta_i = \frac{(2i + N - 1)\pi}{2N} \qquad (6.30)$$

Figure 6-14 shows an example of the arrangement of poles and zeros for a Chebychev II filter. For reference, the circle for the Butterworth poles is also shown.

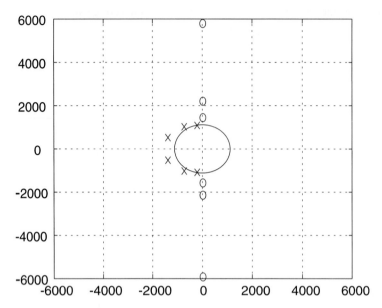

Figure 6-14 Pole and zero locations for example sixth-order Chebychev II filter.

Determining the Order N   Given ε_p, ε_s, Ω_p, and Ω_s, the minimum required order of a Chebychev II filter is the same as for a Chebychev I, namely,

$$N = \frac{\cosh^{-1}\left(\frac{\varepsilon_s}{\varepsilon_p}\right)}{\cosh^{-1}\left(\frac{\Omega_s}{\Omega_p}\right)} \qquad (6.31)$$

Equations 6.5– 6.8 are used to convert design parameters A and δ to ε if necessary.

We'll use the same design parameters as before and design a Chebychev II filter with F_p=1000 Hz, F_s=2000 Hz, A_p=1 dB, and A_s=40 dB.

You know the drill by now—first we translate the filter parameters into the set $\{\Omega_p, \Omega_s, \varepsilon_p, \varepsilon_s\}$. Because nothing has changed in this department, these remain Ω_p=6283 rad/sec, Ω_s=12567 rad/sec, ε_p=0.50885, and ε_s=99.95.

The order N is identical to that found for the Chebychev I case, 4.54, which we rounded up to 5.

The constant α is different; using Equation 6.29, we find α=1.0597.

We can reuse the angles θ_i from the Chebychev I case as N has not changed. Again, the set is {1.885, 2.513, π, 3.770, 4.398}. Now we'll see a problem with that middle, π angle.

Indirect Design Methods

Unlike the prior filter designs, the Chebychev II has finite zeros, which we calculate using Equation 6.27. At least, we'll try! Things are fine for $i=1, 2, 4$, and 5. But as you recall, θ_i for $i=3$ is π; the denominator of the zero expression is the sine of θ_i—leading to a zero at ∞! Actually, this is a reasonable result (though one that your calculator or computer program may balk at). For our purposes, we're just going to ignore that particular zero—it will all "come out in the wash" later when we convert our designs to the z-domain. The finite zeros, then, are $(0+13213j)$, $(0+21379j)$, $(0-21379j)$, and $(0-13213j)$.

The poles are found using Equation 6.28, admittedly a bit messy if you are doing the complex math "by hand." If you're in that position, about the only good thing we can say is that the middle θ_i, source of mild hassle above, has the good manners here to cause the imaginary term to drop off, leaving us with one real pole. The poles are $(-1959.3-7676.4j)$, $(-6594.8-6099.6j)$, $(-9899.4+0j)$, $(-6594.8+6099.6j)$, and $(-1959.3+7676.4j)$.

The expression for $H(s)$ as a ratio of two polynomials is:

$$H(s) = \frac{b_1 s^4 + b_3 s^2 + b_5}{a_0 s^5 + a_1 s^4 + a_2 s^3 + a_3 s^2 + a_4 s + a_5} \quad (6.32)$$

where the coefficients are as follows:

$$a_0 = 1.99 \times 10^{-20}$$
$$a_1 = 5.39 \times 10^{-16}$$
$$a_2 = 7.27 \times 10^{-12}$$
$$a_3 = 6.14 \times 10^{-8}$$
$$a_4 = 3.27 \times 10^{-4}$$
$$a_5 = 1$$
$$b_1 = 1.25 \times 10^{-17}$$
$$b_3 = 7.92 \times 10^{-9}$$
$$b_5 = 1$$

We've already divided both the coefficients of the numerator and denominator, so each is 1 when $s=0$ (i.e., at DC). For a lowpass Chebychev II, the DC gain is always 1.

Figure 6-15 shows the resulting magnitude response, and Figure 6-16 shows the arrangement of poles in the s-plane.

234 Chapter 6 • IIR Filters—Digital Filters with Feedback

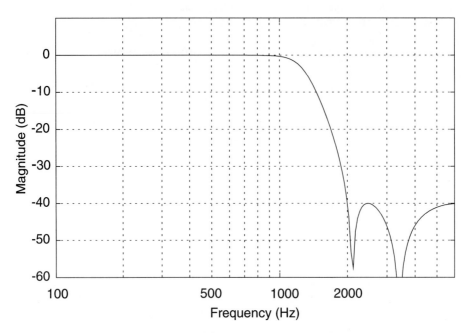

Figure 6-15 Magnitude response of Chebychev II example filter.

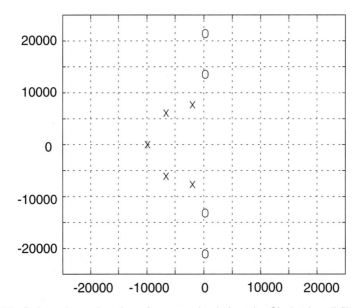

Figure 6-16 Pole and zero locations for example sixth-order Chebychev II filter.

6.4.1.4 Bessel/Thomson—Maximally Linear Phase

Butterworth gives a maximally flat passband while Chebychev I and II filters opt for a sharper transition but give ripple in the passband or stopband, respectively. But we know that the magnitude response isn't the whole story—in the frequency domain we still have phase, and then there are time-domain issues like overshoot, rise time, and so on. Bessel filters, also known as *Thomson filters*, have maximally linear phase for an IIR filter. As we said before, it isn't possible to have exactly linear phase for an IIR filter, but the Bessel filter comes as close as possible. We aren't going to go into the Bessel filter for two very good reasons:

- The wonderful near-linear phase characteristics are lost when the analog filter is converted to a digital filter.
- For most purposes, if linear phase is important, it is just as efficient to use an FIR filter designed for linear phase.

Well, why even mention the Bessel? Recall that no matter what we do, we almost always need both an antialiasing filter and a reconstruction (anti-imaging) filter as part of the A/D and D/A processes. Antialiasing and anti-imaging filters are analog, and the Bessel may be a very appropriate choice for these applications. (After all, what good is a linear FIR filter if your antialiasing filter messes up the phase before it gets to your digital filter?)

Two more Bessel facts: Bessel filters are all-pole, and the near-linear phase takes its toll on the transition width, which is even wider than for a Butterworth filter of the same order.

6.4.1.5 Elliptical/Cauer—Maximally Sharp Transition

If allowing ripples in either the passband (Chebychev I) or stopband (Chebychev II) leads to a sharper transition, why not allow ripple in both? That's what elliptical (or Cauer) filters do. Figure 6-17 shows some sample elliptical magnitude responses. Some of the characteristics of the elliptical filter are:

- Ripples in both passband and stopband.
- Both poles and zeros.
- Maximally sharp transition between passband and stopband.
- Even worse phase than Chebychev I and II.[15]

Because most of the time we'd like to minimize the transition between passband and stopband, the elliptical filter should be a popular filter. We'd love to tell you that it's also very simple to design "by hand," but it's not, and this is one case where mathematical or filter design software is about the only reasonable route. We'll give you the equations for calculating the necessary order N required, which should more than convince you this is one serious filter!

---

15. Perhaps we should say "maximally bad phase."

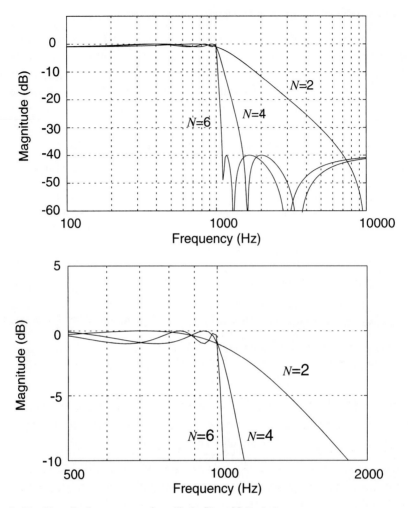

Figure 6-17 Magnitude response for elliptic filter, $N=2, 4, 6$.

Determining the Order N Begin by calculating the "sensitivity factor" k as:

$$k = \frac{\Omega_p}{\Omega_s} \qquad (6.33)$$

Next, find the "modular constant" q:

$$q = u + 2u^5 + 15u^9 + 150u^{13} \qquad (6.34)$$

Indirect Design Methods

where u is defined as:

$$u = \frac{1 - \sqrt[4]{1-k^2}}{2(1 + \sqrt[4]{1-k^2})} \quad (6.35)$$

Then you'll obviously want to calculate a "discrimination factor" D (listen, we're serious):

$$D = \frac{\varepsilon_s^2}{\varepsilon_p^2} \quad (6.36)$$

If you've gotten to this point, we suspect you have far too much time on your hands, but the minimum order is now computed using:

$$N = \frac{\log(16D)}{\log\left(\frac{1}{q}\right)} \quad (6.37)$$

Again, because N must be an integer, this will exceed the specifications a little. The benefit will be in the stopband.

We're going to cheat and use mathematical software to generate an elliptical filter meeting the earlier design specifications. Using the familiar parameters F_p=1000 Hz, F_s=2000 Hz, A_p=1 dB, and A_s=40 dB, and their equivalents Ω_p=6283 rad/sec, Ω_s=12567 rad/sec, ε_p=0.50885, and ε_s=99.95, we just plug in the numbers and go. Equation 6.33 gives a k value of 0.5, Equation 6.35 a u value of 0.01797238325873, and Equation 6.34 a q value of 0.01797238700897. (By the way, q and u vary only beyond the eighth decimal place! Normally we wouldn't bother with this many digits after the decimal point.) The "discrimination factor" D, from Equation 6.36, is 38617, and, finally, the exact N using Equation 6-37 is 3.3178. Looks like N=4 will be fine (whew!).

Excuse us for a moment while we consult our software...

Figure 6-18 shows the resulting elliptic filter magnitude response. With only four poles and four zeros, the elliptic meets the same magnitude response parameters as the Butterworth with eight poles, and the brothers Chebychev with five poles (and four finite zeros for brother number two). Figure 6-19 shows the pole-zero plot in the s-plane.

Indirect Design Methods

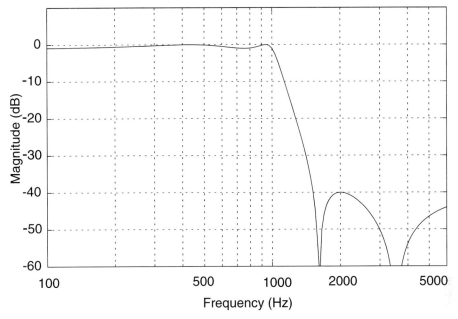

Figure 6-18 Magnitude response for example elliptic filter.

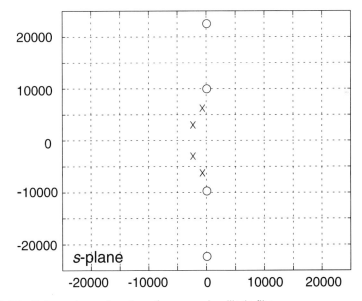

Figure 6-19 Pole and zero locations for example elliptic filter.

6.4.2 Mapping from s to z

Suppose that by using one of the schemes above we have a prototype analog filter design, ready to be turned into a design in the z-plane. Any method we choose for mapping the s-plane to the z-plane should preserve the crucial elements of the s-plane, including:

- The $j\Omega$ axis of the s-plane must map to the unit circle in the z-plane.[16]
- The left-half plane of the s-plane must be mapped to the inside of the unit circle of the z-plane (i.e., stable systems in the s-domain are mapped to stable systems in the z-domain).
- Small negative values of σ should map to locations near (but inside) the unit circle (these are locations where poles and zeros have bigger effects on frequency response).
- Large negative values of σ should map to locations near the origin of the z-plane (less effect).
- In general, small values of Ω should map to locations in the z-plane with small angles.

The relationship between the s and z domains is captured in the equation

$$z = e^{sT_s} \tag{6.38}$$

Unfortunately, when we solve for s, we end up with:

$$s = \frac{1}{T_s}\ln(z) \tag{6.39}$$

where "ln" indicates the natural logarithm (of a complex number, no less).

This might look friendly enough, but we need $H(z)$ to be a polynomial in z; substituting Equation 6.39 for s will give us an equation with logarithms instead.[17] To get around this problem, we can look up some polynomial expansions of the ln function and truncate them at a convenient point; each one will lead to a different s-to-z mapping.[18] Two expansions of $ln(z)$ and the resulting mappings are summarized in Table 6-2.

There are some other mappings, and some folks have proposed taking more than just the first terms of the polynomial expansions (especially for transformations done by software), but by far the most popular method is to use the *bilinear z-transform* (aka the BZT). All methods will introduce distortions in making the mapping; in some cases these distortions limit the use of the mapping to certain cases. The backward-difference mapping, for example, maps the lefthand

---

16. The backward-difference mapping unfortunately doesn't preserve this relationship.
17. In a few pages, we'll discuss the impulse-invariance and matched-z methods, both of which use the relationship $z = e^{sT_s}$; they are limited in the types of filters that may be transformed, however.
18. By the way, some texts provide other interpretations; for example, these mappings can be interpreted in terms of some approximations to integration.

Indirect Design Methods

Transform	Approximation to $ln(z)$	s-to-z
Backward-difference	$\ln(z) = \dfrac{z-1}{z} + \dfrac{1}{2}\left(\dfrac{z-1}{z}\right)^2 + \dfrac{1}{3}\left(\dfrac{z-1}{z}\right)^3 + \ldots$	$s = \dfrac{1}{T_s}(1 - z^{-1})$
Bilinear	$\ln(z) = 2\left[\dfrac{z-1}{z+1} + \dfrac{1}{3}\left(\dfrac{z-1}{z+1}\right)^3 + \dfrac{1}{5}\left(\dfrac{z-1}{z+1}\right)^5 + \ldots\right]$	$s = \dfrac{2}{T_s}\left(\dfrac{1 - z^{-1}}{1 + z^{-1}}\right)$

Table 6-2 Approximations to ln(z) and resulting s-to-z relations.

plane (LHP) of the *s*-plane into a circle of diameter 1 centered at *z*=1/2 (see Figure 6-20). This mapping won't help us at all for bandpass or highpass filters, though it isn't too bad for mapping lowpass filters, where the sampling rate is relatively high with respect to the poles of the system. The BZT does some pretty substantial distortion of its own, too. Figure 6-21 shows the mapping from *s* to *z* using the BZT. However, the ability of the BZT to handle more than just lowpass filters has earned it the position as the mapping of first choice.

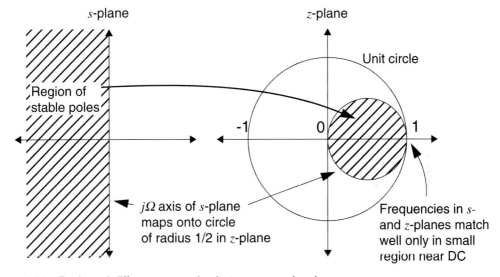

Figure 6-20 Backward-difference mapping between *s*- and *z*-planes.

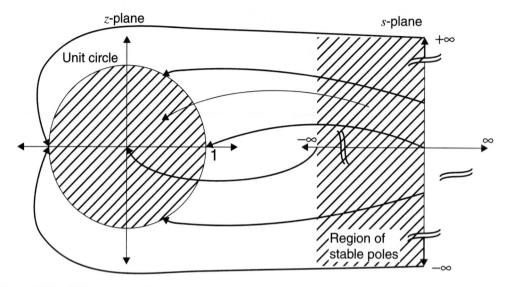

Figure 6-21 Bilinear z-transform mapping from s- to z-planes.

Let's examine the mapping of frequencies in particular. Figure 6-22 shows the relationship between frequencies in the *s*-plane and in the *z*-plane. Low frequencies map fairly linearly to the *z*-plane, but higher frequencies are mapped in an increasingly nonlinear manner, such that even infinitely high frequencies are mapped onto the unit circle. Although we might be able to ignore

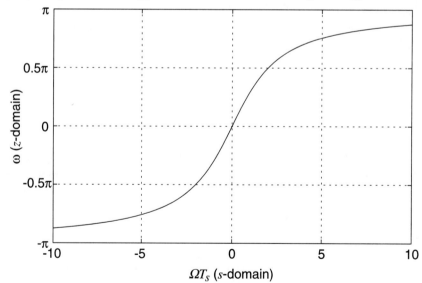

Figure 6-22 Relationship between frequencies in the *s*- and *z*-planes with the BZT.

Indirect Design Methods

this frequency distortion for lowpass filters, what we really need to do is to account for this frequency distortion ("warping") in our design procedure. *Prewarping* is this process.

6.4.2.1 Prewarping Frequencies

Because we know that the BZT will "warp" (distort) frequencies in moving from the *s*-domain to the *z*-domain, the idea behind prewarping is to compensate for this distortion by adjusting the frequencies in the *s*-domain. Some relatively simple algebra gives us the relationship between the frequency in *s* and that in *z* as:

$$\omega = 2\tan^{-1}\left(\frac{\Omega T_s}{2}\right) \quad (6.40)$$

where ω is the frequency (angle) in the *z*-domain (with a range of $-\pi$ to $+\pi$, corresponding to $+/- F_s/2$), Ω is the frequency in rad/sec in the *s*-domain, and T_s is the sampling period ($T_s = 1/F_s$). The inverse relationship is:

$$\Omega = \frac{2}{T}\tan\left(\frac{\omega}{2}\right) \quad (6.41)$$

The indirect design process using the BZT is, then:

1. Determine the sampling rate F_s and calculate T_s.
2. Determine the critical frequencies in ω for the digital filter (e.g., ω_s, ω_p).
3. Prewarp each critical frequency using the relationship in Equation 6.41.
4. Design an analog prototype filter meeting the specifications at these "prewarped" frequencies.
5. Convert the analog prototype filter to a digital filter using the BZT, which will warp the frequencies back to the original ω_s, ω_p, etc.

All *s*-to-*z* mappings involve compromises. The distortion of frequency introduced by the BZT is a drawback, but incorporating prewarping in the design process lets us compensate for this effect. The end result is a complete indirect design method that leverages analog filter design knowledge to help us design digital IIR filters.

6.4.2.2 The BZT

The BZT equation is (from Table 6-2):

$$s = \frac{2}{T}\left(\frac{z-1}{z+1}\right) \quad \text{or} \quad s = \frac{2}{T}\left(\frac{1-z^{-1}}{1+z^{-1}}\right) \quad (6.42)$$

and the inverse relationship (useful for mapping individual poles and zeros) is

$$z = \frac{2 + Ts}{2 - Ts} \qquad (6.43)$$

We can apply the BZT in two ways. First, we can use Equation 6-42 to replace each occurrence of s in some $H(s)$ with an expression in z. A second method is to apply Equation 6.43 to translate each pole and zero individually from the s-domain to the z-domain. The methods are mathematically equivalent, but in practice it is preferable to transform poles and zeros individually, at least for "pencil-and-paper" designs, for two very good reasons.

First, if $H(s)$ is more than a second- or third-order equation, the algebra becomes quite tedious using Equation 6.42. (Even using symbolic math software it can be troublesome, if for no other reason than the sheer length of the intermediate expressions!). Of course, the other route leaves us with a collection of discrete poles and zeros, which we would have to multiply out to give us $H(z)$ as a polynomial—not a pretty task, either.

A second reason is raised by asking, Do we really want $H(z)$ as a monolithic polynomial? As we've mentioned, this decision rests on some implementation issues—specifically, the fact that IIR filters are sensitive to *finite-word-length* (FWL) errors. By approximating filter coefficients with a finite number of bits, we unintentionally affect the roots of the numerator and denominator of the system function—that is, the zeros and poles. The FIR filter, consisting only of zeros, degrades somewhat in performance with this movement of zeros, but is always stable, as the zeros of stable systems may be located anywhere in the z-plane. With an IIR filter, poles near the unit circle might end up outside, leading to an unstable system. IIR filters are also susceptible to other problems stemming from the feedback they require and the use of FWL arithmetic.

We can combat FWL errors by increasing the number of bits used to represent coefficients and data, preserving maximum precision in the arithmetic used, and grouping math operations in ways that minimize the errors. One way of grouping operations is to break up $H(z)$ into second-order sections (SOSs). These SOSs $H_1(z)$, $H_2(z)$, etc., (and possibly a single *first-order section* $H_0(z)$) are cascaded to form the final $H(z)$ as follows:

$$H(z) = H_0(z) \cdot H_1(z) \cdot H(z)_2 \cdot H_3(z) \cdot \ldots \qquad (6.44)$$

where the SOSs are in the form:

$$H_i(z) = \frac{b_{i0} + b_{i1}z^{-1} + b_{i2}z^{-2}}{1 + a_{i1}z^{-1} + a_{i2}z^{-2}} \qquad (6.45)$$

Indirect Design Methods

Each SOS contains two poles and, for our purposes, two zeros. Systems that contain an odd number of poles will also have one (and only one) first-order section.

The SOSs will have real coefficients only if we combine the poles in complex-conjugate pairs. We also may adjust the gain of each SOS for unity DC gain (for a lowpass filter, at least!). SOSs are still affected by FWL errors but to a lesser extent than when $H(z)$ is expressed as the ratio of two polynomials.

These two separate issues—the algebraic tedium of transforming $H(s)$ in polynomial form to $H(z)$ using the BZT and our desire to implement the final filter in terms of SOSs—speak in favor of performing the BZT on individual poles and zeros. Such a choice may be transparent to you or of little consequence if your software allows you to easily move between representations (e.g., polynomial vs. pole/zero). We're going to stick to the easier route, however.

To illustrate the BZT, let's design a digital filter using a Butterworth lowpass filter with the same parameters as we've been using—that is, $F_p=1000$ Hz, $F_s=2000$ Hz, $A_p=1$ dB, and $A_s=40$ dB. We need to know the sampling rate; we'll assume 10 kHz.

Our first step is to prewarp the frequencies. Using $\omega=2\pi F/F_s$ (another way of saying $\omega=\Omega T$), the "digital" frequencies are $\omega_p=2\pi 1000/10000=0.6283$ and $\omega_s=2\pi 2000/10000=1.2566$. Using Equation 6.56, we prewarp these to find the "warped" Ω_p, Ω_s. $\Omega_p=6498.4$ (vs. 6283.2 unwarped) and $\Omega_s=14530.9$ (vs. 12566 unwarped).

Now we can follow the same procedure used for the analog Butterworth filter to generate the poles of $H(s)$. ε_p and ε_s remain 0.50885 and 99.95, respectively. With the new warped Ω_p and Ω_s, N is now 6.56, rounded up to 7 (vs. 8 before). We can now calculate Ω_c, which is 7156.9. By the way, we calculated N and Ω_c based on the warped frequencies instead of using the original, nonwarped frequencies as you might be tempted to do. We'll always do all of the filter design steps using warped frequencies, including finding the order N and any other constants. Recall we found $N=8$ in our earlier Butterworth example using the same unwarped frequencies—one more order than in this case.

As usual, we calculate θ_i, then the positions of the poles for $i=1$ to 7. These are listed in the first column of Table 6-3.

i	Poles in s-plane	Poles in z-plane
1	-1592.5+6977.4j	0.67734+0.54202j
2	-4462.2+5595.5j	0.55387+0.35543j
3	-6448.1+3105.2j	0.49183+0.17515j
4	-7156.9+0j	0.47392+0j
5	-6448.1-3105.2j	0.49183-0.17515j
6	-4462.2-5595.5j	0.55387-0.35543j
7	-1592.5-6977.4j	0.67734-0.54202j

Table 6-3 Poles in the s- and z-plane for an example filter using the BZT.

Next we apply the BZT to each pole, using Equation 6.43. For example, pole $i=1$ in the s-plane is translated to the z-plane as follows:

$$s_{p_1} = -1592.5 + 6977.4j \qquad (6.46)$$

$$z_{p_1} = \frac{2 + T_s s_{p_1}}{2 - T_s s_{p_1}} = \frac{2 + (0.0001)(-1592.5 + 6977.4j)}{2 - (0.0001)(-1592.5 + 6977.4j)} = 0.67735 + 0.54202j \qquad (6.47)$$

We're now ready to construct the SOSs. We'll create SOSs using the poles in the order given above.[19] The "first" SOS, therefore, consists of:

$$H_1(z) = \frac{(1 - z_1 z^{-1})(1 - z_1^* z^{-1})}{(1 - p_1 z^{-1})(1 - p_1^* z^{-1})} \qquad (6.48)$$

where the asterisks indicate the complex conjugates. Of course, the complex conjugate is one of the remaining poles ($i=7$); this SOS has pole $i=1$ and $i=7$, the next SOS will have $i=2$ and $i=6$, and the third $i=3$ and $i=5$. The remaining pole, $i=4$, will form the (only) first-order section. (Once we multiply these terms, we'll have the SOS form given earlier in Equation 6.45.)

19. To minimize FWL errors, we might want to arrange the order of the SOSs somewhat differently in an actual filter—for example, starting with $H_3(z)$, then $H_1(z)$, etc. We'll discuss this later.

Indirect Design Methods

Before we plug the poles into the SOS form, we need to talk about the zeros for this SOS. We noted that the analog Butterworth filter has no finite zeros, and it really makes no difference[20] if we include N zeros at infinity in its system function. (These zeros would lead to terms like $(1-s/\infty)$, which is 1 for any reasonable value of s.[21]) However, you'll recall that the BZT maps the entire $j\Omega$ axis to the unit circle—including ∞! To make things come out right, any time we map a filter from the s-plane to z-plane we must include these zeros at infinity in $H(s)$. From the relationship in Figure 6-21, this means adding zeros at $z=-1$. How many? The total number of zeros and poles in $H(z)$ must be equal. For filters like the Elliptic or Chebychev II that already have finite zeros, you may not need to add any zeros (though you'll recall our example Chebychev II had a zero at infinity); for the Butterworth and Chebychev I, however, we'll need to add N zeros—all at $z=-1$.

We now have everything we need to fill in Equation 6.48. The first SOS is:

$$H_1(z) = \frac{G_1(1-(-1)z^{-1})(1-(-1)z^{-1})}{(1-(0.67735+0.54202j)z^{-1})(1-(0.67735-0.54202j)z^{-1})} \quad (6.49)$$

which, after multiplying out, is:

$$H_1(z) = \frac{G_1(1+2z^{-1}+z^{-2})}{1-1.3547z^{-1}+0.75259z^{-2}} \quad (6.50)$$

So that the gain is unity at DC (i.e., at $z=1$, *not* $z=0$!)[22], substitute 1 for every occurrence of z-anything and take the reciprocal of $H_1(1)$. Note that for a Chebychev I with N even, the gain should be adjusted in one of the SOSs to be $\sqrt{1/(1+\varepsilon_p^2)}$ (the other SOSs should be set to 1). G_1, in our example, is, therefore, 0.09955, which can distributed across the numerator coefficients, like so:

$$H_1(z) = \frac{0.09955 + 0.1991z^{-1} + 0.09955z^{-2}}{1 - 1.3547z^{-1} + 0.7526z^{-2}} \quad (6.51)$$

---

20. At least none that we'll worry about here.
21. Yuck. Jack hates this notation, as ∞ is not a number. But, you get the general idea, we hope.
22. Recall that although $s=\sigma+j\Omega$, z is defined as $z=re^{j\omega}$. For the case $r=1$ (steady state), when $\omega=0$, $z=e^0$ or 1.

In a similar fashion, $H_2(z)$ and $H_3(z)$ are found to be:

$$H_2(z) = \frac{0.08134 + 0.1627z^{-1} + 0.08134z^{-2}}{1 - 1.1077z^{-1} + 0.4331z^{-2}} \quad (6.52)$$

and

$$H_3(z) = \frac{0.07222 + 0.1445z^{-1} + 0.07222z^{-2}}{1 - 0.9837z^{-1} + 0.2726z^{-2}} \quad (6.53)$$

The first-order section, also adjusted for a unity DC gain, is:

$$H_0(z) = \frac{0.2635 + 0.2635z^{-1}}{1 - 0.4729z^{-1}} \quad (6.54)$$

The final expression for $H(z)$, in terms of SOSs, is:

$$H(z) = \frac{0.2635 + 0.2635z^{-1}}{1 - 0.4729z^{-1}} \cdot \frac{0.09955 + 0.1991z^{-1} + 0.09955z^{-2}}{1 - 1.3547z^{-1} + 0.7526z^{-2}} \cdot \\ \frac{0.08134 + 0.1627z^{-1} + 0.08134z^{-2}}{1 - 1.1077z^{-1} + 0.4331z^{-2}} \cdot \frac{0.07222 + 0.1445z^{-1} + 0.07222z^{-2}}{1 - 0.9837z^{-1} + 0.2726z^{-2}} \quad (6.55)$$

Figure 6-23 shows the resulting frequency response of this digital filter.

This cascade of sections can be realized in a series of difference equations as follows. Making sure that the constant term in all denominators is 1 (as it will be if you follow the steps we've outlined), each SOS can be written as a stand-alone difference equation (recall Chapter 4 [Discrete-Time Signals and Systems]), whose input $x(n)$ is the output $y(n)$ of the previous stage. (Again, as a practical matter we may wish to rearrange the order of these SOSs to minimize FWL errors, a matter we discuss later.) Using Equation 6.55:

$y_0(n)=0.2635x_0(n)+0.2635x_0(n-1)+0.4729y_0(n-1)$

$y_1(n)=0.09955x_1(n)+0.1991x_1(n-1)+0.09955x_1(n-2)+1.3547y_1(n-1)-0.7526y_1(n-2)$

$y_2(n)=0.08134x_2(n)+0.1627x_2(n-1)+0.08134x_2(n-2)+1.1077y_2(n-1)-0.4331y_2(n-2)$

$y_3(n)=0.07222x_3(n)+0.1445x_3(n-1)+0.07222x_3(n-2)+0.9837y_3(n-1)-0.2726y_3(n-2)$

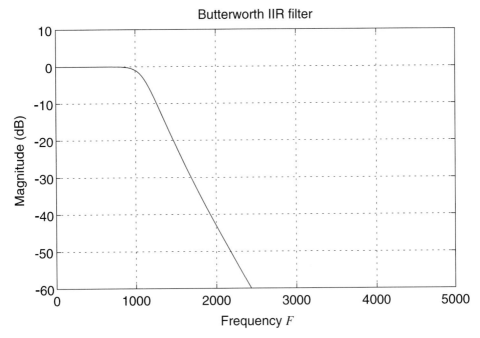

Figure 6-23 Magnitude response of example digital Butterworth filter.

where $x(n)$ is the input to the entire system, $x_0(n)=x(n)$, $x_1(n)=y_0(n)$, $x_2(n)=y_1(n)$, $x_3(n)=y_2(n)$, $y(n)=y_3(n)$, and $y(n)$ is the output of the entire system. The equations, even in SOS form, have some implementation options, so we'll return below to some more issues surrounding implementation.

The indirect method using the BZT can be quite a bit of work, as this example shows. Again, we remind you that there are substantial benefits to IIR filters, most notably, their efficiency compared to FIR filters—the effort is usually worth it! Also, Orfanidis [1996] provides equations that allow you to skip the intermediate stage of determining poles (and zeros) in either the s-plane or z-plane, instead jumping directly to expressions for the SOSs in the z-domain. These equations even include the prewarping of critical frequencies!

6.4.2.3 The Impulse-Invariance and Matched-z Transforms

The *impulse-invariance method* is limited to lowpass and bandpass filters only, and results in an $H(z)$ with an impulse response that is a sampled version of the impulse response of $H(s)$. Approximating the analog impulse response may be important, and the BZT and other transforms do not attempt to preserve the impulse response. In order to use the impulse invariance method, we require the *partial fraction expansion* of $H(s)$, a process that actually isn't too difficult but is a bit much to go into here, as we already have the BZT.

The *matched-z transform* produces an $H(z)$ with poles identical to those for the impulse-invariance method; however, the position of zeros differs between the two. Like the impulse-invariance method, the matched-z is limited to lowpass and bandpass filters, but is easier to perform, as the partial fraction expansion of $H(s)$ is not required. Any textbook will describe these two techniques in greater detail. See, e.g., Proakis and Manolakis [1996].

6.5 Highpass, Bandpass, and Bandstop Conversions

The indirect design methods discussed so far have led to lowpass filters. Although many tasks may, indeed, require lowpass filters, we still need to be able to generate bandpass, bandstop, and (sometimes) highpass filters.

Now, we could have analog filter design equations for lowpass, bandpass, bandstop, and highpass for each type of design (e.g., Butterworth, Chebychev), but it is more common to perform a lowpass to bandpass/bandstop/highpass conversion instead. Furthermore, this conversion can occur in either the s-domain or z-domain, or even as part of the BZT (see Orfanidis [1996]). Table 6-4 and Table 6-5 summarize the two methods.

To convert to	Replace each s by	Where
Highpass	$\dfrac{\Omega_{lp}\Omega_{p}}{s}$	Ω_{lp} =original passband edge frequency Ω_{p} =new passband edge frequency
Bandpass	$\Omega_{lp}\dfrac{s^{2}+\Omega_{u}\Omega_{l}}{s(\Omega_{u}+\Omega_{l})}$	Ω_{lp} =original passband edge frequency Ω_{l} =lower edge of passband Ω_{u} =upper edge of passband
Bandstop	$\Omega_{lp}\dfrac{s(\Omega_{u}-\Omega_{l})}{s^{2}+\Omega_{u}\Omega_{l}}$	Ω_{lp} =original passband edge frequency Ω_{l} =lower edge of stopband Ω_{u} =upper edge of stopband

Table 6-4 Filter conversions in the s-domain.

In which domain should you do the lowpass conversion? If you plan on using a transform other than the BZT, it's best to wait until you have the lowpass filter in the z-domain. Recall that some s-to-z mappings are very bad at mapping high frequencies into the z-domain. However, if you are using the BZT, you can do the lowpass conversion in either domain and end up with the same response.

Although transformations from lowpass to other filter types also introduce some distortion, the magnitude response at the critical frequencies is preserved, and the magnitude response elsewhere is usually reasonable.[23]

---

23. Oppenheim and Schafer [1989] go into greater detail on compensating for the frequency distortions produced by lowpass transformations.

To convert to	Replace each z^{-1} by	Where
Highpass	$\dfrac{z^{-1}+a}{1+az^{-1}}$	ω_{lp} = original passband edge frequency ω_p = new passband edge frequency $a = -\dfrac{\cos\left(\dfrac{\omega_{lp}-\omega_p}{2}\right)}{\cos\left(\dfrac{\omega_{lp}+\omega_p}{2}\right)}$
Bandpass	$-\dfrac{z^{-2}-a_1 z^{-1}+a_2}{a_2 z^{-2}-a_1 z^{-1}+1}$	ω_{lp} = original passband edge frequency ω_l = lower passband edge frequency ω_u = upper passband edge frequency $a_1 = -\dfrac{2\alpha K}{K+1}$ $a_2 = \dfrac{K-1}{K+1}$ $\alpha = \dfrac{\cos\left(\dfrac{\omega_u+\omega_l}{2}\right)}{\cos\left(\dfrac{\omega_u-\omega_l}{2}\right)}$ $K = \cot\left(\dfrac{\omega_u-\omega_l}{2}\right)\tan\dfrac{\omega_{lp}}{2}$

Table 6-5   Filter conversions in the z-domain.

To convert to	Replace each z^{-1} by	Where
Bandstop	$-\dfrac{z^{-2} - a_1 z^{-1} + a_2}{a_2 z^{-2} - a_1 z^{-1} + 1}$	ω_{lp} =original passband edge frequency ω_l =lower stopband edge frequency ω_u =upper stopband edge frequency $a_1 = -\dfrac{2\alpha}{K+1}$ $a_2 = \dfrac{1-K}{1+K}$ $\alpha = \dfrac{\cos\left(\dfrac{\omega_u + \omega_l}{2}\right)}{\cos\left(\dfrac{\omega_u - \omega_l}{2}\right)}$ $K = \tan\left(\dfrac{\omega_u - \omega_l}{2}\right)\tan\dfrac{\omega_{lp}}{2}$

Table 6-5 (Continued) Filter conversions in the z-domain.

6.6 IIR Filter Structures

For reasons already discussed, it is preferable to avoid implementing $H(z)$ as the ratio of two long polynomials. We have emphasized creating SOSs that can be cascaded or placed in series to produce a theoretically identical response as a monolithic H(z) but that exhibit better properties when implemented using FWL coefficients, data, and arithmetic. First, we'll explore implementations based on a series structure of SOSs. It is also possible to express $H(z)$ as the sum of a number of SOSs (versus the product); this leads to a parallel structure that we'll describe in a moment.

6.6.1 The Cascade Structure

Figure 6-24 shows the overall view of a cascade (or series) structure built of first- and second-order sections. (*FOS* stands for "first-order section" in the figure.) Each SOS is of the form:

$$H_i(z) = \frac{b_{i0} + b_{i1}z^{-1} + b_{i2}z^{-2}}{1 + a_{i1}z^{-1} + a_{i2}z^{-2}} \qquad (6.56)$$

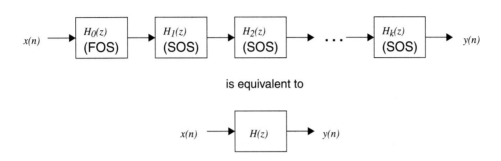

Figure 6-24 Cascade or series structure for IIR filter.

which we can diagram as in Figure 6-25. This structure is known as the *direct form I* structure. However, it is useful to see this equation as:

$$H_i(z) = \frac{b_{i0} + b_{i1}z^{-1} + b_{i2}z^{-2}}{1} \cdot \frac{1}{1 + a_{i1}z^{-1} + a_{i2}z^{-2}} \qquad (6.57)$$

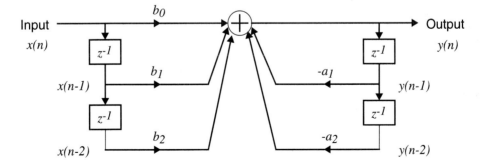

Figure 6-25 Direct form I structure for IIR filter.

or, equivalently, as:

$$H_i(z) = \frac{1}{1 + a_{i1}z^{-1} + a_{i2}z^{-2}} \cdot \frac{b_{i0} + b_{i1}z^{-1} + b_{i2}z^{-2}}{1} \qquad (6.58)$$

Figure 6-26 diagrams this expression using a new sequence variable $w(n)$ to stand for the intermediate value that is no longer $x(n)$ or $y(n)$. As the two sets of delayed values of $w(n)$ are obviously the same, they can be combined as shown in Figure 6-27, a form known as the *canonic*[24] *form* or *direct form II*.

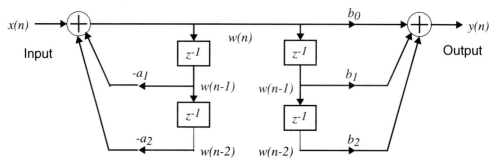

Figure 6-26 An IIR structure based on Equation 6.58.

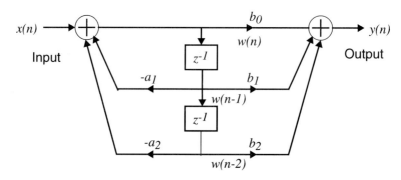

Figure 6-27 Canonic or direct form II structure for IIR filter.

---

24. *Canonic* here means "reduced to the simplest form."

IIR Filter Structures

The canonic structure uses half as much storage for delayed values as does the direct form I structure, with no increase in the number of multiplies or additions. The sections are characterized by the following difference equations:

$$w(n) = x(n) - a_1 w(n-1) - a_2 w(n-2)$$

$$y(n) = b_0 w(n) + b_1 w(n-1) + b_2 w(n-2)$$

In a program, of course, we must explicitly code the movement of the data through each delay. To the operations above, we must add:

$$w(n-2) = w(n-1)$$

$$w(n-1) = w(n)$$

In practice, the three delay locations can be renamed $w0$, $w1$, and $w2$, corresponding to $w(n)$, $w(n-1)$, and $w(n-2)$, as we have no need to reference any but the last three values. Likewise $x(n)$ and $y(n)$ can be simplified to x and y, as we are only concerned with the current values.[25] Therefore, each SOS will resemble the following:

$$w(0) = x - a_1 w(1) - a_2 w(2)$$

$$y = b_0 w(0) + b_1 w(1) + b_2 w(2)$$

$$w(2) = w(1)$$

$$w(1) = w(0)$$

Subsequent SOS stages use the y (outputs) of prior stages as their x (inputs).

We are now in a position to write out the pseudocode for the digital Butterworth lowpass filter of our example:

---

25. The ability to use x and y versus $x(n)$ and $y(n)$ assumes that you are processing samples in real-time (one sample at a time). Block processing requires array storage of multiple input and output perhaps, samples.

```
// Coefficients

b00=0.263538; b01=0.263538;                    // FOS (0)
a01=-0.472925;
b10=0.0994723; b11=0.198945; b12=0.0994723;    // SOS 1
a11=-1.35469;  a12=0.752576;
b20=0.08134;   b21=0.16268;  b22=0.08134;      // SOS 2
a21=-1.10775;  a22=0.433107;
b30=0.0722286; b31=0.144457; b32=0.0722286;    // SOS 3
a31=-0.983662; a32=0.272577;

// variables
x                  // input
y                  // output
w00, w01           // intermediate storage
w10,w11,w12
w20,w21,w22
w30,w31,w32
y0,y1,y2

while not done               // loop as long as there's input

    input x                  // get input value

    // First-order section
    w00=x-a01                // calculate intermediate value
    y0=b00*w00+b01*w01       // output
    w01=w00                  // update delayed values

    // SOS 1
    w10=y0-a11*w11-a12*w12        // calc intermediate value
    y1=b10*w10+b11*w11+b12*w13    // output
    w12=w11                       // update delayed values
    w11=w10

    // SOS 2
    w20=y1-a21*w21-a22*w22        // calc intermediate value
    y2=b20*w20+b21*w21+b22*w23    // output
    w22=w21                       // update delayed values
    w21=w20

    // SOS 3
    w30=y2-a31*w31-a32*w32        // calc intermediate value
    y=b30*w30+b31*w31+b32*w33     // output
    w32=w31                       // update delayed values
    w31=w30

    output y                 // new output value

end while
```

Listing 6-1   Pseudocode for an example digital Butterworth filter.

In the next chapter, we'll discuss strategies for implementing this type of code on a microcontroller, as it's unlikely the floating-point arithmetic we've indicated is available or that a simple translation of this pseudocode into assembly language will result in optimum performance.

6.6.1.1 Ordering and Pairing Poles and Zeros in SOSs

We've hinted before that the order in which we evaluate SOSs may be significant. Even pairing of poles and zeros to form an SOS can be done in a number of ways (for filters with distinct zeros, that is), the choice of which will have some effect on the SNR of the overall filter response. Ordering and pairing are also related to the issue of scaling, which we'll discuss in Chapter 7 (Microcontroller Implementation of Filters).

The following general rules of thumb are fairly easy to apply and are a good start for proper ordering and pairing:[26]

- Pairing: Pair the pole closest to the unit circle with the zero closest to that pole. Repeat until all poles and zeros are paired.
- Ordering: Order the SOSs according to proximity of poles to the unit circle, either in order of increasing or decreasing distance. (Each order minimizes noise in a different way.) If you try only one ordering, evaluate SOSs in order of increasing pole proximities to the unit circle.

The poles and zeros for the current example are: $(0.67734+0.54202j)$, $(0.55387+0.35543j)$, $(0.49183+0.17515j)$, $(0.47392+0j)$, $(0.49183-0.17515j)$, $(0.55387-0.35543j)$, $(0.67734-0.54202j)$. We've plotted these in Figure 6-28 to show locations with respect to the unit circle. (You could also just take the absolute value—i.e., radius—and compare to 1). Based on the rules we've given, we should evaluate the SOSs in the reverse order to which they were calculated—i.e., the SOS we found first, $i=1$, is closest to the unit circle and should be evaluated last. The FOS, the farthest pole from the unit circle, is fine as the first expression evaluated. Because all seven zeros are identical, you are free to match them with whatever poles you like.[27] If we had, instead, used a Chebychev II filter, we might find pairing a finite (s-plane) zero with a nearby pole would increase the filter's SNR somewhat. In practice, the large number of possibilities for pairing and ordering is addressed by optimization software, as it is a nontrivial problem.

6.6.2 The Parallel Structure

The parallel filter structure shown in Figure 6-29 is another arrangement of SOSs. However, these are not the same SOSs as we found for the cascade structure! The poles will be paired into

---

26. Due to Jackson [1986], and summarized in Oppenheim and Schafer [1989].

27. A little mathematical humor, if you will. By the way, since the SOSs are often implemented such that the numerator is evaluated before the denominator, it actually makes sense to shift the denominators down the line by one stage, leaving the zeros where they were. For example, if the SOSs were arranged in a cascade of $N1/D1$ * $N2/D2$ * $N3/D3$, shift the denominators one down to get $N1/D3$ * $N2/D1$ * $N3/D2$. The idea is to use the zeros to reduce the input to the poles and so reduce the dynamic range of the output.

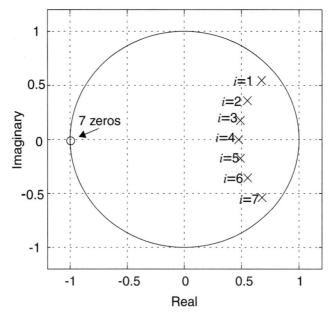

Figure 6-28 Pole-zero plot for digital Butterworth filter example.

sections in the same way (i.e., complex-conjugate pairs, so we have real coefficients), but the zeros in each SOS will be different. This structure is a realization of the following equation:

$$H(z) = H_0(z) + H_1(z) + H_2(z) + \ldots + H_k(z) \tag{6.59}$$

It is necessary to take the partial fraction expansion of $H(z)$ to generate these SOSs (an operation we mentioned in discussing the impulse invariance transform)—not a tricky operation, but an additional step. The advantages of the parallel structure are minimal for coefficient lengths greater than 12 bits (Ifeachor and Jervis [1993]), though, of course, it's unnecessary to worry about the order of evaluation. For our purposes, the cascade structure is a reasonable choice.

6.7 Summary

The primary goal in this chapter has been to determine the filter coefficients for IIR filters. The presence of feedback means the techniques developed for FIR filters do not apply, and a new set of techniques is necessary. We group these into two categories—direct and indirect.

Direct IIR filter design methods include *ad-hoc* ("manual" placement of poles and zeros), and time- and frequency-domain optimization techniques. Placing poles and zeros "by hand" is reasonable for simple systems but difficult for systems that require more than a few poles or

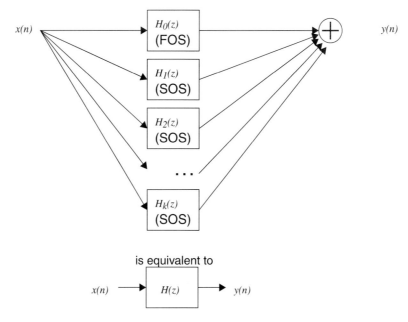

Figure 6-29 Parallel structure for IIR filters.

zeros. Time- and frequency-domain techniques rely on optimization algorithms, and can generate filter coefficients with specialized time or frequency responses within reasonable constraints. However, for standard filter magnitude responses, indirect methods are superior.

Indirect filter design methods involve the following steps: specification of filter parameters in terms of the response of the DT system; translation of the critical frequencies to "real-world" frequencies; design of an optimal analog filter to meet the filter criteria; translation of the analog filter to the DT domain.

The available transformations for going from the CT to DT domains produce varying degrees of distortion and limitations on the types of filters that can be transformed. The most popular transformation, the BZT, produces a "warping" of frequencies that can be precompensated by prewarping the critical analog frequencies.

Finite-word-length errors are reduced by breaking the system function $H(z)$ into either a product or a sum of smaller SOSs (second-order sections). When $H(z)$ is expressed as the product of a series of these sections, the resulting structure is known as a *cascade*; the structure resulting from a summation is a *parallel structure*. Both structures are more easily derived if the individual poles and zeros are preserved in the design process.

The pairing of poles and zeros and the ordering of SOSs in a cascade structure affects the performance of the filter. SOSs with poles near the unit circle should be evaluated last; when possible, zeros should be paired with nearby poles to minimize the maximum magnitude response. In general, optimizing the pairing and ordering of SOSs is a complex task.

Resources

Ifeachor, Emmanuel and Jervis, Barrie, *Digital Signal Processing: A Practical Approach*, New York, Addison-Wesley, 1993.

Jackson, L. B., *Digital Filters and Signal Processing*, Norwell, MA, Kluwer Academic Publishers, 1986.

Orfanidis, Sophocles, *Introduction to Signal Processing*, New Jersey, Prentice Hall, 1996.

Proakis, John G. and Manolakis, Dimitris, *Digital Signal Processing: Principles, Algorithms, and Applications, 3rd Ed.*, Englewood Cliffs, Prentice Hall, 1996.

Rorabaugh, C. Britton, *Digital Filter Designer's Handbook (Featuring C Routines)*, New York, McGraw-Hill, 1993.

Thede, Leslie D., *Analog and Digital Filter Design Using C*, New Jersey, Prentice Hall, 1996.

Software:

A number of programs for IIR filter design are available in source code on the World-Wide-Web. In addition, any filter design programs for analog filters than can produce $H(s)$ can be used if you're willing to then convert to $H(z)$.

The MATLAB Signal Processing Toolbox is an especially nice piece of software, but you can use it without much understanding of what you are doing.[28]

---

28. Be forewarned that many of the toolbox functions that are useful in signal processing seem to have been written by different people. Hence, normalized frequency for one function might use a normalized frequency range of -1 to 1, while another uses $-\pi$ to π. Still another may use -1/2 to 1/2, or even 0 to 2π. This can cause you a lot of grief if you aren't careful with this detail.

CHAPTER 7

Microcontroller Implementation of Filters

This chapter builds on the results of the two filter design chapters, Chapter 5 (FIR Filters—Digital Filters Without Feedback) and Chapter 6 (IIR Filters—Digital Filters With Feedback). Fortunately, there's a lot less math involving complex numbers and such. You should be familiar with binary numbers, two's complement, etc. and be on speaking terms with at least one microprocessor architecture because we need to talk about registers, instruction timing, and so on.

7.1 Overview

In previous chapters, we've seen how a filter specification eventually is turned into a difference equation, a fairly straightforward description of the simple mathematics of a digital filter. Translating the difference equation into C or assembly language does not appear very difficult at first—at most, we're talking multiplies and additions, with a little bit of indexing or data movement mixed in. There are two problems, however, that we face when implementing DSP on microcontrollers. First, we're limited to using fixed-point arithmetic and data. "Multiply and accumulate" (the core of an FIR filter) is really "multiply, accumulate, and see if there were any problems." Second, we're often against the wall on both processor bandwidth and memory space.

In this chapter, we're going to delve into microprocessor architecture, software optimization, and fixed-point arithmetic to see how we can best implement digital filters on a range of architectures. Depending on your application, you may not need to worry about some of these issues, especially if you're able to use a "DSP-friendly" processor. At the other extreme, "DSP-hostile" processors might be all you can use, and a few of the techniques we point out may be necessary to fit your time or space constraints.

Extracting optimum performance is also good fun—maybe too much! There are compromises among performance, robustness, maintainability, portability, and development time that should make us think carefully about how much time and effort we spend optimizing. As we go along we'll try to share our view of how a balance among these performance factors can be struck.

7.2 Architecture Issues

We'll begin by considering the underlying architectures of different types of programmable, single-chip digital processors.[1] Some of these architectures have been designed "from the ground up" to support DSP algorithms, while at the other extreme general-purpose processors may provide no explicit support for DSP operations and, thus, prove far less efficient for DSP applications.

Table 7-1 lists some of the attributes of five general classes of single-chip, programmable digital processors, ranging from dedicated DSP chips to generic microcontrollers. We place the 68HC16, with its explicit support for DSP, in a middle category, but the ordering reflects only the DSP-specific nature of the processor, not its appropriateness for a given application, as each application places different demands on the processor.

The overall performance of a particular digital processor in an application is highly dependent on the total mix of operations required. For many embedded systems, interactions with other hardware plays just as important a role as do the DSP operations. Microcontrollers, with their efficient bit-level control of lines, excel at this type of control, although DSP-specific chips may be less efficient. In one case, a benchmark of the 68HC16Z1 showed it outperforming a DSP-specific chip[2] running at over twice the clock frequency for a particular application that required both DSP and digital control. Of course, there are other applications where a DSP-specific chip would be far more appropriate. Just keep your mind open—and your library full of databooks.

Below we briefly discuss the highlights of each architectural class.

---

1. That is, arithmetic, register, and addressing functions are all performed on the same chip; in many cases the complete system will require external memory, clocks, etc. This is in contrast to the multi-chip designs of not too many years ago, where one chip might generate addresses, one chip (or more, a "bit-slice" approach) perform math operations, and so on.

2. Oh no we're not. Let's just say it was a popular DSP chip, and let it go at that.

Architecture Issues

	DSP	DSP/ Microcontroller combination	DSP w/ microcontroller extensions	Microcontroller w/DSP extensions	Microcontroller
Raw DSP bandwidth	excellent	excellent	excellent	good	poor
Address space	small to medium	small to medium/ medium	small to medium	medium	small to medium
Cost	medium to high	high	medium	low to medium	low to medium
MAC	yes	yes	yes	yes	no
Fast shifter	yes	yes	yes	no	no
Architecture	Harvard/mod-ified Harvard	Harvard & Von Neumann	Harvard/mod-ified Harvard	Von Neumann	Von Neumann
Memory busses	2–3	2–3 for DSP, 1 for micro-controller	2–3	1	1
Circular addressing	yes	yes	yes	yes	no
Saturation/ Overflow	yes	yes	yes	yes	?
Zero-over-head looping	yes	yes	yes	yes	no
Stack	hardware	hardware and memory	hardware (and memory)	memory	memory
FFT address-ing	yes	yes	yes	?	no
Digital I/O	minimal	medium	medium	excellent	excellent

Table 7-1 Single-chip programmable processors for DSP.

	DSP	DSP/ Microcontroller combination	DSP w/ microcontroller extensions	Microcontroller w/DSP extensions	Microcontroller
Peripherals (timers, serial, ADC)	excellent synchronous serial, good timer, some ADC	excellent synchronous serial, good timer	excellent synchronous serial, good timer, some ADC	good to excellent timer, good ADC, minimal to excellent serial (synchronous and asynchronous)	good to excellent timer, good ADC, minimal to excellent serial (synchronous and asynchronous)
Examples	MC56002	MC68356		MC68HC16	MC68HC11

Table 7-1 (Continued) Single-chip programmable processors for DSP.

7.2.1 Single-Chip Programmable Digital Processors

7.2.1.1 Dedicated Digital Signal Processors

DSP chips have an architecture optimized to support DSP operations such as FIR filter evaluation, IIR filters, and FFT routines. In fact, it's relatively common to find no general-purpose registers at all in an architecture—each register is specific to a functional unit (e.g., the multiplier or shifter).

DSP chips "never" use fewer than 16 bits for data and usually have fixed-length, generally 16-bit or longer, instruction words. In many respects, they are architecturally similar to RISC processors, such as in having fixed-length instructions, single-cycle execution of most instructions, etc. You will always find an optimized multiply-accumulate unit, almost always supplemented with a very fast shifter (a so-called barrel shifter). The multiply-accumulate unit has a much wider accumulator than does the basic data path, allowing the inner loop of filters to perform many accumulations without overflowing. The ability to detect overflows is also a necessity, as is the ability to "saturate" (we discuss this later).

While externally, some inexpensive DSP chips may resort to a single bus, internally, a *Harvard* architecture is practically a necessity. Harvard architecture means that the program and data spaces are not only differentiable (as with many processors including the HC16), but are accessed using separate address and data busses. This allows the concurrent fetching of program and data words, though with twice as many addresses and data lines! The Harvard architecture is extended by several DSP chips to more than two busses, allowing fetches from two data spaces and from program space at the same time. (The unified memory space of most microprocessors is called a *Von Neumann* architecture.)

Many DSP chips have a small address space, compared with microprocessors, due in part to the limits a fixed-length program word imposes. The trade-off is between the ability to include an absolute address in an instruction and the desire to keep the instruction word a reasonable length (because each instruction is the same length). Although DSP-specific algorithms are usually quite short (mostly repeating loops), most embedded systems perform other tasks involving large amounts of code and data that can exceed the DSP's memory space. For example, the voice synthesizer Dale designed used a separate microcontroller to run all of the high-level text processing, reserving the DSP chip for the math-intensive DSP chores. (Even so, he was far too close for comfort to both the program and data limits on the DSP chip!)

A major distinction between different DSP chip families is whether they can directly execute floating-point arithmetic. Although we may someday see floating point on nearly every processor (DSP or not), at the moment, floating-point arithmetic is costly—just the logic for handling floating point uses a lot of real estate on a chip. Further, most floating point is done on 32-bit or wider data, and the wider data paths are also more expensive. Most of the references in this book are to the lower-cost, fixed-point DSP chips. In embedded systems, we're usually concerned with cost, but floating-point DSP chips can automatically take care of some of the arithmetic issues that complicate our lives in fixed-point chips and are generally more powerful as well.

DSP chips generally provide fewer addressing modes than do general-purpose processors, but most provide a special addressing mode (bit-reversal) that is useful in performing the FFT. Filters are usually supported by an addressing mode specific to *circular buffers* (discussed later in this chapter). Because so many DSP operations employ tightly coded inner loops, *zero-overhead* looping is common. Thus, instead of taking three instructions to decrement, test a counter, and branch on a nonzero result, the processor can perform these looping operations in the background using special registers.

Many DSP chips use a *hardware stack* instead of data memory for stacks. Usually, a hardware stack is fairly limited, which means you may be seriously limited in terms of nesting subroutine calls. However, hardware stacks are fast and don't clutter the data memory bus. (Some very simple microcontrollers also use a hardware stack, but that's because they have almost no data memory in general.)

In summary, the DSP chip makes nearly all of its trade-offs in favor of optimizing the execution of DSP algorithms. The resulting architecture has a far different "feel" to it—the constraints on register usage and flow of data are similar to what you may have experienced if you first programmed a general-purpose microprocessor (e.g., MC68040) and then a microcontroller. In architectural terms, the DSP chip is far less *orthogonal* than the general-purpose processor (actually, we'd say it's NOT orthogonal at all!), *orthogonality* referring to the ability to use registers in an interchangeable and nondedicated way.

7.2.1.2 DSP/Microcontroller Combinations

One way of combining the advantages of a DSP-specific architecture with the benefits of a general-purpose and/or microcontroller architecture is to, well, have both. This is what Motorola did with the MC68356, which combines a full-fledged 24-bit fixed-point DSP core—the MC56002— with a MC68302, a 32-bit microcontroller core.

The applications for such a device are limited to those that place heavy demands on both types of processing and can afford the overhead of all of this silicon. However, it is a dramatic demonstration of the differences between these two architectures that these two powerful cores could be placed on the same die and used equally.

We should not gloss over the fact that the MC68356 chip designers didn't just cut and paste the two designs next to each other; considerable attention was paid to communications between the two units. This is also an area you should investigate in your own board-level designs. The overhead of communication between processors may not seem like much compared with the heavy-duty number crunching of the DSP algorithms, but this communication takes a potentially large amount of bandwidth.

7.2.1.3 DSP With Microcontroller Extensions

What about just adding some microcontroller-like (and general, microprocessor-like) functionality to a DSP chip? This is exactly the tack some designers have taken, producing chips that support both DSP-intensive processing and a reasonable level of digital control and high-level processing. The DSP56800 family is an example of this approach, aimed squarely at applications that require both (heavy) DSP and microcontroller functionality in a single, low-cost, low-power package. The underlying core is a DSP processor, but bit manipulation, branching, depth of stack, and addressing are greatly enhanced to speed non-DSP processing. To make programming easier, the register set is more orthogonal, at least in comparison with traditional DSP processors. Designs such as the DSP56800 family are likely to continue to appear as applications multiply for DSP in embedded systems (especially portable) and where the elimination of a microcontroller could provide savings in cost, size, and power consumption.

7.2.1.4 Microcontroller With DSP Extensions

What about reversing the recipe above by taking a microcontroller and adding extensions to support DSP? We have such an example with the 68HC16, a 16-bit microcontroller core augmented with DSP-specific registers and instructions.

Again, such a hybrid is influenced by the base architecture, in this case the underlying microcontroller architecture. If we are concerned only about the speed of the inner-loop DSP operations (e.g., MAC), the major constraints are in the nature of data access (i.e., Von Neumann vs. Harvard, support for circular buffers) and the resources devoted to the multiplier and accumulator. The 68HC16 designers did a good job with its multiplier/accumulator; it's about as fast as possible, given the data that has to be moved over a single bus during a MAC cycle. (We'll discuss the MAC instruction in detail in a few pages. Right now, you can see that we need not only to load two words of data—the sample data and coefficient data—but also to adjust pointers to the next sample and coefficient data, to check the loop counter, and so on.) There is support in the 68HC16 for a special case of a circular buffer, and the accumulator has the overflow and saturate features we mentioned earlier. Real-estate limits for the 68HC16 mean making do without a single-cycle barrel shifter. This is not a fatal loss, and it is a necessary trade-off, given the large chip area such a shifter takes.

We'll need to delve more deeply into the 68HC16 when we implement filter code later in this chapter. As we do, you'll see the DSP support in the 68HC16 is in a sort of middle ground—the architecture accommodates, but isn't centered around, the DSP hardware. (For example, the circular buffer addressing is implemented only for the DSP-specific instructions.) It is a microcontroller with very carefully engineered additions that support some, but not all, DSP applications. We'll also see that both IIR and FIR filters can be implemented quite well on the 68HC16.

An alternative approach to that taken by the 68HC16 is to "bolt on" the DSP hardware, leaving the microcontroller core untouched. The DSP side can then run concurrently but is dependent on the microcontroller for data and the sequencing of instructions. This architecture means the core microcontroller can remain 8-bit (or conceivably, even 4-bit), and the DSP side can be a more reasonable 16-bit. Because the DSP is driven completely by the microcontroller, the DSP core is far simpler than a stand-alone DSP chip.

7.2.1.5 Microcontroller

At the opposite end of the spectrum from the DSP chip, we have general-purpose microcontrollers (you may read *microprocessors* here if you like). With more powerful microcontrollers the orthogonality generally increases, making programming somewhat easier. Major architectural issues include the size of internal registers, the type of arithmetic operations supported—especially multiplication—and issues surrounding the throughput of data such as the maximum clock rates and the speed of data movement. The inherent support for useful data formats like signed-fractional representations can be very important.

High-end microprocessors incorporate floating-point processing and techniques borrowed from RISC processors; in some cases, these architectures provide very reasonable execution of DSP routines, as well as excellent support for general-purpose processing. However, general-purpose microprocessors that excel at DSP will almost always be more expensive than chips that specifically target DSP functions; unless you can use the additional functionality, you're paying for a lot of silicon that you aren't using.

Digital cellular phones, wireless communication for notebook computers, and so on are all embedded applications that require hefty amounts of DSP, minimal power consumption, and microcontroller functionality. Embedded systems designers will continue to see new devices become available for DSP and digital control, fueled by this explosion in applications.

7.2.2 Operations

In the discussion of architectures, we've mentioned a few areas of functionality such as computation, addressing, data movement, and so on, that make DSP more or less efficient on a given chip. Let's now examine these areas, keeping in mind the difference equations for an FIR filter:

$$y(n) = h(0)x(n) + h(1)x(n-1) + \ldots + h(N-1)x(n-(N-1)) \qquad (7.1)$$

and the second-order section of an IIR filter:

$$w(n) = x - a_1 w(n-1) - a_2 w(n-2) \qquad (7.2)$$

$$y(n) = b_0 w(n) + b_1 w(n-1) + b_2 w(n-2)$$

7.2.2.1 Computation

Multiply Every time you turn around in DSP, you'll find a need for multiplication, usually constant coefficients multiplying data samples. Unfortunately, multiplication requires either a large number of transistors or a lot of time—many transistors on a chip can produce a multiplication result in a single clock cycle; alternatively, software (or microcode) can use existing resources such as shifts and adds to perform multiplication at a much slower rate. Because DSP algorithms are so dependent on multiplication, for almost any real-time application, you'll need fast hardware multiplication. Slow multiply instructions, or even your own optimized software multiplies, immediately chew up a huge amount of the processing time; just read on to see the lengths to which we will go to optimize 68HC16 programs—and we have decent multiply hardware there!

The MAC Operation Moving up a level in operation, what we usually really want is not just a multiplication but a multiplication followed by an add or *accumulation*. (Recall Equation 7.1). That is, multiply two sets of numbers, accumulating the results. You might also see some opportunities in the IIR filter for this *multiply and accumulate* (MAC) operation. In fact, the FIR filter often consists of a single (repeating) MAC instruction. But accumulating N of these results can potentially overflow the accumulator, a problem that we discuss in a moment.

Before we discuss the mathematical details of the MAC instruction, we comment that MAC instructions usually do far more than multiply and accumulate. Besides multiplying two values together and adding the result to the accumulator, typical MAC instructions also

- Increment (or decrement) a coefficient pointer
- Check to see if the coefficient pointer is still within the coefficient buffer (wrap around if not)
- Increment (or decrement) a data pointer
- Check to see if the data pointer is still within the data buffer (wrap around if not)
- Load a new coefficient value, ready for the next MAC cycle
- Load a new data value, ready for the next MAC cycle
- Decrement a loop counter
- Repeat if the loop counter is not zero (or perhaps -1)

Sure, the multiply-accumulate operation is the centerpiece of the MAC instruction, but if you begin adding up all of the housekeeping issues—incrementing pointers, loading new values, and so on—you see that even with a very quick multiply (or multiply-accumulate), as many as a half-dozen instructions and quite a few clock cycles would be necessary to accomplish these other tasks. For real-time performance, having a multifunction MAC instruction is a necessity, not a mere luxury, and understanding the reasoning behind each function will guide you in crafting code that extracts the maximum performance from the architecture. We'll be spending some time with the 68HC16 MAC and RMAC (repeating MAC) instructions, you can be sure!

Overflow and Saturation It should be clear that the accumulator needs to be wide enough to preserve the entire result of the multiplication—in our case, at least 32 bits wide for a 16-bit by 16-bit multiply. This is so, even though we may eventually only use the 16 most significant bits of the result, as carries from the 16 least significant bits are necessary to preserve the accuracy of the upper 16 bits.

Accumulating more than a few results raises the possibility that an overflow will occur—that is, the result of the addition will exceed the size of the accumulator. There are several ways to address this problem. First, additional bits can be provided in the accumulator to allow intermediate results to be larger than the final result. The 68HC16, for example, provides four additional bits, for a total accumulator size of 36 bits. Although we can access these additional bits, normally we do not—we expect the result to be limited to a specific range (e.g., -1.0 to 1.0); if the result is not, it is usually forced to be in that range!

Adding extra bits to the accumulator doesn't take care of every circumstance. It's still possible for the result of many accumulations to *overflow* the accumulator—that is, cause a carry to be generated for which there is no room in the accumulator. When that happens, the value in the register suddenly appears to have undergone a change in sign (and usually, magnitude). For example, if we are using 8-bit signed integers, adding 0000 0001 (1_{10}) to 0111 1111 (127_{10}) produces the result 1111 1111, or -1_{10} in two's complement. Not exactly graceful behavior, is it?

The usual solution to the overflow problem has two parts. First, a flag is provided to warn that the contents of the accumulator have overflowed into at least one of the extra bits. If this flag is set, we'd better not grab just the 16 bits (or whatever) from the accumulator, as they no longer are the most significant bits of the accumulated result. What to do? The second part of the solution is to *saturate* the result, either "manually" (explicitly in the user's program) or automatically, using a saturate mode of the accumulator.

The saturate and overflow processes work as follows. Logic in the processor monitors the accumulator, setting the overflow flag when necessary. It resets the overflow if results that were temporarily too big drop back down into the acceptable range with further accumulations.[3] Once all of the accumulations are completed, the results can be transferred from the accumulator as usual, assuming the overflow flag is not still set. However, if the overflow flag is set, a saturated value is instead transferred—either the largest positive or largest negative number possible.

3. There may be two overflow flags, as in the 68HC16—one that indicates overflows into the extra bits and one that indicates an overflow "off the end" of the accumulator (even past the extra bits). You can be clever and do some scaling of data and recover an answer if the overflow were into only the extra bits. Once you get the second type of overflow, hang it up. All you learn is that the answer is bigger than the accumulator (you often are also told the sign of that overflow).

Saturation mimics the behavior of an analog circuit when it attempts to process a signal that's "too large"—the op-amp or transistor has only a certain range of output voltages. Ask for more, and the output can go no higher—it has "saturated." If the accumulated result exceeds the capacity of the accumulator, we'd like the same behavior.

Note that saturated values can be used, as is, in further calculations without having to treat them as a special case. By no means are they the same as the positive-infinity, negative-infinity, or "not-a-number" representations that some floating-point encodings allow. When we use signed, fractional representations for fixed-point arithmetic, the range of valid numbers is between -1 and just under 1. The positive saturated value will be just under 1, and the negative saturated value will be -1—both valid numbers. If saturation can be made automatic and if use of "saturated" values is acceptable, then we don't even have to look at the overflow flag.

Addition Obviously, if we don't have an accumulator, we'll have to perform accumulation using a separate add instruction. Addition is typically a fast operation on every microcontroller, so the only issue is the width of the registers. If you can perform only an 8-bit addition, you'll need to use multi-precision adds (adds that include carry bits) to support data sizes of 16 bits or more. Count on additional overhead; multi-precision arithmetic usually takes *more* than twice (e.g., for double-precision) as long as single-precision arithmetic.

Shifting There are no explicit shifts in the FIR or IIR algorithms, but you probably know that arithmetic shifts are equivalent to division by a power of 2 (a right shift) or multiplication by a power of 2 (a left shift). Every microcontroller we've ever seen has included a shift instruction, and for shifts of 1 bit in either direction, these are very fast. (Note that shifts come in two flavors, logical and arithmetic; the arithmetic variety preserves the sign of numbers in division and shifts in zeros when multiplying. Logical shifts usually use the carry bit to allow rotations of data from most significant bit to least, or vice versa. Be sure to use the right kind!)

Shifts of more than 1 bit in either direction are more problematic. A barrel shifter can generate arbitrary shifts in either direction (in a single instruction cycle), but these take a lot of space on the chip. Nonetheless, most DSP chips include barrel shifters because shifting is very useful for scaling data. Looking at the instruction set won't tell you whether you have a barrel shifter, as some microprocessors use microcode to repeatedly cycle data through a simple shifter, shifting 1 bit at a time. Check the instruction timing to be sure. The 68HC16 has a 1-bit-at-a-time shifter; even so, it will form a crucial part of our IIR routines later, in place of some simple multiplication and division.

Sines, Cosines, and All That Other Stuff Part of the "magic" of DSP is that equations for FIR and IIR filters are just multiplication and addition—no need for inverse sines or taking the log of something. At least, that is, in the actual filter evaluation (recall the inverse hyperbolic sines of Chapter 6 (IIR Filters—Digital Filters With Feedback) when we designed some analog filters!). There are still times, though, when you need to evaluate sine or some other functions. Because such functions are not central to DSP, you will almost always have to do this in software—fixed-point processors usually support only one "exotic" function in hardware, A-

Architecture Issues

law or μ-law companding.[4] We discuss implementing functions as an entirely separate issue in Chapter 11 (Synthesizing Signals).

7.2.2.2 Addressing

The next three issues—addressing, data movement, and registers—are all concerned with the flow of data through the processor. All three elements are interrelated; a bottleneck in just one will have a dramatic impact on the performance of the whole DSP system.

The FIR filter clearly requires addressing for both coefficient and data storage. Most microcontrollers allow this, and most will allow convenient indexed addressing—that is, using the value of a register as an offset from a constant address. The major issues are how fast the data is accessed with this form of addressing, and how long it takes to increment—actually, let's say "adjust"—the index to point to the next data. In the case of the 68HC16, we have both fairly quick indexed addressing and the ability to increment the index register easily and quickly (the AIX or AIY instructions). We also have three separate index registers (X, Y, Z); some architectures have only one or two, which erodes performance because the same register must be used for both data access and coefficient access. Ouch! But, with three index registers, we should be set, right?

Circular Buffers A special type of data structure central to DSP is the circular buffer. What's usually happening in DSP is that we'll have two arrays of data to multiply together, one of which consists of prior input or output values. Let's take a concrete example of an FIR filter with 10 coefficients. We could allocate an array of length 10 words to hold these values—say, $x[0]$ through $x[9]$ (i.e., representing $x(n)$ through $x(n-9)$ in the FIR equation). Assuming we have a MAC instruction as outlined earlier, we'd be all set for the first time through. What happens when it's time for the next sample, though? Well, we'd need to move each word down the array by one space—that is, $x[9]=x[8]$, $x[8]=x[7]$, and so on—a total of nine moves. This would really eat up the processor's time, especially for FIR filters of a more realistic length, say, 50 or more coefficients.

Circular buffers take the approach that data never moves—pointers do. The same 10 words are allocated, but we also add a pointer that always points to the next available spot in the buffer. Because the circular buffer is really just simulated in regular "straight" memory, we'll need to look at the values of pointers used to read and write the buffer to make sure we "wrap around" to the start (or end) if we try to move past the end (or start) of the buffer. There's no required direction for the buffer to be filled; in our example FIR code, we'll fill the circular buffer starting at the "top" and moving down so that the MAC instruction increments an index register as it moves up through the buffer. Table 7-2 shows the contents of the buffer as we add samples (the new samples are in bold). Note the "wrap" that occurs as the pointer is decremented from 0 to -1, then is adjusted to the top. A similar wrap occurs from 9 to 0 on increments.

---

4. As you may recall, this is because some CODECs use companding to compress linear data to 8 bits. Some DSPs can automatically perform the conversion to and from linear when connected to such a CODEC.

Pointer	Input	Buffer contents									
		0	1	2	3	4	5	6	7	8	9
9	a	0	0	0	0	0	0	0	0	0	**a**
8	b	0	0	0	0	0	0	0	0	**b**	a
7	c	0	0	0	0	0	0	0	**c**	b	a
6	d	0	0	0	0	0	0	**d**	c	b	a
5	e	0	0	0	0	0	**e**	d	c	b	a
4	f	0	0	0	0	**f**	e	d	c	b	a
3	g	0	0	0	**g**	f	e	d	c	b	a
2	h	0	0	**h**	g	f	e	d	c	b	a
1	i	0	**i**	h	g	f	e	d	c	b	a
0	j	**j**	i	h	g	f	e	d	c	b	a
9 (wrap!)	k	j	i	h	g	f	e	d	c	b	**k**
8	l	j	i	h	g	f	e	d	c	**l**	k
7	m	j	i	h	g	f	e	d	**m**	l	k
6	n	j	i	h	g	f	e	**n**	m	l	k
5	o	j	i	h	g	f	**o**	n	m	l	k

Table 7-2 Example circular buffer operation (decrementing pointer).

Some processors, including the 68HC16, support circular buffers automatically. Often, there are restrictions on both the alignment of the buffer in memory and the length of the buffer. These restrictions simplify the addressing hardware. Simply inhibiting "carries" past a certain bit position when adding to the index register effectively implements circular addressing if the buffer is a power of 2 in length (i.e., 2^B) and the lowest B bits of the starting address are all zeros. The circular addressing on the 68HC16 has both of these restrictions, meaning large circular buffers can sometimes take up to twice as much storage as they really need (e.g., a buffer of length 130 bytes requires 512 bytes when it is a circular buffer) and that buffers may need special attention to start at appropriate addresses.[5]

---

5. If you're feeling quite up against the wall and have no built-in circular addressing, you could modify your address decoder to do hardware circular addressing on external memory! Not for the faint of heart, but if you already have programmable logic doing the decoding, just flip an output bit on your processor to signal the logic to "ignore" upper address bits.

7.2.2.3 Data Movement

Data movement issues includes the width of the external data bus, the speed and flexibility of bus timing, and issues of data alignment. Even if the internal registers of a microprocessor are 16 bits wide, if all data transfers take place over an external 8-bit wide data bus, the flow of data will be suboptimal. But even with a 16-bit external data bus, mistiming of the bus can wreak havoc. When Dale was writing the FIR example in this book, he kept getting extremely long execution timings. Often the filter couldn't finish processing one sample before the next was received.[6] The answer was that he had neglected to check how fast the 68HC16 was accessing external memory. When reset, the 68HC16 slows bus timing way down, slow enough to accommodate glacially slow memory. Among the many registers of the 68HC16 are some that tell the 68HC16 how many wait states are used when external memory is accessed, and once he put in the right numbers, the filter ran as fast as it should. (To accommodate slow memory, the default behavior of the 68HC16 is to include lots of wait states. Once you're running, you can change the bus timing to be as fast as your memory.) By the way, the 68HC16 is a rather sophisticated microcontroller; rarely will you be able to adjust the bus access times for external memory—all of the timing is usually fixed to the processor clock rate.

Even when bus accesses are fast and wide, most microcontrollers still allow 8-bit memory accesses, and some allow unaligned 16-bit memory accesses—that is, memory accesses that require 1 byte from the "end" of one word and the other byte from the "start" of the next word. Although allowing unaligned accesses maintains compatibility with older code, you can get in big trouble if your 16-bit data is not aligned on an even memory address. Instead of one 16-bit transfer, your 16-bit data may require two transfers, one to get the most significant byte, the second for the least significant byte. You'll need to review your microcontroller data book to see whether this is an issue (it is for the 68HC16) and, if it is, pay strict attention to properly starting all 16-bit data on the proper address boundaries. Also make sure you're using the bus to its full capacity—move 16 bits at a time if you can, even if the underlying data are in 8-bit bytes.

7.2.2.4 Registers

What do we need of the registers? Obviously, we'd like the width to be sufficient to allow quick movement of data. While 16-bit microcontrollers are very popular and common, the truth is that 8-bitters will continue to have a niche for a long time to come. Unfortunately, 8 bits doesn't provide a lot of dynamic range; for any appreciable DSP, you'll need to use multiprecision arithmetic, which means twice as many data moves in and out of registers.

Not all registers are created equal—we've referred to the orthogonality of register sets before, that is, the ability to use registers interchangeably, versus the very dedicated registers of many DSP-specific processors. In general, increases in orthogonality make the processor easier to program and a bit more efficient (as you need to move data from one register to another less often), but this usually comes at a dollar cost. Where the real problems come are with low-end microcontrollers; here, needing to move data from one register to another may be so common

---

6. Yes, this would fall into the category of "bad things for DSP programs to do."

that it eats up a substantial amount of processing time. (We pointed out earlier the usefulness of having several index registers to avoid having to swap first in one index, then another.)

Needless to say, if the core of the filter you hope to implement can't be reasonably implemented with the existing registers, arithmetic, and logic units, you'll have to look to another processor. Our example system, a 68HC16 running at 16.77 MHz, has 1,677 clock cycles to process samples at a 10 kHz sampling rate. Spend those cycles however you like, but you can't ask for more. And, like buying a house or car, all the fine-print-costs—whether it's "title and destination charges" or "loading data into the X register"—end up costing you plenty. Leave room in your budget for the overhead of moving data, scaling, and the inevitable addition of function your application will include before you ship it out the door.

7.3 Programming Issues

7.3.1 Languages

7.3.1.1 Assembly

Assembly language provides a friendly way to generate the machine language instructions and data that we want. This control is absolute and carries with it both the ability to extract the maximum performance from a processor and the great need for attention to detail.

We assume that you've had some exposure to assembly language programming and will concentrate on higher-level issues in DSP applications. Here are some hints for assembly language programming in DSP:

- Thoroughly study the underlying architecture of a new processor to understand its strengths and weaknesses. If you're new to the 68HC16, for example, check out the address extension registers (they allow 20 bit addresses), the addressing modes using the IX and IY index registers, and the register usage of the MAC and RMAC instructions, just to name a few features of interest. Be on the lookout for bottlenecks—for example, many microcontrollers may have only a single index register.

- Become familiar with your software tools, especially your *macro preprocessor* (if any). Macros let you build debugged sections of code that you can use almost like subroutines. Instead of the overhead of a subroutine call, however, macros expand in place, a trade-off of more code space for better speed. Macros can make your code easier to read, but document your macros well so that others (or, dare we say, even yourself) can use or debug them. If your assembly doesn't have a macro processor, consider using the preprocessor for a C compiler.

- Acknowledge the need for "ugly" source code if you absolutely need a section of fast object code. As an example, we wanted to generate a variable number of shifts in some of the macros of IIR example appearing later in this chapter. The number of shifts could vary from one filter to another. We ended up with some pretty "ugly" source code to test for the number of shifts—literally, "if shifts equals 1, generate 1 shift; if shifts equals 2, generate 2 shifts," and so on. The end object code contains just the shifts, of course, without any hints of all the behind-the-scenes machinations. The source looks a little clunky, but we wanted to use the same macro for any second-order section, rather than hard code the number of shifts each time. (As a general observation, the moment you hard code some aspect of your program, you'll find a need to change it. If you anticipate changing any aspect of your system—coefficients, sampling rates, etc.—writing defensively and generally will help you out tremendously!) An alternative would be to have a loop generated in the object code itself, but for the core of a DSP routine, that would be wasteful. (The distinction here is between assembly-time and run-time evaluation. More on this later.) Bite the bullet, document the "ugly" code, and move on. Elegant code is a delight to write, but the code has to run in the cycles we have, or it isn't useful.

- Initially, pepper your code with both assembly-time and run-time checks. For constants, use conditional assembly directives to check that constant values are in range or make sense. You can't have "-2" right shifts if you're generating only right shifts. Check overflow flags or add debugging macros to check that values are within range. Consider running your system at a slower sampling rate to accommodate extra code for checking if you won't otherwise have the bandwidth.

- Though your initial assembly-language programs will be short, there are many advantages to splitting up larger assembly-language programs and using a linker. (For small programs, you can get by with *absolute* assembly, wherein you must specify the starting address of code and data explicitly. Code written without absolute addresses is *relocatable*.) Some of your subroutines may even warrant creating a library, which can become a toolbox of routines for your processor. Linkers allow you to program in modules, establishing clear and limited connections among routines and forcing you to explicitly state the interrelationships of data and program.

- Major projects will require familiarity with the full set of software tools, but start with the simplest set of tools possible. Your first job is to understand the processor, then the assembler, then the tools for downloading and running the code. Add other tools as necessary, and always, always, *steal whenever possible*.[7] Use sample assembler command lines, batch files, "make" files, whatever you can find to avoid worrying about more than one thing at a time. Existing code and, especially, code written for the same hardware as you have, already makes the assembler happy, the linker happy, the downloader happy—well, you get the picture. Start there, then modify, if possible, or at least cut and paste.

- For the 68HC16, with its formidable array of configuration registers, having examples was crucial for our programming. This is true of most modern microcontrollers or microprocessor systems with extensive peripherals. Once you see how one person did something, you can delve into the hardware manuals in greater detail to understand the background. (And if you want your code to be robust, you can't skip that understanding, either.)

- A minor hint—don't blindly trust example code. Programmers make mistakes. If you think a byte-wide write to the timer-initialization register should really be word-wide,

---

7. Uh, in the sense of "reusing existing code," rather than, say, walking off with the office laser printer.

check it out. Application engineers are happy to clarify hardware issues when the person on the other end has done her or his homework. Understand why something was done strangely—perhaps it will teach you something new about the architecture or correct a long-standing error.
- Code the core. Don't let too much time go by before you run a "sanity check" on your choice of processor. Sit down with your databook and make sure the core of the DSP routine will execute *well* within your clock cycle budget.[8] This will also tell you other information. Often, the core will define—or at least strongly hint at—other parts of your program. An important example occurs with data formats; in the 68HC16 (and most other 16-bit fixed-point DSP chips), a signed fractional data format is used extensively. In other cases, a MAC instruction might place limits on memory locations for coefficients and data. What index registers get clobbered? Can you arrange the order of samples in memory to minimize data transfers? Building at least part of your program from the inside out can minimize painful (i.e., slow) interfaces at the heart of the data flow through your program.

7.3.1.2   Compilers

The C language is by far the most popular compiled language for microcontrollers. And for good reason—the language constructs map pretty well to many architectures; excellent compilers are available over a wide range of processors; C is fairly standardized, and so on. This isn't to say that C is the perfect embedded-systems language, or that C and DSP are a perfect match—it isn't, and they aren't. But somehow any arguments about languages must come down to what is available, and C (followed at a distance by BASIC) is probably with us for some time to come. What we have to say about compilers, however, is fairly general.

First and foremost, compilers are amazing pieces of software, especially for embedded systems. Some microcontrollers and microprocessors have a half dozen or more different companies competing for the best speed or code density. Some of the generated code is quite impressive, the result of years of compiler research. Dale's voice synthesizer required a few very fast moves of blocks of data, and he thought that he could whip out a quick assembly language routine that would run far faster than the compiler's code. Wrong-o. The compiler knew all about moving data efficiently with just a single pointer register—the trick was to grab as many bytes as possible into internal registers, then swap the pointer to point to the destination. Meanwhile, Dale had merrily been reading a word, loading the pointer, writing a word, loading the pointer, and so on. Learn from your compiler!

Next, compilers let you write at a much higher level. This means increased programmer efficiency, fewer mistakes (reliability), and easier debugging and maintenance. Any time a tool can handle the details, it lets us put more effort into the big picture. Use your brains where *you* make the most difference and, whenever possible, delegate!

8. You can take that "well" as meaning robustly and meeting all your performance criteria, as well as meaning that it uses far fewer than the number of cycles available. On this latter point, if you're using more than 75% of your cycles on core code, beware. Overhead (support routines, communications, etc.) will always get you. You won't believe the amount of overhead until you have the whole program coded. Give yourself some leeway.

Another advantage of compilers is portability. For most applications, we value the portability of the target code—that is, we could take a different compiler and, using the same source, generate code for one processor that would run the same as on the original processor. However, for DSP-intensive applications, we value much more the ability to test code on another platform without regard for timing (non-real-time). It is incredibly useful to have the same code execute on your PC platform as is running on your embedded system, even if the PC code runs much more slowly (or faster). If the same data can be fed into both, you can verify the behavior down to the last bit. Even when parts of the embedded system code have been replaced by assembly language, you can verify the operation against "known good" (well, "assumed good") C code.

Proprietary extensions are common in C compilers for microcontrollers. Among the most common are language extensions to deal with special memory constraints and features such as internal RAM, fast addressing modes, bit-addressable memory, data versus program memory, and so on. Other instructions may generate code that takes advantage of special instructions. Examples include the MAC or bit-manipulation instructions. These extensions come at a cost—they reduce portability and require more familiarity with the architecture. Because you'll be studying the architecture, the latter issue isn't much of a problem, and portability can often be achieved by using conditional compilation (in C, "#if" preprocessor directives). You can often realize enormous time savings by directing the compiler to generate code that exploits fast memory, limited-range data types (e.g., byte values vs. words, unsigned vs. signed), or fast instructions.

Compiled code does have an overhead—it's not uncommon for code generated by compilers to be less efficient at calling subroutines, manipulating data, performing computations, and so on. Much of this overhead is due to the compiler's having generated code that is general, safe, and defensive. Although we do need safe code, it's often possible to drop some of the generality and defensiveness of compiler-generated code in situations in which speed is critical and the situation is well defined. For example, a general-purpose multiply could be changed to a series of shifts in certain circumstances, trading an increase in speed for a restriction to multiplication by a power of 2. This could lead to a 1-byte argument instead of 2-byte, which might avoid using a stack to pass arguments, and so on. The question is whether the resulting code is robust, maintainable, and a good use of programming effort.

Finally, compilers aren't perfect at generating the most efficient code for a given section of source code. First, the language itself may not have any way of expressing operations or data types available to a specific processor. For example, C currently doesn't have a "multiply-and-accumulate" construct, nor a "signed-fractional" data type. In other cases, it may be difficult for a compiler to make a very specific mapping from source code to a special purpose instruction—for example, setting a bit, or perhaps a for/next loop that could be expressed in a repeating MAC instruction. In this case, the source code can express the desired operation, but the compiler may not be able to generate very specific code and, instead, may generate more generic—and slower—code. While we'll continue to see C and other languages include both proprietary and

Programming Issues

standardized extensions to address these mapping issues,[9] for the moment, there are realms in which compilers can't help us.

7.3.1.3 Mixing C and Assembly Language

What To Code in Assembly An obvious strategy in some applications is to use both C and assembly language—C for the non-time-critical symbolic processing and assembly when time counts. It is often clear where your program needs the assembly and where it can remain in C. Obvious places for assembly include the core of DSP routines, operations on large numbers of data, interrupt service routines (interrupt service routines sometimes have too much overhead when written in C), and, yes, even block moves. (By the way, Dale did manage to out-code the compiler on that block move, but it was close.)

Your intuition can mislead you, though. Mixing C and assembly is a balancing act between your development time and the execution time of the code; the way most projects are budgeted, you probably don't have time to overanalyze a single line of code. A few techniques to identify the bottlenecks—thus, sections of code that are candidates for coding in assembly—include:

- Examine the assembly language output of your compiler. Most embedded systems compilers have an option to produce an intermediate assembly language listing. Although there are many uses for this listing, you can certainly pick out a few routines that you suspect take a lot of time and see how good the compiler is at generating code. (You'll want an instruction summary card for your processor close at hand to see how many clock cycles each instruction takes. You might be able to code up a quick utility to append clock cycles to the end of each line of the assembly language listing file.)
- Simulate the code on a system that keeps track of the number of cycles executed. Start at a high level, identifying the subroutines where a lot of time is being spent. (Temporarily change the code to bypass interrupts or other hardware-related aspects that are not supported or could mess up the timing.) Run the timing several times to make sure the simulator is giving you the proper timings. Then divide and conquer. Out of many lines in a routine, you may need to worry about only a few. Many modern simulators will let you work with the source code, displaying the appropriate lines of C instead of the object code, and/or display objects with their symbolic value instead of less meaningful numbers values. You should be able to home in on small sections of code where optimization would be useful.
- Lacking a simulator, consider compiling the code on a PC and using the wide range of tools available there. The code generated for a PC microprocessor will vary quite a bit from that for your target processor, so you'll need to take the results with a grain of salt. You're looking for the "semi-obvious," things that may have escaped your attention the first time through the code but emerge as secondary consumers of clock cycles.

---

9. There are many folks trying to address the problems with C and number crunching. For example, see Ombres [1996].

- Using output pins (ports) to provide timing information is an extremely useful and inexpensive technique. For example, on the MC68HC16Z1EVB, the evaluation board for the 68HC16Z1, many general-purpose input/output pins are available. A single "BSET" (bit set) instruction can output a "high" on a particular pin to indicate the start of a routine while a "BCLR" brings the same line "low" later on. The amount of code and number of clock cycles added is minimal, and a garden-variety oscilloscope then provides a nice graphical picture of the timing. (This is the technique we used to find the memory timing problem discussed earlier.)

In-circuit emulators are a high-end solution. The code runs in real time and can interact with hardware in the rest of your system. Usually nice source-level debugging is possible. Some recent processors from Motorola (e.g., 68HC16), Intel, Hitachi, Philips, and other manufacturers now include a background debugging mode that provides many emulation functions with almost no additional hardware—this makes in-circuit emulation much less expensive. Although not a full-blown emulator, background debugging mode is an extremely powerful tool, not only for the optimization stage but as a learning and debugging tool, too.

Your particular program will have its own mix of time-critical code, but you shouldn't attempt to optimize 100% of your program, except in rare instances. Even when you are writing the entire program in assembly, you've probably noticed the difference in your coding style between time-critical and noncritical routines. Generally speaking, time-critical code usually has far more conditions assumed (e.g., registers loaded with specific values, flags or modes set, input or output values within a specific range, etc.) and more interrelationships among instructions; thus, this code is more "fragile" than noncritical code. Noncritical code is based on few assumptions, and produces clear object code using clear source. (If your object code is "clean," with a minimum of assumptions and "tricks," you'll probably have fewer bugs to worry about and an easier time of debugging those you do have.) Coding tight loops often means taking advantage of side effects of instructions or carefully coding instruction sequences to preserve register values or flags. Normally, we might take a more conservative, safe approach. For example, we might see that a flag was set earlier that would still be valid (maybe the carry bit was set by some operation). However, we might throw in an explicit test, just so we don't have to rely on the permanence of the intervening code. Optimized code, as elegant and efficient as it can be, may also be very difficult to modify, or even understand, when you come back to it in a few days.

How much of your program can you afford to optimize? If you enjoy programming, it's wonderful fun. But it's slow work, the resulting code hard to modify, and so on. An oft-quoted "90/10" rule gives the ballpark figure of 10% of the program uses 90% of the processor time—a generalization, but given the nature of FIR and IIR filters with their repeated inner loops, this may not be a bad rule of thumb. Keeping 90% of the code as clear and high-level as possible maximizes the code's portability, robustness, maintainability, and your coding efficiency, while, if properly chosen, the 10% that's optimized is responsible for the majority of the processing. Table 7-3 summarizes the main points.

High-level language (e.g., C)	Assembly language
+ Portable	- Not portable
+ Easier to debug	- Harder to debug
+ Faster to code	- Slower to code
+ Easier to maintain	- Harder to maintain
- Sometimes slower code	+ Fastest code, if properly written
- Often larger code size	+ Smallest code size
- Not viable on all microcontrollers	+ "Only" choice for small microprocessors
- Good compilers are often expensive (but what's your time worth, really?)	- Low-cost assemblers often available

Table 7-3 High-level languages versus assembly language.

Interfacing Assembly Routines and C Let's assume that you're able to code much of your program in C and have identified some sections that should be rewritten in assembler to speed things up. How does the C compiler pass parameters to subroutines? On the stack? In registers? What registers can you use and what registers must you preserve? Can you use the floating point unit? You'll have to get answers to all of these questions in order to start writing replacement routines.

Your best bet, once you've read the manual, is to let the compiler do the work for you. Each compiler has its own techniques for passing arguments, preserving register values, and so on. This varies not only from one compiler to another but also with the memory model used. So, instead of starting from scratch, write a dummy routine in C that uses the same number and types of parameters, the same types of memory access, some local variables (of different types), and has a variety of references to symbols. Run this through the compiler to generate an assembly-language listing. (Usually, this is a compiler option.) You'll then have a skeleton showing you how to deal with all of the issues that could trip you up.[10] Modify the innards as needed, paying careful attention to any changes that affect the interface to the rest of the program. In particular, the limited resources of microcontrollers often lead compilers to generate very specific code for parameter passing. Changing the number of parameters or their types can lead to very different orderings or arrangements of parameters. (If you do need to change the number of parameters, either understand the compiler thoroughly or use a dummy C routine to generate a new skeleton to double-check the parameter-passing method.)

Let's look at an example showing one compiler's parameter-passing convention. Two source files are shown in Listing 7-1—a main routine that calls a simple two-parameter function *add*. Listing 7-2 is a merged listing showing both the original C source and the generated assem-

10. Once you've generated this skeleton file, though, be sure to delete or rename the original C source so that you don't accidentally recompile it, overwriting either your desired object file or, worse yet, your modified assembly-language file!

bly language. (Please note that we[11] added the comments on the assembly language lines!) We see that the compiler passes arguments using both a register (register *D*, holding variable *a*) and the stack (variable *b*). The return value from the function is in register *D*. (Note that most compiler manuals document the calling convention in great detail; it is rarely necessary to analyze the code as we've done here.)

File c7cmain.c:

```
/* c7cmain.c
   example C routine showing parameter passing
   4/30/97 rce
*/

unsigned int add(unsigned int a, unsigned int b);

void main(void)
{
   unsigned int a,b,c;

   a=0x55;
   b=0x71;
   c=add(a,b);
}  /* main() */

/* c7cmain.c */
```
File c7cmain.c:

File c7cadd.c:

```
/* c7cadd.c
   example C routine showing parameter passing
   4/30/97 rce
*/

unsigned int add(unsigned int a, unsigned int b)
{
   unsigned int c;
   c=a+b;

   return(c);
}  /* add() */

/* c7cadd.c */
```

Listing 7-1 Example C source for parameter-passing analysis.

Example 68HC16 compiled C code using the Introl-C16 Compiler version 3.07

```
Generated via:
        cc16 -l -q c7cmain.c
        imerge c7cmain.c c7cmain.lst >c7cmain.img
        cc16 -l -q c7cadd.c
        imerge c7cadd.c c7cadd.lst >c7cadd.img
(Comments on assembly code lines added by authors!)
```

Listing 7-2 Assembly language generated from code in Listing 7-1.

---

11. Well, Dale. Jack fell asleep after the first line.

Programming Issues

```
File c7cmain.img:
*             /*       c7cmain.c
*                      example C routine showing parameter passing
*                      4/30/97 rce
*             */
*
*
*             unsigned int add(unsigned int a, unsigned int b);
*
*             void main(void)
*             {
    6 00000000                    main:   fbegin
    7 00000000    3430                    pshm    z,k        ;save IZ;EK,XK,YK,ZK
    8 00000002    3ffa                    ais     #-6        ;allocate 6 more bytes
                                                             ;on stack (var a,b,c)
    9 00000004    276f                    tsz                ;ZK:IZ pts to start of
                                                             ;allocated bytes
                                                             ;z+0=a, z+2=b, z+4=c
*                      unsigned int a,b,c;
*
*             a=0x55;
   11 00000006    37b50055                ldd     #85        ;0x55 (in decimal!)
   12 0000000a    aa00                    std     0,z
*             b=0x71;
   14 0000000c    37b50071                ldd     #113       ;0x71 (in decimal)
   15 00000010    aa02                    std     2,z
*             c=add(a,b);
   17 00000012    3401                    pshm    d          ;push variable b
   18 00000014    a500                    ldd     0,z        ;reg D = variable a
   19 00000016   >fa000000                jsr     add        ;call routine
   20 0000001a    aa04                    std     4,z        ;store result in
                                                             ;variable c
*             }        /* main() */
   22 0000001c    3f08                    ais     #8         ;restore stack ptr
   23 0000001e    3506                    pulm    z,k        ;pull IZ;EK,XK,YK,ZK
   24 00000020    27f7                    rts                ;return
   25 00000022                            fend
   26                                     import  add
   27                                     end

c7cmain.s16            Introl C compilation of 'c7cmain.c'
              Section synopsis

    1 00000022  (          34) .text                         ;34 (0x22) bytes of code
                                                             ;in this section

c7cmain.s16            Introl C compilation of 'c7cmain.c'
              Symbol table

.text     1 00000000 | add    I  0 00000000 | main    E  1 00000000

c7cmain.s16            Introl C compilation of 'c7cmain.c'
              Symbol cross-reference

.text                  *4
add                    19       *26
main                   *6
*
*             /* c7cmain.c */

File c7cadd.img:
*             /*       c7cadd.c
*                      example C routine showing parameter passing
*                      4/30/97 rce
```

Listing 7-2 (Continued) Assembly language generated from code in Listing 7-1.

```
*        */
*
*        unsigned int add(unsigned int a, unsigned int b)
    6 00000000             add:      fbegin
    7 00000000    3430               pshm      z,k         ;save IZ;EK,XK,YK,ZK
    8 00000002    3401               pshm      d           ;save reg D
    9 00000004    3ffe               ais       #-2         ;allocate 2 more bytes
                                                           ;on stack (var c)
   10 00000006    276f               tsz                   ;ZK:IZ pts to start of
                                                           ;allocated bytes
                                                           ;z+0=c
*         {
*                    unsigned int c;
*                    c=a+b;
   12 00000008    a50c               ldd       12,z        ;variable b
   13 0000000a    a102               addd      2,z         ;variable a
                                                           ;(old value of D)
   14 0000000c    aa00               std       0,z         ;save as variable c
*
*                    return(c);
   16 0000000e    3f04               ais       #4          ;restore stack ptr
   17 00000010    3506               pulm      z,k         ;pull IZ;EK,XK,YK,ZK
   18 00000012    27f7               rts                   ;return
*         }          /* add() */
   20 00000014                       fend
   21                                end

c7cadd.s16                 Introl C compilation of 'c7cadd.c'
           Section synopsis

    1 00000014 (          20) .text                        ;20 (0x14) bytes this
                                                           ;section

c7cadd.s16                 Introl C compilation of 'c7cadd.c'
           Symbol table

.text      1 00000000 |  add      E   1 00000000

c7cadd.s16                 Introl C compilation of 'c7cadd.c'
           Symbol cross-reference

.text                *4
add                  *6
*
*         /* c7cadd.c */
*
```

Listing 7-2 (Continued) Assembly language generated from code in Listing 7-1.

We had to get out the CPU16 Reference Manual to follow all of the stack-related movements, since, for example, it's not clear that "ldd 12,z" refers to the stacked value of variable "*b*." See Table 7-4 for a picture of the stack prior to executing line 12 of the "add" function. We've also noted stack locations pointed to by the *IZ* register and the stack pointer (usually abbreviated as *SP*—no relation to signal processing).

Another approach that is sometimes supported by compilers is to use *in-line assembly*. You embed assembly language code in your C code; the compiler then includes this code verbatim in the object code. Benefits include the fact that you eliminate the overhead of subroutine calls and parameter passing altogether. This could add up to substantial savings if many calls are saved.

Stack address (decimal)	Stack contents
98	register IZ
96	address extension register: EK, XK, YK, ZK
94	variable c
92	variable b
90 (old IZ)	variable a
88	second parameter to "add" routine: copy of variable b (pushed by "main," line 17)
86	program counter (PC) (pushed by "jsr add")
84	condition code register (CCR) (pushed by "jsr add")
82	register IZ (pointing to address 90)
80	address extension register: EK, XK, YK, ZK
78	first parameter to "add" routine: (register D) copy of variable a (pushed by "add," line 8)
76 (←IZ)	"add" function variable c
74 (←SP)	(next available stack location)

Table 7-4 Analysis of stack during parameter passing.

On the other hand, your code will be very tightly coupled to the surrounding compiler-generated object code; you'll need to assure that the location of data and the usage of registers won't be changing over the life of the code (including changes to the compiler as it is upgraded). If you do choose this route, consider using conditional compilation (e.g., "#ifdef," "#else," "#endif") to select between the optimized in-line code and the same code written in C; you can then easily switch between the two to see if the assembly-language code produces the same results (except for timing).

7.3.2 Data Representations

Although the number of bits available for encoding a value is important, equally important is the representation we choose. Table 7-5 shows some sample values encoded as unsigned integers, signed integers, and signed fractions. Each representation uses the same number of bits (say, 16 for many of our applications) but vary in range and resolution.

Of particular interest is the signed fractional representation. The most common signed fractional format among fixed-point processors is known variously as *Q15* or *1.15* and places an implied decimal point between bits 15 and 14, as shown in Figure 7-1. Adding two numbers in this format is no problem; the result is a single 16-bit number, also in 1.15 format. Multiplying, however, produces a result in 2.30 format (Figure 7-2). This isn't convenient to deal with because we normally arrange computations so that all values are between -1 and 1 (i.e., we don't need the -2– 2 range of 2.30). In addition, we would eventually need to round the results of multiplies to 16 bits, which would require a (32-bit) shift (to get to 1.31 format), followed by a rounding transfer of the most significant word. It's common, therefore, to have multipliers automatically perform the right shift on the result of multiplies, assuming that 1.15 format operands are used. The 68HC16 operates this way. Thus, caution[12] is needed when coding integer multiplies—your result might be off by a factor of 2.

7.3.2.1 Conversions

To convert a decimal fraction to the 1.15 format, use the following rules:

Positive values ($0 \leq x \leq 1\text{-}2^{-15}$): Multiply x by 32768, round to the nearest integer, convert to 16-bit binary value. For example, 0.63859 times 32768 is 20925.31712, rounded is 20925, and converted to binary is 0101 0001 1011 1101. (This is $51BD in hexadecimal.)

Negative values ($-1 \leq x < 0$): Take the absolute value of x, multiply by 32768, subtract 1, round to the nearest integer, convert to 16-bit binary value, and invert all the bits (i.e., change ones to zeros and zeros to ones). For example, -0.32485 is converted by multiplying 0.32485 by

---

Numeric Conventions

Assembler documentation will spell out the conventions used for different number bases. Usually, decimal is the assumed default and hexadecimal (base 16) is preceded by a "$." Binary for the Motorola assembler is indicated by a "%" prefix, and octal (base 8) by a "@." The C language uses a different set of prefixes (e.g., *0x* for hexadecimal), and suffix conventions are also common. In this book, we use "$" for hexadecimal and "%" for binary (when it's not obvious), though we sometimes break up binary numbers into groups of four for ease of reading. (This is not allowed in actual source code.) And we just don't use octal anymore (sigh).

Dale used to give tours at a laboratory where the memory modules were labeled with their addresses in hexadecimal. Sometimes people would comment on how expensive the "$4000" and "$6000" memory modules were...

---

12. And why would we know this? Actually, we've been caught the other way around. One fixed-point DSP chip has a flag that determines whether this shift takes place. One particular program we wrote always kept the multiplier in integer multiply mode (i.e., no shift). But a thoroughly tested sine routine from the manufacturer would not work. Those lousy application engineers! Posting code with bugs! It turned out that the routine assumed the multiplier was set to fractional mode (sheepish grin). Well, now you know better. Watch your assumptions, and comment your code.

Representation	Range	Resolution	Examples	
			Decimal	Binary
16-bit unsigned	0 to +65535	1	0	0000 0000 0000 0000
			1	0000 0000 0000 0001
			32767	0111 1111 1111 1111
			32768	1000 0000 0000 0000
			65535	1111 1111 1111 1111
16-bit signed	-32768 to +32767	1	-32768	1000 0000 0000 0000
			-32767	1000 0000 0000 0001
			-16384	1100 0000 0000 0000
			-8192	1110 0000 0000 0000
			-1	1111 1111 1111 1111
			0	0000 0000 0000 0000
			1	0000 0000 0000 0001
			8192	0010 0000 0000 0000
			16384	0100 0000 0000 0000
			32767	0111 1111 1111 1111
16-bit fractional	-1 to +0.999969	$2^{-15} =$ 0.0000305	-1.0	1.000 0000 0000 0000
			-0.5	1.100 0000 0000 0000
			-0.25	1.010 0000 0000 0000
			-0.000031	1.111 1111 1111 1111
			0	0.000 0000 0000 0000
			0.000031	0.000 0000 0000 0001
			0.25	0.010 0000 0000 0000
			0.5	0.100 0000 0000 0000
			0.999969	0.111 1111 1111 1111

Table 7-5 16-bit unsigned, signed, and fractional representations.

32768 to get 10644.6848, minus 1 is 10643.6848, rounded is 10644, converted to binary is 0010 1001 1001 0100, and inverted is 1101 0110 0110 1011.

Reversing the process is relatively simple. The most significant bit (msb) tells you the sign of the number:

If the sign bit is 0: Convert the binary number to decimal, divide by 32768. For example, 0011 0101 0111 1001 in decimal is 13689 and divided by 32768 is 0.417755.

If the sign bit is 1 (that is, the sign bit is negative): Invert the binary bits, convert the binary number to decimal, add 1, divide by 32768, and multiply by -1 (i.e., make it negative). For example, 1010 1000 0110 0100 inverted is 0101 0111 1001 1011, converted to decimal is 22427, plus 1 is 22428, divided by 32768 is 0.684448, and with a negative sign, -0.684448.

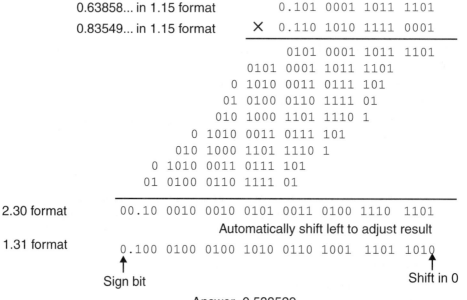

Figure 7-1 Bit position values for 16-bit unsigned, signed, and 1.15 fractional formats.

Figure 7-2 Multiplying two numbers in 1.15 format.

The "subtract 1" and "add 1" business is a consequence of using two's complement instead of one's complement. (Two's complement makes the addition logic easier and avoids two representations for zero.)

By the way, most assemblers will accept signed decimal integers and do the two's complement conversion for you. Just multiply your coefficient by 32768 (regardless of the sign) and round. At some level it's just a question of whether you prefer using the actual bit patterns (usually expressed in hexadecimal, with each group of four bits replaced by a single hexadecimal character) or the scaled decimal value. Sooner or later, it will be helpful to be able to interpret the actual bit patterns, but there's really no reason to have everything in hex in your source. Do whatever is clearest to you (and document everything well!).

7.3.3 Optimizing for Speed

Now for the nitty-gritty—optimizing DSP code for speed. Let's review a few ground rules before getting lost in Programming Heaven[13]:

Optimize correct programs—that is, verify your program before optimizing it.

Optimize as little of the program as possible; keep as much code in C or clean and clear assembly as possible.

Document what you do, especially if you rely on obscure behavior (e.g., little-known side effects).

When you finally do venture out into the "big blue room" (i.e., outdoors), look both ways before crossing the street. Really.

7.3.3.1 Inline Functions

Minimize the number of subroutine calls you make. Each call costs an enormous amount of time because of the need to stack the program counter. Don't forget the cost of the return from subroutine operation, either. Instead of calling a subroutine, use a macro to place a copy of the subroutine "in-line" with your code. This costs more memory, but for short, frequently called subroutines, the cost is usually worth it. (We use this technique in the IIR example later—it saved about 20% in clock cycles, and we used the routine only a few times.)

7.3.3.2 Strength Reduction

This is just a fancy name for replacing one general operation with another mathematically (or functionally) equivalent one. A common and important example is replacing multiplication and division by arithmetic (not logical!) shifts. x divided by 2 is the same as $x >> 1$ (right shift by 1 place). Other mathematical strength reductions are also possible, for example:

Modular arithmetic: Finding the remainder of a division by a constant power of 2 can be done with a simple "AND" operation. For example, 9 mod 4 is 1001 AND 0011 where 0011 is 4-1 or 3. The mask for the AND operation is one less than the divisor.

Multiplication by small integers: Consider repeated adds and/or shifts. x times 6 is $((x<<1)+x)<<1$. Actually, this is just a special case of a software multiplication, and you end up doing about one add and shift for each 1 bit in the constant.

7.3.3.3 Relative and Short Jumps and Branches

In a sense, this is also a "strength reduction" in that it replaces a general-purpose (long) branch with a shorter one. The issue is both code space and speed, as jumps or branches are often available that have a restricted range for destinations but execute faster and/or require fewer bytes.

---

13. Or Hell—take your pick.

Usually, the assembler manual will indicate the circumstances under which the assembler uses short jumps and when longer jumps are used. This can be far down on your list of optimizations.

7.3.3.4 Constant Folding

Constant folding is a special case of "if you can do something at assembly time rather than run-time, do it." In this case, evaluate any expressions involving constants as far as you can. For example, 3 times *y* times 4 should be implemented as 12 times *y*. (Don't laugh! We've seen code that very carefully multiplied two constants together, in assembly language, no less. That was a lot of effort wasted.)

Make sure you understand the concepts of "compile-time" (or, in this case, "assembly-time") versus "run-time" with respect to when calculations are performed. For example, compare the following two pieces of code in 68HC16 assembly language:

```
        ldaa    #HALF+QUARTER       ;load HALF plus QUARTER
        staa    howmuch             ;store in variable
```

and

```
        ldaa    #HALF               ;compute HALF plus
        adda    #QUARTER            ;QUARTER
        staa    howmuch             ;store in variable
```

In the first case, the *assembler* adds two constants, the result of which is a constant in the object code. In the second case, the *microcontroller* performs the addition—which isn't necessary if both values are constants. Most modern assemblers support an extensive set of mathematical and logical operations in expressions. This lets you generate complicated expressions at assembly time, versus having to hand-calculate the constants (and risk forgetting to update them if you change any constants) or—shock and horror—calculate the constants at run-time.

7.3.3.5 Constant Propagation

Look for variables that are assigned constant values and never change. Is it quicker to use a constant directly or to read a memory location to retrieve that constant? In most cases, you should use the constant directly—the so-called immediate addressing mode, wherein the instruction carries the constant with it. Variables should vary! If they don't, then they aren't, and you should consider making them constants.

7.3.3.6 Coding Conditionals

Check out any instruction set and you'll find a big difference between the execution times of a conditional branch that is taken and one that is not. For example, most processors have a "branch if not equal" (BNE, or some such) instruction that conditionally branches if the "zero" flag is not set. Often, the difference in execution time is a factor of 2 or more. For the 68HC16, six cycles are required if the branch is taken, two cycles if not. In certain cases you can arrange code so that a branch is more often not taken. Of course, loops generally take the branch all but the last time, but for nonlooping cases, you can save cycles by coding for the more common condition.

7.3.3.7   Common Subexpression Elimination

Occasionally, you'll find expressions that are common to several different equations. For example, in one equation you might have $k=14.755*(b*d)+17$ and in another $m=u/(b*d)$. As long as the value $b*d$ doesn't change between these two equations, you can calculate this expression once, save it, then reuse it. Remember to look for equivalent expressions such as $b*d$ and "$d*b$," $b(c+d)$ and $b*c+b*d$.

7.3.3.8   Register Versus Memory Storage

Though microcontroller architecture varies wildly, it's almost universally true that reading and writing data to a register is faster than to memory (internal or external). If you like, there's a shorter "address" for the registers than for memory in general, but there are also architectural reasons for the difference in speed. Registers should be used as often as possible to store variables, as long as doing so isn't costly during subroutine calls or other situations in which register values must be saved and restored. If you can't hold all variables, prioritize so you place the most active variables in registers.

By the way—watch out on the 68HC16 for the MAC instruction. It uses nearly every register in the chip!

7.3.3.9   Loop Unwinding

Here's a very important optimization. Loop unwinding involves expanding a loop into a linear piece of code. For example, a "for/next" loop with 10 iterations is expanded into 10 copies of the code from inside the loop, one after another. This eliminates both the iteration counting and the branching, both of which can be big overhead items in architectures without zero-overhead looping (discussed next). You might be able to use macros and conditional assembly to keep the source code looking nice, but this is brute force and the cost is code space. Speed versus space—the trade-off never ends.

7.3.3.10   Zero-Overhead Looping

The best of all worlds is to have zero-overhead looping. In this case there is no overhead cost to keep track of the iteration or to branch at the bottom of the loop. For the tight inner loops of DSP routines, zero-overhead looping is crucial. The RMAC of the 68HC16 implements a zero-overhead loop as one of its many operations. This function is generalized on other DSP chips to allow any series of instructions to be inside the loop. In these cases, there are usually dedicated registers that hold the loop count, and a way of indicating where the loop ends. In the 68HC16, because there's only a single instruction in the loop, only a single count is needed. This count is stored in the E register.

7.3.3.11   Index Registers

Addressing modes are a very important feature of any processor architecture. For efficient DSP, you should look for processors with efficient indexed addressing and a decent supply of index registers. The 68HC16 has both a wide range of indexed addressing modes (with 8-, 16-, and 20-

bit offsets) and a set of three different index registers, nice resources for moving data through the processor quickly.

Indexed addressing generates an address from the value of a register plus (optionally) a constant offset. If simple, quick arithmetic is allowed on the index registers, this drastically cuts down on the time it takes to step through arrays of data. On smaller microcontrollers, you may find only increments or decrements of the index registers; ideally, we'd like to have the ability to step by words (2 bytes at a time), and occasionally more. The *AIX, AIY,* and *AIZ* instructions of the 68HC16 add 8- or 16-bit (signed) values to the three index registers in two or four clock cycles, about as fast as we could hope for.

Concentrate on using index registers instead of less efficient ways of calculating addresses. In a sense, index registers allow us to factor out the addition of the address of the start of arrays, an operation that we'd otherwise have to perform for each new address.

7.3.3.12 Hoisting Variables

Examine the code inside loops to see what operations can be moved outside the loop and performed only once. For example, if you have a loop like this:

```
for i=1 to 100; a[i]=b[i]+j*k; next i
```

you'll want to move the computation of *j*k* outside the loop.

7.3.3.13 Self-Modifying Code

So, feeling pretty lucky? Not satisfied with merely writing fast code? What about code that changes itself?

Actually, self-modifying code is usually a way around bottlenecks in addressing and, when used at all, is used very sparingly. We're not talking about code that somehow decides to throw in a few shifts or branches on its own; at most, we're changing 1 or 2 bytes, the data part of some preexisting instruction. As an example, you might be out of index registers, so you figure out where in memory to write an address such that you overwrite just the address part of some other instruction. Or you might replace the immediate data of an instruction with a new value. Of course, there are a few assumptions here. First, that the program is running in RAM, as writes to ROM or EPROM won't have any effect (the ROM stands for "read-only memory"). If program and data spaces are separate, writing to program RAM might involve some time-consuming operations. Second, there's the assumption that the code is well documented, as it would be a real problem to debug code that was changing during execution if you didn't expect this. Third, the assumption that the programmer is very familiar with the processor architecture and is sure that any pipelining, caches, or other issues won't mess up the intended operation.

The bottom line is whether you need this kind of headache. Unless you're against the wall, be very careful about using self-modifying code.[14]

---

14. If you absolutely must see an example of self-modifying code, there's an example in the application note "Using M68HC16 Digital Signal Processing to Build an Audio Frequency Analyzer," Motorola document AN1233/D. Why was self-modifying code used? To avoid using an index register.

7.3.3.14 Addressing Modes

More "know your architecture." Addressing modes play a crucial role in determining code efficiency and size. In general, the smaller the range of memory accessed, the smaller and faster the instruction. In most cases, there is a speed difference between accessing internal RAM (or registers) and external memory, regardless of the addressing mode. Internal RAM or registers may also be accessible using special addressing modes that are both small and quick to execute. The final image of a microcontroller's memory space, then, is a patchwork of sections with different degrees of accessibility. Even within the same memory chip, you may have a small region that's quite quick to access (e.g., the 256-byte *zero-page* of some microprocessors).

It will be useful to review the addressing modes supported by your microcontroller, checking the instruction summary to see how the different modes vary in terms of the memory space and execution time for instructions. (If you pick a very popular instruction, say, a "move to register A" instruction, you'll be able to directly compare most addressing modes.) For example, on the 68HC16, a "load accumulator a" or LDDA instruction takes two cycles when the data to be loaded follows the instruction (the "immediate" addressing mode), versus six cycles using an indexed or other form of addressing.

Table 7-6 shows the different addressing modes of the 68HC16 to give you an idea of a higher-end microcontroller. Addressing modes for the 68HC11, a popular 8-bit microcontroller that preceded the 68HC16, are marked with a ✓. Not shown is the special circular buffer addressing mode of the MAC and RMAC instructions.

7.3.3.15 Change Pointers, Don't Move Data

In case we didn't convince you in our discussion of circular buffers, *don't move data, change pointers.* We're talking about microprocessor data busses with very limited bandwidth. Buffers of sample data, like those associated with FIR filters, are much too long to shift once per sample. When we look at the IIR filter, we see the designers of the 68HC16 saw that even the movement of sample data for IIR filters could be a problem and gave us a compromise between pointers and data movement. (Well, okay, it's data movement, but very efficient data movement.) Use circular buffers, modulo addressing, even circular buffers with lengthy pointer adjustments—do anything to avoid moving data.[15]

7.3.3.16 Tables Versus Calculated Functions

If you have lots of memory, you can trade some for speed. Use look-up tables to avoid actually calculating values. Even an interpolation between table values can be quicker than evaluating—in real time—some complicated formula. You can do amazing things with look-up tables, perhaps enough to stay with a more modest processor and a bit more memory, rather than upgrade to a faster, more powerful processor. Some processors, such as the MC68HC12 and MC68300, have table interpolation built in.

---

15. Uh, did we mention not moving data?

Addressing mode	Format	Description	Example
Immediate	#<8-bit data> ✓	8 bits of data follows the instruction, usually for small constant data (vs. address).	ldda #$12
	#<16-bit data> ✓	16 bits of data follows the instruction, could be an address.	ldda #$1234
Extended	<20-bit address>	A full 20-bit address follows the instruction. Known as *absolute* addressing in other processors.	jmp $12345
	<16-bit address> ✓	16 bits of address follows instruction. This is modified by the EK page register.	jmp $1234
Relative	<8-bit signed offset> ✓	The 8-bit signed value is added to the current program counter.	bne $20 ;short branch
	<16-bit relative offset>	The 16-bit signed value is added to the current program counter.	lbne $1234 ;long branch
Inherent	<none> ✓	No address is required for many instructions.	inc a

Indexed	<8-bit unsigned offset>, <index reg.> ✓
	 <16-bit signed offset>, <index reg.>
	 <20-bit signed offset>, <index reg.>
		Address is found by adding the value of the index register (X, Y, or Z) to the 8-, 16-, or 20-bit offset. The index register is not changed.	adda $12,X ;8-bit
			 adda $1234,X ;16-bit
			 jmp $12345,X ;20-bit

Accumulator E offset	E,<index reg.>	Same as indexed, but offset is taken from register E.	adda E,X

Table 7-6 68HC16 addressing modes.

Addressing mode	Format	Description	Example
Postmodified index	<16-bit source address>,<8-bit signed destination offset>, X or <8-bit signed source offset>, X, <16-bit destination address>	The effective address is found using the X index register; afterwards, an 8-bit signed offset is added to the X register. Useful for block moves.	movw $1234,$02,X ;from $1234 to address in X, then add 2 to X

Table 7-6   (Continued) 68HC16 addressing modes.

7.3.4 Optimizing for Size

Enough optimizing for speed. Speed kills. We didn't say so before, but we're saying so now. So don't come running to us with some sad story about "Mary in research" or "Bob in systems," because they had it coming. They knew the risks.

Well, we never seem to have too much speed. Memory space is often in short supply as well in microcontroller systems. It's a rare project that has more memory than it needs, and possibly no project has had more than it wants. How can you make the most of the memory you have?

7.3.4.1   Program Size

The major savings in program size comes from "factoring out" common program segments. That's what subroutines are. The problem with subroutines, as we noted earlier, is the overhead of subroutine calls—ten or twenty cycles here, ten or twenty there—pretty soon we're talking real cycles.[16] Minor savings in program size (as well as execution time) can be achieved by making sure your assembler can generate short jumps and branches whenever possible—this is usually a safe proposition. But the total program size can be difficult to reduce.

Depending on the memory hardware, you might consider loading only part of your program at a time. These parts are sometimes called *overlays*. Many systems have well-defined operating states that require only a subset of the entire program to be accessible. For example, a "receiving data" mode might never be active at the same time as a "transmitting data" mode. If you're short of high-speed memory, arrange the program in chunks and leave most of the pieces in low-speed memory, swapping it in when the system changes state. This type of swapping is explicitly supported in several DSP chips with on-chip high-speed memory; they can automatically load this internal memory in milliseconds from an external, inexpensive EPROM. Here the

---

16.   The late U.S. Senate Minority Leader Everett McKinley Dirksen is credited with the comment "a billion here, a billion there, and pretty soon you're talking about real money."

linker and other tools can be important, otherwise you'll need to be tricky with the source code as you'll have multiple sections of code that will look like they belong in the same place.

The high-speed internal RAM of the 68HC16Z1 makes it a nice place to put both program code and data; with only 1K bytes of RAM, you can see that swapping might make sense.

7.3.4.2 Data Size

Know the nature of your data. Pay particular attention to large arrays of data, though your usual options as far as inherent size of data will be only between bytes and words (and sometimes "long words," that is, 32-bit words). It's pretty hard to justify much bit-shuffling even if you have to "waste" 6 bits by storing a 10-bit value in a 16-bit word—the added processing time is just too much. On the flip-side, if you can spare some processing time and have a lot of data, it may make sense to pack data tightly. If you have a barrel shifter, packing and unpacking data across word boundaries will be much, much faster. (But probably never fast enough to justify packing sample data, as you want to be able to cleanly access each sample as fast as possible.)

Breaking up your program into mutually exclusive pieces can help with effective code size (i.e., via overlays); likewise for data. In fact, some compilers can analyze code to determine what variables are never used at the same time, letting the compiler assign more than one variable to the same memory location. This optimization is very important for microcontrollers with internal RAM that can be accessed very quickly (e.g., through special addressing modes). This memory sharing does make your life harder if you need to debug at low levels, and if you assign variables "manually," a mistake could be very hard to find. Nonetheless, the savings can be large enough to consider the technique, whether on a small scale for miscellaneous variables or with a couple of large buffers.

Ping-pong buffers represent an extreme case of variables that might not be in use at the same time, at least in the same way. Ping-pong buffers are used to process data in batches; you can be processing one buffer in its entirety while another is filling up. The processing of the one buffer must be complete before incoming data fills the other; once the incoming buffer is full, the roles of the buffers are swapped and the processed (and, thus, "empty") buffer becomes the "incoming" buffer. Ping-pong buffers often produce very clean solutions to data flow, but before committing to such an arrangement, make sure that the increase in data space is necessary. Perhaps a circular buffer or coupling the sampling and processing functions more tightly would allow the use of a single buffer.

7.3.5 Floobydust[17] on Programming

7.3.5.1 Development System Hardware

---

17. From the *National Semiconductor Audio/Radio Handbook* [1980]: *"'Floobydust' is a contemporary term derived from the archaic Latin miscellaneous, whose disputed history probably springs from Greek origins (influenced, of course, by Egyptian linguists)—meaning here 'a mixed bag.'"* An excellent handbook that is full of application notes and real-world audio and radio issues.

All of this talk about optimizing speed and memory shouldn't obscure the fact that your development platform doesn't need to be slow, or limited in memory, even if the product you're developing does. Buy the fastest version of the processor you'll be using in the ultimate product, then slow the clock down once the code is running. Add extra memory to your development platform, even if you have to wire-wrap a separate board. It will be the best time you ever spent, as extra memory is always useful for saving debugging data and such. Besides, you'll be adding some circuitry to let you read output port pins (if you take our advice, that is). Optimize *final* code, not development code.

7.3.5.2   The Tool Chain— Development System Software

The phrase "tool chain" refers to the series of translators and other software tools used in producing executable object code from source code. We'd like to extend that definition to include tools that take executable object code, load it into the target system, run the code, and provide the programmer with information about code performance. Figure 7-3 shows a generic tool chain for embedded systems programming. Initially, you may not use all of these tools (and most of us will never have our code made into a mask ROM on a chip, either), but, minimally, you need the tools to get from the text editor at the top to execution of code at the bottom.

We've already suggested that you should initially find batch files (or "make" files) that can guide you through the tool chain. Each tool has its own set of command-line flags and default behaviors. Some of these options can be safely ignored, others are crucial to making the code run.

Once you've figured out how to use each tool (e.g., command-line arguments, input and output files required, etc.), it's time to automate the tool chain. For simple projects, a batch file is often sufficient. Remember to have the batch file "bail out" if a tool returns an error. However, if a project involves more than a few source files, you should consider using a *make* utility. You supply a "make" file describing the relationships between files—for example, you can specify that "bob.hex" is made from "bob.obj" by issuing the command "hex bob.obj bob.hex." The make utility uses the time stamp of files to issue only the commands necessary to update the final product. If you edit a single assembly language file, the make utility issues commands to assemble only that one file. Of course, the linker and all the other tools in the chain are invoked as necessary. All in all, make utilities are powerful tools worth getting to know.

Figure 7-3 lists a few outcomes of the tool chain such as code to be simulated or used by an emulator, "burned" into an EPROM, or sent to the factory for use in a masked-ROM version of a chip. There are variations on these themes—EPROM emulators that pretend to be EPROMs (they even plug into the EPROM socket itself), but they connect to your PC and let you change their contents in a second, rather than the minutes required to program an actual EPROM. The *background debug mode* of the 68HC16 lets us load program code into external RAM configured to otherwise look like (EP)ROM. In some cases, you can just download directly into on-chip RAM without the need for external memory. (This is very useful if you have a single-chip processor with enough internal memory. You can download the code from a microcontroller with its own memory, avoiding the need for any external memory for the DSP chip.)

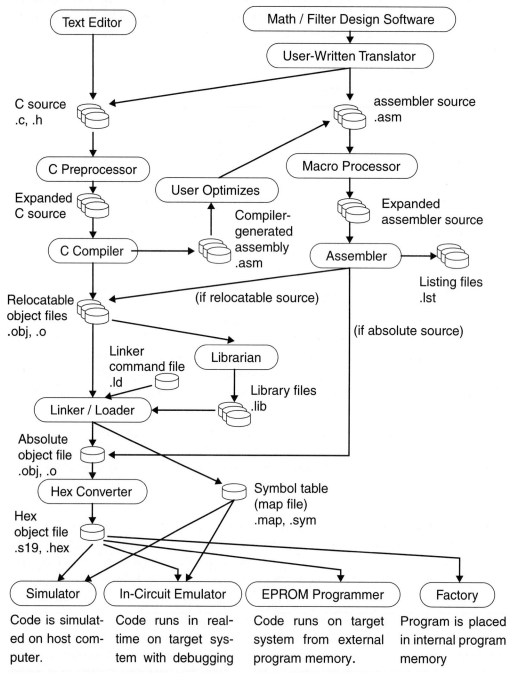

Figure 7-3 A generic "tool chain," from source to executable object code.

Why worry about optimizing the tool chain? Every minute you put into tools is usually paid back, with handsome interest. Automating the tool chain reduces the number of errors, speeds up the code-test-code cycle, and provides documentation so you can later return to the same development setup.

7.4 Finite Word-Length Effects

In Chapter 4 (Discrete-Time Signals and Systems), we found that quantizing sample amplitudes reduced the SNR of the signal being processed. However, quantization of amplitudes does not affect the frequency response of the system—the frequency response is determined by the sampling rate and, of course, the system function.

To realize the system in a processor, we have to quantize the filter coefficients. Unlike quantizing amplitudes, this quantization does change the frequency response (but not so much the SNR—isn't life interesting?). We also need to study the effects of doing arithmetic with fixed-point numbers—for example, the infamous large- and small-scale "limit cycles."

7.4.1 Coefficient Quantization

When we take carefully calculated filter coefficients and mercilessly hack them to fit 16-bit (or smaller!) straightjackets, we also alter the poles and zeros of the system. This movement of each pole and zero in the z-plane may be toward or away from the unit circle, and higher or lower in frequency. In extreme cases, we might even move one or more poles outside the unit circle, producing an unstable system. (Because FIR filters contain no poles, FIR filters are always stable.)

The risk of instability is rather small if 16-bit coefficients are used and the IIR filter is broken down into second-order sections. You can look at the impulse response or run simulations to assure "stable" results. Alternatively, you can find the roots of the denominator of the "quantized" system function (e.g., by using mathematics software); these roots are the poles of the system and must all have a magnitude of less than 1 to be inside the unit circle.

It is important to examine the frequency response of the "quantized" system to make sure the system still meets your design parameters. You don't need to actually implement the system to check the frequency response—you may be able to use your filter design software, even if the software doesn't "know" about fixed-point arithmetic. Using procedures mentioned earlier, you can convert the ideal floating-point coefficients into, say, 1.15 format (i.e., 16-bit fixed point), then back again to floating-point format. Plug these "quantized" floating-point coefficients into your filter design software to generate a frequency response.

Quantizing coefficients to 16 bits of resolution usually won't cause much change in the frequency response, but any change will be for the worse (assuming an optimal design to start with). If your original design had no margins left, you may need to add an additional pole and/or zero to compensate for the movements of the poles and zeros due to coefficient quantization.

7.4.2 Limit Cycles

7.4.2.1 Large-Scale

We've already met the so-called large-scale limit cycles in the discussion of overflow and saturation. Basically, the problem is that without saturation (or without checking the overflow flag "manually"), a very large positive number can, with just one more "straw" added, roll over to become a very large negative number. The output of such a system would contain "spikes" whenever this happens. Such anomalies are quite annoying in audio signals, and even worse for control systems where a spike might cause a failure to control a process. (For example, if you were controlling an industrial oven, rolling over from "very hot" to "very cold" might cause your controller to increase the heat even more.) The solution to large scale limit cycles is a combination of scaling and the use of saturation. Scaling attempts to match the dynamic range of the signals being processed to the internal representations of those signals. Scaling minimizes *clipping* (distortion caused by exceeding the full-scale range of the internal representation) but decreases the SNR (fewer bits represent the signal range). We'll be using the 1.15 format, so we'll normally want signals in the -1 to +1 range. Saturation ensures that if we occasionally have overflow, a value doesn't wrap around to the opposite sign—rather, it is "clamped" or restricted to the valid range.

7.4.2.2 Small-Scale

At the other end of the scale, we have effects due to the limited precision representation of values. Take an IIR filter for example. We know that, in theory, the impulse response never goes to zero, though for many IIR filters the impulse response becomes small quickly. Eventually, the theoretical output of the IIR filter will be less than one quantization level. Because we can't properly represent small numbers, an endless cycle emerges as rounded-off values are fed back into the filter. The output of the filter never goes to zero; instead, a low-level oscillation is set up.

Advanced texts[18] discuss methods of resolving small-scale limit cycles. These methods include using a special form of truncation instead of rounding and making sure that the frequency of the limit cycle of an SOS doesn't inadvertently match the resonant frequency of a later SOS. Small-scale limit cycles, because they are related to the resolution used for storing and processing signals, are much more of a problem for systems with short data lengths; another subtle nudge away from eight-bit architectures, perhaps, and another benefit for FIR filters, which, lacking feedback, don't suffer from small-scale limit cycles.

7.5 FIR Filter Implementation

Let's explore the microprocessor implementation of the FIR filter we designed in Chapter 5 (FIR Filters—Digital Filters Without Feedback). We'll use the BT43367F, a 13-bit, 750-kHz microcontroller with only two registers, both named *A*. We'll also use 500-ns 1K by 4-bit RAM for data storage, and a 16-bit serial ADC for output.

---

18. The usual suspects—Proakis and Manolakis [1992], Oppenheim and Schafer [1989], and so on.

Well, actually the BT43367F isn't a good example. We're always forgetting which register is *A* and which is *A*. Instead, let's use a MC68HC16Z1-based system. We'll use the on-chip ADC for 10-bit samples at 10 kHz and an external 16-bit serial DAC for output. The development environment will be the M68HC16Z1EVB, a simple evaluation board for the 68HC16 that supports background debugging mode (BDM). (The BDM is the mode we mentioned earlier that gives us many of the same features as an in-circuit emulator, but at very low cost.)

The MC68HC16Z1 is quite a sophisticated device. In fact, it's really six different devices, in one package, connected by an internal bus:

- CPU16. This is the central processing unit. All of the registers are found here, as well as the arithmetic and logic unit and the DSP-specific registers and logic.
- System Integration Module. Connects the modules together and supports the operation of the internal and external buses. Also includes a watchdog timer, the system clock, and some input/output ports.
- Analog-to-Digital Converter. A 10-bit ADC with an eight-channel multiplexer. Can be configured to convert multiple channels without intervention by the CPU.
- Queued Serial Module. Provides both synchronous and asynchronous serial communication. Has its own "program" space so it can be programmed to automatically perform sequences of operations.
- General Purpose Timer. Very flexible hardware for timing/counting external events, generating internal events at precise intervals, and generating pulse-width modulated signals.
- Standby RAM. 1K bytes of static RAM for fast internal storage of program code or data.

One price for all of this power is lots of configuration at the start of your program. Sometimes initializing modules like a timer or ADC can be pretty arcane, but if you can exploit an existing program, it's not too bad. In an appendix we list the source code for the examples in this text. These listings might provide a good start if you have similar hardware. We will emphasize, however, the code that is less specific to peripherals and concentrate on code that is accomplishing more "interesting" tasks.

Figure 7-4 shows the programmer's model for the 68HC16 CPU. If you've worked with the 68HC11, some of these registers should look familiar, though perhaps a bit longer than you remember. We're not going to dwell on the 68HC16 architecture except to the extent that it influences our DSP routines. A few of the major points you should be aware of include:

- 20-Bit addresses are often generated using 16-bit values with a 4-bit address extension register prepended. The extension registers *XK, YK, ZK, SK, EK*, and *PK* are generally loaded at the start of the program and left unchanged, but if access to a different 64K page is required, the extension register must be changed.
- Registers *A, B, D*, and *E* are general purpose registers; most logic and arithmetic operations can be performed in these registers. Registers *A* and *B* are the most significant and least significant bytes of register *D*.
- Registers *IX, IY,* and *IZ* provide indexed addressing; limited addition is supported.
- *SP* is the stack pointer, which builds downward in memory.

FIR Filter Implementation

- *PC* is the program counter, which must always have an even address due to the way in which program memory is read.
- *CCR* is the "condition code register" and contains overflow, negative, zero, and carry flags. An interrupt control field is also located here, as well as a bit to enable saturate mode.

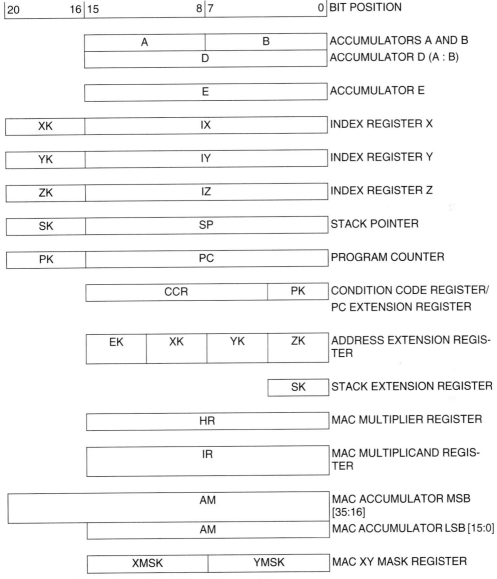

Figure 7-4 CPU16 (e.g., 68HC16) register model.

We'll begin the implementation by converting the FIR filter coefficients into a form we can use with the assembler. For this example, we'll design a filter with the following characteristics:

Passband edge frequency: 2700 Hz

Stopband edge frequency: 3000 Hz

Maximum passband attenuation: 0.5 dB

Minimum stopband attenuation: 40 dB

We'll use the Parks-McClellan program (Chapter 5 [FIR Filters—Digital Filters Without Feedback]) to generate the filter coefficients. (If you're following along, you'll note we had to increase N to meet the design parameters—the estimate from Equation 5-1 was too low.)

Table 7-7 shows the actual coefficient values, followed by the coefficient in 1.15 format in both decimal and hexadecimal. The assembler we're using can perform the two's complement conversion for negative numbers. If you don't mind leaving the coefficients in their scaled decimal equivalents, you don't have to bother with doing the two's complement conversion yourself. Somehow, the hexadecimal format is reassuring to us, but that's just personal preference. And, of course there's no reason you can't use binary, but we usually reserve that for configuration register values that have different bit fields.

	Decimal	1.15 Scaled Decimal	1.15 Hexadecimal
0	-0.010628	-348	FEA4
1	-0.018856	-618	FD96
2	-0.000714	-23	FFE9
3	0.008786	288	0120
4	-0.002800	-92	FFA4
5	-0.008782	-288	FEE0
6	0.008229	270	010E
7	0.006038	198	00C6
8	-0.012383	-406	FE6A
9	-0.001920	-63	FFC1
10	0.016172	530	0212
11	-0.005050	-165	FF5B
12	-0.017625	-578	FDBE

Table 7-7 Coefficients for example FIR filter.

	Decimal	1.15 Scaled Decimal	1.15 Hexadecimal
13	0.014191	465	01D1
14	0.015682	514	0202
15	-0.025147	-824	FCC8
16	-0.008532	-280	FEE8
17	0.036783	1205	04B5
18	-0.005939	-195	FF3D
19	-0.047864	-1568	F9E0
20	0.032174	1054	041E
21	0.057019	1868	074C
22	-0.085798	-2811	F505
23	-0.063057	-2066	F7EE
24	0.311294	10200	27D8
25	0.565166	18519	4857
26	0.311294	10200	27D8
27	-0.063057	-2066	F7EE
28	-0.085798	-2811	F505
29	0.057019	1868	074C
30	0.032174	1054	041E
31	-0.047864	-1568	F9E0
32	-0.005939	-195	FF3D
33	0.036783	1205	04B5
34	-0.008532	-280	FEE8
35	-0.025147	-824	FCC8
36	0.015682	514	0202
37	0.014191	465	01D1

Table 7-7 (Continued) Coefficients for example FIR filter.

	Decimal	1.15 Scaled Decimal	1.15 Hexadecimal
38	-0.017625	-578	FDBE
39	-0.005050	-165	FF5B
40	0.016172	530	0212
41	-0.001920	-63	FFC1
42	-0.012383	-406	FE6A
43	0.006038	198	00C6
44	0.008229	270	010E
45	-0.008782	-288	FEE0
46	-0.002800	-92	FFA4
47	0.008786	288	0120
48	-0.000714	-23	FFE9
49	-0.018856	-618	FD96
50	-0.010628	-348	FEA4

Table 7-7 (Continued) Coefficients for example FIR filter.

Before going further, we should first check that the coefficients rounded to 16 bits will still produce an acceptable frequency response. (We could also double-check the linearity of the phase.) If you have the resources, it is also nice at this point to simulate the finite word-length arithmetic. Figure 7-5 shows results of using both 16-bit coefficients and simulated 16-bit by 16-bit multiplication (with 36-bit accumulator). It doesn't look like we have any problems from finite word-length issues, so we can move on to incorporating the coefficients in our program.

Listing 7-3 shows the file we'll include in our FIR program. We set up our mathematics software so that after generating the coefficients, a "script" we wrote generates the file—including comments. You'll want to do something similar, rather than copying coefficients manually. Time spent building tools like this are almost always worth the effort, as you'll almost always need a few iterations. (Actually, *you* might not want the iterations, but usually someone does...)

Listing 7-3 includes two constants used by the main program. The first is *numtaps*, the number of taps or filter coefficients. The second constant, *bsize*, gives the size of the smallest possible circular buffer that can be used for the filter. The number of coefficients, 51, is greater than 32 (2^5), so we need to go to the next power of 2, which is 64. (More on the circular buffer issue in a moment.) Writing the code so that these two constants could be defined in this "include" file lets us change the filter without having to make changes to the main program. If

FIR Filter Implementation

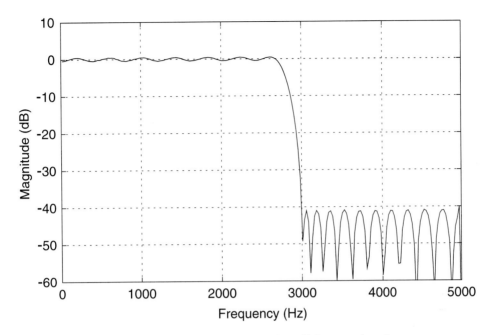

Figure 7-5 FIR response using simulated quantized coefficients and math.

```
*              Coefficient file fircoef1.asm
*              Generated 18-Nov-96
*              Fpb=2700 Hz, Fsb=3000 Hz, Fs=10000 Hz
*              Ap= 0.5 dB, As=40 dB
*
numtaps  equ        51              ;number of coefficients
bsize    equ        64              ;next power of 2 higher
f_coefs:                            ;filter coefficients
         dc.w       -348            ;h(  0)=-0.010628
         dc.w       -618            ;h(  1)=-0.018856
         dc.w        -23            ;h(  2)=-0.000714
         dc.w        288            ;h(  3)= 0.008786
         dc.w        -92            ;h(  4)=-0.002800
         dc.w       -288            ;h(  5)=-0.008782
         dc.w        270            ;h(  6)= 0.008229
         dc.w        198            ;h(  7)= 0.006038
         dc.w       -406            ;h(  8)=-0.012383
         dc.w        -63            ;h(  9)=-0.001920
         dc.w        530            ;h( 10)= 0.016172
         dc.w       -165            ;h( 11)=-0.005050
         dc.w       -578            ;h( 12)=-0.017625
         dc.w        465            ;h( 13)= 0.014191
         dc.w        514            ;h( 14)= 0.015682
         dc.w       -824            ;h( 15)=-0.025147
         dc.w       -280            ;h( 16)=-0.008532
         dc.w       1205            ;h( 17)= 0.036783
         dc.w       -195            ;h( 18)=-0.005939
         dc.w      -1568            ;h( 19)=-0.047864
         dc.w       1054            ;h( 20)= 0.032174
         dc.w       1868            ;h( 21)= 0.057019
```

Listing 7-3 Coefficient file for FIR program.

```
        dc.w      -2811      ;h( 22)=-0.085798
        dc.w      -2066      ;h( 23)=-0.063057
        dc.w      10200      ;h( 24)= 0.311294
        dc.w      18519      ;h( 25)= 0.565166
        dc.w      10200      ;h( 26)= 0.311294
        dc.w      -2066      ;h( 27)=-0.063057
        dc.w      -2811      ;h( 28)=-0.085798
        dc.w       1868      ;h( 29)= 0.057019
        dc.w       1054      ;h( 30)= 0.032174
        dc.w      -1568      ;h( 31)=-0.047864
        dc.w       -195      ;h( 32)=-0.005939
        dc.w       1205      ;h( 33)= 0.036783
        dc.w       -280      ;h( 34)=-0.008532
        dc.w       -824      ;h( 35)=-0.025147
        dc.w        514      ;h( 36)= 0.015682
        dc.w        465      ;h( 37)= 0.014191
        dc.w       -578      ;h( 38)=-0.017625
        dc.w       -165      ;h( 39)=-0.005050
        dc.w        530      ;h( 40)= 0.016172
        dc.w        -63      ;h( 41)=-0.001920
        dc.w       -406      ;h( 42)=-0.012383
        dc.w        198      ;h( 43)= 0.006038
        dc.w        270      ;h( 44)= 0.008229
        dc.w       -288      ;h( 45)=-0.008782
        dc.w        -92      ;h( 46)=-0.002800
        dc.w        288      ;h( 47)= 0.008786
        dc.w        -23      ;h( 48)=-0.000714
        dc.w       -618      ;h( 49)=-0.018856
        dc.w       -348      ;h( 50)=-0.010628

* End of file *
```

Listing 7-3 Coefficient file for FIR program.

we chose instead to modify the constants by hand, sooner or later we'd be sure to forget to update the main program, a recipe for unhappiness.

Listing 7-4 shows excerpts from the main program. (The complete program is listed in Appendix 4.) Notice how much of the program is influenced by the nature of the RMAC instruction—which registers it uses, how data is accessed, the format of data, and so on.

```
        ...
        include 'fircoef1.asm'  ;filter coefficients
        ...
bmask   equ     %01111111       ;mask for modulo addressing
*       Variables (in internal RAM)
        org     $0000           ;(in bank f, actually)
inbuff  ds.w    bsize           ;input buffer storage
topbuff equ     *-2             ;addr of last entry (starting point)
inptr   ds.w    1               ;input buffer pointer
        ...
*       Initialize variables
        ldd     #topbuff        ;start at top of buffer
        std     inptr
*       set up "modulo" addressing registers (circular buffer)
        ldd     #bmask          ;modulo addressing mask
                                ;XMSK (coefs) =$00 (modulo disabled)
                                ;YMSK (data) = bmask (modulo enabled)
        tdmsk                   ;(transfer D -> mask)
```

Listing 7-4 Excerpts from the main FIR program.

FIR Filter Implementation

```
*         Main loop
WAITLOOP:
          wai                         ;sit here until interrupt
                                      ;(minimum and consistent interrupt latency)
          bra       WAITLOOP

* Interrupt service routine, called at sample rate
irq_sample:
*         acknowledge the interrupt
          ...
*         read the ADC, put value in register D
          ...
*         store sample in circular buffer
          ldy       inptr
          cpy       #inbuff           ;will decrement past start of buffer?
          bgt       nowrap
          ldy       #topbuff+2        ;start at top again (+2)
nowrap:
          aiy       #-2
          sty       inptr
          std       0,y               ;store actual data
                                      ;(this is the lowest valid location
                                      ;in buffer)
*         Evaluate filter
          clrm                        ;clear AM (the MAC accumulator)
          lde       #numtaps-1        ;number of times (-1) to do RMAC
          ldx       #f_coefs          ;start of coefs
                                      ;no need to load y—already set
          ldhi                        ;load H and I registers using ix,iy
          rmac      $2,$2             ;repeating MAC instruction:
          tmer                        ;transfer AM to E (saturated if necessary)
*         output data to DAC
          ...
          rti                         ;end isr
          ...
```

Listing 7-4 (Continued) Excerpts from the main FIR program.

Let's start at the center, the RMAC (repeating multiply and accumulate) instruction. RMAC *xo,yo* performs the following sequence of operations:

Do:
1. Multiply the 16-bit value in H by the 16-bit value in I
2. Add the result to accumulator AM
3. Add the 4-bit signed value xo to X, but allow only carries subject to $XMSK$
4. Add the 4-bit signed value yo to Y, but allow only carries subject to $YMSK$
5. Load the data pointed to by X into H
6. Load the data pointed to by Y into I
7. Subtract 1 from E

Until $E < 0$

Let's take each of these operations in turn.
1. RMAC must start with valid data loaded in H and I. A special-purpose instruction, LDHI, loads H and I using the X and Y registers, and is usually found just before the RMAC instruction (as here). Note that after multiplying, RMAC loads H and I with the next data to be multiplied (steps 5 and 6 below).

2. The product of *H* and *I* are put in 1.31 format and added to the contents of the 32-bit accumulator, *AM*. However, we must explicitly clear *AM* prior to using the RMAC instruction by using CLRM. Note, too, that we must enable saturation by setting the *SM* (saturate mode) bit in the condition code register (*CCR*).

3. The 68HC16 supports a limited type of circular buffer based on limiting the number of bits affected by an increment of an index register. Two special registers, *XMSK* and *YMSK*, contain 1s in bit positions that are allowed to change as the circular buffer is addressed. *XMSK* and *YMSK* are loaded at the same time from the (16-bit) *D* register using a special instruction, TDMSK. This limits *XMSK* and *YMSK* to 8 bits each; thus, the largest buffers supported by this address mode are 256 bytes, or 128 words, long. Here, we've disabled *X* modulo addressing by making *XMSK* equal $00; because the coefficient array is always accessed starting at the first word, there is never any wrap-around, and, therefore, no need for modulo addressing. The 4-bit signed constant *xo* is just added to *X* to form the next address. (Because each entry is two bytes long, *xo*, the increment, is +2.)

4. The *Y* register does use modulo addressing because we're accessing sample data and we must be able to start anywhere in the buffer and wrap around to the start. In this program, *YMSK* is set to *bmask*, a constant generated earlier in the program from *bsize* (defined in the filter coefficient file). In this case, *bmask* is 127 or, in binary, 0111 1111. (Be careful to count bytes, not words—the buffer is 64 words long, but 128 bytes long.) The increment (*yo*) is also +2, but both *xo* and *yo* can range from -8 to +7 bytes.

5. RMAC fetches the next value for *H* using the *X* register. For the FIR filter we could have *X* pointing to either the data or coefficient arrays—both *X* and *Y* have the same functionality. In this case, *X* is pointing to the coefficient array.

6. In the same way, the next value of *I* is fetched using the *Y* register, which points to the data array.

7. Finally, the count in register *E* is decremented by 1. *E* is tested to see if it is negative. Therefore, we need to load *E* with one less than the number of samples and coefficients. (See, for example, the "lde #numtaps-1" line in the listing.)

Once the RMAC instruction is complete, data is transferred from *AM* in one of several ways—rounded to the most-significant 16 bits, truncated to the most significant 16 bits, or the entire 36 bits transferred to a set of registers. We're interested only in the most significant 16 bits, and rounding is appropriate, so we've used the TMER instruction to transfer data to the *E* register. From there we'll send it out to the DAC using the Queued Serial Module.

The remaining code of the listing places incoming samples into the sample buffer. Note that the buffer is filled from the top down, so the most recent sample is the first to be multiplied by the RMAC instruction.[19] We do have to worry about moving past the beginning of the buffer, thus the inclusion of code that adjusts the buffer pointer to the start of the buffer. By the way,

---

19. Of course, because the coefficients of a linear-phase FIR filter are symmetric about the center, you have some flexibility in the order of multiplication.

FIR Filter Implementation

because we've set up the buffer according to the 68HC16 modulo addressing rules, you might see a faster way of doing this addressing (e.g., using a bit mask).

There is a nagging question concerning the mismatch between the number of coefficients and the size of the circular buffer. The short answer is that while we unfortunately use the entire 64 words of the circular buffer for storage, we actually use only the most recent 51 words (samples) in our calculations. There's no speed penalty for using the mismatched buffer size, just a loss in memory.[20]

If memory—especially data memory—is an issue, we could rewrite the FIR routine in two parts to capitalize on the symmetry of the coefficients. Recall that for a linear phase FIR filter, the coefficients are symmetric around the center—thus, we can get by with storing only the first half. One RMAC instruction can be used to handle half of the multiply-accumulates, stepping forward through the (shortened) coefficient array. (We would obviously need to change the count we gave to RMAC in register E.) Then, probably with suitable adjustments to the coefficient pointer, a second RMAC performs the remaining multiply-accumulates, this time stepping backward through the coefficient array (i.e., *xo* would be -2, *yo* would still be +2). Note that this isn't the "linear-phase" structure we described in Chapter 5 (FIR Filters—Digital Filters Without Feedback), as we're not minimizing the number of multiplies. We're just adding a little to the execution time to "switch directions" in the middle of the filter to save some coefficient storage. The linear-phase structure doesn't make much sense for the 68HC16 since the RMAC is so powerful and fast. On microcontrollers without great multiplication support, though, you might consider the linear-phase structure to see if it would save some cycles.

The source code reveals that the sampling in this program is *interrupt-driven*. You might recall we earlier discussed several other alternatives, including making the timing dependent on the execution of instructions, setting up the ADC to sample regularly, and polling a timer. These are all options with the 68HC16. However, program-dependent timing is hard to maintain (and write!); the ADC in this case can't be set to sample precisely at 10 kHz (though, for example, 10.49 kHz is possible), and polling a timer leads to irregular sampling intervals. (In Chapter 4 [Discrete-Time Signals and Systems] we discussed how variations in the sampling interval result in a decrease in the SNR. Polling a timer can introduce over a dozen clock cycles of difference between sample intervals. If the jitter time is 400 nsec—a little conservative—the SNR of a full-scale sinewave would drop to around 40 dB.)

We can also incur some variation in the interrupt latency; the two main issues are the relationship of the interrupt signal to the microprocessor's execution cycle and the operations executed by the microprocessor at that moment. Interrupts are usually sensed at periodic intervals (often a small multiple of the clock period); thus, interrupts that are asynchronous to the microprocessor clock will have a bit of jitter to them. Once an interrupt is sensed, the microprocessor may have to complete the execution of an instruction that takes many more cycles to complete or may even be in a mode where interrupts are disabled—the ultimate jitter!

---

20. We were going to write something about getting older, but we forget what it was.

In this program, we avoid both problems using simple techniques. First, the source of interrupts is the on-chip general-purpose timer—which is synchronous with the processor. Second, we place the chip in a wait mode (using the "wai" instruction) so that each interrupt is processed with the same amount of delay. Note that we really don't care how long the interrupt latency is (within reason), only that this time is the same from sample to sample. Were we not able to put the chip in a wait mode, we would experience variation in the interrupt latency as different instructions were interrupted during different samples.

The evaluation of the filter occurs in the interrupt service routine. There are other valid arrangements. For example, the interrupt service routine can insert samples into a buffer, while a foreground task can extract the samples at its leisure, evaluate the filter, then put samples into an output buffer to be sent by another interrupt service routine. The actual structure will be influenced by the other operations your program must perform—evaluating the filter in the interrupt service routine is only one option. Of course, no structure in the world will help you if your filter evaluation per sample takes longer, on average, than the sampling period.

How fast does this code run? The RMAC instruction itself takes 12 clock cycles per coefficient (plus 6 cycles overhead); if both program and data were in fast memory, the core of this filter would therefore take about 37 μsec.[21] Running the program in the appendix, which toggles pins to allow the different sections of code to be timed, about 54 μsec are required for the interrupt service routine to execute. (Not only is other code executed in the interrupt service routine, but running the program in external memory[22] adds time, too.) At a 10-kHz sampling rate, this is just over 50% of the available processor bandwidth.

By the way, what is the actual frequency response of the system? (Not that we would want to break with tradition and suggest that anything in this book is actually *practical*...) We hooked up a function generator and digital multimeter, both controlled remotely by a PC, and found the response shown in Figure 7-6. (The ideal response is also plotted.) Granted, this was not a low-distortion sine wave generator, nor did we spend days designing low-noise analog input and output stages, but even so, it looks like the final filter performs in the real world very similarly to our "paper-and-pencil" design. Success!

The point of this book isn't to teach you how to program the 68HC16. Our point is to use the 68HC16 as an example microcontroller to see how the core of an FIR digital filter can be implemented in an efficient way. This code isn't fully optimized, though we've tried to point out some general directions. You should begin to see how the architecture and available instructions

---

21. This is with a 16.78-MHz processor clock—newer versions of the 68HC16 can run at higher speeds. At 16.78 MHz, each clock cycle is 59.6 nsec. As we write this, the 68HC16 is available at speeds up to 25 MHz, close to 50% faster than the speed used here.

22. This is getting pretty technical, but on the 68HC16 you can access memory using "fast termination," "0-wait state," or up to 13 wait states. All of the instruction timing in the chip manuals is based on running code and accessing data in the fastest of these, "fast termination" or "two-cycle" memory. 0-wait state memory requires an additional cycle (a total of 3), and each additional wait state adds a cycle. Our programs here are running in external, 0-wait state memory, and the data is accessed in internal, fast-termination memory.

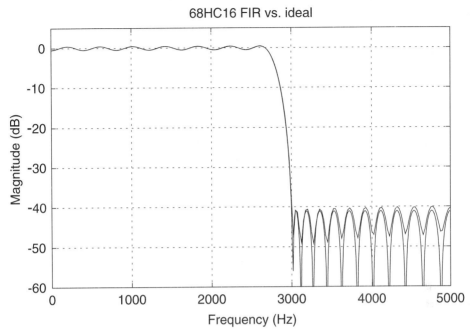

Figure 7-6 Magnitude response of actual and ideal FIR filter.

deeply influence how a DSP routine is written for a microcontroller and, specifically, how an FIR filter is implemented in one architecture.

But wait—there's more to digital filter implementation. The IIR is a much trickier beast to code, as we'll want to use second-order sections. We'll also open a can of worms when we examine the coefficients needed for the task.

7.6 IIR Filter Implementation

As you recall from Chapter 6 (IIR Filters—Digital Filters With Feedback), we generally implement IIR filters as a cascade of second-order sections (SOSs), also known as *biquads*. Figure 7-7 shows the structure in the canonic or direct form II structure. This implements the following equation (the G term assures that the filter gain is what we want in the passband).

$$H_i(z) = \frac{b_{i0} + b_{i1}z^{-1} + b_{i2}z^{-2}}{1 + a_{i1}z^{-1} + a_{i2}z^{-2}} \cdot G \qquad (7.3)$$

We face two issues concerning fixed-point computation with this filter implementation. First, the values for x, y, and w must be in the range of the numeric representation. (We don't have to worry about intermediate results like $b_1w(n-1)$, as this value will be directly added to the

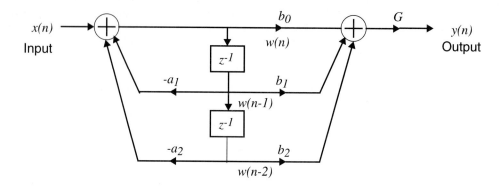

Figure 7-7 Canonic or direct form II structure of a second-order section for an IIR filter.

accumulator, which has extra bits to handle temporarily large results.) Using 16 bits and a 1.15 format, this range is $-1 \leq x < 1$. Second, the filter coefficients must be represented in 16 bits, again in the same -1 to 1 range. Neither of these conditions is necessarily met by the coefficients found for the example IIR filter in Chapter 6 (IIR Filters—Digital Filters With Feedback). (You'll note we conveniently avoided both issues in the FIR example—the coefficients are less than 1, and there's no intermediate storage in the FIR structure used.) We need to modify the SOS to incorporate data scaling and to develop a recipe that assures both conditions are satisfied without sacrificing the system's SNR.

We begin by introducing three additional gain terms—$G1$, $G2$, and $G3$—and replacing the filter coefficients with modified coefficients $A1$, $A2$, $B0$, $B1$, and $B2$. The new structure is shown in Figure 7-8. Let's take each of our requirements and see how we can generate appropriately modified coefficients and gain terms. As an example, we'll use the lowpass Butterworth filter we designed earlier and will assume that the numerator and denominator coefficients are those polynomials resulting from multiplying the simple zero and pole factors, respectively. The poles and zeros, along with the SOSs, are shown in Table 7-8.

Our first job is to calculate a scaling factor G such that $y(n)$, the output of this section, always has an absolute value less than or equal to 1 for any sequence of inputs $x(n)$, all of which are less than or equal to 1. We know that for a lowpass Butterworth filter, the maximum gain of 1 occurs at $\omega=0$. (For a Chebychev I with N even, the point of maximum gain (1) is elsewhere, while the gain at $\omega=0$ is $1-\delta_p$.) Frequency $\omega=0$ corresponds to $z=1$, so the required G is the reciprocal of $H(z)$ evaluated at $z=1$. For $H_1(z)$, G is $1/((1+2+1)/(1-1.3547+0.7526))$, or 0.09947. Table 7-9, where we'll record the rest of our calculations, shows the remaining gain values for the other SOSs.[23] We'll put the G factor into the value of G_1, which will also carry some additional factors.

This satisfies our need for $|y(n)|$ to be less than 1; and for completeness, we'll explicitly note that $|x(n)|$—either the actual input to the filter (e.g., ADC), or the output of the previous cascaded section—must also be less than 1.

IIR Filter Implementation

i	$H(z)$	Poles	Zeros
0	$H_0(z) = \dfrac{1 + 1z^{-1}}{1 + (-0.4729)z^{-1}}$	$0.47392 + 0j$	-1
1	$H_3(z) = \dfrac{1 + 2z^{-1} + 1z^{-2}}{1 + (-0.9837)z^{-1} + 0.2726z^{-2}}$	$0.49183 + 0.17515j$ $0.49183 - 0.17515j$	$-1, -1$
2	$H_2(z) = \dfrac{1 + 2z^{-1} + 1z^{-2}}{1 + (-1.1078)z^{-1} + 0.4331z^{-2}}$	$0.55387 + 0.35543j$ $0.55387 - 0.35543j$	$-1, -1$
3	$H_1(z) = \dfrac{1 + 2z^{-1} + 1z^{-2}}{1 + (-1.3547)z^{-1} + 0.7526z^{-2}}$	$0.67734 + 0.54202j$ $0.67734 - 0.54202j$	$-1, -1$

Table 7-8 Poles and zeros of example lowpass Butterworth.

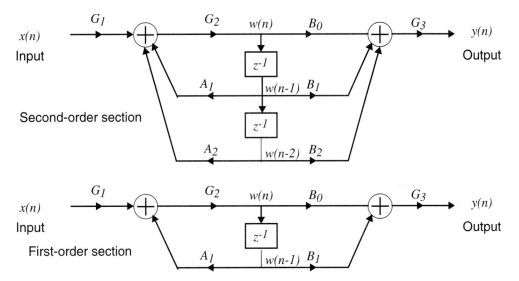

Figure 7-8 Modified structures for first- and second-order sections.

23. A brief aside on calculations. We're not here to overwhelm you with numbers, so we've tried to restrain ourselves, numerically speaking, usually recording only the first 4 or 5 digits after the decimal point. However, in practice, you are well advised to keep the values stored with as much precision as possible, perhaps in your calculator's memory or in your mathematics software. This is particularly important in calculating the coefficients of IIR filters. We'll worry enough about the effects of having a limited number of bits to store the coefficients— no need to make matters worse by starting off with coefficients that have fewer than 16 bits of resolution! (16 bits of resolution corresponds to about 5–6 digits past the decimal point.)

There is one more important data array to worry about—$w(n)$. (By the way, this will be a different value than the $w(n)$ for the unscaled SOS.) How can we guarantee that $|w(n)|$ will never be larger than 1? By scaling the input $x(n)$ by an appropriate factor s_1.

There are three popular ways of calculating s_1, depending on how adventurous you feel.

The l_1 *norm*: Assumes the worst-case input. There's no chance for overflow, ever. Because it scales down the input the most, it degrades the SNR the most.

$$s_1 = \sum_{k=0}^{\infty} |h(k)| \qquad (7.4)$$

$h(k)$ is the impulse response of the system and is, of course, infinite in length for systems with feedback but very tractable for systems without feedback. And note, when we're talking about "systems," we're talking about the part of the filter that is directly affecting the data at the point in question. Our IIR filter can be broken into two parts, each of which can be analyzed separately. One side has feedback, and the other side has no feedback. Thus, to scale the value of $w(n)$, we'll be dealing with a system with feedback, but, to scale $y(n)$, we can just treat the system as one without feedback. The l_1 scaling factor is also known as the *broadband* scaling factor, in contrast to the factor below.

The l_∞ *norm*: Takes the opposite tack from l_1 and assumes that all input is relatively narrowband—perfect sinusoids, in fact. All we do is look for the frequency where there is maximum gain:

$$s_1 = \max(|H(\omega)|) \quad \text{over} \quad 0 \leq \omega \leq \pi \qquad (7.5)$$

l_2 *norm*: The least amount of scaling of all, this norm is based on the energy in the input and system function; like the l_1, it is found from the impulse response as:

$$s_1 = \left[\sum_{k=0}^{\infty} h^2(k) \right]^{1/2} \qquad (7.6)$$

A closed form solution exists for the SOS:

$$s_1 = \frac{1}{\sqrt{(1 - a_2^2) - a_1^2 \left(\frac{1 - a_2}{1 + a_2} \right)}} \qquad (7.7)$$

IIR Filter Implementation

Arranged in order, l_1 is the largest scaling factor (most scaling), followed by l_∞, and l_2 is the smallest (least scaling). The trade-off is between distortion (or, if saturation is not used, large-scale limit cycles!) with not enough scaling, or a reduction in the SNR of the overall system if too much scaling is done. In practice, l_∞ is a reasonable choice, especially for systems with saturation, as the 68HC16 has. (With saturation, even if things do get too big, we avoid large-scale limit cycles. There's just some distortion—clipping—in the output.)

Using our example, we know the maximum gain is at $\omega=0$ or $z=1$, so s_1 (using l_∞) is

$$s_1 = \max(|H'(\omega)|) = \frac{1}{1 + (-1.3547) + 0.7526} = 2.5132 \qquad (7.8)$$

(That is, just the denominator of $H(z)$ evaluated at $z=1$. We ignore the feed-forward part of the filter—the numerator—as it doesn't affect the value of $w(n)$. Therefore, we've used the notation $H'(\omega)$ and not $H(\omega)$, above.)

We place this scaling factor in G_1 to scale the input and in G_3, so that the overall filter gain remains the same.

Now that we have $w(n)$ properly scaled, it's time to turn to the coefficients. Right away we have a big problem in the a coefficients: a_1 is -1.3547, outside the allowable range. Without going into details, we can scale the a coefficients by choosing a scaling factor s_2, such that dividing each coefficient produces a coefficient within the acceptable range. We also must compensate elsewhere in the system for this scaling—adding a $1/s_2$ term in G_1 and s_2 in G_2.

The b coefficients likewise require scaling. We don't have to worry about the value of $y(n)$ for reasons we'll discuss in a moment, but all the same, let's use the l_∞ norm to calculate a scaling factor s_4 as if we were concerned about $y(n)$. In our example, l_∞ is just the numerator of $H(z)$ evaluated at $z=1$, which gives us a result of 4. (Again, $z=1$ is the point of maximum gain, which makes sense as the zeros are all at the "other end," at $z=-1$.) Because all of our SOSs have zeros at $z=-1$, this scaling factor is the same for each SOS. We also need to compensate for this scaling by adding an s_4 term in G_3.

Let's pause for a moment and summarize the new coefficients:

$$G_1 = \frac{1}{s_1 s_2} \qquad G_2 = s_2 \qquad G_3 = s_1 s_4 G \qquad (7.9)$$

$$A_1 = \frac{-a_1}{s_2} \qquad A_2 = \frac{-a_2}{s_2} \qquad (7.10)$$

$$B_0 = \frac{b_0}{s_4} \quad B_1 = \frac{b_1}{s_4} \quad B_2 = \frac{b_2}{s_4} \tag{7.11}$$

If multiplication is significantly slower than shifting, it pays for us to make the various scaling factors powers of 2 if possible. On the other hand, depending on the architecture of the microcontroller, other benefits such as saturation may tip the scales toward using true multiplication for the scaling. Shifting by 2 can result in coarse scaling, though much of the time the necessary scaling (such as from s_2) will be in a reasonable range, say 2 or 4. We're going to go with shifting for our example here; it's the best choice for most microcontrollers and not a bad choice for the 68HC16.

We have one more refinement to make, and that is to get rid of the gain term G_3 by including this term in the b coefficients. This makes sense, given the fact that we would like to saturate the result of the scaling by G_3 anyway, and it will often be the case that G_3 times each of the b coefficients will still produce coefficients with an absolute value less than 1. (By the way, if you've followed the math, this completely cancels out the s_4 scaling factor from before.) Figure 7-9 shows the final structure of the SOS and FOS stages, the coefficients of which are found using the final set of equations below:

$$G_1 = \frac{1}{s_1 s_2} \quad G_2 = s_2 \tag{7.12}$$

$$A_1 = \frac{-a_1}{s_2} \quad A_2 = \frac{-a_2}{s_2} \tag{7.13}$$

$$B_0 = s_1 G b_0 \quad B_1 = s_1 G b_1 \quad B_2 = s_1 G b_2 \tag{7.14}$$

Table 7-9 summarizes these values for the FOS and for each SOS, using powers of 2 for the scaling factors s_1 and s_2 (thus, G_1 and G_2 are powers of 2).

With these changes, our SOS looks quite different from where we started, but by matching coefficients and the range of data to the underlying hardware, we should realize much higher efficiencies.[24]

24. Two more side comments about IIR coefficients. First, sometimes you'll see coefficients arranged such that *B0* is equal to 1—eliminating a multiplication. This isn't hard to do if you can put some scaling elsewhere (e.g., output). Second, some IIR code for DSP chips hard codes the $a1$ coefficient so that even though it is greater than 1 (and cannot be represented in 1.15 format), a separate addition is performed to get the correct results. By hard-coding the coefficients, however, you lose some flexibility to change the filter coefficients later.

IIR Filter Implementation

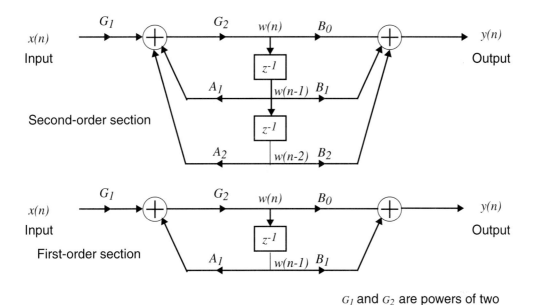

G_1 and G_2 are powers of two

Figure 7-9 Final structure of first- and second-order sections.

7.6.1 An Example IIR Implementation

Now that we have reasonable numbers, let's implement an IIR filter using the BT43367F—er, MC68HC16Z1.

Following the same idea used in the FIR example, we break the program into two pieces—a general-purpose IIR filter program and a filter-specific coefficient file. Again, we've written a "script" to automate the generation of the coefficient file. Listing 7-5 shows the coefficient file.

The (modified) coefficient values are expressed in hexadecimal this time. Choose whatever is most comfortable for you (but be sure you're getting the correct values assembled). The coefficients are generated in a very specific order to be compatible with the IIR routine. Each FOS or SOS appears in the order of evaluation, which is also a product of the structure of the filter routine.

Note that in this file we're generating constants that are stored in memory (using the DC.W directive) as well as those that are associated with symbolic names (using the EQU directive). The two types of constants have different roles to play, even though they're intermixed in the coefficient file.

For the FIR filter, we had to worry about only the number of coefficients. For the IIR filter, there are a few more concerns. Specifically, we need to know whether there's an FOS present or not, the number of SOSs, and the scale factors $G1$ and $G2$ for each section.

	Poles in z-plane	Original coefficients	Scaled coefficients
FOS0	$0.47392+0j$	$a_1=-0.472925$ $b_0=1$ $b_1=1$ $G=0.263538$	$A_1=0.472925$ $B_0=0.527075$ $B_1=0.527075$ $G_1=0.5$ $G_2=1$
SOS1	$0.49183+0.17515j$ $0.49183-0.17515j$	$a_1=-0.983662$ $a_2=0.272577$ $b_0=1$ $b_1=2$ $b_2=1$ $G=0.0722286$	$A_1=0.983662$ $A_2=0.272577$ $B_0=0.288915$ $B_1=0.577829$ $B_2=0.288915$ $G_1=0.25$ $G_2=1$
SOS2	$0.55387+0.35543j$ $0.55387-0.35543j$	$a_1=-1.10775$ $a_2=0.433107$ $b_0=1$ $b_1=2$ $b_2=1$ $G=0.08134$	$A_1=0.553874$ $A_2=-0.216553$ $B_0=0.32536$ $B_1=0.65072$ $B_2=0.32536$ $G_1=0.125$ $G_2=2$
SOS3	$0.67734+0.54202j$ $0.67734-0.54202j$	$a_1=-1.35469$ $a_2=0.752576$ $b_0=1$ $b_1=2$ $b_2=1$ $G=0.0994723$	$A_1=0.677344$ $A_2=-0.376288$ $B_0=0.397889$ $B_1=0.795779$ $B_2=0.397889$ $G_1=0.125$ $G_2=2$

Table 7-9 Original and scaled coefficients for example IIR filter.

```
*          Coefficient file iircoef1.asm
*          Generated 1-Nov-96 by i6x013.m
*          Butterworth Example
*          Lowpass, Fsamp=10000 Hz
*          Fp=1000 Hz, Ap=1 dB
*          Fs=2000 Hz, As=40 dB
*          Order=7
*
*
num_fos   equ     1            ;number of FOSs
num_sos   equ     3            ;number of SOSs
*
```

Listing 7-5 Coefficient file for example IIR program.

```
*
iircoefs:                       ;start of coefficient storage

* FOS0:
* zero at (-1.000000+0j)
* pole at ( 0.472925+0j)
* b0= 1.000000, b1= 1.000000
* a1=-0.472925
* G= 0.263538
            dc.w    $3c89           ;A1= 0.472925
            dc.w    $4377           ;B0= 0.527075
            dc.w    $4377           ;B1= 0.527075
FOS0G1      equ     $1              ;G1= 0.500000

* SOS1:
* zeros at (-1.000000+0j), (-1.000000+0j)
* poles at ( 0.491831+ 0.175154j) and conjugate
* b0= 1.000000, b1= 2.000000, b2= 1.000000
* a1=-0.983662, a2= 0.272577
* G= 0.072229
            dc.w    $7de9           ;A1= 0.983662
            dc.w    $dd1c           ;A2=-0.272577
            dc.w    $24fb           ;B0= 0.288915
            dc.w    $49f6           ;B1= 0.577829
            dc.w    $24fb           ;B2= 0.288915
SOS1G1      equ     $2              ;G1= 0.250000
SOS1G2      equ     $0              ;G2= 1

* SOS2:
* zeros at (-1.000000+0j), (-1.000000+0j)
* poles at ( 0.553874+ 0.355431j) and conjugate
* b0= 1.000000, b1= 2.000000, b2= 1.000000
* a1=-1.107747, a2= 0.433107
* G= 0.081340
            dc.w    $46e5           ;A1= 0.553874
            dc.w    $e448           ;A2=-0.216553
            dc.w    $29a5           ;B0= 0.325360
            dc.w    $534b           ;B1= 0.650720
            dc.w    $29a5           ;B2= 0.325360
SOS2G1      equ     $3              ;G1= 0.125000
SOS2G2      equ     $1              ;G2= 2

* SOS3:
* zeros at (-1.000000+0j), (-1.000000+0j)
* poles at ( 0.677344+ 0.542017j) and conjugate
* b0= 1.000000, b1= 2.000000, b2= 1.000000
* a1=-1.354687, a2= 0.752576
* G= 0.099472
            dc.w    $56b3           ;A1= 0.677344
            dc.w    $cfd6           ;A2=-0.376288
            dc.w    $32ee           ;B0= 0.397889
            dc.w    $65dc           ;B1= 0.795779
            dc.w    $32ee           ;B2= 0.397889
SOS3G1      equ     $3              ;G1= 0.125000
SOS3G2      equ     $1              ;G2= 2
*
*
* end of file *
```

Listing 7-5 (Continued) Coefficient file for example IIR program.

In this example, we use macros to handle the variable number of sections and the scaling factors. There are other approaches that could be employed, including subroutines and "for/next" loops. We chose macros for the following reasons:

- Macros eliminate the overhead of subroutine calls. During the initial design stage, we calculated that each SOS would take about 120 clock cycles; adding a separate subroutine call (10 cycles) and return from subroutine (12 cycles) would have added almost 20% to the execution time. The trade-off is against code space, which is plentiful in this case. A "for/next" loop might have less overhead than a subroutine, but then there's the issue of changing *G1* and *G2* for every iteration.
- Using macros, we can hard-code the exact number of left or right shifts for the *G1* and *G2* scale factors without the need for a for/next loop around the shift. (Yes, this is the "ugly" source code we mentioned earlier when we talked about optimization. Ugly, but clear.)
- Macros can also include or exclude code for the FOS as needed. (If the number of poles is even, there is no need for an FOS.)

All applications have different goals and constraints. As you gain experience in optimizing filter routines, you'll find a number of factors pushing and pulling in different directions—saving data memory, code memory, execution speed, register usage, preserving data precision, clarity of source code, maintainability, portability, and so on. Some folks might argue that architectural features like the MAC instruction more or less determine the code, but there are also a number of secondary decisions, especially for IIR filters.

Let's look at the macro that generates the code for an SOS, the core of this routine. Listing 7-6 shows the macro in the format accepted by the Motorola MASM macro assembler.

```
*
*         sos:   Create a second-order section (SOS) with arguments
*                G1 shift count (e.g. SOS1G1), and G2 shift count (e.g. SOS1G2).
*
*                   G1             G2           W0          B0
*          x(n) >----->----(+)----->--------->---(+)----> y(n)
*                           |                     |
*                           |                   [z^-1]
*                           |         A1          |         B1
*                         +---<--------->---+
*                           |                    |W1                |
*                           |                   [z^-1]
*                           |         A2          |         B2
*                         +---<--------->---+
*                                                W2
*
*
*         (Refer to text for value of G1, G2, and calculation of
*         A1, A2, B0, B1, and B2 from a1, a2, b0, b1, and b2)
*
*         Assumes input is in AM, ix points to W1, iy points to A1,
*         HR and IR are loaded with W1 and A1 respectively.
*         Enable saturation; clear x and y masks.
*
*         Order of variable & coefficient storage is:
*         A1, A2, B0, B1, B2 , next A1, next A2, ..., yk set to page
*         W0, W1, W2, next W0, next W1, ..., xk set to page
*         All variables, coefficients in 1.15 signed format
*
*         Output is in AM, with ix, iy, HR and IR set up for
*         a call to another SOS.
*
```

Listing 7-6 Macro to generate a second-order section.

IIR Filter Implementation

```
*           Cycles: Depends on number of shifts for G1, G2;
*           100 cycles + 4*G1shifts + 4*G2shifts (about 120 cycles typical)
*
*           Example usage:
*
*           lde     (input)             ;e is 1.15 format input
*           tem                         ;e -> am
*           ldx     #w_of_n+2           ;ix points to W1 of SOS #1
*           ldy     #iircoefs           ;iy points to A1 of SOS #1
*           ldhi                        ;load hr with W1, ir with A1
*           sos     SOS1G1,SOS1G2       ;implement SOS#1
*           sos     SOS2G1,SOS2G2       ;SOS #2
*           tmer                        ;output -> e (rounded & saturated)
*           ste     (output)
*
*           Cycles per instruction noted in first column of comments
sos         macro                       ;     G1 shift, G2 shift
            rasrm   \1                  ;4*G1shift  generate right shifts for G1
                                        ;     am=G1*x, hr=(ix)=W1, ir=(iy)=A1
            mac     2,2                 ;12   hr=(ix)=W2,ir=(iy)=A2, am=G1*x+A1*W1
            mac     -4,2                ;12   hr=(ix)=W0,ir=(iy)=B0,
                                        ;     am=G1*x+A1*W1+A2*W2
            raslm   \2                  ;4*G2shift  generate left shifts for G2
            tmer                        ;6    am -> e, rounded & saturated
            ste     $00,x               ;6    W0=new result
            ldhi                        ;8    hr=(ix)=W0 (new!), ir=(iy)=B0 (again)
            clrm                        ;2    am=0
            mac     2,2                 ;12   iz=hr=W0, hr=(ix)=W1, ir=(iy)=B1
                                        ;     am=B0*W0
            stz     $00,x               ;4    W1=W0
                                        ;note hr=old W1, prior to this write!
            mac     2,2                 ;12   iz=hr=W1, hr=(ix)=W2, ir=(iy)=B2
                                        ;     am=B0*W0+B1*W1
            stz     $00,x               ;4    W2=W1 (note again that hr=old W2)
            mac     0,2                 ;12   hr=(ix)=W1, ir=(iy)=next A1
                                        ;     am=B0*W0+B1*W1+B2*W2
            aix     #8                  ;2    ix=next W1
            ldhi                        ;8    hr=(ix)=next W1, ir=(iy)=next A1
                                        ; result is in am
            endm
* Note—The last mac above would like to be 'mac 8,2' which
* would set up hr and ix for the next section without need
* for a separate 'aix #8' and 'ldhi'—half of which is redundant
* since we already have ir/iy correct.  Alas, the mac instruction
* has a 4-bit *signed* increment for x, with a range of -8 to 7.
```

Listing 7-6 (Continued) Macro to generate a second-order section.

The MAC instruction is at the core of this macro. It is functionally identical to the RMAC (discussed earlier), with two exceptions. First, the multiple-accumulate operation is performed only once. Second, the value in H is transferred to the Z register before any other operations by the MAC instruction, allowing this value to be written to a different location. If H contains sample data, this provides a low-overhead way of implementing the data moves at the end of the SOS algorithm. We repeat the pseudocode description of the SOS from Chapter 6 (IIR Filters—Digital Filters With Feedback) here as:

$$w(0) = x - a_1 w(1) - a_2 w(2)$$

$$y = b_0 w(0) + b_1 w(1) + b_2 w(2)$$

$w(2)=w(1)$

$w(1)=w(0)$

Of course, we make these data moves in the course of evaluating the first two expressions, requiring good planning. Note that in the macro we've added comments showing the values of each register at each instruction to avoid clobbering innocent values by mistake.

Having decided to use this efficient movement to update $w(2)$ and $w(1)$, the order of evaluation for the filter is partially determined. In turn, this influences the ordering of the coefficients and data in memory. The ordering of the coefficients and data is also influenced by limits on how the *X* and *Y* registers can be updated by the *xo* and *yo* arguments of the MAC instruction. (You'll notice in comments at the end of the macro that we lament the 4-bit limit to this offset. You might be able to rearrange the coefficient ordering to avoid this problem and save more cycles.)

The scale factors *G1* and *G2* deserve special comment. Recall that these values are required to be in the coefficient file as equated symbols for each SOS (*G1* for the FOS). Actually, we should say that the number of shifts associated with *G1* and *G2* are given, as we use shifts to implement the scale factors. Because the number of shifts varies for each SOS, the shift counts must be passed to the macro as arguments. In the MASM syntax, these arguments are called "/1" and "/2." In the macro listed above, these values are used to invoke the two shift macros "*raslm*" and "*rasrm*," macros that generate zero to four arithmetic shifts (left or right, depending on the macro) of the accumulator. Listing 7-7, below, shows the (ugly!) code for "*rasrm*."

```
*         rasrm:   repeating asrm
*         call with 1 argument that evaluates to 0,1,2,3 or 4,
*         generates 0 to 4 asrm's
rasrm     macro
          ifeq    \1-1
          asrm
          endc
          ifeq    \1-2
          asrm
          asrm
          endc
          ifeq    \1-3
          asrm
          asrm
          asrm
          endc
          ifeq    \1-4
          asrm
          asrm
          asrm
          asrm
          endc
          ifgt    \1-4
          fail    "Too many shifts for rasrm"
          endc
          endm
```

Listing 7-7   Macro to generate right shifts for scaling.

Again, this code avoids using a "for/next" loop and is completely determined by the symbols defined in the coefficient file at assembly time. We never need to modify the code to match

IIR Filter Implementation

a new set of coefficients. The conditional directives ("ifeq") cause the code following the directive to be assembled only if the expression (e.g., "\1-2") evaluates to zero. Note, too, that we added a "cow-catcher" error message to indicate that more shifts were requested than the macro could generate—otherwise the code would fail "silently." Listing 7-8 shows the filter-evaluation section of the main program.

```
*               (get data from ADC)
*               filter
                tem                         ;e -> am[31:16], am[35:32]=am[31],
                                            ;am[15:0]=$00
                ldx     #w_of_n+2           ;start ix pointing at W1 (not W0!)
                ldy     #iircoefs           ;start of coefficients (A1)
                ldhi                        ;hr=(ix)=W1, ir=(iy)=A1

*               Error message if wrong number of FOS's (can only be 0 or 1)
                ifgt    num_fos-1
                fail    "Wrong number of FOS's!"
                endc

                ifeq    num_fos-1
                fos     FOS0G1              ;do one first-order section [macro]
                endc

*               Error message if wrong number of SOS's
                ifgt    num_sos-6
                fail    "Wrong number of SOS's!"
                endc

                ifge    num_sos-1
                sos     SOS1G1,SOS1G2       ;do second-order section [macro]
                endc
                ifge    num_sos-2
                sos     SOS2G1,SOS2G2
                endc
                ifge    num_sos-3
                sos     SOS3G1,SOS3G2
                endc
                ifge    num_sos-4
                sos     SOS4G1,SOS4G2
                endc
                ifge    num_sos-5
                sos     SOS5G1,SOS5G2
                endc
                ifge    num_sos-6
                sos     SOS6G1,SOS6G2
                endc
                                            ;result is in AM

                tmer                        ;transfer AM to E (saturated if necessary)
*               done with filter, output data to DAC
```

Listing 7-8   Excerpt from main IIR example program.

The FOS and SOS macros handle the internal storage, so there is no buffer-handling code at this level, as in the FIR case. Prior to calling the first section (which happens to be an FOS), we need to set up the accumulator *AM* with the sample data, and index registers *X* and *Y* to point to data and coefficients, respectively. Then we need to "preload" *H* and *I* with the values to be multiplied. After that, it's just a matter of stringing together the calls to subsequent SOSs;

because we wrote the SOS macro to set up *X, Y, H, I,* and *AM* properly at the end of the macro, each macro flows smoothly into the next, without a lot of unnecessary data movement. If we haven't hammered that point home yet, here it is again—the core of a DSP routine running on a microcontroller is sensitive to wasted cycles, as the core is usually small and executed many times. Even with a fast MAC, ignoring mundane data movement instructions can result in code that's too slow.

The complete program appears in Appendix 5. Like the FIR program, a lot of code is devoted to the overhead of setting up modules. (In fact, most of the FIR and IIR code is similar.) We also have an additional file containing the macros for the FOS, SOS, and shifts.

Figure 7-10 shows the results of running the program on our evaluation system. In the passband and partially into the stopband, the performance of the system matches the ideal very closely. Further into the stopband, reality raises its head for a look around—minimum stopband attenuation is at around 47 dB, worst case. This is superior to the design parameters, which require 40 dB stopband attenuation, but you can see that some combination of the ADC, our test bench, coefficient rounding, and other finite word-length issues (e.g., small-scale limit cycles) would require additional refinement if we really needed more than 40 dB of stopband attenuation.

This filter executes in about 31 μsec (44 μsec for the entire interrupt service routine). (Running in the fastest memory, the ideal is about 26 μsec.) Each SOS takes about 8.5 μsec, which means that, at 10 kHz, the highest order we could construct would be about $N=20$. The FIR routine we wrote evaluated a filter of order 51 in a little over 50 μsec.

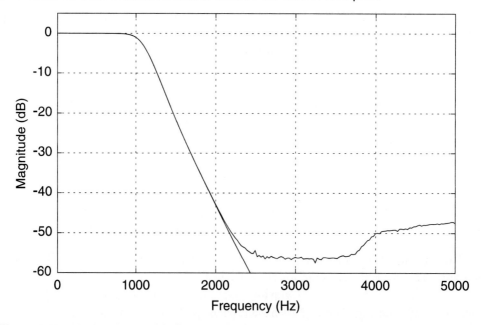

Figure 7-10 Ideal and actual IIR filter magnitude response for example program.

Summarizing the IIR filter implementation, we see that the SOS structure is not very amenable to the repeating MAC instruction and that a carefully choreographed flow of data is necessary. Scaling becomes an issue as coefficient values and intermediate results would otherwise exceed the 1.15 format. For these and other reasons, the core of an IIR routine is generally much more complicated than the simple RMAC used in the FIR filter. Recall that IIR filters usually have fewer coefficients than do FIR filters. This is important, given that, with the 68HC16, IIR filters require four times more computation time per coefficient than do FIR filters.

7.7 Summary

The architecture of a processor strongly affects the efficiency of a DSP algorithm implemented on it. Architectures designed explicitly to support DSP usually include multiply-accumulate instructions, a Harvard memory architecture, a fast shifter, support for circular buffers, saturation/overflow logic, zero-overhead looping, and FFT-specific addressing. Orthogonality is greatest for high-end general-purpose microprocessors and lowest (nonexistent!) in very specialized DSP chips.

Operations central to digital filters include multiplication, accumulation, overflow/saturation, shifting, efficient indexed and circular addressing, and good bandwidth for data movement to and from external memory.

Compiler-generated code has many advantages over assembly language, including increased coding efficiency, fewer bugs, easier debugging, increased portability, easier maintenance, and efficient code.

Assembly language provides the ultimate in control for code and data but at the cost of more time spent coding, more difficulty debugging, less portability, and more difficult maintenance and modification. Optimization of assembly language code exaggerates these effects.

In practice, a minimum of code should be optimized. Optimizing code takes a lot of time and it is usually the case that a small amount of code is executed most of the time. However, microcontrollers require very good optimization to execute DSP routines in real time.

Data can be encoded in a number of formats with varying ranges and resolutions. A popular 16-bit format is Q15 or 1.15, which allows numbers between -1 and just less than 1 to be represented using a sign bit, an implied decimal point, and 15 data bits.

Optimizations for code speed include reducing overhead (subroutine calls, loop unwinding, zero-overhead looping), reducing run-time computation (strength reduction, constant folding, common subexpression elimination, hoisting variables, use of tables), and speeding data movement (constant propagation, register usage, index registers, self-modifying code, addressing modes, and use of pointers).

Optimizations for size include the use of subroutines, program overlays, and reuse of memory for mutually exclusive operations.

The tool chain is the series of software used to produce executable code (and perhaps to run this code) from source code. The complexity of these tools can be tamed somewhat by automating their use with batch or make files.

The use of finite word-lengths for data, coefficients, and arithmetic operations introduces changes in the expected filter behavior. Quantization of coefficients changes pole and zero locations. Finite-word arithmetic introduces both large- and small-scale limit cycles. Saturation/overflow can alleviate large-scale limit cycles at the cost of distortion; scaling reduces large-scale limit cycles at the cost of reduced SNR. Small-scale limit cycles are present only in systems with feedback (IIR filters), and require more analysis to prevent or reduce.

FIR filters are relatively straightforward to implement using the 68HC16 microcontroller. The repeating multiply-accumulate instruction (RMAC) performs the core of the FIR routine with one instruction. Often, the coding of the core of the filter will have a strong influence on the surrounding program structure—for example, location of data, format of data, and so on. Crucial aspects of the 68HC16 architecture used in the FIR filter example include the 1.15 format support for data and circular buffer addressing.

Interrupt-driven input/output has some advantages over other ways of generating sampling intervals. Other options include program loops, generating timing with the ADC, and polling a timer. Variations in interrupt latency can be minimized by placing the processor in a known "wait" mode.

IIR filters are usually constructed using SOSs. To implement such structures on a 16-bit processor requires modification of the filter coefficients and scaling as part of the filter. These scale factors can be powers of 2, allowing shifts to be used instead of multiplication.

The data movement required by IIR filters is less regular than FIR and requires more careful coding. On the 68HC16, an IIR filter takes longer to evaluate, due to additional data movement.

Macros allow code to be generated without the overhead of run-time "for/next" loops, subroutine calls, or "hard-coding" special operations. Separating filter coefficient data from digital filter code allows changes to be made to filters without changing program code.

Resources

Ombres, Denise, "C and C++ extensions simplify fixed-point DSP programming," *EDN*, October 10, 1996, pp. 135–138.

Glenewinkel, Mark, *Using M68HC16 Digital Signal Processing to Build an Audio Frequency Analyzer*, Motorola document AN1233/D, 1995.

Burrus, C. Sidney, McClellan, James H., Oppenheim, Alan V., Parks, Thomas W., Schafer, Ronald W., and Schuessler, Hans W., *Computer-Based Exercises for Signal Processing Using MATLAB®*, New Jersey: Prentice-Hall, 1994.
This is a good companion to any of the traditional texts. Using MATLAB, you can experiment quickly with powerful signal processing tools, including digital filter design software.

Ingle, Vinay K. and Proakis, John G, *Digital Signal Processing Laboratory Using the ADSP-2101 Microcomputer*, New Jersey: Prentice-Hall, 1991.

Van Sickle, Ted, *Programming Microcontrollers in C*, Solona Beach, CA: HighText, 1994.
Not only contains a lot of information on using C with microcontrollers but has practical information on programming the 68HC11 and 68HC16 for a variety of tasks. Includes technical information on the on-chip peripherals for these microcontrollers.

Dr. Dobb's Journal[25], a monthly magazine concentrating on software tools for programmers, has printed a number of articles on filtering, the FFT and wavelets (see next chapter), and other algorithms (ISSN 1044-789X).

All of the following databooks are useful in explaining the 68HC16 (Note: some of these are on the accompanying CD-ROM):

MC68HC16Z1 User's Manual. This includes much of the information found in the manuals below, so it may be all you need initially.

CPU16: (M68HC16 Family) Reference Manual. Detailed information on the central processing unit of the 68HC16 family, including DSP, the background debug mode (BDM), and comparisons with the 68HC11 microcontroller.

ADC: (Modular Microcontroller Family) Reference Manual. Details on the 68HC16Z1's ADC.

GPT: (Modular Microcontroller Family) Reference Manual. The General-Purpose Timer.

QSM: (Modular Microcontroller Family) Reference Manual. The Queued Serial Module, including the Queued Serial Peripheral Interface (QSPI) and SCI (Serial Communications Interface).

SIM: (Modular Microcontroller Family) Reference Manual. The Systems Integration Module, including system clock, internal RAM, external bus interface, watchdog timer, and more.

Hardware:

M68HC16Z1 evaluation board. (May be superseded by other development hardware). Contains M68HC16Z1, 64K of RAM configured as pseudo-ROM for program storage, background debug mode with software, serial port, socket for 16-bit DAC, expansion headers. Assembler, linker, and demonstration code. Includes M68HC16Z1EVB User's Manual.

Software:

MASM Toolset. (May be superseded by other development software.) This includes an assembler, linker, and other tools for programming the Motorola M68HC05, M68HC08, M68HC11, and M68HC16 processors.

Source code examples are also available on the Motorola Freeware bulletin board system and on the Web. (See Preface for locations.)

Evaluation versions of 68HC16 compilers and software development systems are available from several vendors, including Intermetrics Microsystems Software, Inc. and Introl Corp. These programs usually have limits on the maximum number of lines or functions compiled, but the software can be downloaded from the Internet and will give you a good idea of their compiler's capabilities.

Source code for the examples in this book are also available on the World Wide Web and are included on the CD-ROM with this book.

25. One of the few surviving magazines from the early microcomputer days. It was originally named *Dr. Dobb's Journal of Computer Calisthenics & Orthodontia* with a subtitle of "Running Light Without Overbyte" and was published by the People's Computer Company, starting in 1976.

CHAPTER 8

Frequency Analysis

You might want to review Chapter 2 (Analog Signals and Systems) and Chapter 4 (Discrete-Time Signals and Systems). Again, complex numbers come up in this chapter, but any math, as usual, will be fairly simple.

8.1 Overview

So far, we've been concerned with modifying the frequency content of signals using filters. The tools we'll explore in this chapter, primarily the discrete Fourier transform (DFT) and relatives, will let us

- Identify the frequency spectrum of a signal
- Detect the presence of one or more frequencies in a signal
- Potentially reduce the data necessary to send or store a signal
- Accelerate the processing (e.g., filtering) of signals
- Deduce the magnitude and phase response of an unknown system by analyzing its "impulse" response.

We start this chapter with a philosophical question—what do we mean by the frequency content of a signal that has a finite time span? We then need to ask a more practical question, namely, What information do we want from the signal? The first question leads us to some limitations of the tools like the DFT, while the second question requires us to consider which tools or techniques are the most appropriate for the task at hand.

The bulk of this chapter is about a family of DFT algorithms that are lumped under the name *fast Fourier transform* (FFT). First, we'll discuss the background of this transform, then how to use it, and finally, an overview of FFT implementation. On this final point, we'll see that an FFT routine is not trivial to write (especially for a microcontroller!), at least if it is to be computationally efficient. So we have to warn you that if you've been planning to implement a real-time FFT on a 4-bit, 1 MHz microcontroller, you might be in for a little disappointment. Well, make that a *lot* of disappointment, unless you've got some really slow signals. Even the 68HC16 code we present has to make some serious trade-offs to run in anything approaching real-time.

8.2 What Do You Want?

8.2.1 A Philosophical Question

Figure 8-1 is a sequence of 64 samples at a sampling rate of 8 kHz. What should we say is the frequency spectrum of that signal?

The signal looks like a sinusoid at about 150 Hz, so we'd like to say that the frequency spectrum is something like Figure 8-2 . The problem is that we're trying to talk about the spectrum of a *time-limited* signal—that is, the signal we're looking at has a finite number of samples (and, therefore, was taken over a finite length of time). What happens to the signal outside this time? Is the signal zero? Does it repeat?

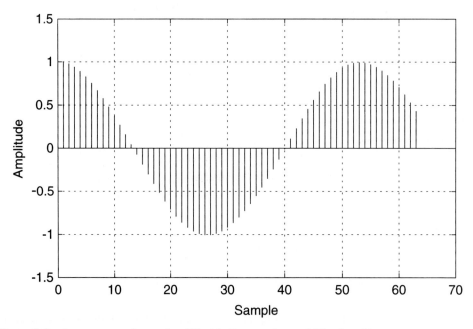

Figure 8-1 A sequence of samples. What is the spectrum of this signal?

What Do You Want?

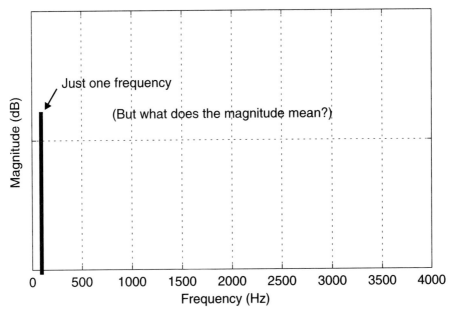

Figure 8-2 An ideal magnitude spectrum for the signal in Figure 8-1.

Normally, we assume that the signal we've sampled continues in "much the same way," that is, with the same frequency spectrum (i.e., "stationary"). Otherwise, we'll also need to include all of the high frequencies that are necessary to create a signal that was zero except when we happened to look at it.

It turns out that we'll have to do more than just "assume" the signal is periodic, at least if we want to use the DFT to process the data. We'll need to window the input data. These windows are the same as those we used before in Chapter 5 (FIR Filters—Digital Filters Without Feedback) to design FIR filters. But the windows introduce their own problems, which we'll need to explore in greater detail.

Before we plunge into the DFT in earnest, let's consider the second question we posed earlier—what information do we really want from the signal? The DFT, even when implemented as the FFT, still takes a lot of processing time. There are often ways to get the information we want with a lot less effort.

8.2.2 Using Individual Filters

If the problem you want to solve is to detect the presence of frequencies within a small number of frequency ranges, one popular solution is to use one or more individual filters. Simple two-pole IIR filters are often sufficient, and their bandwidth can be varied to accommodate narrow or wide bands of frequencies. A simple way of measuring the magnitude of the signal that is passed by one of these filters is to take maximum absolute value of the output.

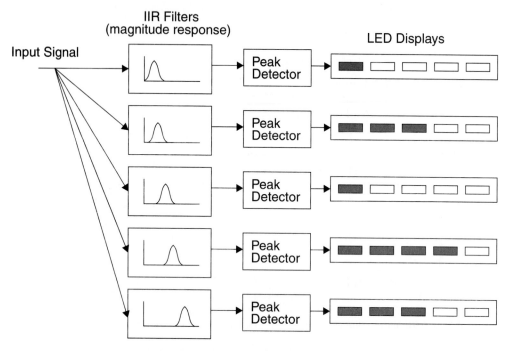

Figure 8-3 Using a bank of filters to analyze frequency content.

The Motorola application note *Using M68HC16 Digital Signal Processing to Build An Audio Frequency Analyzer*[1] uses this approach to analyze an audio signal in real-time. Five two-pole IIR filters, with center frequencies at 125 Hz, 500 Hz, 1 kHz, 4 kHz, and 10 kHz, run in parallel, with their outputs driving five banks of LEDs. For this type of application, five or ten bands is sufficient resolution, and the bandwidths are quite wide, so that the transition from one band to the next is smooth. Figure 8-3 shows an outline of this process.

8.2.3 Looking For a Small Set of Frequencies

Once the number of frequencies of interest becomes large, other processing methods can be used more efficiently. For example, detecting DTMF signals—the familiar Touch-Tone® sounds of telephones—requires eight different frequencies to be detected. It's also useful to assess the energy of the second harmonics of those tones (to avoid false triggers), so 16 total frequencies are of interest. You could use 16 different IIR filters, but a special case of the DFT, the Goertzel algorithm, can be used instead. We'll discuss this briefly in the context of the FFT later.

Filter banks and techniques like the Goertzel algorithm often use less processing time than generating a complete frequency-domain description of a signal. This makes sense, given that

---

1. Application note AN1233/D, 1995. The source code is available on the Motorola website.

we start out knowing a great deal about the frequency content (of interest) of the signal. To fully analyze, compress, or speed processing, however, we need tools like the DFT.

8.3 The Discrete Fourier Transform and Fast Fourier Transform

8.3.1 A Complete Frequency-Domain Description

The input to the DFT is a discrete-time sequence of samples[2]; the output is a discrete-frequency sequence representing the magnitude and phase of the input spectrum at regular frequency intervals. Other transforms like the Fourier transform, the discrete-time Fourier transform, the *z*-transform, and the Laplace transform, all involve either continuous time or continuous frequency or both; they're useful tools (as you've seen!), but not something we can implement on a digital processor, where we can deal only with discrete samples in both time and frequency. In fact, to emphasize this fact, we really should call the *DFT* the *DTDFFT*—the *discrete-time discrete-frequency Fourier transform*. Not only would this name be a bit more descriptive, but it hints at the connection between the DFT and the DTFT (more in a moment).

The DFT produces a "complete" description of the discrete-time input signal within the window—there is no loss of information going from the time domain to the frequency domain, though sometimes the desired information won't be so easy to see in the DFT. The inverse DFT (IDFT or "IDTDFFT") proves there's no loss of information by transforming the magnitude and phase spectrum back into the original input sequence.

We're not going to show you the DFT equations yet. In fact, you probably won't ever have to write a DFT (FFT) routine yourself. It's far more crucial to understand how to *use* the DFT properly, understanding its limits and advantages. Once we've covered DFT usage, we'll give you an overview of the issues involved in DFT (FFT) implementation.

8.4 Using the DFT

The big picture is that the DFT accepts a discrete-time signal and produces a discrete-frequency transform. Below, we discuss first the nature of the input to the DFT, then the issue of windowing and the interpretation of the output. Finally, we'll discuss the issue of frequency resolution, where we'll see the precise connection between the DFT and the DTFT.

8.4.1 Input Data

8.4.1.1 Discrete-Time and Amplitude

Because the input to the DFT is a sequence of time samples, all of the issues we've raised earlier with respect to sampling apply. In particular, antialiasing filters must be used if there are any frequency components higher than the Nyquist frequency present. Quantization of the signal will degrade the signal and must be taken into account.

---

2. By the way, the samples may be complex valued. We'll talk about what that would mean later in this chapter.

We've already addressed these issues in Chapter 4 (Discrete-Time Signals and Systems); the only point we might add is that the FFT requires even a bit more attention to scaling than do filters. This scaling can directly affect the resolution of the output. One trade-off that's sometimes necessary when using fixed-point math is to perform worst-case scaling at each of many FFT stages. This scaling substantially reduces the resolution of the FFT results. We'll discuss this point in more detail when we look at the 68HC16 FFT implementation.

8.4.1.2 Complex Versus Real Data

In this book, we've treated signals, whether continuous- or discrete-time, as real-valued. One thing you'll run into immediately with the DFT is the concept of complex-valued signals. What could this mean?

Complex signals arise in situations where there are two or more independently variable components to a signal. Let's make up an example using radio waves, which have both an amplitude and an orientation (i.e., the oscillation of the waves is in a specific plane). You've seen this orientation—though in a static sense—in TV antennas. In the U.S., TV waves oscillate in a horizontal plane; thus, the signals are best received by TV antennas with a horizontal orientation. There are countries where the oscillation is in a plane that is vertical with respect to the ground. Mount an antenna horizontally in those countries and you'd be disappointed with your reception. Imagine now that a TV station not only varied the signal amplitude (the signal we're used to thinking about), but could also continuously vary the orientation. Two receiving antennas, one horizontal and one vertical, taken together would produce a complex signal—in this case, neatly divided into the "real" (horizontal) and "imaginary" (vertical) components. The information from this complex signal (magnitude and orientation) would be greater than just the amplitude-only signal we'd receive from just one antenna.[3]

Complex-valued signal models are quite common in communications where they make certain types of processing much easier, though in these cases the complex signals are a mathematical convenience, not "real," as our example above was. But, we have to admit that, in practice, you might never have to worry about time-domain complex signals per se. So why do most FFT routines accept complex-valued inputs?

There are two good reasons. First, each stage of the FFT involves multiplying the input by a complex-valued constant. thus producing a complex result. (There are generally $\log_2 N$ stages to the FFT, where N is the number of samples to be processed.) All the variables after the first section therefore must be complex regardless of the input to the very first stage. Second, the same FFT routine is often designed to perform the IFFT by just changing a simple flag. Since the output of the FFT is complex-valued, obviously the input to an IFFT should be able to take complex values. If you use this type of routine, remember to set the imaginary part of each sample

---

3. This is a very contrived example—most radio communication does not vary the orientation of the radio wave oscillations. But we'd have to get into some complicated mathematics to see the power of complex signals in a real example.

value to zero if your input data is purely real. (We'll have a further comment on real inputs with complex-input FFTs later on that can save you some processing time.)

8.4.2 Why Use Windows?

The DFT operates only on discrete-time samples and on only a finite number of them. This is pretty convenient for us mortals, as a finite amount of time is all we have, but it often means we're dealing with a time-limited version of the "actual" signal that we wish to transform. The DFT gets away with using a finite number of samples by assuming that it is processing exactly one period of a periodic signal. It has all the information it needs (or so it thinks) because the real-world signal just repeats the period that the DFT already knows about. The problem, as you can imagine, is that we'll practically never have exactly one period of the real-world signal as input. For example, taking the earlier waveform of Figure 8-1, the DFT "thinks" the actual real-world signal looks like Figure 8-4. Those "glitches" where we've pasted together copies of the original samples will show up in our DFT spectrum as a lot of extra frequency components. Not that we can complain—if you perform an IDFT on this "noisy" data, you'll get the original waveform of Figure 8-1.

Windows let us deal with the conflict between the reality of our data (rarely exactly periodic in our samples) and the DFT's view of the world. Recall we used windows in Chapter 5 (FIR Filters—Digital Filters Without Feedback) to "smooth out" the truncated impulse response of an ideal filter when we created an FIR filter by windowing. Here, we'll use a window to

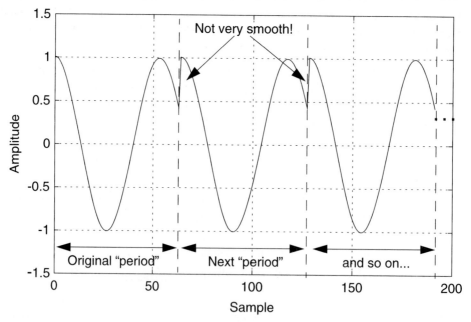

Figure 8-4 The DFT's view of the signal in Figure 8-1.

Using the DFT

smooth out both ends of the input samples to the DFT. By reducing the discontinuities the DFT sees, the DFT spectrum won't be cluttered with the frequency components necessary to generate "glitches" in the time domain.

The "default window" is the rectangular window, which is to say no window at all. This is rarely a good choice, except when the window is very long. Instead, other windows are used, based on two characteristics: first, their ability to resolve closely-spaced frequency components and, second, how well they preserve the relative magnitudes of the frequency components. The shape and length of window allows us to address both these issues.

8.4.2.1 Window Shape: Spectral Leakage

To understand how windowing affects the DFT, we need to make a small digression. A fundamental observation from Chapter 5 (FIR Filters—Digital Filters Without Feedback) is that "convolution in the time-domain is equivalent to multiplication in the frequency domain." We used this idea to create FIR filters. The converse[4] is also true— "multiplication in the time domain is equivalent to convolution in the frequency domain." In practical terms, this means that when we apply a window to a signal, any particular frequency component of the signal will be replaced in the frequency domain by a shifted (and scaled!) version of the window's spectrum. For example, a nice clean "spike" representing a single frequency component will be replaced by the main lobe and side lobes of the frequency spectrum of the window we chose. If there's more than one "spike" (frequency component) in the input signal, there will obviously be some problems as each frequency component's main lobe and side lobes interfere with those of other frequency components. Figure 8-5 shows in general terms a signal with two frequency components and the spectrum that results from windowing. (Be aware, however, that this spectrum is not necessarily what we'll get from the DFT, only what is possible.)

Figure 8-6 shows some common windows in both the time and frequency domains. Note that now we're showing both positive and negative frequencies for the frequency domain of the windows,[5] as "both sides" of the window's spectrum show up in the spectrum of the windowed input. As was the case in designing FIR filters, the magnitudes of the side lobes are nonnegotiable. Once we've chosen a window type (and, thus, shape), the side lobe magnitude is set.

The relative magnitude of the side lobes affects the relative accuracy of the magnitude spectrum, an effect known as *spectral leakage*. Some of the energy at one frequency is "leaking" over and corrupting the magnitude at nearby frequencies. Filters with low side lobe magnitudes, such as the Blackman and Kaiser (β=8.96), reduce spectral leakage, and the Hamming, Hanning, and rectangular windows have relatively larger side lobe magnitudes and thus more spectral leakage.

8.4.2.2 Window Shape and Length: Frequency Resolution

---

4. A kind of sneaker.

5. In the chapter on FIR filter design, we worried about only the positive-frequency spectrum of the windows.

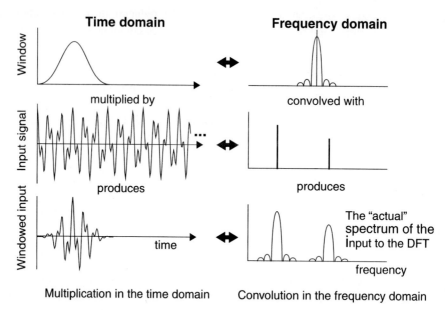

Figure 8-5 The relationship between window shape and a windowed signal's spectrum.

The choice of window type also affects our ability to resolve closely spaced frequency components, or the *frequency resolution* of the spectrum. Frequency resolution is directly related to the width of the main lobe of the window. A wide main lobe will cause nearby frequency components to blend together into an inscrutable "hump," and a narrow main lobe will allow (up to a limit) closely spaced frequency components to be identified. This blending effect is sometimes called *smearing*.

There are two ways we can minimize the width of the window's main lobe. First, for a given window length, each window type has a different main lobe width. Windows such as the rectangular, Hamming, and Hanning have relatively narrow main lobes compared with Blackman or Kaiser (β=8.96) windows. Unfortunately, the first three windows have relatively poor spectral leakage characteristics (i.e., relatively high side lobes).

Our second method for decreasing the main lobe width is by increasing the window length.[6] Recall we adjusted the window length in our FIR filter design to get the right transition width—we're doing much the same here. Even the Kaiser (β=8.96) window can produce good frequency resolution if we use a long enough window. (Watch out, though—this assumes the signal is stationary. See the section later on time-frequency analysis.) Figure 8-7 shows how the

6. In many real-world situations, unfortunately, playing with the length of the window is not an option—other constraints establish the window length. For example, to produce real-time results, the window cannot be arbitrarily long or unacceptable delays will occur. Another common situation is that the signal itself can be treated as stationary over only a relatively short period of time. More on this latter point later in this chapter.

Using the DFT

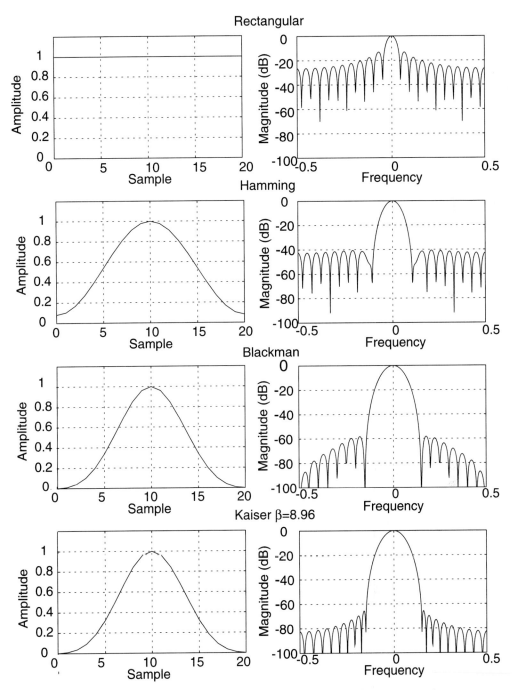

Figure 8-6 Common windows and their time- and frequency-domain representations.

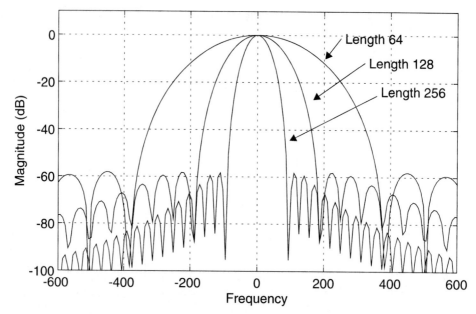

Figure 8-7 Main lobe width for the Blackman window (N=64, 128, 256).

main lobe width decreases with increases in window length for the Blackman window (the frequency scale is given for an 8-kHz sampling rate).

Let's take some actual data and examine the effects of window type and length on the resulting spectra. We'll generate input data according to the following equation:

$$x(n) = \cos\left(2\pi \frac{878}{8000} n\right) + \cos\left(2\pi \frac{960}{8000} n\right) + \cos\left(2\pi \frac{2478}{8000} n\right) \quad (8.1)$$

That is, we're generating a sequence of samples that, for a sampling rate of 8000 Hz, has frequency components at 878 Hz, 960 Hz, and 2478 Hz. The first two frequency components are quite close together compared with the third—we'll want to see how well these two close components are resolved.

Figure 8-8 shows the DFT of the same data windowed by the rectangular, Hamming, Blackman, and Kaiser (β=8.96) windows. Figure 8-9 zooms in on the area of the spectrum where the two frequency components are closely spaced. The superior resolution of the rectangular window is evident, but so, too, is its spectral leakage. Meanwhile, the other windows are having a hard time with the closely spaced components. (Note that we've been a little dishonest in our plots—we actually should be plotting individual *points*, as the DFT produces a discrete frequency output. You can see these points in the second, zoomed plot.)

Using the DFT

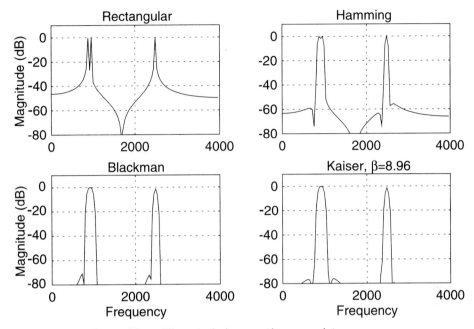

Figure 8-8 The effects of four different windows on the same data.

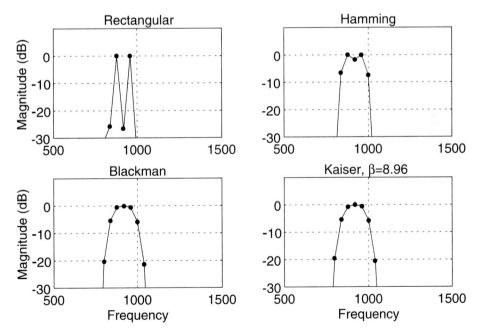

Figure 8-9 Details from Figure 8-8 showing resolution of nearby frequencies.

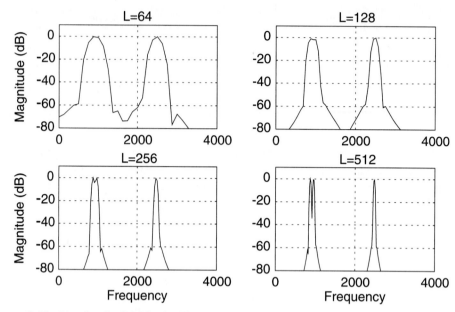

Figure 8-10 Varying the length of a Blackman window.

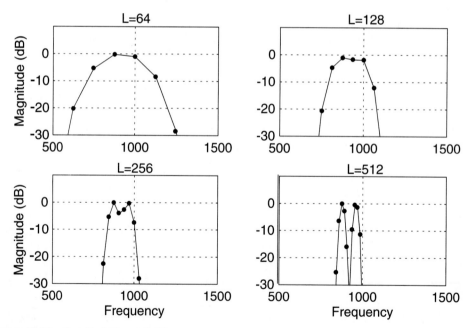

Figure 8-11 Detail of Figure 8-10.

Using the DFT

Now let's look at the effect window length has on frequency resolution. Figure 8-10 shows our same data windowed with a Blackman window, but now we'll vary the window length from 64 to 1024. Figure 8-11 again zooms in on the closely spaced frequency components, where we can see that the hump that ate Cincinnati (and our two frequency components) turns into nicely distinguishable peaks if we increase the window length.

8.4.2.3 Choosing a Window

Table 8-1 summarizes some popular windows, noting the side lobe magnitudes and main lobe widths. The article by Harris [1978] is a widely cited review of windows for use with the DFT, discussing more than 20 different windows. Good window choices tend to emphasize low spectral leakage, though there are other issues we haven't discussed that may be important in some applications. The Blackman and Kaiser windows (e.g., $\beta \geq 8.96$) are good windows to start with.

Window	Side lobe magnitude relative to main lobe peak (dB)	Main lobe width (normalized Hz) (N=window length)
Rectangular	-13	$1.2/N$
Kaiser β=2.12	-19	$1.4/N$
Hanning	-31	$2.0/N$
Kaiser β=4.5	-34	$1.8/N$
Hamming	-41	$1.8/N$
Kaiser β=6.76	-49	$2.2/N$
Blackman	-57	$2.4/N$
Kaiser β=8.96	-66	$2.4/N$
Kaiser β=9.42	-69	$2.4/N$

Table 8-1 Side lobe magnitude and main lobe width of popular windows.

Table 8-1 lists the main lobe width in terms of N, the window length. One point to note is that the main lobe in this case is measured to the -6 dB point (versus, for example, the -3 dB point), because what we're looking for is the closest we can have two frequency components and still be able to distinguish between them. If our criterion for distinguishing two components is that there be a valley of at least -3 dB between the peaks, this point must be at the -6 dB point of the main lobes.[7] (See Figure 8-12.) Because both main lobes will be identical (they're just "copies" of the same window's spectrum), the -6 dB width of the main lobe is also the minimum frequency difference we can have between frequency components and still be able to resolve them as separate peaks.

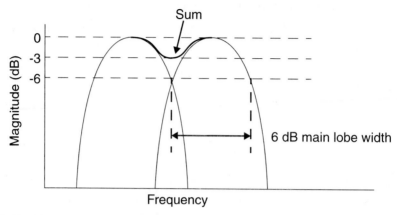

Figure 8-12 Main lobe width and distinguishing nearby frequencies.

Let's look at our example signal and calculate what the *minimum* window length should be if we wish to resolve the two frequency components at 878 Hz and 960 Hz. Converting both quantities to normalized Hz and subtracting, the frequency difference is 0.01025 (normalized Hz). The Blackman window used in Figure 8-11 has a -6 dB main lobe width of about $2.4/N$. The minimum window length is, therefore, $N=2.4/0.01025$, or 234. As you can see in the figure, we were unable to resolve the peaks for lengths of 64 and 128, but the peaks are distinguishable for lengths of 256 and 512. (Why powers of 2 for window length? We'll tell you in just a moment.)

There are two important points about window length. First, we've been assuming that the real-world signal is stationary. If the signal changes frequency content over time, then our window can't be arbitrarily long while at the same time providing a meaningful spectrum. Second, window length and shape modify the frequency spectrum of the signal we give to the DFT. The DFT cannot "undo" any smearing or spectral leakage. The frequency resolution we preserve by our choice of window type and length is the best that the DFT will be able to resolve—no better. However, as we'll see in a few pages, the DFT is not guaranteed to give us all the resolution in its output that we have preserved in its input, as the DFT output is not continuous, but is just a sampled version of the underlying spectrum. (Don't worry if this doesn't make sense right now. We'll come back to it.)

Let's now turn our attention to the output of the DFT algorithm, the DFT itself.

---

7. Watch out—if we *multiplied* -6 dB by -6 dB, as a shortcut we could add -6 and -6 to get -12 dB. But we're *adding* two signals, so we should convert to linear, add, then convert back to dB. This is assuming that we can just add the magnitudes of two complex quantities, however. Depending on the phases involved, -3 dB is the maximum we can get. If the phases are exactly 180° apart, the magnitude will be 0 (linear)!

8.4.3 Interpretation of Output

There are three distinct aspects to the DFT output we could consider. First, the indexing of the output samples—what frequency range do the samples cover, and what symmetries are there in this range (for real signals)? Next, how should we interpret the magnitude of the samples? Finally, we could consider the phase of the output samples.

But we won't. Consider the phase, that is, aside from saying "look—there's the phase." As you well know, phase is necessary to reconstruct the original time-domain signal and carries useful information in many other cases. But at the level of this book, phase is not a critical product of the DFT. Most of the interesting things we want to do with the DFT for analysis have to do with the energy at different frequencies. Fair enough? Let's look at the frequencies of the DFT output samples.

8.4.3.1 Frequency

Our plots so far have had a nicely labelled x-axis in real-world frequency (Hz). However, the DFT produces N output samples. What is the relationship between the DFT output samples and the frequencies they represent?

The relationship is

$$F_k = \frac{kF_s}{N} \qquad k = 0,\ldots,N-1 \qquad (8.2)$$

where k is the index of the output sample. However, you might notice this means the DFT output frequencies range from 0 to just shy of F_s—what happened to Mr. Nyquist and his sampling theorem?

Recall that the spectrum of a sampled signal is periodic. Many chapters ago we saw that for a real (vs. complex) sampled signal, the frequency range 0 to $F_s/2$ was reversed for the range of $-F_s/2$ to 0, and this entire range ($-F_s/2$ to $F_s/2$) was repeated infinitely in both directions in the frequency domain. Well, the DFT just outputs a different period, from 0 to F_s, which is the same length, and (pretty much) the same spectrum.

For a real input of N samples, the DFT produces N complex outputs. However, because of the symmetry of the magnitude and phase spectrum mentioned above, only $N/2$ complex outputs carry unique information. We need to look only at the first $N/2$ outputs for both magnitude and phase information and can completely recreate the input signal from these outputs. (However, if we have a complex signal as input to the DFT, all N complex outputs could carry information.)

Figure 8-13 shows the output of the DFT using 64 samples from Equation 8.1 and a Blackman window. (Note we're using a linear scale for magnitude, not dB, as earlier.) You can see the symmetry of both the magnitude and phase between samples 0–31 and 32–63. Using Equation 8.2, we can convert the indices to real-world frequencies by multiplying each index k by F_s/N (or 8000/64).

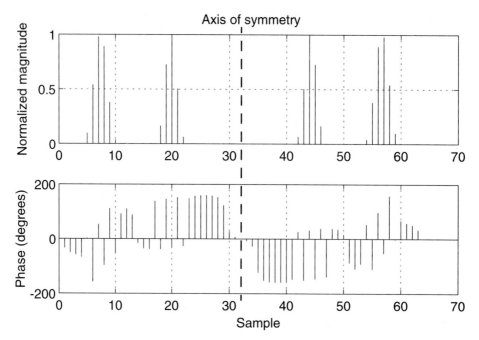

Figure 8-13 Symmetry of magnitude and phase of DFT output.

8.4.3.2 Magnitude (and a Little Bit About Phase)

What about magnitude of the DFT output? Intuitively, the magnitude should be telling us about how much energy is found at a given frequency. The truth is a little trickier.

It turns out that the DFT is a sampled version of the continuous-frequency spectrum we get using the DTFT[8] on the same input. This is *not* to say that we first compute the DTFT and then sample it to produce a DFT (although we could in principle—just not with a computer). Rather, the DFT produces the same values that we would get by sampling the DTFT at those points. Figure 8-14 shows the spectrum generated by the DTFT (the continuous curve) and the discrete samples generated by a DFT, both using 64 samples from Equation 8.1. (By the way, you might compare this figure with the first plot of Figure 8-10, where we lied a bit by "connecting the dots" of the DFT magnitude.)

Let us digress a moment to understand the nature of the DTFT output. The magnitude of the DTFT output is not a statement like "at 45 Hz there is a frequency component with a value of 1.4 volts." Instead, the DTFT magnitude is an *amplitude density*. Instead of units of volts (assuming our original signal was a voltage signal), the units are volts/Hz. We can't ask for the amplitude at a certain frequency, but we can ask over a range of frequencies what the average amplitude is.

---

8. You might recall we used the inverse DTFT back in the FIR filter design chapter to go from an ideal frequency response curve to an impulse response.

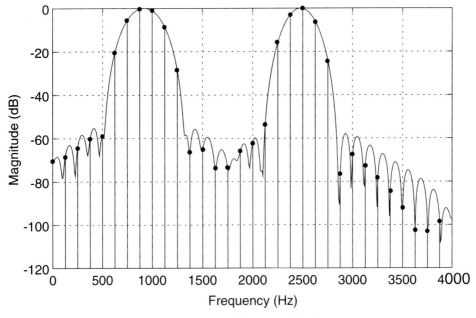

Figure 8-14 DTFT spectrum (solid line) and DFT sample (dots).

If this distinction isn't clear, you might consider a population density map. Such a map doesn't say, for example, that there is exactly one person in every 10 square meters of a particular locale. Rather, it says that if you take a reasonable area, *on average,* you'll find one person for every 10 square meters you've included. In the same way, the amplitude density tells you what the average amplitude is over a range of frequencies. (If the amplitude density is not constant over this range, you'll need to integrate to find the average. If the density is constant, you can just multiply by the frequency width.)

The DFT is a sampled version of the DTFT spectrum but has units of volts—at least it does after you scale the DFT output by multiplying by the sampling rate. (Mathematically, this scaling cancels out the "/Hz" in the amplitude density, giving us straight amplitude.) Mathematics aside, why isn't the DFT output an amplitude density like the DTFT? Because the DFT thinks it's looking at a periodic signal. A periodic signal can have energy only at frequencies that are integer multiples of the fundamental frequency, that is, $k=0, 1, 2, ..., N-1$. No energy can be "in between" the output samples because that wouldn't be consistent with the input samples representing exactly one period of a periodic signal. On the other hand, the DTFT makes no assumption about the periodicity of what it transforms, and so produces a (continuous-frequency) amplitude density output. We will have more to say about the relationship between the DTFT and DFT in a moment.

In practice, it is the *relative* values of the DFT spectrum that are usually of interest, so it's not uncommon to normalize the DFT so that the largest output is unity, and to express the magnitudes in decibels.

Power Spectrum In many cases, we'd much rather know about the relative power at different frequencies. Squaring the magnitude of the DFT produces a power spectrum, with units of watts. The average of many power spectra is known as a *periodogram*. (The DFT comes close to producing a "true" periodogram when N is large.) The output is "normalized" power, as it assumes a resistance of 1 ohm.[9]

8.4.4 Increasing Frequency Resolution

Now that we've waved our hands about the relationship between the DTFT and the DFT, it's time to revisit the question of frequency resolution. We told you that the DFT represents samples of the DTFT, and we should add that this sampling is at N equally spaced points (recall Equation 8-2.) If we want a more detailed (better frequency resolution) output from our DFT, there are two things we can do. First, produce a DTFT with better resolution in the first place (we can't sample what's not there!) and, second, sample the DTFT at more points.[10]

8.4.4.1 Adding Data

The frequency resolution of the underlying DTFT is a function of the shape and length of the window used to window the input. For a given window, increasing the length of the window reduces the width of the main lobe, which leads to better frequency resolution. A longer window implies using more data, so there's obviously a trade-off here between resolution and the ability to process longer windows.

Adding data—or, rather, making sure you have enough data to begin with—is the first step in implementing a DFT-based system. If the resolution is not available in the data (i.e., would not be available in the output of the DTFT for that data), it will not be available in the DFT.

8.4.4.2 Adding Zeros—Sampling the DTFT

It is sometimes the case that the underlying DTFT contains the information we want in sufficient detail, but we don't happen to be sampling it at enough points to get the desired detail in our DFT. Is there some way to increase the "sampling rate" of the DFT to get a better picture of the underlying DTFT? In fact, there is. The process is called *zero padding*, and it involves tacking on some samples with zero value to the input to the DFT. Note that the zeros are appended *after* any windowing is performed—that is, the window is not applied to the zeros also. The N of the DFT increases, which means the DFT outputs are more closely spaced in frequency. However, this doesn't change the underlying DTFT. Which means if you don't have sufficient frequency

---

9. *Power=Voltage^2/Resistance.*

10. In this discussion, it sounds like we're focused on the DTFT of the data. In fact, the DTFT is just the appropriate transform; it's not the DTFT per se that we are concerned with, but the frequency domain information that can be found in the data.

resolution in the DTFT, all you'll get from zero padding is a better picture of the hump that ate your peaks.

Figure 8-15 shows the relationship between window length and zero padding. (We're again using the three-tone signal made up of components at 878 Hz, 960 Hz, and 2478 Hz, and we'll use a Blackman window.) The underlying magnitude spectrum is the same for a given window length, regardless of the amount of zero padding. Therefore, the plots in the upper row all have the same overall "shape," which appears with increasing resolution as the amount of zero padding increases. (The amount of zero padding is just the difference between the DFT length and the window length; for the top row, the zero padding is 0, 128, and 394.) However, increasing the window length (i.e., adding actual data) changes the underlying magnitude spectrum, as you can see by comparing the plots in the rightmost column (DFT length=512). Although all three of these plots have the same DFT length and, thus, are sampling the underlying DTFT at the same "sample rate," the underlying DTFT for each plot is different.

The distinction between window length and zero padding is important to realize. It's quite acceptable to zero pad your data (especially to bring the number of samples to a power of 2, for reasons we'll discuss momentarily), but zero padding increases only the "sampling rate" (in the frequency domain) of the DFT—the underlying spectrum is unchanged. Most important is the fact that the window shape and length determine the frequency resolution of the underlying (DTFT) spectrum.

8.5 Implementing the DFT

8.5.1 Overview

The DFT ranks as one of the most important tools we have in signal processing. It would be hard to overstate its impact in virtually any field of science. However, the DFT in its basic form takes a large number of calculations; were it not for the discovery of much more efficient algorithms for computing the DFT, the DFT would not be as widespread as it is today.

Although this book isn't going to cover the implementation of the DFT in great detail, it will be useful for you to have a basic understanding of the operation of both the DFT and the class of improved DFT algorithms that are known as *fast Fourier transforms* (FFT).

8.5.2 From the DFT ...

8.5.2.1 Computation

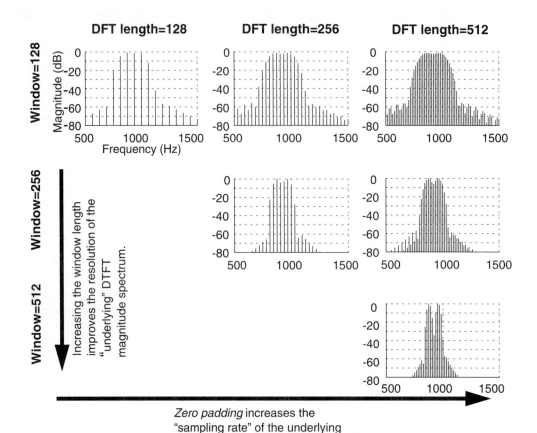

Figure 8-15 Relationship between zero padding and window length.

The DFT equation is rather simple:

$$X(k) = \sum_{n=0}^{N-1} x(n)(W_N)^{kn} \qquad k = 0, 1, \ldots, N-1 \tag{8.3}$$

Implementing the DFT

The complex values $X(0), X(1), ..., X(N-1)$ are the frequency-domain output, and the (possibly complex) values $x(0), x(1), ..., x(N-1)$ are the time-domain input. $(W_N)^{kn}$ are a set of constants known as the *twiddle factors*,[11] defined as follows:

$$(W_N)^{kn} = \left(\cos\left(\frac{2\pi}{N}\right) - j\sin\left(\frac{2\pi}{N}\right)\right)^{kn} \quad k = 0, 1, ..., N-1 \quad (8.4)$$

A little math reveals

$$(W_N)^{kn} = \cos\left(\frac{2\pi kn}{N}\right) - j\sin\left(\frac{2\pi kn}{N}\right) \quad k = 0, 1, ..., N-1 \quad (8.5)$$

It's actually much more common to see $(W_N)^{kn}$ written as:

$$(W_N)^{kn} = e^{-j\frac{2\pi kn}{N}} \quad (8.6)$$

which is mathematically equivalent, but because we have to store the twiddle factors in terms of real and imaginary parts in the microcontroller memory, the form of Equation 8.4 is a bit more useful.

What is the DFT doing? It's basically taking the input sequence $x(n)$ and "comparing" it to a series of sampled sinusoidal waves (the twiddle factors) to see if there is any similarity—that is, any similar frequency component. (This "comparing" is similar to the correlation we discuss in Chapter 9 [Correlation].) Index k steps through different frequencies, and index n steps through the sinusoidal waveform and the input sequence. Because we have no idea of the phase of frequency components, both sine and cosine waves are used (at the same time via complex math)—that way, if a frequency component happens to be exactly out of phase with the cosine wave (and hence, "invisible" when we look for similarities by multiplying the two sequences together), the sine wave will not only "detect it," but allow us to determine the phase.

Equation 8.3 is evaluated once for every DFT output. If we have N outputs, this is a total of approximately N^2 complex multiplications and the same number of complex additions. (Since each complex multiplication requires four real multiplications and additions, the actual number of computations is $4N^2$.) A 512-point DFT requires 262,000 complex multiplications; a 1024-point DFT in excess of a million. An algorithm that requires operations proportional to N^2 is technically known in most circles as "a bad thing."

---

11. Trust us on this—we're not clever enough to have come up with a name like "twiddle factor."

8.5.3   ...To the FFT

The fast Fourier transform has roots in the 1800s as folks were very motivated to find any shortcuts in performing these types of calculations by hand. However, the FFT in the form we use today was first described by Cooley and Tukey in 1965. Since then, a great deal of research has produced variations on the original theme, all of which are generally lumped under the name FFT. Each FFT has different strengths and makes different trade-offs between code complexity, memory usage, and computation. We're going to describe some general features of these algorithms and go into a little more detail on the most popular FFT, the *decimation-in-time* (DIT) FFT.

It will become apparent that the FFT is more involved to implement than the FIR or IIR filters we considered earlier. A full discussion of the implementation issues for FFTs would be quite lengthy, and the fact remains that FFT routines are best "borrowed" rather than written.[12] In addition, the operations central to the FFT are quite different from those of the FIR and IIR filters. Virtually no microcontroller explicitly supports the FFT operations, including the 68HC16. This means that the FFT may not be computed fast enough to be very useful, except in certain limited applications.

8.5.3.1   Symmetry and Periodicity of Twiddle Factors

The route to the FFT begins by noting that there is a lot of symmetry that can be exploited in the set of twiddle factors. Because the twiddle factors involve sines and cosines, this is not hard to imagine—the sine has odd symmetry and the cosine even.[13] A second gain in computational efficiency can be achieved by noting the periodicity of the twiddle factors—again, nothing surprising, as the sine and cosine functions are periodic.

These factors yield some decrease in computation of the DFT based on Equation 8.3. The real gains in efficiency require a quite different approach—divide and conquer.

8.5.3.2   Decimation-in-Time, Decimation-in-Frequency

The FFT capitalizes on the fact that it is possible to split the input to a DFT into two or more parts, perform separate DFTs on each part, then combine the results to get the same answer as if one larger DFT had been performed. (The DFT is a linear operator!) The most common algorithm splits the input into two parts. The savings of this method come from the fact that each DFT is now operating on $N/2$ points—so the number of computations is now $2\times(N/2)^2$, or $N^2/2$. This savings is hardly worthwhile—except what if we took each of those two small DFTs and broke each of those into smaller DFTs? If we continue with this process, we'll end up having broken the single N-point DFT into $\log_2 N$ DFTs of length 2. The final algorithm ends up requir-

---

12.   Almost any DSP textbook will give you an in-depth discussion of FFT implementation and point you to books and papers devoted entirely to the FFT. Another good place to start is the January 1992 issue of *IEEE Signal Processing Magazine*, which includes a piece by J.W. Cooley on "How the FFT gained acceptance."

13.   To say a function has even symmetry means that $f(n)=f(-n)$; that is, the plot of the function for negative values looks like a mirror image of that for positive values. Odd symmetry, when $f(-n)=-f(n)$, is what you'd get if you rotated the positive side of the function 180° around the origin.

ing computation proportional to $(N/2)log_2 N$, a very significant savings over N^2. For a 1024-point DFT, the difference is between 1,048,576 and 5,120. That's right—a million versus five thousand. The difference is even larger as we increase the number of points.

The *decimation-in-time* (DIT) FFT breaks the original input sequence into two smaller parts by putting the even-numbered samples (i.e., those with even "times") into one sequence and the odd-numbered samples into the other sequence. It is also possible to take the first $N/2$ samples for one sequence and the remaining $N/2$ samples for the other—this is the basis of the *decimation-in-frequency* (DIF) FFT. Both take roughly the same number of operations; the major difference tends to be in the order of the input and output (discussed below).

8.5.3.3 Sequence Lengths

The original (and most popular) FFT is known as a *radix-2 FFT*; the initial DFT is broken up into two smaller DFTs, each of these into two smaller DFTs, and so on. The input to radix-2 FFTs must therefore be a power of 2 (e.g., 256, 512, 1024, etc.). That's the reason, for example, why everyone gives benchmark times for 1024-point FFTs, but no one ever mentions 1000-point FFTs. For all but the shortest sequences, if you have a radix-2 FFT routine available, it is almost always quicker to zero pad the data to the next power of 2 and use the FFT rather than use the unmodified DFT.

Other FFTs can be implemented that break the DFT into four- or eight-point DFTs. The computational savings can be helpful, but the jump is nowhere near the savings the radix-2 FFT has over the DFT. There is some increase in the complexity of the program as well, and added restrictions on the length of the input (i.e., a power of 4 or 8).

FFTs also exist that do not require the input to be a power of 2; in general, however, these algorithms don't work well when the length is a number with few factors. (You can't break the big DFT into smaller parts well if the length has only a few factors.) These algorithms are more complicated than the radix-2 FFT and are far less likely to be available for microcontrollers.[14]

Overall, the radix-2 (DIT or DIF) FFT is the most common FFT algorithm used and, thus, the input length to FFTs in practice is almost always a power of 2.

8.5.3.4 Core Operations

The FIR filter has a "core" that involves a single multiplication and accumulation, which is almost always supported with a dedicated MAC instruction in DSP-friendly microprocessors. The core of the FFT is a bit more involved and involves two complex inputs, two complex outputs, a complex multiply, a negation, and two additions. In addition, we require a special form of addressing.

---

14. Cynics might suggest that radix $N \neq 2$ algorithms serve little purpose but to clutter the literature and keep signal processing professors off the streets. To which we reply, isn't that enough justification? Younger readers will not be familiar with the problems many communities faced from roving bands of underemployed signal processing professors prior to the widespread availability of personal computers. The problem was especially bad when mainframes or card punch machines went down, and several communities enforced curfews on signal processing faculty after 8:00 P.M. Let us never return to those dark days.

To make some sense of these operations, we need to look at the structure of the FFT graphically. Figure 8-16 shows the flow of data through an eight-point DIT FFT. In particular, notice the operation that is highlighted—the whole 8-point FFT is made up of three stages of four of these operations. This is the famous FFT *butterfly*, captured and mounted for your viewing in Figure 8-17.

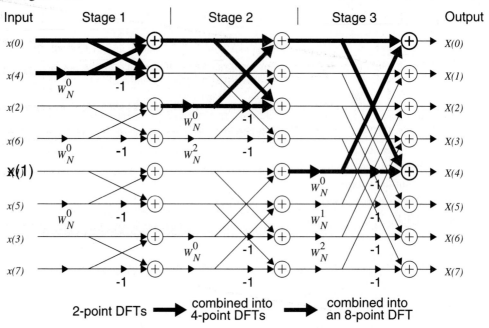

Figure 8-16 Structure of an eight-point DIT FFT.

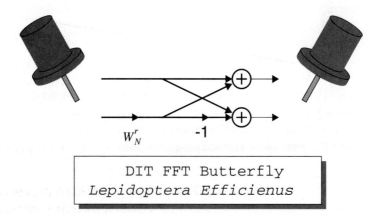

Figure 8-17 The FFT "butterfly."

Butterfly The *butterfly* operation is roughly the MAC of the FFT world. By dividing the DFT into smaller and smaller DFTs, the FFT ends up being composed of *(N/2)log₂N* FFT butterflies. For the radix-2 DIT FFT, there are log_2N stages, each of which accepts N complex inputs and generates N complex outputs. For the eight-point FFT here, there are just three stages. Within each FFT there are $N/2$ butterflies, each of which involves just a single complex multiplication by a twiddle factor, a negation, and two complex additions (or, if you prefer, one complex addition and one complex subtraction).

Because the butterflies combine the results of smaller DFTs to make a larger DFT, it's reasonable to ask "where's the actual 2-point DFT?" The two-point DFT is identical to the butterfly, though it's even a little less complicated since the twiddle factor for the first stage, W_N^0, is just 1. You can speed your code up somewhat by coding certain steps like this as special cases, but most generic FFTs don't. Keeping each stage the same results in a compact FFT.

Bit-Reversal Addressing The butterfly operation is buried deep within the FFT algorithm—it may never flutter its wings across your desktop. But an important side effect of the FFT is the unusual ordering it imposes on either the input or output data. This is visible in Figure 8-16, where the input is shown in what appears a quite scrambled order. In fact, the ordering has a simple underlying basis—if you express the index of the sample in binary form, you just reverse the order of the digits. For example, sample 4 has a 3-bit (binary) address of 100. Reverse this to get 001 and, sure enough, sample 4 is the second sample in the input. (Remember that 000 is the first.) Table 8-2 shows you the idea.

Original index		Bit-reversed index	
Decimal	Binary	Binary	Decimal
0	000	000	0
1	001	100	4
2	010	010	2
3	011	110	6
4	100	001	1
5	101	101	5
6	110	011	3
7	111	111	7

Table 8-2 Bit-reversal addressing (3 bits wide).

The DIT FFT requires that the input be processed in bit-reversed order. This scrambling might be your responsibility, but it is often included in the first part of the FFT code.

The DIF FFT produces an output that is bit-reversed while accepting a normally ordered input. Perhaps for historical reasons, this form is less common, but there are no major differences between the two otherwise. Again, the unscrambling might be built into the FFT code you're using, so bit-reversal may be transparent to you.

As a rule, microprocessors do not support bit-reversed addressing, but this is a relatively common feature of DSP chips. Just by setting a flag, you can increment or decrement an index register as if you were accessing a normal array, but the address hardware automatically bit-reverses the address for you.

If you do need to bit-reverse, consider using a look-up table rather than shifts or other operations. Other bit-reversing strategies are often discussed in Internet newsgroups (e.g., *comp.dsp*); some might be very useful on microprocessors. Also note that creating a bit-reversed array involves swapping pairs of values, so you can do the bit-reversal "in place."

8.5.3.5 Scaling

The FFT butterfly performs one complex multiply and two complex additions. There is no opportunity to use a special long accumulator to avoid temporary scaling problems with the FFT, so for each butterfly we must address the potential problem of overflow. There are four common methods to prevent (or address) overflow in the FFT. We discuss them (in bit-reversed order?) below.

Unconditional Input Scaling First, we can avoid overflow within the FFT by unconditionally scaling the input. Because each stage of the FFT can add at most 1 bit to the data flowing through, we can just right shift the input by the number of stages of the FFT (which is $log_2 N$). For a 1024-point FFT, our data must be shifted by 10 bits. If we're using 16-bit data, this leaves us with 6 bits of data—not very much resolution. On the positive side, the code for the FFT can run at maximum speed because there is no possibility of overflow.

An alternative way of looking at unconditional input scaling is to examine the number of additional bits that must be accommodated in the output. These additional bits are then known as *guard bits* (somewhat akin to the 4 extra bits in the 68HC16 accumulator), but the end result is the same—additional space is provided for the result of the calculations to grow.

Unconditional Scaling Per Stage Instead of scaling the input to account for every stage, we can scale at each stage by 1 bit. That is, we shift the input by 1 bit to the right, perform the first FFT stage, right shift that result by 1, perform the second stage, and so on. This will have better accuracy than unconditional input scaling, but the bottom line is that we are still scaling unconditionally, without regard for the actual results of calculations.

Block Floating-Point (Conditional Scaling Per Stage) Block floating-point is a "bit" smarter than unconditional scaling. The results of each FFT stage are examined to determine whether all the results of that stage should be scaled. For example, if all of the results of the first FFT stage have a magnitude less than 1 (in 1.15 format), no scaling is necessary. If just one has an overflow, however, we scale each of the results.

Block floating-point requires a way of keeping track of the total amount of scaling that was done, but this bookkeeping may be tolerable. We get better accuracy, as we ordinarily are scaling far less often than with the unconditional scaling methods.

Floating-Point Using floating-point numbers takes block floating-point to its extreme. Floating point numbers, after all, are just a sign, a magnitude, and a scaling factor. In block floating-point, the scaling factor was common to all the values at a given stage; using floating-point values, the scaling factor is specific to each value.

The problem with floating-point representations is that they require much more computation than fixed-point numbers (or much more hardware). Hardware floating-point (often a "floating-point unit," or FPU), practically speaking, is a luxury item that has not yet found its way into even medium-priced microcontrollers. On the other hand, there is nothing inherently complicated about performing simple floating-point mathematics. If your application does not require real-time performance, floating-point performed in software may be quite viable. (Floating-point libraries are available for almost any microcontroller, including Motorola's.)

The issue of scaling is primarily an issue of speed—if time is not a concern, multiple-precision or floating-point math can give you whatever resolution you need. Many applications simply don't require fast evaluation of an FFT. For example, let's take a hypothetical industrial application where we want to monitor the frequency components of vibrations from some piece of machinery to detect wear. The sampling rate might be tens of kHZ or more, but the underlying signal is not going to change much in the course of minutes (or even hours or days, perhaps). The application might sample at high speed for a moment, capturing perhaps 1024 points or more in a few tenths of a second, but then can spend hundreds of seconds, if necessary, calculating the FFT. This assumes that there is some time lag between the onset of the detectable condition and any catastrophic failure.

8.5.3.6 In-Place Algorithms

An important implementation issue is the fact that at any time, no more than N complex values are required only for an FFT implementation. As Figure 8-16 shows, each FFT stage needs to reference only the N values that are the output of the prior stage, and it generates only N output values itself. Note, too, that the two inputs to each butterfly are used be only that butterfly, and the two outputs of each butterfly are specific to that butterfly as well—no other butterflies affect a given butterfly's output values.

This observation is used to produce *in-place* FFT algorithms—FFTs that use the same memory for both input and output (and all intermediate calculations). Although this memory reduction may not be very important in systems with large amounts of undifferentiated memory, the issue is very important for microcontrollers with on-chip RAM. First, this may be the only memory in the system—for example, the 68HC16Z1 contains 1K bytes of RAM. Second, and potentially more important, on-chip memory often offers the fastest possible access times—an important factor in the overall FFT performance.[15] For these reasons, an in-place FFT is preferable.

8.5.4 The Goertzel Algorithm

The FFT produces a complete discrete-frequency spectrum of the input signal. There are cases in which the detection of only a limited number of frequencies is of interest; the Goertzel algorithm is a kind of semioptimized DFT that is more efficient for these cases.

The general idea behind the Goertzel algorithm is to compute only the needed DFT components and to take advantage of the periodicity of the twiddle factors. Complex math is avoided by grouping complex conjugate pairs of the twiddle factors. The final structure of the Goertzel algorithm (for M different frequencies) is of M different second-order IIR filters. The feedback part of each filter is evaluated for every sample, but the feedforward part needs to be evaluated only every Nth sample. Each second-order filter requires only one real multiply (one coefficient is -1) and two additions—a total of $M(N+1)$ multiplies compared to the FFT's $N\log_2 N$. Thus, the Goertzel algorithm is preferred when M is less than $\log_2 N$.

The Goertzel algorithm has many appealing characteristics, but the design involves a somewhat involved process of identifying optimal sampling rates and lengths for N. Application notes from some manufacturers[16] can guide you through the process if you determine the Goertzel algorithm is appropriate for your application.

8.6 Implementation of the FFT on the 68HC16

Listing 8-1 shows one implementation of an FFT on the 68HC16. This routine was written by Derrick Klotz, a Motorola field application engineer, and was modified by Alex Upchurch of Crosstek Microsystems Ltd. (North York, Ontario). Although the FFT and 68HC16 are definitely not a match made in heaven, these engineers have managed to code a 1024-point FFT that runs in about 65 msec—with other overhead, that's about 10 spectra per second.

```
************************************************************************
*                                                                      *
*                                                                      *
*          fft1024.asm     -   1024 Point Fast Fourier Transform       *
*                              for the MC68HC16                        *
*                                                                      *
*          Written by:         Derrick H.J. Klotz                      *
*                              Field Application Engineer              *
*                              Motorola Semiconductor, Canada          *
*                              4000 Victoria Park Avenue               *
*                              North York, ONT  M2H 3P4                *
*                                                                      *
```

Listing 8-1   1024-point FFT example code.

---

15. The 68HC16Z1 can access high-speed external memory as fast as the internal memory using the "fast termination," or "two-cycle" mode).

16. *Analog Devices* has several application notes on using the Goertzel algorithm to process DTMF signals.

```
*       Modified by:                                                        *
*                                                                           *
*                               Alex Upchurch                               *
*                               Crosstek Microsystems Ltd.                  *
*                               53 Belgrave Ave.                            *
*                               North York, Ontario                         *
*                                                                           *
*       This is a modification of the MC68HC11 FFT presented by:            *
*                                                                           *
*                               Ron Williams                                *
*                               Department of Chemistry                     *
*                               Ohio University                             *
*                               Athens, OH 45701                            *
*                                                                           *
*       Which was a modification of the MC6800 FFT presented by:            *
*                                                                           *
*                               Richard Lord                                *
*                               Byte Magazine, pp. 108-119                  *
*                               February 1979                               *
*                                                                           *
*       Additional references were:                                         *
*                                                                           *
*               Motorola Application Note APR4/D REV 2                      *
*                       Implementation of Fast Fourier Transforms           *
*                       on Motorola's DSP56000/DSP56001 and                 *
*                       DSP96002 Digital Signal Processors.                 *
*                       Guy R.L. Sohie                                      *
*                       Digital Signal Processor Operation                  *
*                                                                           *
*               Motorola's Dr. BUB DSP Bulletin Board                       *
*                       Motorola DSP Operations                             *
*                       6501 William Cannon Drive W.                        *
*                       Austin, TX 78735                                    *
*                                                                           *
*               Introduction to Digital Signal Processing                   *
*                       J.G. Proakis & D.G. Manolakis                       *
*                       Macmillan Publishing Company                        *
*                       New York, 1988                                      *
*                                                                           *
*                                                                           *
*       This program listing was originally compiled to execute on          *
*       Motorola's M68HC16Z1EVB Evaluation Board.                           *
*                                                                           *
*                                                                           *
*       This program listing is offered as an example exercise in           *
*       progress.  It is not considered to be complete nor can its          *
*       correctness be guaranteed.  Any suggestions or other help           *
*       is welcomed and greatly appreciated.      - DK                      *
*                                                                           *
*****************************************************************************
*****************************************************************************
*                                                                           *
*       Revision History                                                    *
*       01.00   17 May 1994     Original                                    *
*                                                                           *
*****************************************************************************
```

Listing 8-1   (Continued) 1024-point FFT example code.

```
*****     Constants    *****
yes       equ     0
no        equ     1

*****     Data Section  *****
          org     data_seg

kount0            rmb     2       ; 16-bit counter 0
kount1            rmb     2       ; 16-bit counter 1

*****     Program Section  *****
          org     prog_seg

**********************************************************************
*
*         Fast Fourier Transform Subroutine
*
*      This subroutine will perform a real-data fast fourier transform.
*         This calculation is done "in place" upon a input sample array
*         pointed to by IZ upon entry.
*
*      In order to optimize this subroutine for speed of execution, self-
*         modifying code techniques have been utilized to take advantage of
*         the fastest addressing modes possible.  For this reason, the
*         entire FFT subroutine must reside in RAM.
*
*      Input/Output sample array data organization:
*
*                 (IZ    )    ->   Ar
*                 (IZ + 2)    ->   Ai
*                 (IZ + 4)    ->   Br
*                 (IZ + 6)    ->   Bi
*                 (IZ + 8)    ->   Cr       ...etc
*
*      Since the incoming data is considered to be entirely real, all
*         imaginary values (Ai, Bi, Ci, etc) are initialized as zero.
*
**********************************************************************
*
*      Subroutine entry point.  Do basic initialization first.
*
*      Note that IZ points to the start of sample array upon entry.  In
*         addition, the bank number of the sine table must be in XK.
*
_fft1024
fft1024   equ     (_fft1024-iram_code_begin)+iram_code_start
          ldab    #page(iram_start)
          tbyk                    ; point YK to 'iram_bank'
          ldy     #pass-_fft1024  ; IY -> primary pass control routine    4
t_eval    set     fft_top-pass    ; calculate 8-bit offset to 'fft_top'
          stz     t_eval,y        ; save start of sample array            4
          clrd                                                            2
          tdmsk                   ; disable modulo addressing             2
*
**********************************************************************
*
```

Listing 8-1   (Continued) 1024-point FFT example code.

Implementation of the FFT on the 68HC16

```
*       Special case for first two passes of FFT for real incoming data.
*
*       Note that no scaling is performed by the first two passes,
*         therefore, incoming data should already be scaled by four (ie.,
*         divided by four).  This is automatically performed by the
*         windowing input sample subroutine "window_in" in conjunction with
*         the scaled sample window (ie., the sample window has been scaled
*         by four).
*
t_eval  set     temp0-pass          ; calculate 8-bit offset to 'temp0'
        ldaa    #<(fft_size/8)      ; working with sets of 8 samples each    2
        staa    t_eval,y            ; set up pass iteration counter          4
*
*       Fundamental calculations are:
*
*                   Ai = Bi = Ci = Di = 0
*
*                   Ar" = (Ar+Br) + (Cr+Dr)
*                   Ai" = (Ai+Bi) + (Ci+Di) = 0
*                   Br" = (Ar-Br) + (Ci-Di) = (Ar-Br)
*                   Bi" = (Ai-Bi) - (Cr-Dr) = -(Cr-Dr)
*                   Cr" = (Ar+Br) - (Cr+Dr)
*                   Ci" = (Ai+Bi) - (Ci+Di) = 0
*                   Dr" = (Ar-Br) - (Ci-Di) = (Ar-Br)
*                   Di" = (Ai-Bi) + (Cr-Dr) = (Cr-Dr)
*
_Ar     equ     0
_Ai     equ     2
_Br     equ     4
_Bi     equ     6
_Cr     equ     8
_Ci     equ     10
_Dr     equ     12
_Di     equ     14
*
pass12
        lde     _Ar,z               ; ACCE = Ar                              6
        ldd     _Br,z               ; ACCD = Br                              6
        ade                         ; compute (Ar+Br)                        2
        ste     _Ar,z               ; save partial result for Ar"            6
        sde                         ; subtract twice to                      2
        sde                         ;   compute (Ar-Br)                      2
        ste     _Br,z               ;::::::::::::::: Br" = (Ar-Br)           6
        ldd     _Dr,z               ; ACCD = Dr                              6
        ste     _Dr,z               ;::::::::::::::: Dr" = (Ar-Br)           6
        lde     _Cr,z               ; ACCE = Cr                              6
        sde                         ; compute (Cr-Dr)                        2
        ste     _Di,z               ;::::::::::::::: Di" = (Cr-Dr)           6
        ste     _Bi,z               ; Bi" = (Cr-Dr)                          6
        negw    _Bi,z               ;::::::::::::::: Bi" = -(Cr-Dr)          8
        ade                         ; add twice to                           2
        ade                         ;   compute (Cr+Dr)                      2
        ted                         ; ACCD = (Cr+Dr)                         2
        lde     _Ar,z               ; ACCE = (Ar+Br)                         6
        ade                         ; compute (Ar+Br) + (Cr+Dr)              2
        ste     _Ar,z               ;::::::::::::::: Ar" = (Ar+Br) + (Cr+Dr) 6
```

Listing 8-1 (Continued) 1024-point FFT example code.

```
            sde                         ; subtract twice to                              2
            sde                         ;   compute (Ar+Br) - (Cr+Dr)                    2
            ste     _Cr,z               ;:::::::::::::::: Cr" = (Ar+Br) - (Cr+Dr)       6
            clrw    _Ai,z               ;:::::::::::::::: Ai" = 0                        6
            clrw    _Ci,z               ;:::::::::::::::: Ci" = 0                        6
*
*
*                   Ei = Fi = Gi = Hi = 0
*
*                   Er" = (Er+Fr) + (Gr+Hr)
*                   Ei" = (Ei+Fi) + (Gi+Hi) = 0
*                   Fr" = (Er-Fr) + (Gi-Hi) = (Er-Fr)
*                   Fi" = (Ei-Fi) - (Gr-Hr) = -(Gr-Hr)
*                   Gr" = (Er+Fr) - (Gr+Hr)
*                   Gi" = (Ei+Fi) - (Gi+Hi) = 0
*                   Hr" = (Er-Fr) - (Gi-Hi) = (Er-Fr)
*                   Hi" = (Ei-Fi) + (Gr-Hr) = (Gr-Hr)
*
_Er         equ     16
_Ei         equ     18
_Fr         equ     20
_Fi         equ     22
_Gr         equ     24
_Gi         equ     26
_Hr         equ     28
_Hi         equ     30
*
            lde     _Er,z               ; ACCE = Er                                      6
            ldd     _Fr,z               ; ACCD = Fr                                      6
            ade                         ; compute (Er+Fr)                                2
            ste     _Er,z               ; save partial result for Er"                    6
            sde                         ; subtract twice to                              2
            sde                         ;   compute (Er-Fr)                              2
            ste     _Fr,z               ;:::::::::::::::: Fr" = (Er-Fr)                  6
            ldd     _Hr,z               ; ACCD = Hr                                      6
            ste     _Hr,z               ;:::::::::::::::: Hr" = (Er-Fr)                  6
            lde     _Gr,z               ; ACCE = Gr                                      6
            sde                         ; compute (Gr-Hr)                                2
            ste     _Hi,z               ;:::::::::::::::: Hi" = (Gr-Hr)                  6
            ste     _Fi,z               ; Fi" = (Gr-Hr)                                  6
            negw    _Fi,z               ;:::::::::::::::: Fi" = -(Gr-Hr)                 8
            ade                         ; add twice to                                   2
            ade                         ;   compute (Gr+Hr)                              2
            ted                         ; ACCD = (Gr+Hr)                                 2
            lde     _Er,z               ; ACCE = (Er+Fr)                                 6
            ade                         ; compute (Er+Fr) + (Gr+Hr)                      2
            ste     _Er,z               ;:::::::::::::::: Er" = (Er+Fr) + (Gr+Hr)       6
            sde                         ; subtract twice to                              2
            sde                         ;   compute (Er+Fr) - (Gr+Hr)                    2
            ste     _Gr,z               ;:::::::::::::::: Gr" = (Er+Fr) - (Gr+Hr)       6
            clrw    _Ei,z               ;:::::::::::::::: Ei" = 0                        6
            clrw    _Gi,z               ;:::::::::::::::: Gi" = 0                        6
*
*       Cycle through array until complete.
*
            aiz     #32                 ; offset sample array pointer                    2
```

Listing 8-1 (Continued) 1024-point FFT example code.

Implementation of the FFT on the 68HC16

```
                dec     t_eval,y        ; check if done (temp0)                      8
                lbne    pass12          ; loop back if not                           6,4
*
****************************************************************
*
*       Core Fast Fourier Transform for passes 3 to 10.
*
*       Initialize control variables.
*
t_eval  set     cellnbr-pass    ; calculate 8-bit offset to 'cellnbr'
        ldd     #fft_size/8     ; get number of remaining cells              4
        stab    t_eval,y        ; save in 'cellnbr'                          4
t_eval  set     delta-pass      ; calculate 8-bit offset to 'delta'
        ldd     #256            ; angle increment is 256                     4
        std     t_eval,y        ; save in 'delta'                            4
t_eval  set     pairs-pass      ; calculate 8-bit offset to 'pairs'
        ldd     #4              ; number of pairs is 4                       4
        std     t_eval,y        ; save in 'pairs'                            4
t_eval  set     pairoff-pass    ; calculate 8-bit offset to 'pairoff'
        ldd     #16             ; distance between pairs is 16 bytes         4
        std     t_eval,y        ; save in 'pairoff'                          4
t_eval  set     sinbank-pass    ; calculate 8-bit offset to 'sinbank'
        txkb                    ; get sine table bank number                 2
        stab    t_eval,y        ; save in 'sinbank'                          4
        aiy     #twdtbl-pass    ; IY -> top of twiddle table                 2
*
*       Primary pass control.
*
*       The next instruction will load ACCA with the current cell count.
*       Note that this variable is implemented as self-modifying code in
*       RAM to take advantage of the faster immediate mode addressing.
*
pass     ldaa    #0              ; get the current number of cells           2
cellnbr  equ     *-1             ; number of cells in current pass
t_eval   set     cellctr-twdtbl  ; calculate 8-bit offset to 'cellctr'
         staa    t_eval,y        ; save in 'cellctr'                         4
*
*       The next instruction will load IZ with the starting address of
*       the input sample array.  Note that this variable is implemented
*       as self-modifying code in RAM to take advantage of the faster
*       immediate mode addressing.
*
         ldz     #0              ; IZ -> start of sample array              4
fft_top  equ     *-2             ; starting address of sample array
*
*       Cell calculation control.
*
cell     ldx     #sintbl         ; IX -> top of sine table                  4
t_eval   set     sinptr-twdtbl   ; calculate 8-bit offset to 'sinptr'
         stx     t_eval,y        ; initialize sine table pointer            4
*
*       The next instruction will load ACCD with the current pair count.
*       Note that this variable is implemented as self-modifying code in
*       RAM to take advantage of the faster immediate mode addressing.
*
         ldd     #0              ; get current number of pairs per cell     4
```

Listing 8-1 (Continued) 1024-point FFT example code.

```
pairs     equ     *-2               ; number of pairs per cell
t_eval    set     pairctr-twdtbl    ; calculate 8-bit offset to 'pairctr'
          std     t_eval,y          ; save in 'pairctr'                          4
          bra     r2bfly            ; start butterfly calculations               6
*
*
*     Program variables not buried in the code are located here to take
*     advantage of the fact that IY will always be pointing to the
*     twiddle factor look-up table.  This allows the faster indexed
*     addressing mode to be used whenever possible.
*
twdtbl                              ; twiddle factor look-up table:
          rmb     2                 ;   cos(w)
          rmb     2                 ;   sin(w)
cellctr   rmb     2                 ; cell counter                          +4
pairctr   rmb     2                 ; pair counter                          +6
temp0     rmb     2                 ; temporary holding register 0          +8
temp1     rmb     2                 ; temporary holding register 1          +10
*
*
*     Radix-2 Butterfly Kernel.
*     Fundamental calculations are:
*
*           Ar' = Ar + Br(cos(w)) + Bi(sin(w))
*           Ai' = Ai - Br(sin(w)) + Bi(cos(w))
*           Br' = Ar - Br(cos(w)) - Bi(sin(w)) = 2Ar - Ar'
*           Bi' = Ai + Br(sin(w)) - Bi(cos(w)) = 2Ai - Ai'
*
*     All results are divided by 2 to prevent overflows (auto scaling).
*     This yields the following:
*
*           Ar" = Ar' / 2          Br" = Br' / 2
*           Ai" = Ai' / 2          Bi" = Bi' / 2
*
*     Routine input conditions:
*       ACCA = pair counter
*       IX -> sine table
*       IZ -> sample array
*       IY -> twiddle table .......... format:    cos(w)
*                                                 sin(w)
*
*     Note that 'r2bfly' is the initial entry point for this routine.
*
*
*     Set up new data pointer.
*
r2bloop   aiz     #4                ; IZ -> next real value                      2
*
*     The next instruction will load IX with the sine table pointer.
*     Note that this variable is implemented as self-modifying code in
*     RAM to take advantage of the faster immediate mode addressing.
*
          ldx     #0                ; IX -> current sine table index             4
sinptr    equ     *-2               ; sine table pointer
*
```

Listing 8-1 (Continued) 1024-point FFT example code.

Implementation of the FFT on the 68HC16

```
*              The next instruction will add the current angle increment to IX.
*              Note that this variable is implemented as self-modifying code in
*              RAM to take advantage of the faster immediate mode addressing.
*
               aix    #512            ; IX -> new sine table index             4
delta   equ    *-2                    ; angle increment
t_eval  set    sinptr-twdtbl          ; calculate 8-bit offset to 'sinptr'
               stx    t_eval,y        ; save new sine table index              4
*
*
*              Butterfly entry point.
*
r2bfly
*
*              The next instruction will establish the sine table bank number.
*              Note that this variable is implemented as self-modifying code in
*              RAM to take advantage of the faster immediate mode addressing.
*
               ldab   #0              ; get sine table bank number             2
sinbank equ    *-1                    ; sine table bank number
               tbxk                   ; XK = sine table bank number            2
               ldd    512,x           ; get cos(w)                             6
               std    0,y             ; save cos(w)                            4
               ldd    0,x             ; get sin(w)                             6
               std    2,y             ; save sin(w)                            4
*
*              Calculate Ar" via two MAC instructions.
*
*              First, get Ar into the MAC unit.
*
               lde    0,z             ; ACCE = Ar                              6
               tem                    ; AM = Ar                                4
*
*              Then set up a pointer to Br.  Note that this instruction will make
*              XK:IX = ZK:IZ.
*
               tzx                    ; IX -> Ar                               2
*
*              The next instruction will add the current pairs offset to IX.
*              Note that this variable is implemented as self-modifying code in
*              RAM to take advantage of the faster immediate mode addressing.
*
               aix    #512            ; IX -> Br                               4
pairoff equ    *-2                    ; current distance between pairs
*
*              And load the MAC unit registers.
*
               ldhi                   ; HR = Br, IR = cos(w)                   8
               stz    t_tmp0,y        ; save IZ in 'r2tmp0'                    4
*
*              Here's what the first MAC will do, step by step:
*
*              Before MAC instruction:
*
*                     AM = Ar
*                     IX -> Br
```

Listing 8-1 (Continued) 1024-point FFT example code.

```
*              IY -> cos(w)
*              HR =  Br
*              IR =  cos(w)
*
*         After MAC instruction:
*
*              ED = HR * IR
*              AM = AM + (HR * IR) ....... AM = Ar + Br(cos(w))
*              IX = IX + 2 ................ IX -> Bi
*              IY = IY + 2 ................ IY -> sin(w)
*              IZ = HR
*              HR = [IX] ................. HR = Bi
*              IR = [IY] ................. IR = sin(w)
*
          mac      2,2                ;                                    12
*
*         Here's what the second MAC will do, step by step:
*
*         Before MAC instruction:
*
*              AM =  Ar + Br(cos(w))
*              IX -> Bi
*              IY -> sin(w)
*              HR =  Bi
*              IR =  sin(w)
*
*         After MAC instruction:
*
*              ED = HR * IR
*              AM = AM + (HR * IR) ....... AM = Ar + Br(cos(w)) + Bi(sin(w))
*              IX = IX + 2 ................ IX -> [Bi + 2]
*              IY = IY + 0 ................ IY -> sin(w)
*              IZ = HR
*              HR = [IX]
*              IR = [IY]
*
          mac      2,0             ; AM = Ar'                              12
*
*         The next instruction will restore IZ with its previous value.
*         Note that this variable is implemented as self-modifying code in
*         RAM to take advantage of the faster immediate mode addressing.
*
          ldz      #0              ; restore IZ                             4
r2tmp0    equ      *-2             ; temporary IZ storage during MACs
*
*         Round off the result and save:  Ar" = Ar' / 2
*
          asrm                     ; AM = Ar' / 2 = Ar"                     4
          tmer                     ; ACCE = AM (rounded) = Ar"              6
          ldd      0,z             ; ACCD = Ar                              6
          ste      0,z             ; save Ar"                               6
*
*         Calculate Br" with simple subtractive technique:
*                 Br' = 2Ar - Ar' ....... Br" = Br' / 2
*                                             = Ar - (Ar' / 2)
*                                             = Ar - Ar"
```

Listing 8-1   (Continued) 1024-point FFT example code.

Implementation of the FFT on the 68HC16

```
*
               nege                       ;  ACCE = -Ar"                              2
               ade                        ;  ACCE = Ar - Ar" = Br"                    2
               ste     t_tmp2,y           ;  save Br" temporarily in 'r2tmp2'         6
*
*      Calculate Ai" via two MAC instructions.
*
*      First, get Ai into the MAC unit.
*
               lde     2,z                ;  ACCE = Ai                                6
               tem                        ;  AM = Ai                                  4
*
*      Then set up a pointer to -Br.
*
               aix     #-4                ;  IX -> Br                                 4
               negw    0,x                ;  IX -> -Br                                8
*
*      And load the MAC unit registers.
*
               ldhi                       ;  HR = -Br, IR = sin(w)                    8
               stz     t_tmp1,y           ;  save IZ in r2tmp1                        4
*
*      Here's what the first MAC will do, step by step:
*
*         Before MAC instruction:
*
*             AM  =  Ai
*             IX  -> Br
*             IY  -> sin(w)
*             HR  =  Br
*             IR  =  sin(w)
*
*         After MAC instruction:
*
*             ED = HR * IR
*             AM = AM + (HR * IR) ....... AM = Ai - Br(sin(w))
*             IX = IX + 2 ............... IX -> Bi
*             IY = IY - 2 ............... IY -> cos(w)
*             IZ = HR
*             HR = [IX] ................. HR = Bi
*             IR = [IY] ................. IR = cos(w)
*
               mac     2,-2               ;                                          12
*
*      Here's what the second MAC will do, step by step:
*
*         Before MAC instruction:
*
*             AM  =  Ai - Br(sin(w))
*             IX  -> Bi
*             IY  -> cos(w)
*             HR  =  Bi
*             IR  =  cos(w)
*
*         After MAC instruction:
*
```

Listing 8-1   (Continued) 1024-point FFT example code.

```
*               ED = HR * IR
*               AM = AM + (HR * IR) ....... AM = Ai - Br(sin(w)) + Bi(cos(w))
*               IX = IX + 2 ............... IX -> [Bi + 2]
*               IY = IY + 0 ............... IY -> cos(w)
*               IZ = HR
*               HR = [IX]
*               IR = [IY]
*
         mac      2,0                 ; AM = Ai'                                  12
*
*       The next instruction will restore IZ with its previous value.
*       Note that this variable is implemented as self-modifying code in
*       RAM to take advantage of the faster immediate mode addressing.
*
         ldz      #0                  ; restore IZ                                 4
r2tmp1   equ      *-2                 ; temporary IZ storage during MACs
*
*       Round off the result and save:  Ai" = Ai' / 2
*
         asrm                         ; AM = Ai' / 2 = Ai"                         4
         tmer                         ; ACCE = AM (rounded) = Ai"                  6
         ldd      2,z                 ; ACCD = Ai                                  6
         ste      2,z                 ; save Ai"                                   6
*
*       Calculate Bi' with simple subtractive technique:
*               Bi' = 2Ai - Ai' ....... Bi" = Bi' / 2
*                                           = Ai - (Ai' / 2)
*                                           = Ai - Ai"
*
         nege                         ; ACCE = -Ai"                                2
         ade                          ; ACCE = Ai - Ai" = Bi"                      2
         ste      -2,x                ; save Bi"                                   6
*
*       Get and save Br".
*
*       The next instruction will load ACCD with the value of Br".
*       Note that this variable is implemented as self-modifying code in
*       RAM to take advantage of the faster immediate mode addressing.
*
         ldd      #0                  ; ACCD = Br"                                 4
r2tmp2   equ      *-2                 ; temporary storage for Br"
         std      -4,x                ; save Br"                                   6
*
*       Check if finished with this cell via pair counter.
*
t_eval   set      pairctr-twdtbl      ; calculate 8-bit offset to 'pairctr'
         decw     t_eval,y            ; decrement pair counter                     8
         lbne     r2bloop             ; loop back if not finished                6,4
*
*       Advance the cell count and check if finished with this pass.
*       Note that IX already points to Ar in the next cell.
*
         txz                          ; IZ -> Ar                                   2
t_eval   set      cellctr-twdtbl      ; calculate 8-bit offset to cellctr
         dec      t_eval,y            ; decrement the cell count                   8
         lbne     cell                ; loop back if not finished                6,4
```

Listing 8-1   (Continued) 1024-point FFT example code.

Implementation of the FFT on the 68HC16

```
*
*       Increase the cell size and check if done.
*
t_eval   set     cellnbr-twdtbl    ; signed 16-bit offset to 'cellnbr'
         lsr     t_eval,y          ; divide the number of cells by two       8
         beq     done              ; exit if result is zero                  6,2
t_eval   set     pairs-twdtbl      ; signed 16-bit offset to 'pairs'
         aslw    t_eval,y          ; double the number of pairs              8
t_eval   set     pairoff-twdtbl    ; signed 16-bit offset to pairoff
         aslw    t_eval,y          ; double the distance between pairs       8
t_eval   set     delta-twdtbl      ; signed 16-bit offset to 'delta'
         lsrw    t_eval,y          ; divide the angle offset by two          8
         lbra    pass              ; loop back for next pass                 6
done     rts                       ; return
fftend
*
*
t_tmp0   set     r2tmp0-twdtbl     ; calculate 8-bit offset to 'r2tmp0'
t_tmp1   set     r2tmp1-twdtbl-2   ; calculate 8-bit offset to 'r2tmp1'
t_tmp2   set     r2tmp2-twdtbl-2   ; calculate 8-bit offset to 'r2tmp2'
```

Listing 8-1   (Continued) 1024-point FFT example code.

8.6.1 Design Issues

Although the code is far from simple, we can pick out some of the issues that Klotz and Upchurch had to deal with in a real-world implementation such as real data input, self-modifying code, use of pointers, optimization of the stages of the FFT, scaling, and use of tables. We'll also discuss some of the code needed to support an FFT routine. (This routine is also available on-line—see Preface.)

8.6.1.1   Real Data Input

First, we note that the program uses only real data as input. Thus, the first two stages ("passes") can omit some calculations that involve only the imaginary components, and the calculations themselves can be hardcoded, rather than requiring the more complicated logic of stages 3–10. (Stage1 and 2 processing begins at the label "pass12.")

Note, however, that we still use a data buffer with enough room for complex data—that is, two words (a real followed by an imaginary) per sample, rather than a buffer just big enough for the 1024 real-valued samples. This is because the FFT here is an "in-place" algorithm that replaces the values in the buffer with the results of each stage of processing. Because the later stages involve complex results, the buffer has to accommodate complex values, even though we won't use them as input. (We don't need to explicitly set the imaginary components to zero, however; this is done in the code.)

8.6.1.2   Self-Modifying Code

This FFT routine uses self-modifying code to increase speed of execution. We can examine the instruction at the label "pass" to see what kind of time savings are involved.

The "ldaa #0" instruction loads the accumulator A with the immediate value of 0, taking two cycles (as long as it is executing in fast memory). What if we stored this value, which needs to change, and used a more conventional load from an absolute address? This would take six cycles (three times as long)—and that's only if the memory we were accessing were also fast (no wait state). In fact, any other addressing mode takes six cycles for the "ldaa" instruction. 16-bit loads, such as "ldx," take four cycles for an immediate load and six cycles for other addressing modes.

So how is the code modified, especially given that it is being moved to RAM from its original home in ROM? Immediately following the "ldaa" line is a line assigning the value "*-1" to the label "cellnbr." That is, "cellnbr" contains the address (in ROM) of the immediate data argument of the preceding "ldaa #0" instruction. Note that, about 20 lines earlier, some clever calculations produce an 8-bit offset to "cellnbr" from "pass," then use this offset in conjunction with the pointer *IY* (which is earlier set to point to the RAM location of the code) to store a new 8-bit value in place of the "#0" argument of the "ldaa" instruction.[17]

At the start of the routine, by the way, you'll find the machinations that calculate the address in RAM where the code is copied, "fft1024." If you think about it, this is necessary since the assembler thinks the code is going to be run from ROM (the address equated with the label "_fft1024"), the normal state of affairs for the 68HC16Z1 evaluation board. There are also some "games" you can play with some linkers (e.g., making multiple passes) to make this address issue a little cleaner at the assembly-language level, but then you have the complexity of yet another tool (the linker) to deal with. Even if this code looks a bit complicated, all the complexity is at one level—the assembler.

8.6.1.3 Use of Pointers

You'll note that the code is very careful about conserving and using pointers, again mostly for speed considerations. For example, store instructions benefit slightly from using an 8-bit indexed offset addressing mode (four cycles versus six cycles), so several variables are bunched together to allow this slightly faster addressing mode to be used. (See "twdtbl," "cellctr," etc.)

The MAC instruction has the dubious distinction of using every register in the chip—including the *IZ* pointer register. More self-modifying code is used to quickly save and restore the *IZ* contents during the MAC instructions as the *IZ* pointer is used throughout the FFT routine. Alternatives, such as stacking *IZ*, would take many more instruction cycles, though some alternatives might be much easier to understand!

8.6.1.4 Optimization of the FFT Stages

As we mentioned, the first two stages of the FFT have been implemented separately from stages 3–10. In part, this is due to the use of real data as input; however, the first two stages also use simple twiddle factors that we don't need to look up in the twiddle factor table. Although hard-coding the first two stages complicates the code, the savings in cycles is worthwhile.

---

17. Do we really have to tell you that you have to know your instruction set pretty well to be able to go in and change the argument inside an instruction? Not every processor has immediate data so easily accessible. Beware, too, of systems that have separate program and data space.

8.6.1.5   Scaling

Earlier, we discussed the need for scaling and the common methods for preventing scaling problems in the FFT. Although any of the methods are possible on the 68HC16, if we're looking for high performance, the only real contenders are unconditional input scaling and unconditional scaling per stage. (The other two options are conditional scaling per stage—block floating-point—and floating-point.)

The FFT routine in Listing 8-1 uses a hybrid approach, with unconditional input scaling (by a factor of 4) on the input, then unconditional scaling per stage for the remaining eight stages. Because the input is a maximum of 16 bits wide and we lose one bit per stage unconditionally, the final results will have only 6 bits of resolution, at best.

The only way to get better results is to resort to conditional scaling, the extreme example of which is performing all math in floating-point. Even conditional scaling per block would take a considerable toll on the performance of FFT code on the 68HC16. This is a fundamental trade-off between speed and resolution that we can't avoid, at least in an architecture like the 68HC16, that was never intended to be an FFT engine.

8.6.1.6   Use of Tables

Chapter 11 (Synthesizing Signals) will discuss subroutines for calculating trigonometric functions (such as the twiddle factors), but as we know exactly what twiddle factors we need beforehand, a table is a far faster approach. Recalling Equation 8.5, the twiddle factors, in rectangular form, can be found from a simple sine and cosine table. In fact, we need only a sine table if we use a trigonometric identity such as $cos(x)=sin(x+\pi/2)$. ($\pi/2$ is equivalent to an increment of the table index of 512, an operation you can see just below the "sinbank" label in the code.)[18]

8.6.2   Support routines

To actually use the FFT routine, we must supply additional routines—data input (sampling), windowing, and bit-reversed ordering prior to the FFT—and magnitude calculation afterward (if desired).

8.6.2.1   Sampling

All of the sampling issues we've mentioned before apply to the sampling of data for processing by an FFT. Proper antialiasing, a precise and appropriate sampling rate, and good resolution are all important if you expect accurate output data. You can also implement simple automatic gain control (AGC) at this stage by keeping track of the maximum (absolute) input value and scaling all the samples so the largest value is just shy of the largest representable value in the register size you are using (e.g., 16-bit). Of course, any conversion that's necessary from the ADC encoding to the 1.15 format (for example) should also be done here.

8.6.2.2   Windowing

18.   The sine table and additional files are available electronically.

The next step is to window the samples using an appropriate window such as the Blackman or Kaiser (e.g., β≥8.96) windows. The window is real-valued and will be calculated at as many points as samples in the data buffer (i.e., 1024). Because this is only 2K bytes (1024 x 16-bits), a look-up table again makes a lot of sense; the table is precomputed and stored in 1.15 format along with the program in ROM. (If you have enough RAM and it's faster than your ROM—not an uncommon situation—you can copy the table from ROM to RAM and gain some speed.)

8.6.2.3 Bit-Reversed Ordering

Because the FFT implemented in Listing 8-1 is the DIT FFT, the input data needs to be in bit-reversed order. Recall that this ordering can be done via swaps, so it is also an in-place algorithm. As the 68HC16 lacks bit-reversed addressing, we have to look at a few options for generating the bit-reversed addresses.

The first option is to use rotate instructions to build up a bit-reversed address one bit at a time. For example, if register D contained the normal address and register E were zero, repeating the following sequence 10 times would produce a bit-reversed address in register E:

RORD

ROLE

This is some 40 cycles (much more if you use a loop); not exactly a speedy operation, but not too bad in terms of memory usage (at most, 40 bytes at 2 bytes/instruction).

Another option is to use a look-up table. It's another 2K bytes (1024 addresses by 16-bits), but considerably faster than the bit-shifting above. The comments we made earlier for other tables apply.

Note that you will usually be dealing with words, not bytes, and also that your buffer may not start at address 0. The actual address you need will likely be of the form base + (bit_reversed_offset)*2.

Depending on the flow of your program, you could place data samples into the data buffer using the bit-reversed address to begin with—this would avoid a second pass through the data performing the bit-reversal.

8.6.2.4 Magnitude Calculation

Once the FFT has processed the data, the results will be complex-valued and stored in rectangular format. Most of the time we'll eventually want the magnitude information, which you'll recall is:

$$|x(i)| = \sqrt{x_r^2(i) + x_i^2(i)}$$

where $x_r(i)$ is the real component and $x_i(i)$ the imaginary component of sample i.

Calculating the square root is not very easy, but if we instead just calculate the sum of the squares, we'll end up with $x^2(i)$—the power spectrum. As long as you keep the difference between the magnitude and power spectra in mind, the power spectrum often is quite acceptable.

This is especially true if you convert the results to decibels, where the difference is only a scale factor.

8.6.3 Additional Comments on the 68HC16 Code

Lest you get the wrong idea, the basic DIT FFT algorithm can be written in a few dozen lines of C or FORTRAN; it is only as the time constraints become tighter and our computing resources more limited (e.g., no floating-point math) that the complexity starts building. Now, perhaps you can appreciate our comments earlier that the FFT is a routine we much prefer to reuse, rather than write from scratch. However, at the same time, it should be apparent that to get high performance, the FFT must be tightly tuned to not only the processor it will be run on, but to the entire system, especially to the memory available for program and data storage. The files that accompany this book include an example program showing how to interface to this FFT routine, including the necessary bit reversal. (We measured 70 msec for a 1024-point FFT on our hardware.)

8.7 The Inverse DFT/FFT

The Inverse DFT (IDFT) takes a finite set of discrete frequency samples (i.e., magnitude and phase—or real and imaginary components—the output of a DFT) and produces a finite discrete-time sequence. The output is actually just one period (exactly) of an infinitely long discrete-time sequence, according to the IDFT. (Recall that the DFT believes its input is exactly one period also, so there is symmetry here.)

As we mentioned, the Inverse FFT (IFFT) is computationally very similar to the FFT. The only changes that need to be made are in the sign of the twiddle factors and the need to divide the output by N when you're done. (Isn't the division by N curious?) Often, the same code is used to implement both the FFT and IFFT—a flag that you set determines what sign of twiddle factor to use and whether to divide at the end.

Note that if you feed the IFFT purely real data, you've in essence given a phase of zero for each component. If you are intending to modify a discrete-time sequence by first performing an FFT, modifying the results and using an IFFT to convert back to discrete-time, you'll need to preserve the phase or end up with a signal that has different time-domain characteristics (i.e., different wave shape).

8.8 Time-Frequency Analysis

Up to this point we've been assuming the signal that we want to analyze is stationary—that is, it doesn't matter when we look at it, the frequency content is always the same.[19] The problem is that there are many signals of interest that change over time. Time-frequency analysis is an

---

19. We're being a little loose here with terminology—stationarity comes in a few flavors (strong sense, weak sense, and chocolate-cinnamon mocha) and is a bit more involved than our hand-waving statement above indicates.

attempt to reconcile the changing nature of these signals with the most prevalent engineering tools, most of which assume stationary signals.

The following discussion is rather brief; for our purposes in this book, time-frequency analysis illustrates some interesting points, rather than representing a major area of microcontroller-based DSP application.

8.8.1 Multiple FFTs

One approach that can be applied to nonstationary signals is to perform multiple FFTs on the discrete-time sequence of interest. A single FFT is not suited to analyzing a sequence where there are changes in frequency content, so instead we break up the sequence into shorter segments during which it is assumed that the signal is "more" stationary. Because it's unlikely that frequency components will have the good manners to change instantaneously at exactly the right times, overlapping windows are used. So a given sample might be included in two or more FFTs.

8.8.1.1 Time Resolution Versus Frequency Resolution

The problem with using multiple FFTs comes down to resolution in the time and frequency domain. If we hope to have good frequency resolution, we'll need long windows—which implies using a longer input sequence. But to have a good idea of *when* a frequency component is present, we need to have fairly short windows. This trade-off means that each application will have a different "optimum" window length. Even the same signal (e.g., speech) will often be analyzed with different window lengths, depending on the relative importance of the time resolution versus the frequency resolution.

8.8.2 Wavelets

Wavelets are a fairly recent development in the DSP field. Instead of using sines and cosines as the "basis functions" (whatever those are), a different family of functions is used that gives the wavelet transform the ability to provide information about *when* a frequency component is present in the window.

In addition to the time-frequency properties of the wavelet transform, a particularly efficient implementation of the wavelet transform (Mallat's pyramid or tree algorithm) requires computation proportional (approximately) to N, compared with the FFT's $Nlog_2N$. Transforms based on the wavelet transform are not as common as the DFT (FFT), though there is good reason to believe wavelets will become more popular over time.

8.9 Miscellaneous Topics

Believe us when we say that we've only scratched[20] the surface of the DFT. Before we leave the topic, though, we'd like to just briefly mention a few DFT-related issues.

---

20. Or *defaced*, take your pick.

Miscellaneous Topics

8.9.1 Fast Convolution

Recall the famous "convolution in the time domain is equivalent to multiplication in the frequency domain" theorem? When we combine this fact with the following three observations,

- FIR filters operate by convolving the input with a set of coefficients
- Convolution takes time proportional to N^2
- the FFT takes time proportional to $N\log_2 N$

we see an opportunity to speed up evaluation of long FIR filters. The idea is that we process the input sequence using an FFT ($N\log_2 N$ operations), multiply the discrete-frequency domain output by a desired frequency-domain response[21] (about N operations), then convert back to the discrete-time domain using an IFFT (again, $N\log_2 N$ operations). Time-domain convolution requires time proportional to N^2; the FFT-based convolution requires time proportional to about $N(1+2\log_2 N)$. That difference can make FIR filter evaluation by FFT—a process known as *fast convolution*—more efficient for FIR filters past a certain length. (We've glossed over some issues in how you must handle data that are much longer than the FFT length, but this is the general idea.)

On some DSP-friendly chips, fast convolution is more efficient than normal convolution for lengths as low as $N=32$. However, for microcontrollers with FFT-unfriendly architectures, the crossover point is much, much higher. The crossover point depends on how efficient the FFT can be made (for the same resolution as the FIR filter), but on a microcontroller, we'd work pretty hard to see if standard convolution wouldn't work first, even if that means using a higher clock speed (and spending more money). Also recall that the linear-phase characteristic of the FIR filter comes at a substantial cost—perhaps an IIR filter would work instead. Fast convolution is a useful tool, but probably not a mainstay of microcontroller DSP.

8.9.2 Other Transforms

Our laundry list of transforms is pretty long: Fourier transform, discrete-time Fourier transform, discrete Fourier transform, Laplace transform, and the z-transform. And the discrete wavelet transform. Well, we have two more to briefly mention—the discrete cosine and discrete Walsh transforms.

8.9.2.1 Discrete Cosine Transform

The *discrete cosine transform* (DCT) is essentially just the real part of the DFT.[22] The DCT is useful in data compression applications. For example, a complicated waveform (perhaps a video signal) can be processed with the DCT to produce a discrete-frequency sequence. For many sig-

---

21. If you like, approximately the FFT of the filter coefficients, but you'll recall we got those coefficients from an ideal frequency response function to begin with—just cut out the middle man and use the ideal frequency response here.

22. It is necessary that the input data have even symmetry (usually by mirroring the data) for the DCT. A discrete sine transform, basically the same idea but the imaginary part only of the DFT, requires an odd symmetry (also easy to produce by mirroring the data properly).

nals, the discrete-frequency spectrum—or select parts of it—can be transmitted with far fewer bits than the original signal. An inverse DCT at the receiving end produces a discrete-time sequence that approximates the original.

The use of purely real math makes this transform appealing for a number of applications involving data compression, but the DFT is more appropriate for analysis of signals.

8.9.2.2 Discrete Walsh Transform

Here's a transform tailor-made for digital processors. The *discrete Walsh transform* (DWT) uses rectangular waveforms instead of cosine and sine waves. Table 8-3 gives the values of the waveforms used for a DWT of length 8.

WAL(n)	Index							
	0	1	2	3	4	5	6	7
0	1	1	1	1	1	1	1	1
1	1	1	1	1	-1	-1	-1	-1
2	1	1	-1	-1	-1	-1	1	1
3	1	1	-1	-1	1	1	-1	-1
4	1	-1	-1	1	1	-1	-1	1
5	1	-1	-1	1	-1	1	1	-1
6	1	-1	1	-1	-1	1	-1	1
7	1	-1	1	-1	1	-1	1	-1

Table 8-3 Discrete Walsh transform waveforms (length eight).

The discontinuous nature of the DWT functions is an advantage in analyzing waveforms that themselves have discontinuities, but a disadvantage for waveforms that lack discontinuities (the DFT, with its sinusoids, is better there). Like wavelets, the DWT may emerge over time as a much more popular tool for certain applications, but, at the moment, it is far less popular than the ubiquitous FFT.

There are many other transforms currently available, and it is almost certain that the discovery of both new transforms and ways of efficiently applying existing transforms will continue. Much of this research could be applicable to microcontroller applications—the DWT is an example. The magazines, journals, and on-line newsgroups we've mentioned before are good sources of up-to-date coverage of these developments. In addition, a number of tutorials on both established and emerging topics in spectral analysis are available at various worldwide web sites—use key words like "tutorial" or "primer" in your search.

8.10 Summary

Frequency-analysis techniques can be used to identify the frequency spectra of a signal, detect one or more frequencies, reduce the data necessary to send or store a signal, and/or accelerate the processing of a signal.

The discrete Fourier transform provides a complete discrete-frequency description of a discrete-time signal. A collection of very efficient algorithms for implementing the DFT is known as the *FFT*.

In some applications, a complete frequency spectrum is unnecessary. A small collection of IIR filters may identify a handful of frequencies, or a special version of the DFT—the Goertzel algorithm—can be used to handle more frequencies.

The signals we process are necessarily time-limited. The DFT calculates a spectrum assuming its input is exactly one period of an infinitely long (sampled) signal. Because it is usually not the case that the sample sequence is exactly one period (or that the real-world signal is periodic), the data must be windowed.

The DFT (and FFT) generally accept complex values. Although complex-valued signal models are useful in communication applications, it is often the case that the DFT is used to process real values.

The window used to preprocess the data prior to a DFT operation determines both the accuracy of the magnitudes of the output of the DFT, as well as the spectral resolution. Though the accuracy of the magnitudes (spectral leakage) is a function only of the window type, spectral resolution is a product of both window type and the length of the window. (The degradation of spectral resolution is called *smearing*.)

The DFT output is a sampled version of the DTFT of a given input. Zero-padding the data prior to computing the DFT increases the resolution with which the underlying DTFT is sampled, but does not change the DTFT spectrum. Only by adding additional data (i.e., increasing the window length) or selecting a different window type (shape) will the underlying spectrum change.

The DFT requires on the order of N^2 complex operations to compute. By dividing the DFT into smaller DFTs and these, in turn, into smaller DFTs, the FFT calculates the DFT using on the order of $N\log_2 N$ operations. Two popular implementations are known as the *decimation-in-time* and *decimation-in-frequency* algorithms. Both require that the input be an integer power of 2 in length.

The common FFT algorithms require that either the input data or output data be shuffled in "bit-reversed" order. DSP chips may support bit-reversed addressing in hardware, eliminating the need for a separate reordering routine.

The FFT requires proper scaling to avoid overflows in the intermediate results. Two types of unconditional scaling (a single scaling at the input stage or smaller scaling at each intermediate stage) or a conditional scaling method (per stage) allow the use of fixed-point arithmetic. Floating-point arithmetic eliminates the need for scaling but is very hardware-intensive.

The FFT can be programmed such that intermediate results and the output can be placed in the same memory as the input. These are known as *in-place algorithms* and can substantially increase microcontroller performance.

The Goertzel algorithm can detect the presence of a number of frequencies without calculating the entire spectrum as the DFT does, and in less time than the FFT. It is implemented as a number of IIR filters that are only partially evaluated for each sample.

The inverse discrete Fourier transform performs the opposite function of the DFT, taking a discrete-frequency spectrum and producing a discrete-time signal. A signal (either in the time or frequency domain) is preserved if processed by first one, then the other transform. The IFFT is computationally very similar to the FFT, requiring only that the sign of some coefficients be changed and a division by N be performed at the end.

Time-frequency analysis is applicable to signals whose frequency content is not constant over time (i.e., the signal is nonstationary). Using shorter windows for the FFT can indicate more clearly changes in frequency over time but decreases spectral resolution in any given frame. Multiple, overlapping FFTs can be used, but other transforms, like the discrete wavelet transform, are especially well equipped to deal with time-frequency analysis.

The efficiency of the FFT can be exploited to speed up convolution; this process is known as *fast convolution*. This method is viable only above a certain signal length N, which varies dramatically from processor to processor, depending on their architecture.

The discrete cosine transform and discrete Walsh transform are but two transforms that address special applications. The DWT has very simple coefficients and is hence fairly amenable to digital processors but the DWT is not well-suited to transforming data that is continuous in amplitude.

Resources

Higgins, Richard J., *Digital Signal Processing in VLSI*, New Jersey: Prentice-Hall, 1990.

Harris, F. J., "On the use of windows for harmonic analysis with the discrete Fourier transform." *Proceedings of the IEEE*, 66(1), 1978, pp. 51–84.

Personal Engineering and Instrumentation News, a monthly magazine, often has articles on frequency analysis issues.

Bruce, Andrew, Donoho, David, and Gao, Hong-Ye, "Wavelet Analysis," *IEEE Spectrum*, October 1996, pp. 26-35. A good overview of wavelet-based analysis. Contains listings of both commercial and freely available wavelet tools.

CHAPTER 9

Correlation

Correlation is not a mathematically complicated operation; it just involves summing some products. We usually don't have to worry about complex values, either! Chapter 2 (Analog Signals and Systems) through Chapter 4 (Discrete-Time Signals and Systems) will be helpful, and we'll mention convolution (discussed in Chapter 5). The DFT (FFT) is described in Chapter 8 (Frequency Analysis); it plays a small role in this chapter. The FBI investigates incidents of copying videotaped movies. Never stick your head or hands out the window of a moving vehicle.

9.1 Overview

This chapter is about a simple operation called *correlation*. Correlation quantifies the "similarity" between two signals (crosscorrelation) or between two segments of the same signal (autocorrelation). Correlation is closely related to convolution; both involve the same "sum of products" operation that is usually implemented by a MAC instruction and hardware.

Correlation is a very useful DSP tool; in this chapter, we'll see correlation used to detect signals that otherwise seem drowned out by noise, to determine the time delay between signals, to determine the impulse response of a system (and, thus, system function) without using an impulse, to design FIR filters, to find the periodicity of a signal, and even to estimate the shape of an unknown periodic signal in the presence of noise. This is by no means a complete list of the applications for the correlation operation, but it gives some idea of the potential of this operation in microcontroller applications.[1]

We start this discussion by examining the crosscorrelation operation, followed by autocorrelation. We then discuss pseudo-noise (PN) signals, which often occur in applications that use correlation. To close this chapter, we look at signal averaging, a technique related to correlation.

9.2 Crosscorrelation

9.2.1 Definition

Crosscorrelation quantifies the similarity of two signals. We're going to use discrete-time signals, but the concept is similar for continuous-time signals.[2] The correlation operation multiplies the two signals point by point, then sums the products. Then one of the signals is shifted one sample, and the process is repeated. This shifting continues (in theory) until there is no overlap between the signals. In equation form, we say the crosscorrelation between signal x_1 and signal x_2 is:[3]

$$r_{12}(l) = \sum_{n=0}^{N-1} x_1(n) x_2(n+l) \qquad l = 0, 1, \ldots N \qquad (9.1)$$

where N is the number of elements in both x_1 and x_2, and l is known as the *lag* or (time) *shift*. (The shift can also be spatial rather than temporal. It just depends on the source of the signal.)

Figure 9-1 shows a numerical example of crosscorrelation. Signal x_1 is shown along the top, and signal x_2 is shown being shifted to the left as the values of r_{12} are calculated. (Zero values are shown beyond the end of x_2.) As we would hope, the crosscorrelation reaches a maximum when there is strong agreement between the two signals—in this case, the match of the sequence {-2, 0, 2} that occurs in both signal segments.

Figure 9-2 shows the same crosscorrelation, but graphically. Below each shifted pair we've plotted the pointwise product of the two signals—r_{12} is the sum of these products. For example, at lag=1, we add the quantity -2 (at sample 2) and -4 (at sample 4) to get -6, the same result we computed numerically.

1. Let the weaseling begin. Actually, this chapter is about "short-term correlation," and, as usual, we'll avoid some of the sticky issues that would be involved in a thorough treatment.

2. By the way, we've again "connected the dots" on many of the plots in this chapter because they look better that way, but as we're talking discrete-time signals, you know by now that the plots would be drawn more accurately using discrete points.

3. In some books, you will also see crosscorrelation expressed with the lag subtracted. It just depends on what definition you have of *lag* (be sure you know what definition an author uses when you read other texts!). It is also common to see a $1/N$ factor prepended to the expression; a full explanation would take us a bit afield but has to do with whether the signal is an energy or power signal. For reasonable sampling rates, using $1/N$ is a "matter of taste." We'll omit this factor, since it will make other definitions in this chapter a bit clearer. (See Jack's speech book [Deller, *et al.*, 1993] for a discussion, "On the Role of '$1/N$' and Related Issues" in Chapter 4.)

x_1:	0	0	-2	0	2	0	0	0	0	0	0	0	lag	r_{12}
x_2:	0	0	2	1	-1	-2	0	0	-2	0	2	0	0	-6
	0	2	1	-1	-2	0	0	-2	0	2	0	0	1	-6
	2	1	-1	-2	0	0	-2	0	2	0	0	0	2	2
	1	-1	-2	0	0	-2	0	2	0	0	0	0	3	4
	-1	-2	0	0	-2	0	2	0	0	0	0	0	4	-4
	-2	0	0	-2	0	2	0	0	0	0	0	0	5	0
	0	0	-2	0	2	0	0	0	0	0	0	0	6	8
	0	-2	0	2	0	0	0	0	0	0	0	0	7	0
	-2	0	2	0	0	0	0	0	0	0	0	0	8	-4
	0	2	0	0	0	0	0	0	0	0	0	0	9	0
	2	0	0	0	0	0	0	0	0	0	0	0	10	0
	0	0	0	0	0	0	0	0	0	0	0	0	11	0

Figure 9-1 Crosscorrelation (numeric example).

The definition in Equation 9.1 depends on the amplitudes of the signals being compared as well as the density of samples. We can rid the analysis of this dependence by computing a *normalized crosscorrelation sequence*, which we'll call ρ_{12}. The normalized crosscorrelation varies

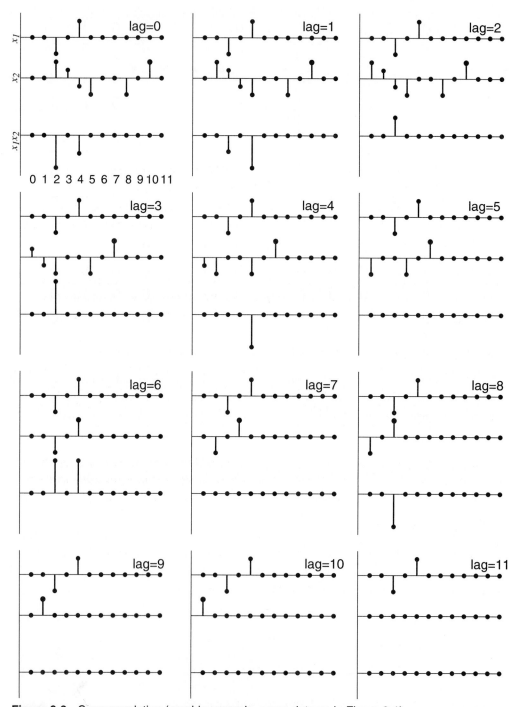

Figure 9-2 Crosscorrelation (graphic example, same data as in Figure 9-1).

between 1 (the two signals are exactly similar, i.e., 100% correlation) and -1 (the two signals are exactly opposite or antiphase):

$$\rho_{12}(l) = \frac{r_{12}(l)}{\sqrt{\sum_{n=0}^{N-1}[x_1(n)]^2 \sum_{n=0}^{N-1}[x_2(n)]^2}} \qquad (9.2)$$

The quantity on the bottom is related to the energies of the two signals and can also be stated in terms of the individual "autocorrelations" of the two signals (described later).

Note that -1 indicates a "negative" similarity—the two signals are 180° out of phase (antiphase). A normalized crosscorrelation of 0 at lag l indicates no similarity at all between the two signals at this spacing.

Both the crosscorrelation and the normalized crosscorrelation generate values for up to N different lags or shifts between the two signals. In some cases, we'll use values generated at a large number of lags; in other cases, it might be understood that the single crosscorrelation value of two signals is the maximum crosscorrelation value of all reasonable lags. What are reasonable values of lags?

Without any adjustments to the expressions we've given, you can expect good results for lags of no more than 5%–10% of the length of the signals. Beyond this point, the decreasing overlap between the two signals is not reflected in the equations, so you can't accurately compare correlation values with lags outside this range. In general, you should avoid calculating lags that are more than 50% of the signal length. (Not to restate the obvious, but the more data you have, the better results you get.)

In a related vein, we might ask how long our signals should be. As a rule of thumb, the signal length in time should be at least five times longer than the period of the lowest frequency component of interest. If N is the length in samples, F_s the sampling rate, and F_0 the lowest frequency of interest, $N/F_s \geq 5/F_0$.

In practice, some care must be taken in directly implementing Equation 9.1 in software. The index value $n+l$ extends beyond the end of x_2 for any positive values of lag and includes negative values when the lag is negative. If we need to compute only positive lags (x_2 advanced with respect to x_1), for example, we can rewrite the equation as

$$r_{12}(l) = \sum_{n=0}^{N-1-l} x_1(n)x_2(n+l) \qquad l = 0, 1, \ldots N \qquad (9.3)$$

which ensures that we don't exceed the range of valid samples for x_2.

9.2.2 Correlation and Convolution

If we compare the expressions for crosscorrelation:

$$r_{12}(l) = \sum_{n=0}^{N-1} x_1(n)x_2(n+l) \qquad l = 0, 1, \ldots N \qquad (9.4)$$

and convolution:

$$y(n) = \sum_{m=0}^{n} x(m)h(n-m) \qquad (9.5)$$

we see that they have the same basic structure. Convolution just "reverses" one of the sequences (actually, either x or h could be reversed). Is there more to this similarity? The answer is found in the matched filter, though as we'll see in a moment, the "filtering" it performs is very specific.

The relationship between convolution and correlation is also helpful in terms of implementation. Not only can we use the MAC operation of most DSP processors, but we can also use the FFT to perform fast correlation, just as the FFT can be used to perform fast convolution. We'll discuss this later when we talk about implementation issues.

9.2.3 Applications

We'll describe four applications that use crosscorrelation. The first, matched filters, detects a known time-domain signal that may be buried in a noisy signal. Measuring the "time of flight" or time delay between two signals is the next application, and is crucial in sonar, radar, and earth sciences. Crosscorrelation can be used in image processing where signals represent images. And finally, crosscorrelation provides a way of determining the system function of a system without the practical problems involved in trying to create and use an impulse signal.

9.2.3.1 Detection of a Known Signal in Additive Noise

Let's say that we have a noisy channel over which someone is trying to send us a particular signal—for example, the situation of Figure 9-3. The "desired" signal is buried under so much noise that we can't even see the signal in the time domain, but we do happen to know what the signal—if it is sent—should look like. What can we do?

Because we know what the desired signal looks like when it is not corrupted by noise, we can compute the crosscorrelation of the desired signal with the received noisy signal. If the signal we're looking for is present, we should see a strong peak in the crosscorrelation values. Fig-

Crosscorrelation

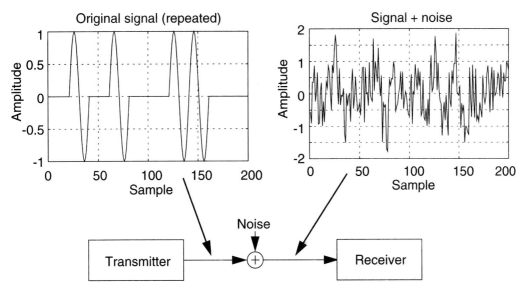

Figure 9-3 Sending a signal over a noisy channel.

ure 9-4 shows the desired signal—a single cycle of 500 Hz sine wave—and the results of crosscorrelating this signal with the noisy received signal. We see peaks in the normalized crosscorrelation where the sinusoid is buried in the noisy signal. However, there are other peaks due to the correlation of the desired signal with the noise in the received signal. Therefore, we have to set a threshold carefully. If the threshold is too high, we risk not recognizing a valid correlation; but if the threshold is too low, noise may create false positives. A threshold is selected based on the properties of the received signal, length of the desired signal, costs associated with false alarms and misses, and other parameters.[4]

Although we used sinusoids in this example, correlation works regardless of the waveshape. For example, while your transmitter may be sending a rectangular wave, the limited bandwidth of the channel (say, a cable) may round off edges, produce ringing, etc. Your "desired" signal can then be this distorted waveform. A very common application of correlation is in processing a binary bit stream, where a (rectangular) pulse represents a "1" bit. The only major constraint is that the signals to be correlated must not have a bias (that is, a nonzero average). Preprocessing to remove a bias is very easy to arrange.

What we've just created is a *matched filter*. This filter is a bit different from the type of filtering we've done before in this book as it is not designed to be explicitly frequency-selective in a simple passband/stopband way. Instead, it is matching the time-domain waveform of the desired signal with the incoming signal. Since we expect noise to be uncorrelated to our signal,

---

4. Many of the advanced signal processing texts in the list of references go into the mathematics in greater detail; some background in probability is usually necessary.

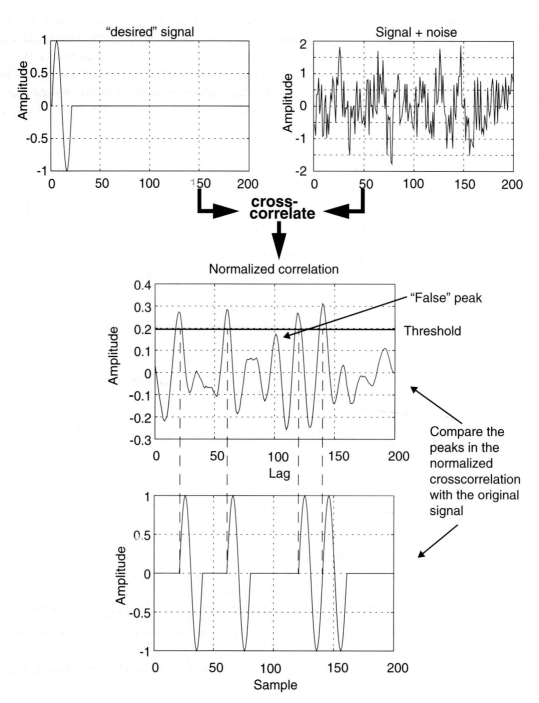

Figure 9-4 Crosscorrelation used to detect a signal in additive noise.

we should see very low crosscorrelation values when the desired signal is not present and large values when it is.[5]

We just saw the similarities between the equations for convolution (which we used in implementing FIR filters) and crosscorrelation. Convolution is just correlation with one of the sequences reversed. Is there some deeper connection between FIR filtering with convolution and matched filtering?

The connection is pretty straightforward. Recall that we generally designed our FIR filter coefficients[6] based on an ideal filter magnitude response (ignoring phase). Certain frequencies would be attenuated and certain frequencies passed. But there is nothing (except mathematical pain and misery, the two mainstays of professorial entertainment) that prevents us from generating the filter coefficients from an arbitrary magnitude and phase frequency response. So let's take the desired time-domain signal and transform that into the frequency domain (keeping both magnitude and phase) to get the desired frequency response for the filter. Now use whatever method (e.g., windowing) to get the impulse response and, thus, the filter coefficients.

We could do that, but it's not necessary. That impulse response would be just a time-reversed copy of the desired signal. The two transformations, from the time-domain to frequency domain and back, cancel out, so we can skip those steps.

By the way, it is certainly permissible to have multiple matched filters running at the same time, each "looking for" a different signal. At any given moment, at most one of the matched filters should then indicate a match (assuming that only one desired signal can be present at a time).

9.2.3.2   Determination of Time Delay

Let's take a slightly different scenario. Say that, instead of a signal being present or not, we know that the signal is always present. However, we don't know how much time shift there is between the desired signal and the input signal. This is the problem in sonar, radar, and other processes that determine distances by calculating the time delay between a reference signal and its echoed version. Figure 9-5 shows the situation.

We are interested in the time lag (delay) between the transmitted signal and the returned signal (the correlation magnitude may also be of interest, but it is of secondary importance). However, we will have a problem if we transmit a signal that repeats with a relatively short period. If the delay between signals is greater than the signal period, we can't tell for certain what the actual delay is. For example, if we're sending sound through air, the signal will travel at about 340 m/sec. If the reflecting object is 100 meters away (200-meter round trip), the transmitted signal can't have a period of less than 200/340 or 0.6 seconds. If the transmitted signal has a period of 0.5 seconds, the crosscorrelation indicates a delay of only 0.1 seconds. We'll discuss a

---

5. This matched filter business (in the simple form described here) works optimally only if the noise is "white," a subject we discuss elsewhere. Colored noise is a whole different can o' worms.

6. FIR filter coefficients=(truncated) impulse response.

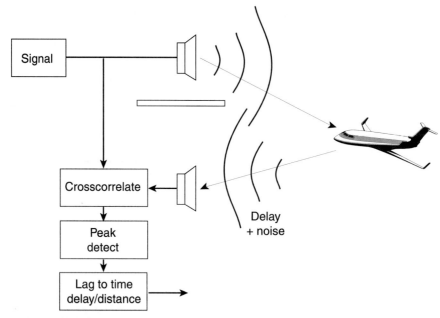

Figure 9-5 Application of crosscorrelation to determine distance.

type of signal that works well for this application later in this chapter. Let's first look at an example showing how crosscorrelation can be used to determine signal delay.[7]

Figure 9-6 shows a signal consisting of +1 and -1 values chosen at random (our reference), the delayed and noisy reflected signal, and the results of crosscorrelating the two signals. (Be aware we're "connecting the dots" here; the signals are actually discrete-time.) Though visually it's difficult to see our original signal in the noisy returned signal, the crosscorrelation of the original and reflected signal shows a very clear peak at a lag of 10 samples. Since the speed of sound in air is about 340 m/sec, a sampling rate of 10 kHz implies a delay of 0.001 seconds between the signals, or a distance of 0.034 meters (round trip). The one-way distance is half that, or about 17 cm.[8]

9.2.3.3 Feature Detection in Image Processing

Crosscorrelation has uses in image processing where it quantifies the similarity between features of two images. Here the "lag" between signals is not in time but in space (e.g., pixels). The correlation operation needs to be extended to two dimensions, but the underlying idea (a sum of products) is similar.

---

7. The example here is for illustrative purposes. Texts such as Higgins [1990] go into greater detail on optimum methods for measuring time delays between signals, using correlation.

8. Such distances might be observed in a medical diagnostic application using ultrasound, but hopefully not in detecting the location of an aircraft!

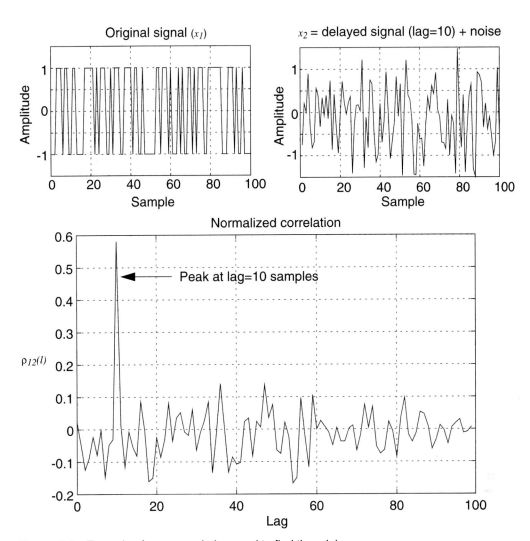

Figure 9-6 Example of crosscorrelation used to find time delay.

9.2.3.4 Determination of the System Function or Impulse Response

Another use of crosscorrelation is to determine the system function of an analog system.[9] As you know by now, we could find the system function by taking the impulse response of the system and using an appropriate transform (perhaps the DTFT, if the impulse response is discrete-time) to find the system function.

---

9. Discrete-time systems are not a problem—the discrete-time impulse is easily produced and is also easily processed by discrete-time systems.

The problem with using an impulse to probe an analog system is that you either must use a strong impulse or risk losing the impulse response in ambient noise. But an impulse of a large amplitude may "overload" the system (i.e., cause the system to be nonlinear).

The crosscorrelation-based technique avoids this problem by probing the system with a "noisy" signal and crosscorrelating this signal with the system output. The result of the crosscorrelation is an approximation to the impulse response $h(t)$ of the unknown system. Again, a DTFT can be used to find the (continuous) system function. Unlike using an impulse, we've kept the system well within its linear range, and by using long sequences, we can generate as accurate an impulse (or system function) as we need.

Application to System Identification and FIR Filter Design In fact, the crosscorrelation technique just described can be used to design an FIR filter according to an optimality criterion known as *least square error*. Suppose that we can provide the input and see the output of a system that has an unknown impulse response. (Recall that having an unknown impulse response is the same as saying we don't know the system function for the system.) (See Figure 9-7.)

We'll assume this system has a finite impulse response—that is, $h(n)$ is zero for n greater than some M. If we can determine $h(n)$, the impulse response of the system, we can then either model the system (perhaps use the system model in another problem or study the characteristics of this system as it is) or create an FIR filter (system) that acts the same. If our interests are the former, we're doing *system identification*, although if we're out to create an FIR filter, this is just another type of FIR filter design. The problem and the solution are the same, regardless of what name we give it.

So how do we find a good $h(n)$? One simple approach involves comparing the outputs of both the unknown system $y(n)$ and an FIR system $\hat{y}(n)$ for the same $x(n)$ input. When the out-

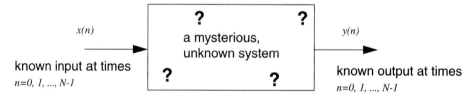

Figure 9-7 Determining the system function of an unknown system.

put of the FIR system—with an impulse response of $h(n)$—has a minimum total squared error compared with the unknown system, we've found our h(n). The error criterion to be minimized is:

$$\text{error} = \sum_{n=0}^{N-1}[y(n) - \hat{y}(n)]^2 = \sum_{n=0}^{N-1}\varepsilon^2(n) \qquad (9.6)$$

That is, we will attempt to minimize the square of the error between the outputs.[10] For obvious reasons, this is called the *least-squared error* method for designing the filter. The idea is diagrammed in Figure 9-8.

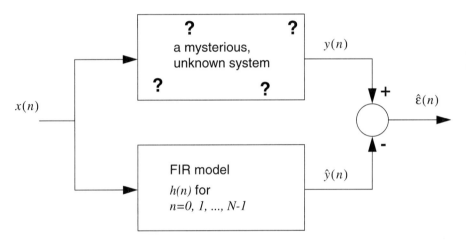

Figure 9-8 Finding h(n) for an unknown system using the least-squared error method.

The impulse response h(n) that meets this criterion is found by solving the set of *normal equations*:

$$\sum_{n=0}^{N-1} h(n) r_{xx}(n-k) = r_{xy}(k) \qquad (9.7)$$

---

10. We square the errors so that positives and negatives don't cancel one another.

for $k=0, 1, ..., N-1$, where $r_{xy}(k)$ is the crosscorrelation between x and y, and r_{xx} is the autocorrelation of x. (We discuss autocorrelation shortly).

9.2.4 Implementations

Implementation issues for crosscorrelation are almost identical to those for FIR filter evaluation by convolution. In fact, you can use the same FIR filter routine by just reversing the order of the "coefficients."

The role of "coefficients" is usually taken by the reference signal. The reference signal might be from a table or, more commonly, calculated in real-time.

9.2.4.1 Fast Correlation

Just as the FFT can be used to implement a fast version of convolution, a similar trick can be played to perform fast correlation. The only difference is that the complex conjugate of the FFT of one of the signals is used to account for the time reversal of one of the signals with respect to convolution. The crosscorrelation of two sequences x_1 and x_2 can, therefore, be calculated as:

$$r_{12}(l) = IFFT[FFT(x_1)FFT^*(x_2)] \quad (9.8)$$

where l is again the lag. Crosscorrelation calculated in this way is more efficient for large values of N for which the $\frac{3N}{2}\log_2 N$ operations of the two FFTs and one IFFT are fewer than the N^2 operations using the correlation equation directly.

Just as we noted in the our discussion of fast convolution, the relative inefficiency of the FFT compared to direct sum of products (i.e., RMAC) on the 68HC16 means you need fairly large N (over 100, likely) to make the FFT route more efficient. However, if $N=1024$, then N^2 is over a million, and $N\log_2 N$ about ten thousand. There are definitely applications for fast correlation!

9.3 Autocorrelation

9.3.1 Definition

Autocorrelation is identical to crosscorrelation, except that the two signals compared are segments of the same signal. When we use autocorrelation, we're looking for or identifying some periodicities in a signal, or at least similarities at certain lags. The notation for autocorrelation is the same as for crosscorrelation, except that only one signal is used instead of two different ones:

$$r_{xx}(l) = \sum_{n=0}^{N-1} x(n)x(n+l) \quad l = 0, 1, ...N \quad (9.9)$$

We can also calculate a normalized autocorrelation in the same manner as we did for crosscorrelation.

9.3.2 Applications

Crosscorrelation requires that we know quite a bit about the nature of the "underlying" (desired) signal that we're processing. The unknown information is "when" and possibly "how strong." There are many cases in which we do not know much about the signal and so can't apply crosscorrelation. What would we crosscorrelate the received signal with? However, it may still be possible to extract information about the signal using correlation between different segments of the same signal. The next few sections discuss using autocorrelation to determine periodicity, recover the waveform of a repeating signal, find the energy of a signal, and produce a filter that predicts the output of a system using past output values.

9.3.2.1   Determination of the Period of a Signal

If a signal is periodic, we would expect to see a peak in the autocorrelation values at a lag corresponding to the period of the signal. Let's take a signal we've seen before and examine the autocorrelation values.

Figure 9-9 shows a segment of the vowel sound "ahhh"— the sound, you may recall, that a person might make upon falling into a body of water such as the Crab Haul Creek sometime in March. The normalized autocorrelation values are shown below, with obvious peaks at lags of 0, 90, 180, and so on. Knowing that the sampling rate is 10 kHz, the period corresponding to a lag of 90 is 9 msec, or a frequency of about 110 Hz. In this case, this frequency is the pitch of the speech signal.[11] The result from autocorrelation agrees with a visual inspection of the original signal, though the technique of autocorrelation to determine periodicity will also work for signals whose periodicity is less obvious (e.g., buried in noise).

We expect a peak of 1 at a lag of 0—this is just the signal correlated with itself, so we'd better get 1 or we've made a mistake.[12] The other major peaks are just integer multiples of the first, as shifting by two complete periods also should give us high correlation. The downward trend we noted on the plot is a result of the shorter and shorter overlap between the two signal segments. (Recall the 5-10% overlap rule of thumb.) It's not too difficult to compensate for this effect by modifying the normalized crosscorrelation equation to sum only sequences in the

---

11. This is not necessarily the best way to find the pitch of a speech signal. Strangely enough, we have a book at hand that goes into a bit of detail on this and other speech processing issues. It's by some guys named John R. (Jack) Deller, John G. Proakis, and John H. L. Hansen, and is called *Discrete-Time Processing of Speech Signals* (Prentice Hall/MacMillan, New York, 1993). We cannot say enough about this book, and we suspect those around you would be delighted to receive several copies as birthday gifts. (This is a pitch of a different sort altogether.)

12. These values assume the correlation values are normalized. Also, note that $r(0) \geq |r(l)|$ for all values of l. Also also, (don't you wish you could ignore these footnotes?), we feel compelled to point out that $r_{xx}(0) = \sum x^2(k)$, which is equal to the *energy* in this segment of signal $x(n)$. Sometimes you'll see an added factor of $1/N$, which gives you a quantity called something like *average energy*. (The exact terminology will vary.)

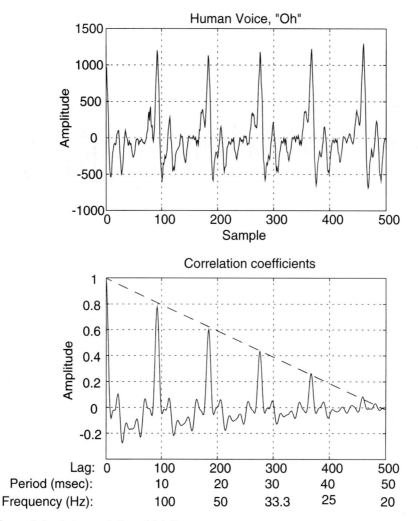

Figure 9-9 Autocorrelation of "oh."

denominator over the range that actually overlaps. However, you can see that at the logical extreme—say, a lag of 95, when you only have 100 samples—your correlation is based on very few samples. The results will be much more susceptible to noise.

9.3.2.2 Estimation of Periodic Signals

As we'll see later in this chapter, if we know the period of an unknown periodic signal, we can sometimes use a technique called *signal averaging* to recover the original waveform. We just saw that autocorrelation can determine the period of a signal even in the presence of noise; if we have enough periods available to process, it is possible to follow up the autocorrelation step with a crosscorrelation of the input signal with a signal consisting of periodic impulses (of the same period). Recall that the correlation operation sums the product of the two signals at a given delay. If we have multiple periods of the "unknown" signal and a second signal with properly spaced impulses, for any given lag we'll be just "averaging" the multiple periods together. As the delay is incremented along, the uncorrupted waveform will emerge as the normalized crosscorrelations, assuming that the noise is "random" and has a mean of zero.

This technique is mathematically "neat," but the lengths of N required make the direct implementation of this technique memory intensive. Because the crosscorrelation part of the technique is identical to signal averaging, it usually makes more sense to implement signal averaging directly. The memory requirements are much less (and independent of the number of periods over which you can collect data).

9.3.2.3 Linear Prediction

Autocorrelation plays an important role in a technique known as *linear prediction*. The idea behind linear prediction is to create a system that, given past values of a signal, can make accurate predictions about future values of that signal. This predicting system could be an FIR "filter," which means our job is to find the impulse response $h(n)$ that does the "best" job of predicting.

Why might such a system be useful? Linear prediction finds applications in a number of fields, notably, contemporary speech and image coding and recognition. Here, linear prediction lets us reduce the amount of information we need to describe a speech signal. Instead of sending every sample, we can describe the predictor with about a dozen parameters and then describe a simple input signal to be processed (again, using just a few parameters). Because speech is relatively unchanging ("stationary") on the scale of 10–20 msec, we need to update only the predictor coefficients and input signal parameters 50–100 times a second. The combination of a simple input signal (perhaps a pulse train or noise) and the filter produces a speech signal that is similar to the original speech.

Our task is to find a set of coefficients for an FIR predictor such that, on some "training" signal, we minimize the error between the value estimated by the predictor and the actual value. (See Figure 9-10.) Once we have $h(i)$, we can use this in our actual application—where we won't have access to the original signal. (If we did, there wouldn't be much point to the predictor!)

What criterion should we use to pick the best predictor? We'll use the least-square error, similar to the approach taken to FIR filter design earlier in this chapter. The approach is diagrammed in Figure 9-11.

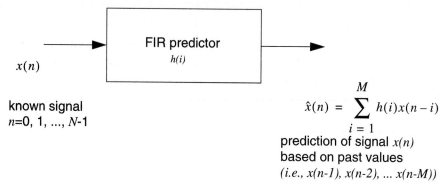

Figure 9-10 A linear predictor system.

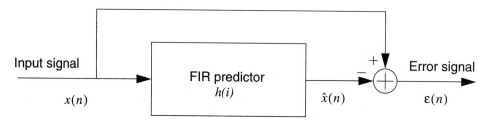

Figure 9-11 Calculating the "error" in a linear predictor.

In equation form, we want to minimize:

$$\sum_{n=0}^{N-1} \varepsilon^2(n) = \sum_{n=0}^{N-1} [x(n) - \hat{x}(n)]^2 \qquad (9.10)$$

To actually find $h(n)$, we can use a set of M normal equations, similar to those used in the FIR filter design problem:

$$\sum_{n=0}^{M-1} h(n) r_{xx}(n-j) = r_{xx}(j+k) \qquad (9.11)$$

for $j=0, 1, ..., M-1$, and where r_{xx} is the autocorrelation of signal $x(n)$. In practice, there are shortcuts to solving this set of equations, such that doing real-time linear predictive coding (LPC) is

Pseudo-Noise (PN) Signals

now commonplace—in fact, digital cellular phones are based in part on linear prediction (with a lot of other enhancements besides).

Obviously, we haven't given you enough information to actually do linear prediction, but since we were in the neighborhood, we thought we would at least sketch out the subject. Linear prediction is another one of those techniques that has been extremely useful in practical signal processing; like the FFT, linear prediction is not a great match to the processing power of current microcontrollers, though the steady progress in microcontrollers means it is just a matter of time before linear prediction in real-time is quite viable on microcontrollers.[13]

9.3.3 Implementation

Aside from the obvious memory savings of having only a single signal, autocorrelation is similar to crosscorrelation in terms of implementation, including with respect to the use of fast correlation. (Only one FFT is required, though, instead of two.)

9.4 Pseudo-Noise (PN) Signals

In the earlier "sonar" example, we saw a use for a signal with a very long period. Actually, what we want is a sequence that has close to zero (auto)correlation except at lag 0 (and also has no DC bias). Although truly random noise[14] fits this bill, for practical reasons we look to sequences known as *pseudo-noise* (PN) or *pseudo-random bit sequences* (PRBS). These signals have the appropriate correlation and bias properties and can also be relatively easily generated in software.

Generating seemingly random bits or numbers is surprisingly nontrivial. We'll have more to say in Chapter 11 (Synthesizing Signals) when we discuss the topic of random numbers in more detail, but for correlation applications, a particular PRBS technique is used called the

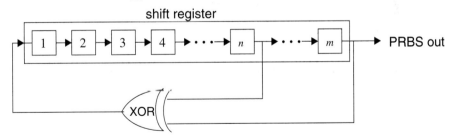

Figure 9-12 Structure of the feedback shift register (FSR).

---

13. Actually, we should say the analysis part of linear prediction is not a good match—the synthesis part could be just a garden-variety FIR filter, at which the 68HC16 is quite adept.

14. In this section, we're using some terms like *random*, *noise*, and *the* without much rigorous definition. In truth, we're dancing on the edge of some fairly heavy-duty probability and math here; take what we're saying as a first approximation.

Pseudo-Noise (PN) Signals

tapped or *feedback shift register (FSR) generator*. Not only is this technique fairly easy to implement in software on a microcontroller, but it is quite simple to implement in hardware as well. The basic structure of the FSR generator is shown in Figure 9-12.

The FSR operates by taking two or more bits from the shift register, combining them using an "exclusive-or" function (several texts also call this a "modulo-2 addition" function), and using the result as the input to the shift register. Although you can take any bit as the output, it is traditional to use the last bit of the shift register as the output. The arrows in the diagram show the movement of bits within the shift register—they move one bit position (to the right) every clock cycle. Shift registers are available as ICs (e.g., 74HC164, an eight-bit shift register), should you need a hardware version; nearly any digital processor has a right and/or left logical shift instruction.

9.4.1 Characteristics of PN Sequences

Several characteristics of PN sequences generated with an FSR are worth noting. First, the autocorrelation of the sequence is essentially zero (except at a lag of 0 of course!). Second, there are an equal number of 0s and 1s generated over the entire period (and over any reasonably long part of a period), so we don't need to worry about a DC bias if we map the 0s to -1s.[15] Third, if the taps are properly chosen, an m-bit register can produce a PRBS with a length of 2^m-1 bits before the sequence repeats (such an arrangement is known as a *maximal-length shift register*).[16] Fourth, the configuration shown requires that the shift register start off with a nonzero value (any value will do fine).

Table 9-1[17] describes maximal-length shift registers for some useful bit lengths. Not every combination of taps or register lengths will give a maximal-length sequence and, as you can see, registers that are exactly multiple bytes long require more than one tap.

By the way, Donald Knuth (admired by programmers everywhere for his three-volume *The Art of Computer Programming*) warns us not to fall into the trap of making up k-bit random numbers using k sequential bits from a feedback shift register. The resulting numbers are not as satisfactory as those from the techniques we discuss in Chapter 11 (Synthesizing Signals). (Just to confuse matters, though, a very powerful pseudo-random number generator is based on running a bunch of FSR generators in parallel! More later.)

15. Okay, actually there's one fewer zero than ones, as we don't allow the shift register to be all zeros. But this is negligible over the sequence lengths we use. And, just to be up front, in fact the autocorrelation of a PN sequence (other than at lag 0) is $-1/N$ rather than zero. But again, for N of any reasonable length, this is nothing to worry about.
16. We don't get a sequence length of 2^m because we can't have a situation in which the shift register is all zeros.
17. Table based on Horowitz and Hill [1989] and Knuth [1981].

Bytes	m (bits)	n (tap(s))	sequence length (2^m-1)
1	7	6	127
1	8	4, 5, 6	255
2	15	14	32,767
2	16	4, 13, 15	65,535
3	23	18	8,388,607
3	24	17, 22, 23	16,777,215
4	31	28	2,147,483,647
5	39	35	5.498×10^{11}
6	47	42	1.407×10^{14}
7	55	31	3.603×10^{16}
8	63	62	9.223×10^{18}

Table 9-1 Maximal-length shift registers for selected bit lengths.

9.4.2 Software Implementation of Feedback Shift Register Generators

Before presenting a 68HC16 subroutine to implement a FSR, we need to say a little about shift instructions in microprocessors, as there's more than one way to shift data on most microcontrollers. Four common types of shifting instructions are:

- Left shift. The bits are shifted one position to the left, the most significant bit is placed in the carry bit and a 0 is shifted into bit 0. This instruction may be called either *arithmetic* or *logical*. In the 68HC16, this is known as Arithmetic Shift Left (ASL).
- Signed right shift. The bits are shifted one position to the right, the least significant bit placed in the carry bit, and the value of the most significant bit is retained. This preserves the sign of two's-complement signed values. This is known in the 68HC16 as *Arithmetic Shift Right* (ASR).
- Logical right shift. The bits are shifted one position to the right, the least significant bit placed in the carry bit and a 0 bit shifted into the most significant bit. This is known as *Logical Shift Right* (LSR) on the 68HC16. (Can be simulated by doing a right rotate with the carry bit clear if you don't have an LSR instruction.)
- Left and right rotate. The carry bit value is used as both input and output to the shift process (obviously, as input first). Shifts across multiple bytes are easily realized with rotates, or (with additional operations), rotations of data within a byte. This is known as *Rotate Left/Right* (ROL/ROR) in the 68HC16.

Figure 9-13 shows these four forms of shifting instructions.

Pseudo-Noise (PN) Signals

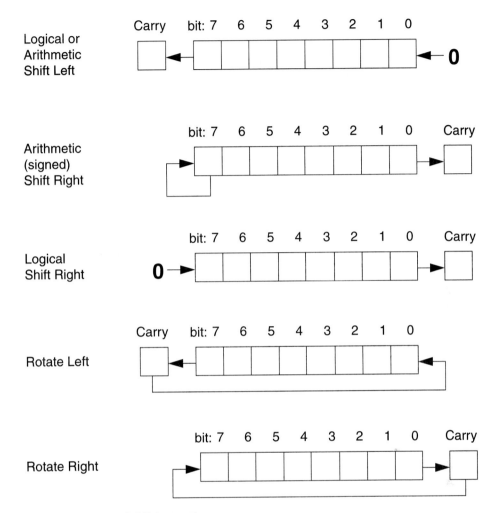

Figure 9-13 Forms of shift instructions.

For the feedback shift register, the left or right rotate instruction should be used, as we'll always need to supply the bit being shifted in, as well as to preserve the bit being shifted out (either as output data or as data to be shifted into the next higher byte/word). Whenever possible on the 68HC16, shifts should be performed using word-length shift instructions (e.g., ROLW) rather than byte-wide, as both take the same amount of time.

Listing 9-1 shows a subroutine written for the 68HC16 that implements a 31-bit-long FSR generator using one of the entries from Table 9-1. This sequence has a length of 2^{31}-1 (about 2 billion), which would take about 6 hours to send at 100K bits per second. Extending this to 48 bits should be more than sufficient if 2^{31} isn't quite long enough for you.

```
*       Feedback Shift Register Pseudo-Random Bit Sequence Generator
*       Usage:
*       Initialize fsr_w0 and/or fsr_w1 to any non-zero value.
*       Then call rand_fsr to get a single pseudo-random bit
*       returned in the carry bit.
*       Implements feedback shift register of length 31 with
*       tap at 28th bit; maximal length, length is 2^31.
*       (Note—28th bit is bit 27; 31st bit is bit 30.)
*       Uses A register
*       60 cycles (38 if in-line)

rand_fsr:                          ;(10 for jsr)
        ldaa    fsr_w1             ;(6) A=bits 24:31 of shift register
                                   ;(watch out—byte access of word!)
        rola                       ;(2)
        rola                       ;(2)
        rola                       ;(2)
        eora    fsr_w1             ;(6) bit 30 of A = (bit30) XOR (bit27)
        rola                       ;(2)
        rola                       ;(2) carry bit= (bit30) XOR (bit27)
        rolw    fsr_w0             ;(8) shift carry into shift register
        rolw    fsr_w1             ;(8) continue
        rts                        ;(12)
; Note that the final rolw shifts the old bit 31 into the carry bit.
; This is just a delayed version of bit 30, and saves some instructions.
```

Listing 9-1 FSR generator routine for 68HC16.

The first thing to notice is that nearly one-third of the cycles for this subroutine (22 out of 60) are spent calling or returning from it! If you really require speed, you might consider putting this code in the routine that needs it, rather than having it as a separate routine.

Next, notice that because the "exclusive-or" (EORA) instruction is bit-wise, there is no interaction between other bits and the two bits we're XORing together. Thus, there's no need to "mask" the quantities (for example, using "AND #$04" to get just bit 27). Let the other bits fall where they may—we're just interested in setting the carry bit to the "exclusive-or" of bits 30 and 27. (By the way, these are the 31st and 28th bits of the shift register.)

Finally, to save instructions, we haven't bothered to shift bit 30 out to the carry bit; instead, the last ROLW instruction has shifted the old bit 31 into the carry bit, which we return as the result. This bit 31 is a delayed version of bit 30, so all we're doing is delaying the bit stream by two sample times.

(Not shown in this code fragment is the storage required—you'll need to allocate two words in RAM for storage of "*fsr_w0*" and "*fsr_w1*," and you must make sure that these are not both zero. We used "seed" values of "*fsr_w0*=$5678" and "*fsr_w1*=$1234" in both the 68HC16 code and in the C program described below.)

As we coded these and other examples involving random sequences, it was crucial to have a way of testing the output, as a visual inspection wouldn't tell us if we had made a mistake. If we trust the algorithm, an excellent approach is to code the same routine in another (preferably, high-level) language, and to compare the results. We used the C program shown in Listing 9-2 to verify the 68HC16 FSR routine. We compared the first 1024 bits generated by the 68HC16 FSR routine with the output of the C program running on a PC. Ideally, the high-level implementation should be written as clearly and explicitly as possible—there's no need to optimize it for performance (in most cases), and you need to be able to see intermediate results easily. (In our C program, for example, we've split calculations into multiple lines, and avoided any "tricks" in coding conditionals.) The best case is when you can get hold of some "known good" code, or at least some known good output.

```c
/* fsr.c
   Pseudo-Random Bit Sequence by Feedback Shift Register
   Implements an FSR of length 31 with feedback at 28th bit (bit 27).
   This is maximal length FSR with a length = 2^31 (>2 billion)
   12/28/96 rce
   (Compiled under Borland C; unsigned long must be >=32 bits)
*/

#include <stdio.h>
#include <stdlib.h>

unsigned long sr;       /* shift register */

int rand_fsr(void);

int main(void)
{
   unsigned int c;
   unsigned int i;

   printf("\n\n  FSR:   \n");
   sr=0x12345678L;         /* initialize same as 68hc16 code */
   for(i=0;i<1024;i++)
   {
      if( (i & 0x1f)==0)
         printf("\n");      /* new line every 32 values */
      c=rand_fsr();
      printf("%1d ",c);     /* space after each bit */
   }
   return(0);
}  /* main() */

int rand_fsr(void)
{  /* Returns a single bit in lsb using FSR
      Not fast—designed to return same values as hc16 code */

   unsigned long a;
   unsigned int c;

   a= (sr<<3) ^ sr;       /* bit 30 = bit 30 XOR bit 37 */
   a=a & 0x40000000L;     /* bit 30 */
   if( a==0L)
      a=0L;
   else
      a=1L;

   if( (sr & 0x80000000L)==0L)
```

Listing 9-2   FSR generator routine in C.

```
        c=0;
    else
        c=1;

    sr=(sr<<1) | a;

    return(c);
}   /* rand_fsr() */

/* fsr.c */
```

Listing 9-2 (Continued) FSR generator routine in C.

Listing 9-3 shows the output from the C program (identical to the 68HC16 output) in case you want to check your code output manually. (A good-file compare utility is invaluable and can be used as follows: Capture the output of your C program to a text file; use a terminal emulation program to likewise capture the output of the 68HC16 code as an ASCII file. Edit the two files to eliminate any headers or trailing text, then perform a line-by-line file compare. Any differences between the two outputs will be shown, but you'll also see where there are similarities.)

```
0 0 0 1 0 0 1 0 0 0 1 1 0 1 0 0 0 1 0 1 0 1 1 0 0 1 1 1 1 0 0 0
0 0 0 0 1 1 1 0 0 1 0 1 1 0 1 1 1 0 0 1 0 1 1 0 1 1 1 0 0 0 0 0
0 1 1 1 1 0 0 1 0 0 0 0 1 1 1 0 0 1 0 0 0 0 1 1 1 1 0 0 1 1 1
0 0 1 1 0 0 0 1 0 1 1 1 1 0 0 0 1 0 1 1 1 0 1 1 0 1 1 1 1 0 1
0 1 1 1 0 1 0 1 0 0 1 1 1 1 0 1 0 1 1 0 0 0 0 0 1 0 1 0 1 1 0 1
1 0 1 1 1 0 0 1 1 0 1 0 1 1 0 0 1 1 0 0 1 0 1 1 1 0 0 0 0 0 0 0
1 1 1 0 1 0 0 1 1 0 0 1 0 1 0 0 1 0 1 0 1 1 1 1 0 0 0 0 1 1 1 1
0 1 0 0 1 0 1 0 0 1 1 1 1 0 0 0 1 0 1 1 1 1 0 1 1 1 1 0 1 0 1 0
0 0 1 1 0 0 1 1 0 1 1 1 1 0 1 1 1 0 1 1 0 0 1 1 0 1 1 1 0 1 1 1
0 1 0 1 0 0 0 1 0 1 0 0 1 1 0 0 0 1 0 1 0 0 0 1 1 0 0 1 1 0 1 1
1 0 1 1 0 1 1 0 0 1 0 1 1 1 0 1 1 0 1 1 1 0 1 0 1 0 0 0 1 1 0 0
0 0 0 0 1 0 0 1 0 1 1 0 0 0 0 1 1 0 1 1 1 0 1 1 1 0 1 1 0 0 0 0
1 0 0 0 0 1 0 0 1 1 0 0 1 1 0 0 0 1 1 0 0 1 1 0 0 0 1 1 1 0 0 1
0 1 0 0 0 1 0 1 0 0 1 0 1 1 1 1 0 1 0 1 0 1 1 1 1 1 1 0 0 1 1 0
1 1 0 1 1 1 1 1 0 1 0 1 0 1 1 1 1 1 0 1 0 0 0 0 1 1 0 1 0 0 0 0 0
0 1 0 0 1 0 1 1 1 1 0 1 0 1 0 1 0 1 1 0 0 1 0 1 0 0 0 0 1 0 0
0 0 1 0 1 0 1 0 1 1 1 1 1 1 1 0 0 1 0 0 1 1 0 1 1 0 0 1 0 1 0
1 1 1 1 1 0 1 0 0 0 0 0 1 1 0 0 0 0 1 0 0 0 0 1 0 0 1 1 1 0 1 0
0 1 0 1 0 1 0 0 1 1 0 1 1 0 1 0 0 1 0 1 0 0 0 1 1 1 0 1 0 0 0 1
1 1 1 0 0 1 0 0 0 0 0 1 0 0 0 1 1 0 1 1 1 1 1 0 1 0 1 1 1 1 0 1
1 0 0 0 1 0 0 1 0 0 1 1 1 1 0 0 0 0 1 0 0 1 0 1 1 0 1 0 1 0 0 0 1 1
1 0 0 0 0 0 0 1 1 1 1 1 1 0 0 0 0 1 0 0 0 0 1 1 1 0 1 1 1 1 1 1 1
0 0 0 1 1 1 0 0 0 1 1 1 0 1 0 0 1 1 1 1 0 0 1 0 0 0 0 1 1 1 1
1 1 1 1 1 1 1 1 1 0 1 0 0 1 1 0 0 1 1 0 0 0 1 0 1 1 1 0 0 0 0 0
0 0 0 0 0 1 0 1 0 0 1 0 1 0 1 0 1 1 1 0 1 0 1 1 1 0 0 0 0 0 0 0
0 1 0 1 1 0 0 0 1 1 1 1 0 1 1 0 1 1 0 1 0 1 1 1 0 0 0 0 1 0 1
0 0 1 1 1 1 1 0 0 1 0 0 0 0 0 0 1 1 0 1 1 1 1 0 1 0 1 1 0 0 1
1 0 0 1 1 0 0 0 1 0 0 0 0 1 1 0 0 0 1 0 1 0 1 1 0 0 1 0 1 0 1 0
1 0 1 1 1 0 0 1 0 1 1 0 1 1 1 0 1 1 1 0 0 1 0 0 1 1 1 1 1 1 1 0
1 1 1 0 0 1 0 0 0 0 1 1 0 0 1 1 1 0 0 0 0 1 1 0 0 0 0 0 1 0 0 1 1
1 0 0 0 1 0 1 1 0 1 0 1 1 1 1 0 1 1 0 1 1 0 1 0 0 0 1 1 1 1 1
1 0 1 0 0 0 1 1 0 1 0 0 1 0 0 0 0 0 0 1 0 1 1 1 1 0 0 0 1 0 1
```

Listing 9-3 Output of FSR generator routine.

9.5 Signal Averaging

Correlation, whether crosscorrelation or autocorrelation, works because when we add a large number of random values together, we should get a result quite close to the average value of those numbers (times N, the number of values). This is the "law of large numbers." Because we've been assuming the noise we experience has a mean value of 0 (no DC bias), any time we add up either random values or the product of random values, we expect those values to drop out. Signal averaging exploits this same phenomenon to allow us to extract a relatively clean waveform using many noisy copies of that waveform.

We begin by assuming that we either know the period of an otherwise unknown waveform or can find the period (e.g., via autocorrelation). The idea is to add up many copies of the noisy waveform exactly in phase. The noise should, over enough periods, cancel out; however, the underlying waveform will always be adding "in phase."

As an example, we take a simple periodic waveform and add some noise, as shown in Figure 9-14. We assume that we know the period and can thus average any number of periods to produce a new waveform. Figure 9-15 shows the results of averaging 5 and 100 periods of the noisy signal. The improvement in the recovered waveform with increasing periods is obvious; in terms of the SNR, signal averaging increases the SNR by $\sqrt{N}$ (in linear terms, not dB).

Signal averaging requires sufficient storage for an entire period (or at least the part of the waveform you're interested in), and each sample must have enough range to handle the sum of the N waveforms. Thus, if you have 12-bit samples and want to use signal averaging with $N=100$, you'll need more than $12+\log_2 100$ bits; that is, at least 19 bits, plus some extra to account for noise.

9.5.1 Stimulus/Response

We've mentioned two situations where signal averaging is useful—when the period of a signal is given (e.g., 10 msec) and when we can determine the period of a signal (e.g., by autocorrelation). However there is another important situation in which we either supply or are given the actual original signal. In this case, the signal need not be periodic; we just need to be able to know when the signal will repeat.

There are many cases where the same device both generates a "stimulus" and processes the "response." For example, "impulses" generated by small explosions can be used to probe the earth's crust. The shock-absorber system of a car might be tested by simulating bumps or potholes at specific times and measuring the movement of the car's frame. And so on. The important thing is that if we know when the stimulus occurs, we can average multiple responses *coherently* (in phase). The same improvements are possible as were noted for signal averaging. (In both cases, we know when the signal will repeat—whether because we know the period of some external source or because we, ourselves, determine the timing.)

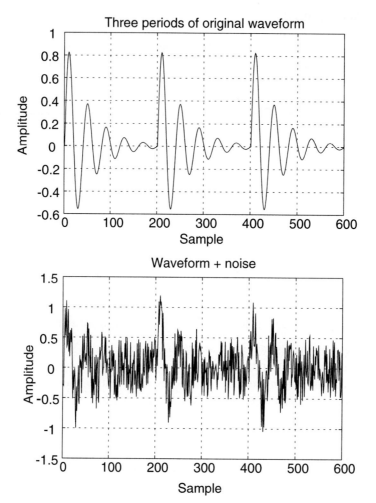

Figure 9-14 A periodic waveform and with noise added.

9.6 Summary

Correlation quantifies the similarity of signals. The correlation of two signals is the sum of the products of the two signals as one is shifted in time with respect to the other. Correlation is a function of delay (or [time] shift/lag) of one of the signals; sometimes a summary correlation figure is taken to be the maximum correlation between two signals across all lags.

Correlation between two different signals is called *crosscorrelation*.

Correlation between two segments (or the same segment) of the same signal is called *autocorrelation*.

Summary

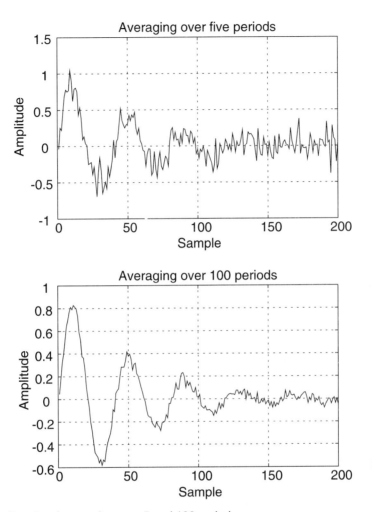

Figure 9-15 Results of averaging over 5 and 100 periods.

The normalized crosscorrelation produces a value between +1 (100% correlation) and -1 (antiphase relationship). A normalized crosscorrelation of 0 indicates no similarity between the two signals at that lag.

A good rule of thumb is to limit the maximum shift or lag to 5–10% of the signal length; more than this shift produces correlations that cannot be compared meaningfully.

As a rule of thumb, the signal length in time should be at least five times longer than the period of the lowest frequency component of interest.

Correlation and convolution are closely related (convolution reverses one of the signals in time) and so can be implemented in similar ways on microcontrollers. This includes the use of the FFT to speed correlation, a process known as *fast correlation*.

A few common applications for crosscorrelation include matched filtering, time of flight (time delay) calculation, image processing, filter design, and determining the system function of an unknown system.

In matched filtering, an input signal is crosscorrelated with a copy of the desired signal. A maximum output occurs when the input signal correlates highly with the desired signal.

Time of flight calculations crosscorrelate a copy of the signal transmitted (e.g., sonar) with a received version. Maximum correlation occurs at a lag corresponding with the round-trip delay the signal experienced. Using the speed of propagation of the signal, the distance traveled can be calculated.

Images or areas of images can be crosscorrelated to determine areas of similarity between two different scenes or the same scene separated by time.

The system function of a system can be established without using an impulse by crosscorrelating a "noisy" input to the system with its output. Such a technique avoids complications that can be caused by probing the system with an impulse.

Autocorrelation's principal applications involve determining the periodicity of a signal. In combination with a type of crosscorrelation or equivalently—signal averaging—autocorrelation can also be used to determine the periodicity and waveshape of an unknown signal.

The autocorrelation of a signal at lag 0 is equivalent to the normalized energy of the signal. The normalized autocorrelation of a signal at lag 0 is always 1.

Using autocorrelation, a linear predictor can be created that predicts future values of a signal using an FIR filter.

Pseudo-noise sequences, also known as *pseudo-random bit sequences*, are very useful in correlation applications, as they have zero correlation at every lag except lag 0. They are easily generated in hardware or software by means of an appropriate tapped or feedback shift register generator.

Signal averaging is a signal processing technique distinct from correlation but sometimes used in conjunction with correlation. Signals that repeat, though not necessarily periodically, if averaged exactly in phase with each, produce a waveform with increased SNR. The SNR increases as $\sqrt{N}$. Signal averaging is applicable for periodic signals when the period is known or can be determined (e.g., via autocorrelation). Signal averaging is also viable for nonperiodic signals if the interval between repetitions is known or if the analyzing device itself generates the "stimulus" signal to a system.

Resources

Deller, John R., Proakis, John G., and Hansen, John H. L., *Discrete-Time Processing of Speech Signals,* New York: Prentice-Hall/Macmillan, 1993.

Higgins, Richard J., *Digital Signal Processing in VLSI*, Englewood Cliffs, NJ: Prentice Hall, 1990.

Knuth, Donald E., *The Art of Computer Programming, Vol. 2: Seminumerical Algorithms, 2nd Edition*, Reading, Mass.: Addison-Wesley, 1981.

CHAPTER 10

Changing Sampling Rates

We'll use many of the topics of Chapter 4 (Discrete-Time Signals and Systems), and have need for a few FIR filters (Chapter 5 [FIR Filters—Digital Filters Without Feedback]). One example will use an RC filter, which you might remember from Chapter 3 (Analog Filters). Not much in the way of math here—sorry.

10.1 Overview

In this chapter we look at techniques for changing the sampling rate of a discrete-time (or digital) signal. The topic of changing sampling rates is part of a field known as multirate signal processing, and we'll be scratching only the surface here. However, some of these techniques are potentially quite important in embedded DSP applications.

We've hinted in earlier chapters that we might be able to reduce some of the hardware filter requirements by changing the sampling rate. Changing sampling rates can also reduce the computation necessary for some DSP operations like filtering or spectral analysis. In still other cases we're *required* to change sampling rates, in order to match two digital signal sources that have different rates.

We'll begin with applications for *decimation* (decreasing the sampling rate) and *interpolation* (increasing the sampling rate), then look at how to implement each of these operations. In each case, efficient methods result in computational savings. We then will discuss combining these two operations to change the sampling rate by noninteger (but rational) ratios.

Before we begin, one question you might have is "Why don't we just use a DAC and ADC to change sampling rates? With the proper filter between them (a combination anti-imaging/antialiasing filter), there shouldn't be any problems with arbitrary sampling rate conversion." In practice, this would be a fairly expensive solution, at least compared with a DSP-based technique. In addition, all of the ills of DACs and ADCs—noise, nonlinearities, missing codes, etc.—would corrupt the signals. With the techniques in this chapter, you should be able to keep your signal in the digital realm until you're done processing—a cheaper, higher quality solution.

10.2 Applications

Why change the sampling rate of a digital signal? Among the most important reasons are the following:

- the sampling rate of a signal can be altered to match a (fixed) sampling rate required by a system;
- interpolation can reduce the need for anti-imaging, or $sin(x)/x$ compensation, and decimation can reduce the need for antialiasing;
- decimation can reduce the amount of computation necessary to process a signal.

Let's take these one at a time.

10.2.1 Matching Fixed Sampling Rates

A contemporary example of mismatched sampling rates is that between compact disks (CD) (44.1 kHz) and digital audio tape (DAT) (48 kHz). For example, to convert a digital signal from a CD to store on a DAT, we need to change the sampling rate by 48 kHz/44.1 kHz. (We'll discuss this case a little later.) A simpler example is in storing a telephone conversation on DAT—here, if we use an 8 kHz sampling rate for the phone call, we would need to interpolate by a factor of 6 to get the 48 kHz of the DAT. Cases in which the increase or decrease in sampling rate is an integer factor are easier to implement, but when external standards or devices dictate the sampling rate, sometimes we have to implement more complicated conversions.

10.2.2 Reducing Input/Output Hardware Filters

In Chapter 4 (Discrete-Time Signals and Systems), we discussed antialiasing filters on the input to a DSP system, and anti-imaging and $sin(x)/x$ compensation filters on the output. As you know by now, a major factor in the complexity of a filter is the transition width that's required. For both input and output, if the sampling rate is increased, the requirements of the filters are effectively reduced. For high-volume products, the savings in using simple RC filters instead of costly active filters, for example, can be substantial; we'll take you through one such design example. You might also recall that the $sin(x)/x$ effects that were produced by the DAC's "zero-order hold" behavior are much less of a problem if the bandwidth of interest is concentrated at relatively low frequencies where the $sin(x)/x$ effect is minimal—which will be the case if our sampling rate is relatively high. So this is another area of potential savings. Note that we'll *decimate* if we've used a high sampling rate on the input and *interpolate* to use a high sampling rate on the output.

10.2.3 Reducing Computation

Decimation can reduce the amount of computation required for processing both bandpass and baseband signals. Recall that bandpass signals have their frequency content starting at some nonzero frequency, and baseband signals' spectra begin at 0 (DC), or at least fairly close. In the case of bandpass signals, appropriate (bandpass) filtering followed by decimation can dramatically reduce the amount of computation necessary for filtering or spectral analysis. For example, if we are interested in only a small range of high frequencies, the FFT would require the calculation of values at the same resolution for all parts of the spectrum. By decimating,[1] however, we effectively shift the frequency range of interest down to the baseband and can just take the FFT over this limited range.

Baseband signals may also have a lot of "unused" bandwidth. By decimating, we can trim some of the "fat" off the signal, which will speed processing. In addition, it is often more efficient to do lowpass filtering in stages. For example, an FIR filter that passes only the very lowest frequencies will have many more coefficients than an FIR filter, where the passband is closer to the Nyquist range. (More on this when we talk about decimation.)

The main point is that if we've somehow ended up with a digital signal that doesn't use the entire "available" bandwidth, we can sometimes be more efficient by appropriately decimating the signal.

10.3 Decimation

We'll start with decimation, because it's a bit easier to implement than interpolation.

A first approach might be to just take every Dth sample—in effect, resampling the signal. To decimate by a factor of 2, we would take every other sample (D=2). Or for a factor of 10, we would take every tenth sample. The problem with this approach is that just as when we sample continuous signals, we can suffer aliasing. Any frequencies present in the original digital signal will be mapped *somewhere* in the decimated signal's spectrum.

The solution is an antialiasing filter, though we'll use a *digital* filter as we're sampling a discrete-time, not continuous-time, signal. Now the question is, What type of filter—IIR or FIR?

For decimation filters, the FIR is used almost exclusively; we'll see why in a moment. Not only can we get linear phase, but with the 68HC16, we get pretty good performance. The decimation process is shown in Figure 10 1.

10.3.1 Efficient Decimation

An important thing to notice in this simple decimation scheme is that we use only every Dth output of the FIR filter. Because FIR filters have no feedback, there's no reason to calculate any output values we don't use—which is D-1 of every D input samples! So we have to compute the

---

1. This is identical to the purposeful undersampling we've mentioned before, just that we're doing it to a digital signal, rather than analog. Decimation is almost always preceded by an appropriate filter, except in this under-sampling case.

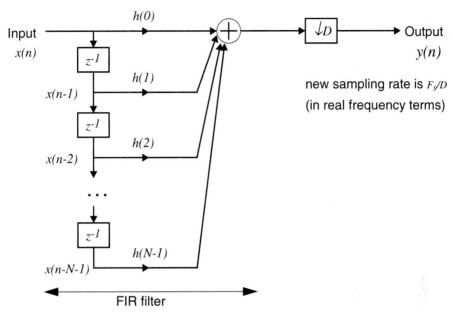

Figure 10-1 The decimation process.

output of the FIR filter only once per output value, and we also need to update the input array only once per output value. If the FIR filter has N coefficients where N is an integer multiple of D, this updating can omit any testing for buffer wrapping and just write a block of D new values quickly (e.g., check for buffer wrap-around at start or end of block write.)

If we instead use an IIR filter for antialiasing, we would need to calculate all of the feedback values for each new input value (though we could postpone doing the feedforward operation until we needed the output like we did for the FIR). For large values of D, the advantage goes to the FIR filter, especially for our FIR-friendly 68HC16. (And, of course, the IIR wouldn't have linear phase, but you knew that.)

10.3.2 Buying More Time With Buffers, and Decimating in Stages

A few more observations about decimation. Normally we need to evaluate an FIR filter in one sample period—otherwise, the samples $x(n$-$1)$, $x(n$-$2)$ and so on would change before we were done calculating the output value. But we evaluate this particular FIR filter once for every D samples. A strategy that gives you more time to evaluate the FIR filter is to precede this FIR filter with a first-in/first-out (FIFO) buffer that can hold D samples. You can now spend up to D sample periods evaluating the FIR filter, then once you have the output, you can shift in the next D samples into the x array. The samples keep arriving at the normal sample rate, but they stack up in the FIFO buffer until the FIR filter is ready for new data.

You can often realize great computational savings by breaking the decimation into stages. This is because the required antialiasing filters of multiple stages, even when added together, are almost always much smaller than one big monolithic filter, because the transition width is much wider for each of these filters. So instead of decimating in one step by a factor of $D=12$, you can instead cascade two or more stages—for example, $D=4$ followed by $D=3$, or $D=2, 2, 3$, or some other factorization of 12. Most advanced texts (e.g., Proakis and Manolakis [1992]) have the details of finding the optimum arrangement of stages. This technique can also be applied to interpolation, with similar savings.

10.4 Interpolation

Interpolation increases the sampling rate of a digital signal, producing I output samples for each input sample. Just as we had to deal with aliasing with decimation, we have to worry about "imaging" when we perform interpolation, which means we need to use a digital "anti-imaging" filter. We'll use an FIR for this filter too, but in a different way and for slightly different reasons than in the decimation case.

10.4.1 Interpolate/Filter Structure

Figure 10-2 shows the starting point for interpolation. The box marked $\uparrow I$ adds I-1 zeros between each incoming sample,[2] then this new signal is filtered to remove the high frequency images (recall Chapter 4 [Discrete-Time Signals and Systems]). We need a gain of I at the end as we've spread out the energy of the incoming signal over I-1 zero samples.

We can design this digital anti-imaging filter using the same criteria as we would use for analog anti-imaging filters. Figure 10-3 shows an example magnitude spectrum of a discrete-time signal, the spectrum after the sampling rate is increased (by adding zeros), the required anti-imaging filter, and, finally, the magnitude spectrum of the interpolated signal.

Notice the big difference between the transition width an (external, analog) antiimaging filter would need for the original signal versus that for the interpolated signal. The smaller that transition width, the more complex the filter. Of course, the *digital* anti-imaging filter has to handle that original, narrow transition width, but this is a reasonable approach if we have the extra instruction cycles to spend.

10.4.2 Efficient Interpolation

Unlike decimation, an interpolation process must evaluate the FIR filter for each sample. However, most of the data values going into the filter are zero (I-1 out of every I values, in fact), so we should be able to figure out a way of speeding things up. In fact we can. The technique is known by the scary-sounding name of *polyphase filtering*, but it just involves knowing what coefficients will be multiplied by nonzero data and doing only those multiplications. Of course,

---

2. Why not just duplicate the samples instead of sticking in zeros? You wouldn't need the gain stage, but you'd need to take into account the $sin(x)/x$ effect of this digital equivalent to the zero-order hold.

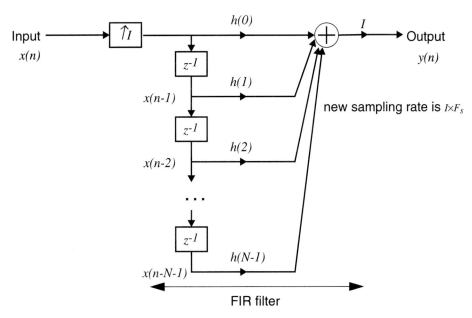

Figure 10-2 The interpolation process (naive approach).

to make things efficient, we'll want to use the RMAC instruction of the 68HC16, but the final result isn't too complicated.

Let's construct a table showing the values of the input array for a simple case. We'll let I, the interpolation factor, be 3, and we'll assume that the FIR filter has 11 coefficients. Table 10-1 shows input and the output for a number of cycles, where x_n stands for one of the original samples and h_n stands for a filter coefficient. The pattern of multiplications that emerges is complicated only by the fact that N/I isn't an integer, so the last multiplication we're expecting sometimes "drops off" the end.

A useful way of looking at the pattern of multiplications is to concentrate on adjacent time periods when the original samples are the same. Time periods 9, 10, and 11 are like this—original samples x_3 through x_0 are the only samples used. The coefficients are $\{h_0, h_3, h_6, h_9\}$ for time period 9, then $\{h_1, h_4, h_7, h_{10}\}$ for period 10, then $\{h_2, h_5, h_8, 0\}$ for period 11. The interpolation filter acts as though there were three filters—each with the same inputs (just the original signal values, no added zeros) but with different coefficients—and that are evaluated one after the other. This is where the term *polyphase filtering* comes from—the idea that you could replace the one FIR filter with a lot of smaller FIR filters, each with a different set of coefficients and each evaluated once every I times.

To implement the polyphase filter, we don't have to write I different filter routines—we just change coefficients in one routine. This single filter, evaluated once for every output sample, takes as input the original, noninterpolated signal, and changes coefficients for each new output sample. The quantity N/I is the length of this filter; at sample n, the first coefficient used is n

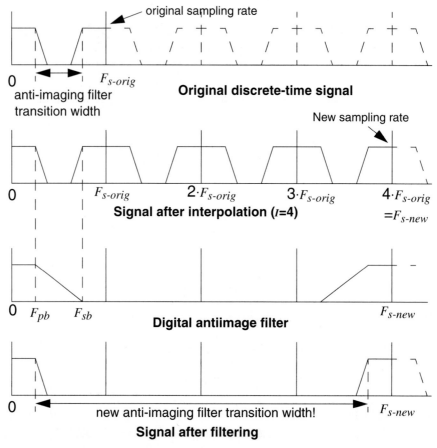

Figure 10-3 Spectra of signals during the interpolation process.

modulo I, and the increment between coefficients is I. If N/I is not an integer, you'll either need to vary the length of the filter (for one or more cases, the length will be[3] *floor(N/I)*), or add zeros to the filter coefficients. Instead of performing the modulo division, you can just use a counter that counts down from I; properly ordering your coefficients (in "scrambled" order) makes stepping through them in the correct order trivial. We'll illustrate these techniques in the sample code below.

10.4.3 Interpolation Example

Let's demonstrate interpolation on the 68HC16 for the following problem. Say we have a signal, already in digital form after we've done some other processing, and we'd like to convert this back to analog form. However, we'd like to use a simple (one-pole) RC filter on the DAC output,

---

3. The function "floor" returns the largest integer less than or equal to the argument.

Interpolation

Input	\multicolumn{11}{c}{Filter coefficients, h_n}	Output
	0	1	2	3	4	5	6	7	8	9	10	
0	x_0	0	0	0	0	0	0	0	0	0	0	$h_0 x_0$
1	0	x_0	0	0	0	0	0	0	0	0	0	$h_1 x_0$
2	0	0	x_0	0	0	0	0	0	0	0	0	$h_2 x_0$
3	x_1	0	0	x_0	0	0	0	0	0	0	0	$h_0 x_1 + h_3 x_0$
4	0	x_1	0	0	x_0	0	0	0	0	0	0	$h_1 x_1 + h_4 x_0$
5	0	0	x_1	0	0	x_0	0	0	0	0	0	$h_2 x_1 + h_5 x_0$
6	x_2	0	0	x_1	0	0	x_0	0	0	0	0	$h_0 x_2 + h_3 x_1 + h_6 x_0$
7	0	x_2	0	0	x_1	0	0	x_0	0	0	0	$h_1 x_2 + h_4 x_1 + h_7 x_0$
8	0	0	x_2	0	0	x_1	0	0	x_0	0	0	$h_2 x_2 + h_5 x_1 + h_8 x_0$
9	x_3	0	0	x_2	0	0	x_1	0	0	x_0	0	$h_0 x_3 + h_3 x_2 + h_6 x_1 + h_9 x_0$
10	0	x_3	0	0	x_2	0	0	x_1	0	0	x_0	$h_1 x_3 + h_4 x_2 + h_7 x_1 + h_{10} x_0$
11	0	0	x_3	0	0	x_2	0	0	x_1	0	0	$h_2 x_3 + h_5 x_2 + h_8 x_1$
12	x_4	0	0	x_3	0	0	x_2	0	0	x_1	0	$h_0 x_4 + h_3 x_3 + h_6 x_2 + h_9 x_1$
13	0	x_4	0	0	x_3	0	0	x_2	0	0	x_1	$h_1 x_4 + h_4 x_3 + h_7 x_2 + h_{10} x_1$
14	0	0	x_4	0	0	x_3	0	0	x_2	0	0	$h_2 x_4 + h_5 x_3 + h_8 x_2$
15	x_5	0	0	x_4	0	0	x_3	0	0	x_2	0	$h_0 x_5 + h_3 x_4 + h_6 x_3 + h_9 x_2$
16	0	x_5	0	0	x_4	0	0	x_3	0	0	x_2	$h_1 x_5 + h_4 x_4 + h_7 x_3 + h_{10} x_2$
17	0	0	x_5	0	0	x_4	0	0	x_3	0	0	$h_2 x_5 + h_5 x_4 + h_8 x_3$
18	x_6	0	0	x_5	0	0	x_4	0	0	x_3	0	$h_0 x_6 + h_3 x_5 + h_6 x_4 + h_9 x_3$
19	0	x_6	0	0	x_5	0	0	x_4	0	0	x_3	$h_1 x_6 + h_4 x_5 + h_7 x_4 + h_{10} x_3$

Table 10-1 Analyzing a simple interpolation example.

but not to have less than 40 dB attenuation of images. The signal has a bandwidth of 300 Hz and a sampling rate of 1 kHz, and the DAC can handle any reasonable sampling rate. We also require less than 1 dB of passband deviation from this stage forward.

Before we begin, we can use formulas from earlier chapters to find that we'd need a five-pole Chebychev I filter for anti-imaging if we don't interpolate—and that's ignoring any $sin(x)/x$

and phase issues. (We'll assume that the phase shift from an RC filter is acceptable—if not, that's another area to examine.)

Working backward, if we determine the RC filter parameters (i.e., the constant *RC*), we can find the necessary interpolation factor *I*. We know for a one-pole RC filter:

$$|H(jw)| = \frac{1}{\sqrt{1 + \omega^2 (RC)^2}} \qquad (10.1)$$

If we allocate half of the 1-dB passband deviation to the RC filter, we can use Equation 10.1 to determine *RC*. Substituting $\omega = 2\pi F_{pb}$ (F_{pb}=300 Hz) and setting the equation equal to 0.5 dB (0.944061 linear), we find RC=0.000186488.

Now we can ask what value of *I* gives 40 dB of stopband attenuation at the edge of the first image to be rejected. This edge is at $I \cdot F_s - F_{pb}$, and it's probably easiest to just try some numbers. Don't forget the *sin(x)/x* effect from the DAC's "zero-order hold" behavior—it helps us a lot in meeting the stopband requirements! The magnitude response of the zero-order hold is:

$$|H_{\sin x / x}(F)| = \frac{\sin(\pi F T_s)}{\pi F T_s} \qquad (10.2)$$

(You can use $\pi F/F_s$ in place of $\pi F T_s$.)

We wrote a program to calculate the stopband attenuation for various values of *I*; the results are in Table 10-2. Interpolating by a factor of 6 takes the stopband edge frequency out far enough for the combined efforts of the RC filter and *sin(x)/x* effect to total more than 40 dB. Now we've got *I*. (By the way, notice that the *sin(x)/x* effect exceeds the 1-dB passband deviation for the case of *I*=1. But for higher values of *I*, the attenuation is far less and can be ignored for *I*=6.)

Independent of the interpolation factor, we know the parameters of the FIR anti-imaging filter we need—the passband edge frequency is 300 Hz, the stopband edge frequency is 1000-300, or 700 Hz, we'll use up our remaining 0.5 dB of passband ripple (the RC filter has 0.5 dB deviation) and also require 40 dB of stopband attenuation. (This last item, stopband attenuation, still has to be (around) 40 dB, because the RC filter doesn't do much attenuating at this frequency.) Using the methods of earlier chapters (we'll use the Parks-McClellan program), we calculate an estimated *N*=21, but actually generating the coefficients and plotting the response shows we need at least *N*=25. The program we wrote for the 68HC16 requires *N/I* to be an integer; we can either add zeros (which is what we did), or else tighten up the filter specifications and "use up" more coefficients (there's no reason not to!). The stopband attenuation is -43 dB for *N*=25 coefficients, and we added five zeros to bring *N* to 30. *N/I* is then 5, the length of the filter.

Interpolation

I	Passband attenuation due to $sin(x)/x$ (dB)	Stopband edge frequency for RC filter (Hz)	Stopband attenuation of RC filter (dB)	Stopband attenuation of $sin(x)/x$ (dB)	RC + $sin(x)/x$ stopband attenuation (dB)
1	-1.3	700	2.2	8.7	10.9
2	-0.32	1700	7.0	15.4	22.4
3	-0.14	2700	10.4	19.2	29.6
4	-0.08	3700	13.0	21.9	34.9
5	-0.05	4700	15.0	24.0	38.9
6	-0.04	5700	16.6	25.6	42.2

Table 10-2 Stopband attenuations for interpolation factors I=1 to 6.

This is all the information we need to implement the interpolation. The program is listed in its entirety in Appendix 5; we'll note some of the major points here.

First, we scrambled the order of the filter coefficients so we can just step through the coefficients in the order they're used. Because we chose to use a fixed-length filter for all the filters, this means adding a zero to the end of some filters' lists of coefficients. (You might think because the RMAC instruction lets us specify the step size for the pointers, we could leave the coefficients in normal order. But the range is only 4-bits signed, which is -8 to +7; because we need to step by words, this is a maximum step size of three words. Not big enough for this case and too restrictive for a general-purpose routine we might want to reuse in different cases.)

The next major issue is the length of the filters. You can perform your own analysis, but we saw the opportunity for only a slight savings if we had varying filter lengths (from N/I not being an integer). The issue is whether coding to avoid a possible extra multiply outweighs the cost of doing a multiply by a zero coefficient. On a microcontroller other than the 68HC16, however, the savings might be bigger. If you do code for varying lengths, you might look at embedding the filter length in the coefficient array to avoid the need for another pointer, versus putting the lengths in a separate array.

(A minor point in this example code is how we update the pointer to the coefficient array, stepping from the coefficient values for one filter to the values for the next. We've coded this as a separate addition, when you'd think we could just use the value of *IX* that has been properly adjusted by the RMAC instruction. We ran into some differences in how this code executed when debugging, so we put in the explicit add just to be safe.)

We set the timer to interrupt at the fast sampling rate—$I \cdot F_s$, or 6 kHz here—and use a counter to determine whether we should take a new input (once every I times). The program always calculates a new output value on every interrupt. By the way, you'll note that we don't output the value just calculated—instead, we output the value calculated at the prior interrupt. And we don't have multiple instruction paths before we read the ADC either. In both cases, we want to avoid introducing variation in the input or output sampling period (i.e., jitter). Coding like this adds an additional delay of one sample period to the system, but it means we can code much of the routine without worrying about inconsistent timing.

Just for fun, we added a bit of a kluge to the code to allow you to disable the interpolation and see the raw input passed to the output. (Not a bad way to begin writing a program, either.) An external switch is sensed through an unused I/O port documented in the program listing.

Using the RC constant computed above, you should be able to construct an appropriate RC filter using a 180K resistor and 0.001 µF capacitor on the DAC output.

10.4.4 Interpolation Results

The results from interpolation are dramatic. Let's follow an example sinusoid through the interpolation process. (Although these figures were generated from simulations, the same filter coefficients and interpolation factors as those used in the 68HC16 demonstration program were employed, so the results are similar.)

A 117-Hz sinusoid is shown in Figure 10-4a, with the 1 msec (F_s=1 kHz) sample times marked. Figure 10-4b shows the output we would get if we sent the sampled signal directly to a DAC. The RC-filtered output is shown in Figure 10-4c. The filtering helps a little, but it's clear that there's a lot of high-frequency content in the output.

Interpolating the original signal, we create the much smoother signal, shown as it would be output from a DAC in Figure 10-4d. If we've done our work correctly, the remaining high-frequency components should largely be eliminated by the RC filter. Figure 10-4e shows that we met our goals. We've replaced a fairly complicated five-pole filter with a single-pole RC filter without degrading the output.

10.5 Rational Interpolation/Decimation

So far we've discussed changing the sampling rate by integer factors—decimating by 4, or interpolating by 10, for example. In this section, we explore combining interpolation and decimation to change sampling rates by rational[4] factors—say, interpolating by 48/44.1 (the CD-to-DAT problem).

10.5.1 Original Structure

The idea key to changing sampling rates by rational factors is to express the interpolation factor in terms of the ratio of two whole numbers, interpolating first by the numerator, then decimating by the denominator. Figure 10-5 shows the idea for interpolating a signal sampled at 1 kHz by a factor of 2/3.

Interpolation should always be done first. If you decimate first, you can lose frequency content. Also, you should express the two factors in their "lowest" form (that is, with no common factors). As an example, interpolation by 4/6 would produce an intermediate sampling rate of 4 kHz, instead of the 2 kHz that interpolation by 2/3 produces, though both produce the same results.

10.5.2 Combining *D* and *I* Filters

Because both the interpolation anti-imaging filter and the decimation antialiasing filter are low-pass filters operating at the same sampling rate, they can be combined into one filter, as shown in Figure 10-6. This single filter must meet the most restrictive of either filter specification, but this is more efficient than having two separate filters.

This single filter can also incorporate the efficient interpolation and decimation methods we just discussed. On the input side, a polyphase implementation replaces explicit interpolation by properly sequenced filter coefficients, while on the output side, the filter is evaluated only at the final output sampling rate. Thus, we need not actually evaluate the filters at an extraordinarily high sampling rate, or allocate memory for the intermediate signal. Orfanidis [1996], among other texts, discusses the mechanics of this process in detail.

As an example of interpolation by a rational factor, let's examine the CD-to-DAT problem. Multiplying (48 kHz)/(44.1 kHz) by 10 gives us an interpolation factor of 480/441, and dividing by 3 (the only common factor) gives the ratio 160/147. The intermediate sampling rate will be 160×44.1 kHz, or 7.056 MHz(!), so efficient interpolation and decimation are obviously necessary. (For fun, you might want to calculate the filter order required. Then perform the same analysis if the CD sampling rate were 44.0 kHz instead, a seemingly negligible difference. Can you reach any conclusions about the choice of 44.1 kHz as the CD sampling rate?)

Other techniques exist for performing multirate processing, for both the problems we've described here and others. Start with the usual suspects; if the general DSP texts don't go into enough detail for you, there are any number of texts just on multirate processing. (See, for example, Vaidyanathan [1990].)

Also, for some applications you should be aware that hardware interpolators are available that are optimized for sample rate conversions over a wide range of frequencies.

---

4. *Rational* numbers are numbers that can be expressed as the *ratio* of two whole numbers. 355/113 is a rational number, for example. Plenty of expressions are not, such as π or $\sqrt{2}$, just to name two famous examples.

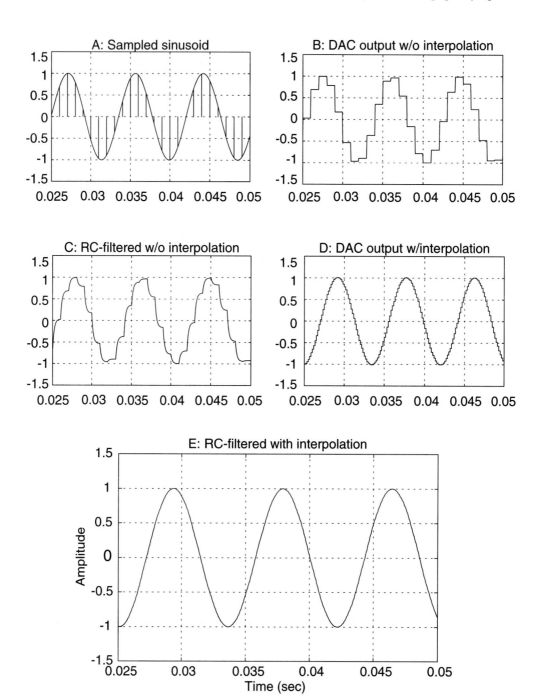

Figure 10-4 Waveforms from interpolation example.

Summary

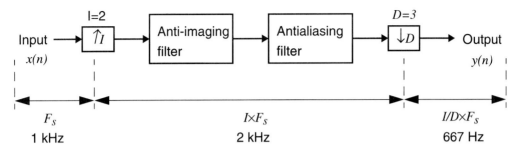

Figure 10-5 Rational interpolation.

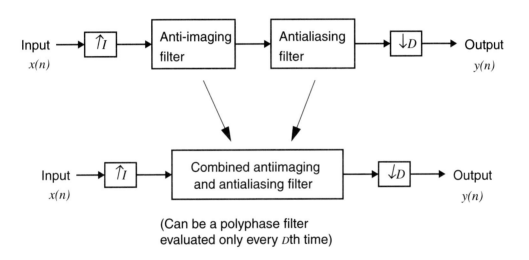

Figure 10-6 Combining antiimaging and antialiasing filters in rational interpolation.

10.6 Summary

Interpolation is the process of increasing the sampling rate of a digital signal; decimation decreases the sampling rate.

It is useful or necessary to change the sampling rate of a digital signal for three main reasons:
- to reduce the requirements for hardware filters on either input or output
- to reduce the amount of computation required (e.g., filtering, spectral analysis)
- to match a different (fixed) sampling rate of another system.

Decimation by an integer factor D is accomplished by first removing frequencies that would otherwise be aliased, then taking every Dth sample to produce a new digital signal.

When an FIR filter is used as the antialiasing filter for decimation, it is not necessary to evaluate the filter except at every Dth sample time. This increases the efficiency of decimation considerably and makes FIR filters preferable for decimation applications.

Decimation by large factors can require high-order FIR filters. Decimating in stages can reduce the order of the antialiasing filters considerably.

Interpolation by an integer factor I is accomplished by inserting I-1 zeros between each sample, then removing high-frequency images with an anti-imaging filter.

When an FIR filter is used as the anti-imaging filter for interpolation, efficient polyphase filtering can be used. This method avoids multiplication of zero input data values.

Interpolation and decimation can be combined to produce rational interpolation factors. In this case, interpolation by an integer factor (the numerator) is first performed, followed by decimation by an integer factor (the denominator). The anti-imaging and antialiasing filters can be combined into one filter, which can be implemented as a polyphase filter whose output is evaluated once every D samples.

Resources

Analog Devices, *Digital Signal Processing Applications Using the ADSP-2100 Family*, Englewood Cliffs, NJ: Prentice-Hall, 1990.

Orfanidis, Sophocles J., *Introduction to Signal Processing*, New Jersey: Prentice-Hall, 1996.

Proakis, John G. and Manolakis, Dimitris G., *Digital Signal Processing: Principles, Algorithms, and Applications*, 3rd Edition, New York: Macmillan, 1995.

Vaidyanathan, P. P., "Multirate Digital Filters, Filter Banks, Polyphase Networks, and Applications: A Tutorial," *Proceedings of the IEEE*, vol. 78, January 1990, pp. 56–93.

Vaidyanathan, P. P., *Multirate Systems and Filter Banks*, New Jersey: Prentice-Hall, 1993.

CHAPTER 11

Synthesizing Signals

A background in multivariate astronumerology might be helpful if you are kidnapped by superstitious aliens, but this chapter is largely devoid of math beyond algebra. Well, a bit of probability comes up in talking about random numbers, but nothing you need to know beforehand. We will mention the pseudo-noise generators of Chapter 9 (Correlation), which is also where you can see a good application of random numbers. No other chapters are really necessary to understand this material.

11.1 Overview

So far in this book, we've looked at systems that modify the frequency content of signals (filters) and methods that analyze signals for frequency content (the DFT) or detect specific time-domain signals. What about creating systems that generate signals—from our friend the sine wave to various flavors of noise?

We have two major topics to cover in this chapter—creating functions that are deterministic (like $sin(x)$ or x^3+4x^2) and creating functions that are described using probability ("random numbers"). Both of these areas sound pretty simple at first—perhaps look up a simple equation for the "sine" function, then code it. Or code the sine function incorrectly and end up with a random number generator. But you probably have an inkling that this simplicity just isn't so—and right you are. It is surprisingly difficult to generate "good" random numbers, and functions like the sine call for some tricks and trade-offs to get reasonable accuracy at the speeds we need.

We'll begin with random numbers.

11.2 Random Numbers

11.2.1 Why Do We Want Random Numbers?

Though they have a valuable role in accounting, economics, and marketing,[1] random numbers also have some important roles to play in DSP. First, we saw that the analysis of pseudo-noise bit sequences using correlation makes it possible to determine the time lag between two such signals. In addition, random numbers are useful in simulating noise, though we'll often use this in laboratories, rather than in the field. Finally, there are many signals that we may wish to synthesize that include various degrees of "noise," for example, in speech synthesis. (Outside of DSP, random number generators can create behavior that appears less predictable—say, in a game—or can be the foundation of algorithms that conduct searches or process data in ways that may be more efficient than deterministic algorithms.)

11.2.2 What Do We Want?

The first thing to figure out is what we want of our "random numbers." Intuitively, we might want each number to be unrelated to any that come before or after ("independence")[2] and for any number (or range) to have an equal chance of occurring over some period of time ("uniform distribution").[3]

Because we are using a microprocessor to generate numbers, we can't produce "truly" random numbers. The microprocessor is a very nonrandom device—which is fortunate, as we depend on it to execute our millions of instructions per second completely deterministically. However, it is possible to generate sequences of numbers that "look" unrelated to each other and have a uniform distribution but are produced using deterministic algorithms. Such numbers are called *pseudo-random* numbers.

Because this isn't a book on probability or "stochastic processes" (whatever that means) we're going to wave our hands a bit about what "looking random" means. Basically, we're going to ask that our sequences have a long period (as they are all going to repeat), that the distribution of values be either uniform or some other specific distribution, and that there be no "obvious" relationships between a number and its predecessors.

1. Well, we suspect that there is some type of accounting that uses numbers related to real-world events, but this remains unconfirmed.
2. We're doing some heavy hand-waving here. As usual, there are a lot of technicalities we're going to ignore as we're dancing on the edge of some serious math and probability theory.
3. Sometimes we need other nonuniform distributions or want a sequence of such numbers, interpreted as a noise signal, to have certain spectral characteristics. However, we can obtain the most important nonuniform distribution the *Gaussian* or *normal* distribution by clever manipulation of independent uniform random numbers. In addition, we can manipulate the spectra of sequences using filters. We'll discuss this later in this chapter.

11.2.3 Generating Pseudo-Random Numbers

There are many algorithms for generating pseudo-random numbers. The surprising thing is that it is difficult to find algorithms whose output is reasonably random. Lest you think this is purely academic, a great deal of research in physics and other sciences depends on using huge quantities of random numbers; if the characteristics of the pseudo-random number generator (*PRNG* for short) are not good statistically speaking, the research results would be questionable. It so happens that a popular PRNG, RANDU, which was used for nearly a decade by most of the computer systems in the world, didn't meet many statistical tests of randomness. According to Knuth, "Its very name RANDU is enough to bring dismay into the eyes and stomachs of many computer scientists!"[4]

Our applications on microcontrollers are likely far less critical and probably could work with practically any PRNG. We're going to give you two different PRNGs, both with good properties, that represent two extremes in terms of period length but with fairly efficient execution. These are the "linear congruential generator" and the "R250 algorithm." We also discuss techniques for producing even "more-random" random numbers, should your faith in pseudo-randomness be a little weak.

11.2.3.1 Linear Congruence

The *linear congruence generator* (LCG) is probably the most widely used random number generator. Because it involves a single multiplication, addition, and modulo division, it is quite simple to implement and is reasonably fast. It is also very easy to create a bad LCG—the RANDU function mentioned earlier is an LCG with a "bad" set of constants.

Knuth devotes the first 177 pages of volume 2 of *The Art of Computer Programming* to random numbers, and the LCG receives the bulk of the attention among various methods. If you really want to know about random numbers, you should start with that book. What follows is our brief summary.

The linear congruence algorithm is:

$$X_{n+1} = (aX_n + c) \text{ modulo } m \qquad (11.1)$$

The modulo operation just returns the remainder after division by m. For example, 17 mod 4 is 1, and 17 mod 9 is 8. The constants a, c, and m are chosen to produce both a long period and good statistical characteristics of the sequence.

To generate the next number, the LCG depends on only the prior number in the sequence. The sequence can therefore be no longer than m. (This also means you can start the LCG with the same "seed" and get the same sequence—a very useful property!) However, poor choices of a and c can produce much shorter sequences, which means that not every possible value will be generated. We discuss some necessary conditions below.

---

4. Knuth [1981], p. 104.

Modulo division in general is not a strong point for microcontrollers. However, if m is a power of 2, we can just pick off the $\log_2 m$ bits of a result to get modulo m. There's another reason we want m to be a power of 2—we may need random numbers with uniform distribution over 0 to 2^m-1. If we instead had a distribution over 0 to some arbitrary value, we'd need to scale the random numbers.

Having picked m as a power of two, the following are necessary (but not sufficient) conditions to pick good values of a and c:

- Pick m as large as possible. (Here, we assume m is a power of 2.)
- Pick $a=4n+1$, where n is an odd number.
- Choose c to be any odd number.

In addition, it may be a good idea to make n and c relatively prime (that is, they have no factors in common).

We emphasize that these are necessary conditions for the LCG constants, and the actual performance of the LCG must be verified in a number of ways to be confident that the LCG is "good." We're going to use some values that others have reported as "good" rather than getting sidetracked, but if you need to use a different m or a for some reason, the criteria above will likely be sufficient.

Implementing the Linear Congruence Generator The code of Listing 11-1 shows one implementation of an LCG for the 68HC16. The constants meet the criteria given above and, further, are reported to produce an LCG with good statistical characteristics (though we have not confirmed that). The m is 65536, or 2^{16}, which lets us avoid the explicit modulo division. The value of a is 25173, which is $(7\times29\times31)4+1$. The choice of c is arbitrary at 13849 (which is 11×1259, the latter a prime).[5]

```
...
*          Constants
LCG_A      equ      25173           ;(7*29*31)4+1
LCG_C      equ      13849           ;11*1259
;LCG_M     equ      65536           ;implemented via modulo arithmetic

*          Variables (in internal RAM)
...
lcg_x      ds.w     1               ;seed/random number
...
*          Linear Congruent Pseudo-Random Number Generator
*          Usage:
*          Call rand_lcg_seed with D=16-bit seed
*          Then call rand_lcg to get 16-bit pseudo-random numbers
*          Uses D,E registers
*          52 cycles (30 if in-line)
rand_lcg_seed:
```

Listing 11-1 A linear congruence generator (LCG) routine for the 68HC16.

---

5. Actually, this value for c was apparently picked using the relation $c=(1/2 - \text{sqrt}(3)/6)m$, which has some mild justification in the study of LCGs. However, Knuth suggests that though this probably doesn't hurt, any odd value would do just fine.

```
        std     lcg_x
        rts

rand_lcg:                       ;(10 for jsr)
        ldd     lcg_x           ;(6)
        lde     #LCG_A          ;(4)
        emul                    ;(10) E*D -> E:D
        addd    #LCG_C          ;(4) (2 cycles if C is 8-bit const)
        std     lcg_x           ;(6) lcg_x= (aX+c) mod 2^16
        rts                     ;(12)
```

Listing 11-1 A linear congruence generator (LCG) routine for the 68HC16.

The LCG routine given is a fairly compact source of long-period pseudo-random numbers. Note that the overhead of calling the routine is 22 cycles versus the 30 cycles of the actual routine! It might make sense to "in-line" the code if you need additional speed.

Unlike the feedback shift register of a prior chapter, you can start this algorithm off with a "seed" value of zero. Any value is acceptable; it will take 65536 calls before the same value is repeated.

The output of the routine is a full 16 bits. However, all LCG routines, and this one in particular, have least-significant bits that are much less "random" than higher bits. If you need fewer than 16 bits, you should take these bits starting from the MSB, not from the lowest bits.

Listing 11-2 shows the C language version of this same LCG. If both are started from the same seed, you'll get the same output sequence. The first ten numbers generated (in hexadecimal) are: 3619, be66, 79f7, 431c, 3665, efa2, cae3, 7978, 7af1, 4a1e.

```c
/* rndlcg
   Linear Congruential Method
   12/28/96 rce
*/

#include <stdio.h>
#include <stdlib.h>

#define LCG_A 25173L
#define LCG_C 13849L
#define LCG_M 65536L

static unsigned int lcg_x = 0;   /* default value */

unsigned int set_seed(unsigned int sd)
{
   return lcg_x = sd;
}

unsigned int randlcg(void)        /* returns a random unsigned integer */
{
   unsigned long x;

   x= (LCG_A * lcg_x) + LCG_C;
   lcg_x= (unsigned int) (x % LCG_M);
   return(lcg_x);
```

Listing 11-2 A linear congruence generator (LCG) routine in C.

```
}   /* randlcg() */

void main(void)
{
   int i;

   printf("LCG:\n");
   set_seed(0);
   for(i=0;i<10;i++)
      printf("%04x ",randlcg());
   printf("\n");
}   /* main() */

/* randlcg.c */
```

Listing 11-2 (Continued)A linear congruence generator (LCG) routine in C.

11.2.3.2 R250 and Relatives

The R250 algorithm is a more recent PRNG, but it has become quite popular in research communities where a large number of nonrepeating pseudo-random numbers must be generated efficiently. The period of this PRNG[6] is an astounding $2^{250}-1$, or 1.81×10^{75}. In addition, it uses only addition. Well, it uses only "modulo-two addition," better known to us programmers as exclusive-or ("XOR"). Thus, R250 is quickly calculated on almost any processor.

The only disadvantage of the R250 algorithm is its need for an array of 250 variables, each as wide as the width of the random numbers being generated. Also, we need to initialize this array in a special way, but the initialization isn't too horrendous.

The R250 algorithm works by XORing the data generated 250 calls ago with the data generated 147 calls ago. Because XOR is a bit-wise operation, in effect we are computing N feedback shift registers (FSRs—see Chapter 9 [Correlation]), each of length 250, in parallel. N is the number of bits we're operating on—which, in our example code, will be 16. So the period is just like that of an FSR, which is $2^{250}-1$. In fact, if we had continued the table of FSR taps, we would have an entry for length 250, with a tap at 147 (or equivalently, 103).[7]

With a "shift register" length of 250, we can't efficiently shift the actual data, so we instead use pointers. The major inefficiency of the R250 algorithm is the need to adjust these pointers. Because the array length is not a power of 2, we can't just mask the pointer to wraparound. (Recall modulo addressing.) Thus, on the 68HC16, the R250 code we wrote takes longer than the LCG, even though the LCG is doing multiplication and R250 requires only XOR.

Again, like the FSR, we can't start the algorithm with all zeros. Maier [1991] writes that it is necessary to initialize the 250 array elements such that, when viewed as a matrix, the array is

---

6. Maier [1991] gives a period of 2^{249}, but we think this is a typo. However, it's pretty unlikely that the disparity will make much difference in any application.

7. You are also free to construct your own R250-like algorithm with fewer elements. If the number of elements is a power of 2 (e.g., 128), you'll need more than one feedback tap. (See the discussion of the FSR generator.) However, the cost for using more than one feedback tap may be more than offset by the simplification of the pointer arithmetic.

linearly independent. This is accomplished by filling the array with more-or-less random numbers,[8] then for each *j* = 0 to *N*-1, picking a different number in the array, setting the *j*th bit to 1 and all the bits to the left to 0. Any *N* numbers can be chosen; Maier's implementation picks the fourth and every eleventh entry after that to modify; to remain consistent, our algorithms will do the same.

Implementing the R250 Algorithm on the 68HC16   Listing 11-3 shows our R250 routine for the 68HC16, along with some storage and initialization code.

```
...
*               Variables (in internal RAM)
                org     $0000           ;(in bank f, actually)
                                        ;Variables for R250
r_index ds.w    1                       ;index to buffer
r_buff  ds.w    250                     ;buffer (word wide)
r_cnt   ds.w    1                       ;misc counter
r_diag  ds.w    1                       ;init variable
r_zeros ds.w    1                       ;init variable
...
*               R250 Pseudo-Random Number Generator
*               Usage:
*               (Optional: Set up rand_lcg with call to rand_lcg_seed)
*               Must call rand_r250_init to initialize array
*               Then call rand_r250 to get 16-bit pseudo-random numbers
*               Uses Y,Z,D registers
*               80 cycles (58 if in-line)
*               (rand_r250_init uses Y,D,E registers)
*
*               Note that the indexing is in word increments!

rand_r250_init:
                                        ;Fill array with random numbers
                ldy     #$0000          ;(assumes yk pts to buffer RAM)
rr250i1:
                jsr     rand_lcg
                std     r_buff,y
                aiy     #$02
                cpy     #500            ;(250*2)
                blt     rr250i1
                                        ;now make numbers linearly independent
                                        ;(see text for discussion)
                ldy     #6              ; (3*2)
                ldaa    #16
                staa    r_cnt
                ldd     #$8000
                std     r_diag
                ldd     #$ffff
                std     r_zeros
rr250i2:
                ldd     r_buff,y
                andd    r_zeros
```

Listing 11-3   An R250 routine for the 68HC16.

8. You can use pseudo-random numbers from an LCG to initialize the array. This is the approach we take in the example code that follows.

```
            ord     r_diag
            std     r_buff,y
            aiy     #22             ; (11*2)
            lsrw    r_diag
            lsrw    r_zeros
            dec     r_cnt
            bne     rr250i2

            ldd     #$0000          ;index=0
            std     r_index
            rts

rand_r250:                          ;(10 for jsr)
            ldy     r_index         ;(6) y=index
            tyz                     ;(2) z=j
            aiz     #206            ;(4) (note—positive 16-bit value)
            cpz     #500            ;(4) j=(index+103) mod 250
            blt     rr2501          ;(2/6)
            aiz     #-500           ;(4) (note negative)
rr2501:
            ldd     r_buff,y        ;(6) new=buff(index) xor buff(j)
            eord    r_buff,z        ;(6)
            std     r_buff,y        ;(6) buff(index)=new
            aiy     #2              ;(2) index=(index+1) mod 250
            cpy     #500            ;(4)
            blt     rr2502          ;(2/6)
            ldy     #$0000          ;(4)
rr2502:
            sty     r_index         ;(6)
            rts                     ;(12)
```

Listing 11-3   (Continued) An R250 routine for the 68HC16.

The execution time is 80 cycles (58 without the "jsr/rts" overhead), compared with 30 cycles for the LCG routine earlier. It is debatable whether this additional time, code space, and RAM space is worth the longer period for noncritical applications, but in some applications, the R250 algorithm may be appropriate and useful.[9]

Relatives   Structurally, the R250 algorithm is similar to the "additive" algorithm discussed in Knuth [1981], but, as there is carry propagation to adjacent bits during addition, the additive algorithm also "mixes up" the separate FSRs that in the R250 algorithm are running in parallel. Knuth indicates this might be the best PRNG for practical purposes, the only problem being a lack of theory supporting its apparent randomness.

11.2.3.3 Randomizing Random Numbers

The field of PRNGs is not as well developed as some folks would like. Given the decade-long reign of RANDU (the infamous LCG with horrible properties), you can't blame them for a bit of hesitation about LCG or other PRNGs. Knuth and others have suggested a few methods for producing "super" random numbers using the outputs from two different PRNGs. One method

---

9. Complete 68HC16 source code that prints the first 1024 R250 numbers (via the serial port) as well as a C language implementation of the R250 algorithm are available electronically, as discussed in the Preface.

involves an array of length N that is filled with numbers from one PRNG. The second PRNG is used to select elements (at random, obviously) from the array. As an element is selected (and output), it is replaced by another random number from the first PRNG.

Again, this is probably more than you'll ever need to know about random numbers, but if you really do need some "serious" random numbers, be sure to do your homework. For the rest of us, the LCG is a good source of everyday random numbers on the 68HC16, and the R250 algorithm a heavy-duty source that may also be more efficient for microcontrollers without good multiplication.

11.2.4 Nonuniform Amplitudes and Frequencies

11.2.4.1   Normal (Gaussian) Distribution

After going through all kinds of pain to get uniformly distributed values from our PRNGs, we've got to tell you that much of the time in DSP we actually need numbers with a *Gaussian* or *normal distribution*. This means that instead of expecting the random numbers to have an equal probability of having any value, values at either extreme of the range are far less likely. For example, if we plot the number of occurrences of values in a sequence of pseudo-random numbers that has a normal distribution, we'll see a plot with a "bell" shape like Figure 11-1.

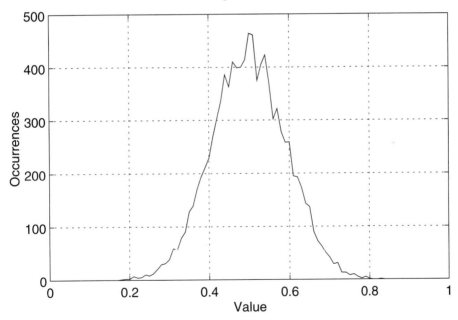

Figure 11-1   An example of a "normal" distribution. (This is actually an approximation—see text.)

Normal distributions are, well, pretty normal. Many naturally occurring physical processes—and that specifically includes most "noise" in the natural environment—have a normal distribution, and so we often need our PRNGs to have normally distributed values if we are seeking to simulate this noise.

There are several routes you can take to create a normal distribution from a uniform distribution. (See, for example, Press [1986].) Many involve computationally taxing operations like computing a logarithm or a square root, but a relatively "quick and dirty" method is based on the famous *central limit theorem*. If you add up independent random numbers of any distribution (all the same distribution, though), the distribution of that sum approaches a normal distribution. On a practical level, this means we can take a sufficiently large number of independent random numbers, sum them, and have a new random number whose value would occur with a probability governed by a normal distribution. In fact, we generated Figure 11-1 by adding 10 independent (pseudo-) random numbers, each with uniform distribution, to produce a single, "pseudo-normal" random number. We did this repeatedly, then plotted the number of occurrences of each value to produce the plot. For most noncritical applications, about 10 numbers should suffice; for critical applications, you may want to compare more complicated techniques or use additional numbers.

For *uniformly* distributed random numbers, the parameter of interest is the range (or equivalently, the mean or average). The *normal* distribution is characterized by a *mean* and a *variance* (σ^2), where the variance is a measure of how much variation or spread there is to the normal curve. The *standard deviation* is the square root of the variance, and might be given instead of the variance. (On a normal plot, the standard deviation is the length from the mean to the point on the curve where the curve changes from convex to concave.) To produce pseudo-random numbers with a normal distribution and a specific mean μ and variance σ^2, you can use the following procedure:

1. Generate N uniformly distributed independent random numbers in the range 0 to 1 from a PRNG. N is typically ≥ 10.

2. Sum the N generated numbers.

3. Subtract $N/2$ from the sum. (This step normalizes the mean to zero.)

4. Multiply the result by $\sqrt{12\sigma^2/N}$. (This step produces the desired variance. Be careful not to overflow the result.)

5. Add μ to the result. (This step produces the desired mean.)

This procedure produces a single number; sequences of such numbers will have the desired mean and variance.

As a special case, if 12 numbers are summed and the value 6 subtracted from this sum, the result has a distribution that is approximately normal with a mean of 0 and a variance (and standard deviation) of 1. (This is equivalent to setting $\mu=0$ and variance $\sigma^2=1$.)

11.2.4.2 Spectral Characteristics

What about the frequency spectrum of this "noise" we're generating? If necessary, the same IIR (or FIR) filters from earlier chapters can be applied to "shape" the magnitude frequency spectrum of a noise source.

Engineers have named some common noise types according to their frequency spectra[10]:

- white noise: constant power density over a frequency range.
- pink noise: power density decreases 3 dB/octave, which means each octave (power of 2) in frequency contains the same amount of power.

A recent "comp.dsp" newsgroup posting listed definitions for red, green, blue, grey, brown, and black(!). However, it's unlikely you'll need to worry about these exotic colors of noise.

11.3 Functions

We've mentioned any number of times that digital filters and much of DSP in general require nothing more than multiplication and addition. Yet, there are applications involving DSP where we must occasionally calculate a sine, evaluate a polynomial, or generate a complex waveform. These techniques fall into the field of "numerical methods."

11.3.1 Polynomials

Polynomials occur both as natural expressions of some functions and as approximations to functions. Because polynomials can be computed with just multiplication and addition, they are usually easy to evaluate, even on microcontrollers.

Polynomials in the form:

$$y = c_0 + c_1 x^1 + c_2 x^2 + \ldots + c_n x^n \tag{11.2}$$

at first look like a good fit to the RMAC operation of the 68HC16—that is, a sum of products. However, the powers of x must be calculated beforehand; possible, but a complication. (It is possible to calculate the powers of x as they are needed, but the register set of the 68HC16 is a bit too small to do this efficiently.)

The polynomial above can be rearranged using Horner's method so that multiplication is by x, not by higher powers. For example, a third-order polynomial can be arranged like so:

$$y = [(c_3 x + c_2)x + c_1]x + c_0 \tag{11.3}$$

---

10. Perhaps by analogy with the light spectrum.

The problem is that we no longer have a sum of products; the intermediate results are multiplied by x. On the 68HC16, this means either making do with 16-bit intermediate results, or resorting to multiple-precision math. We've taken the former approach below in one of our sine implementations.

If you really must calculate long polynomials quickly, Knuth [1981] and Press [1986] describe methods that involve "adapting" the coefficients; the trade-off is fewer multiplications for more additions. For example, a fifth-order polynomial can be calculated with four multiplications and five additions, and a sixth-order using four multiplications and seven additions. This optimization may be necessary on architectures where eliminating multiplication can have a major effect. The cost of multiplication is minimal on the 68HC16.

11.3.2 Sine, Cosine, and Some Others

Because the sine, cosine, and other trigonometric functions are so familiar, it may be easy to forget that they are *transcendental* functions—they can't be expressed as a finite number of additions, subtractions, multiplications, divisions, or raising quantities to a power. None of the elegant trigonometric identities shows us a polynomial with a finite length, nor will one ever be found. Instead, we must depend on approximations; the range over which the approximations are accurate leads us to several different implementations. In this section, we'll look at the sine function and how we can compute it with maximal speed, maximum accuracy, and minimal code size.

11.3.2.1 Polynomial Approximations

The sine function can be expressed as the following infinite power series expansion:

$$\sin(x) = \frac{x^1}{1} - \frac{x^3}{3!} + \frac{x^5}{5!} - \frac{x^7}{7!} + \dots \quad (11.4)$$

where x is in radians, and "!" registers not our surprise, but the factorial operation.[11] Because this polynomial skips even powers of x, we can fold the negative signs into the coefficients to end up with the same form as Equation 11.2.

Before you rush off to implement this, you must realize that this equation, although mathematically accurate, requires more and more terms to provide reasonable accuracy as you increase the value of x. Crenshaw [1992a] notes that two terms are needed when x is less than 15°, but seven terms when x approaches 180° (for 16 bits of accuracy).

We can avoid evaluating lengthy polynomials by restricting the range of the input. Realizing that $\sin(x)=-\sin(x-180°)$ for $x>180°$, and $\sin(x)=\sin(90°-x)$ for $90°<x<180°$, we find we need to evaluate only $\sin(x)$ for a range of 0–90° (i.e., 0 to $\pi/2$ radians). Because the same accuracy

---

11. To compute the factorial, just multiply each of the integers from 1 to N together. Thus, 4! is 4×3×2×1, or 24.

Functions

issues occur in the cosine approximations, we can do a bit better by using the sine approximation only over the range 0–45° and using the cosine for the range 45°–90°. (Note that in this range, sin(x)=cos(90°-x).)

You should expect by now that just truncating the infinite series (for example, using only a finite number of terms from Equation 11.4) isn't always the best thing to do. Recall that truncating didn't seem to work well for FIR filter coefficients or FFT windows. In fact, there are optimizations that can be done on the polynomial coefficients to minimize the error over the desired range. One method is based on the use of Chebychev polynomials, and the coefficients we'll be using in our example program have been so adjusted.

Using the above techniques, we need about three terms for sine approximation, and two terms (plus a constant) for cosine approximation. There's some added baggage in terms of figuring out the correct range (and negating the output if necessary), but we get decent accuracy (11 bits) using just the two polynomials.

Listing 11-4 shows the polynomial-based sine function in 68HC16 assembly language. Note that we also included a cosine function—at the expense of just one more instruction! (Recall that cos(x)=sin(x+π/2))

```
...
*               Constants
                                        ;cosine
C0      equ     $0fff                   ;0.999999842
C2      equ     $b117                   ;-4.931917315
C4      equ     $3ef6                   ;3.935127589
                                        ;sine
S1      equ     $3243                   ;3.141576918
S3      equ     $ad5a                   ;-5.165694407
S5      equ     $27c3                   ;2.485336730

*               Variables (in internal RAM)
                org     $0000           ;(in bank f, actually)
origx           ds.w    1               ;sine routine
x1              ds.w    1               ;sine
xsq             ds.w    1               ;sine
...
*               Sinep (Cosinep)
*               Usage:
*               Call with angle in pirads in D.
*               ($0000=0 degrees, $8000=180 degrees, $ffff=(almost) 360 degrees)
*               Returns with sine in 1.15 format in D.
*               Maximum error is +0, -16 counts
*               Verified against C program sinep.c.
*               Uses Chebychev-optimized coefficients from J. Crenshaw,
*               "Getting the Sines Right," Embedded Systems Programming,
*               February 1992, pgs 30-50.
*               132-152 cycles, depending on angle
*               (110-130 cycles w/o jsr/rts), +4 for cosine
*               Uses D,E registers
cosinep:
                addd    #$4000          ;(4) cos(x)=sin(x+ pi/2)
sinep:                                  ;(10 for jsr)
```

Listing 11-4 Polynomial-based sine (and cosine) routine for the 68HC16.

```
                            ;Determine quadrant & adjust if necessary
        std     origx       ;(6) see if should negate output
        andd    #$7fff      ;(4) remove hi bit
        cpd     #$4000      ;(4) if x > $4000, adjust x
        ble     sine1       ;(6/2)
        negd                ;(2)     x = -x
        andd    #$3fff      ;(4)     mask
sine1:
                            ;if in range 0..45 degrees, use sine eq.
                            ;else, use cosine(90-x) eq.
        cpd     #$2000      ;(4) if x < $2000, use sine eq
        bge     sine2       ;(6/2)
;Calc sin(x) = x*s1 + x^3*s3 + x^5*s5
;Use Horner's method to get sin(x) = ((x^2*s5 + s3)x^2 + s1)x
        std     x1          ;(6) save x
        tde                 ;(2) calc xsq=x*x
        fmuls               ;(8) E:D=1.31 result
        ste     xsq         ;(6)
                            ; x^2*s5
        ldd     #S5         ;(4) s5 is in 4.12 format
        fmuls               ;(8) E:D=4.28 format
                            ; x^2*s5 + s3
        adde    #S3         ;(4) 4.12 + 4.12 -> 4.12 format
                            ; (x^2*s5 + s3) * x^2
        ldd     xsq         ;(6) 1.15 format
        fmuls               ;(8) E:D=4.28 format
                            ; (x^2*s5 + s3) * x^2 + s1
        adde    #S1         ;(4) still 4.12 format
                            ; ((x^2*s5 + s3) * x^2 + s1) * x
        ldd     x1          ;(6) 1.15 format
        fmuls               ;(8) E:D=4.28 format
                            ;now left shift E:D 3 bits to 1.31
        asld                ;(2) into carry
        role                ;(2) into e
        asld                ;(2) into carry
        role                ;(2) into e
        asld                ;(2) into carry
        role                ;(2) into e
        ted                 ;(2) result is in 1.15 format
        bra     sine3       ;(6)
sine2:                      ;if range 45...90 degrees, use cosine eq.
;sin(x)=cos(90-x) over range x=45..90
;Calc cos(x)= C0 + x^2*C2 + X^4*C4
;Use Horner's to get (x^2*C4 + C2)x^2 + C0
        subd    #$4000      ;(4) calc 90 deg - x
        negd                ;(2) (actually, calc x-90 & negate)
        std     x1          ;(6) save x
        tde                 ;(2) calc xsq=x*x
        fmuls               ;(8) E:D=1.31 result
        ste     xsq         ;(6)
                            ; x^2*c4
        ldd     #C4         ;(4) C4 is in 4.12 format
        fmuls               ;(8) E:D=4.28 format
                            ; x^2*c4 + c2
```

Listing 11-4 (Continued) Polynomial-based sine (and cosine) routine for the 68HC16.

Functions

```
            adde    #C2         ;(4) 4.12 + 4.12 -> 4.12 format
                                ; (x^2*c4 + c2) * x^2
            ldd     xsq         ;(6) 1.15 format
            fmuls               ;(8) E:D=4.28 format
                                ; (x^2*c4 + c2) * x^2 + c0
            adde    #C0         ;(4) still 4.12 format
                                ;now left shift E 3 bits to 1.15
            asle                ;(2)
            asle                ;(2)
            asle                ;(2)
            ted                 ;(2) result is in 1.15 format
                                ;fall thru
sine3:
                                ;Negate if in quadrants III or IV
            tstw    origx       ;(6)
            bpl     sine4       ;(6/2)
            negd                ;(2)
sine4:
            rts                 ;(12)
```

Listing 11-4 (Continued) Polynomial-based sine (and cosine) routine for the 68HC16.

This function is much simpler than it might be due to our use of yet another angular measurement—the *pirad*. In this system, "$0000" is the angle 0°, "$8000" is 180°, and "$ffff" is just shy of 360°. The nice thing about using pirads is that you can add a constant to any angle and the proper wrapping automatically occurs. For example, adding 270° + 180° = 450° - 360° = 90°; in pirads the same math (using 16-bit math) is "$c000 + $8000 = $4000" (=90°). In addition, it's easy to determine what mappings apply to a given angle. You can interpret pirads as 1.15 numbers[12] that, multiplied by π, give you the angle in radians.

A few comments on the code: First, the sine and cosine approximations we've used are valid for positive and negative values of x, though we're using only positive values (for example, 0°–45°, rather than -45°–45°). You might be able to exploit this fact to speed up the routine a little. Next, note that we've used Horner's method to evaluate the polynomial. An alternative is to compute the powers of x and store them in an array, then use RMAC to create a sum of products with potentially more accuracy. Just be sure to use the proper multiply instruction (either MAC or FMULS), as x is in a 1.15 format, and simple multiplication will yield a result in 2.30 format. Fractional multiplication will perform the shift automatically, yielding an answer in 1.31 format.

Finally, in addition to using 1.15 format for the angular measure, 4.12 format is used for the coefficients (as the coefficients have a range larger than -1 to 0.999). Keeping track of the format of the results of multiplies, shifts, and adds can be tricky—we've added a lot of comments to track the format of the results in the code.

(Work through a few simple examples if you have trouble following all of the changes in formats. As a starting point, adds require similar formats and produce no change in format. The result of normal multiplication is found by adding the integer part of both formats and the frac-

---

12. Recall the notation 1.15 means that in the 16-bit number (add 1+15), there is an implied decimal point after the most significant bit. As the MSB is the sign bit, a 1.15 number can hold values between -1 and 0.9999695.

tional parts separately—thus, a number in 1.15 multiplied by a number in 4.12 format produces a result in 5.17 format. Fractional multiplication (supported on the 68HC16 as FMULS and the MAC/RMAC instructions) automatically applies a single left shift, so 1.15 by 4.12 would yield 4.28.)

11.3.2.2 Table-Based Approximations

The sine function has nice symmetries that are crucial to the polynomial-based routine above. We saw how reducing the range of the angle let us use fewer terms of the polynomial approximation. Can we take the idea of reducing the angle to an extreme? That is, could we use only first-order polynomials if we only look at very tiny ranges of angles?

Actually, the logical extreme is zero-th order polynomials—that is, just look up a constant based on the angle. In other words, a simple look-up table. In fact, that would make a lot of sense if you needed the sine of only a limited number of angles. (FFT routines often precompute the sine and cosine values they need). Most of the time, the number of angles is too large to justify storing each value in a table.

One step back brings us to first-order polynomials. Because a first-order polynomial describes a line, what we're doing is approximating the sine function (or any other function) with a series of straight lines. The only question is how many lines are sufficient to approximate the sine function with the accuracy we want. Crenshaw [August 1995] gives the following expression:

$$\varepsilon = \frac{\Delta x^2}{8} \qquad (11.5)$$

where ε is the error, and Δx is the increment, in radians, between table entries. If we restrict our table to the range 0–90° using the techniques from earlier, to get 15 bits of accuracy requires roughly 100 entries. We'll bump to this 128 entries (a power of two), so that we can use modulo arithmetic and shifting instead of messy multiplication and division.

Using the table of 128 entries, sin(x) can be computed as:

$$\sin(x) = s(i) + \frac{m(s(i+1) - s(i))}{step} \qquad (11.6)$$

where i is the index to the table s, $step$ is the increment between table entries, and m tells us where between the table values x actually falls.

To make the answer come out right and exploit the powers of 2 involved, we'll measure $step$ in terms of the number of counts between table entries. Thus, for our 128-value table, covering a range of "$0000" to "$3fff" pirads (0–90°), $step$ is 16384/128 or 128 counts per table entry. The table index i is just the integer part of $i/128$, or $x\!>\!>\!7$ (i.e., x right shifted by 7). And we'll

Functions

calculate $s(i+1)-s(i)$ in terms of the number of counts as well—more a question of interpretation than of anything we need to do. By the way, the step size and the divisor for the table index are both 128 in this example, but this is just coincidence.

Listing 11-5 shows the table-based sine routine (and again, cosine tossed in for free). We've shaved the execution time down to 100 cycles (vs. a minimum of 130 for the polynomial-based routine), and now the error is -2 counts—about 14 bits.

```
*         Sine (Cosine)
*         Usage:
*         Call with angle in pirads in D
*         ($0000=0 degrees, $8000=180 degrees, $ffff=(almost) 360 degrees)
*         Returns with sine in 1.15 format in D.
*         Maximum error is +0, -2 counts; verified against C program sine.c.
*         Uses a table with only entries from 0 to pi/2; maps
*         other quadrants to this range.
*         100-104 cycles (78-82 w/o jsr/rts), +4 for cosine
*         Uses D,E,X registers
*         Assumes XK points to ROM
cosine:
          addd     #$4000           ;(4)  cos(x)=sin(x+ pi/2)
sine:                                ;(10 for jsr)
                                     ;Determine quadrant & adjust if necessary
          std      origx            ;(6)  see if should negate output
          andd     #$7fff           ;(4)  remove hi bit
          cpd      #$4000           ;(4)  if x > $4000, adjust x
          ble      sine1            ;(6/2)
          negd                      ;(2)    x = -x
          andd     #$3fff           ;(4)    mask
sine1:
                                     ;Do table lookup (range is 0...$4000,
                                     ;or 0 to pi/2, *inclusive*
          tde                       ;(2)  calc index to table
; Either do 7 shifts (14 cycles), or use fact that D register is
; made of A & B registers.  Saves 6 cycles.
          asld                      ;(2)    i= x>>7
          tab                       ;(2)
          clra                      ;(2)
          asld                      ;(2)
          addd     #s_table         ;(4)
          xgdx                      ;(2)

                                     ;Calculate interpolation part first,
                                     ;y= ((x & $7f) * sine_dx[i]) >> 7
          ande     #$007f           ;(4)  e= x & $7f
          ldd      >(sine_dx-s_table),x ;(6)  d= sine_dx[i]
          emuls                     ;(8)  e:d = (x & $7f) * sine_dx[i]
; Now, could do 7 shifts (14 cycles), or, knowing that e is in range
; 0..127, and sine_dx is in range 0..402, the result will be only in
; register D, which is made of A & B.  The shift by 7 will ignore
; all but msb of B, so save it in carry bit.
          aslb                      ;(2)  get bit 7
          tab                       ;(2)  moves bits 8-15 to 0-7
          ldaa     #$00             ;(2)  clear upper byte (leave carry)
          rold                      ;(2)  rotate in bit 7
```

Listing 11-5 Table-based sine and cosine routine for the 68HC16.

```
                              ;Add the sine_x table value
        addd    >(sine_x-s_table),x  ;(6)

                              ;Negate if in quadrants III or IV
        tstw    origx         ;(6)
        bpl     sine2         ;(6/2)
        negd                  ;(2)
sine2:
        rts                   ;(12)
```

Listing 11-5 (Continued) Table-based sine and cosine routine for the 68HC16.

You'll note that we're using two tables here—*sine_x* and *sine_dx*. *Sine_dx* contains the precomputed differences $s(i+1)-s(i)$; you can explicitly compute these values instead for very little extra execution time and save 256 bytes of memory.

We also used some ugly optimizations to avoid some annoying multiple shifts. To shift by 7 bits using normal shift instructions would normally cost 14 cycles (two cycles per shift); however, the *D* register of the 68HC16 can also be manipulated as two separate registers, the *A* and *B* registers, using fewer cycles for the same result. Your own code will need to choose between clarity, flexibility, code size, and speed, as this code is very specific to the table and step size chosen.

There is nothing that says you can't pick a method in between the polynomial and table versions—using a second-order polynomial could dramatically reduce the number of table entries but still produce accurate results. To avoid the logic that maps all the angles into a restricted range, the table could be extended to cover all angles. It may take a number of tries to find the fastest method for the accuracy and code size goals you have.

By the way, all these bit manipulations and various numeric formats can get extremely confusing. Our approach was to first code the routines in C on a personal computer. The quick compile cycle, excellent debugging, and familiarity made it easy to get the routines up and running. We could also easily compare the results of our routines with the sine function of the compiler's library. After this, we could use the C language source as a guide in writing the assembly. Tricky bugs (the "ldaa #$00" in the table-driven sine function was originally a "clra" but wiped out the carry bit!) were easy to catch as we could step through our C routine and compare the intermediate results with the emulation system results. This approach is highly recommended!

11.3.2.3 Using IIR Filters

Another technique you can use to generate sinusoidal signals is to construct an IIR filter with its poles on the unit circle. Note that this isn't the same thing as a general-purpose sine function, but perhaps you need to generate only a sine wave. As you know by now, putting poles on the unit circle is not something that is normally done. Generally, it will be impossible for the filter to not either decay or grow exponentially. But perhaps you can live with a slight exponential decay; the IIR filter might be a compact solution. (You need to "excite" the filter with an impulse to get it started, but after that it needs no input.)

11.3.2.4 Other Functions

Because mathematicians and others have needed to perform numerical calculations for hundreds of years and could use only human brain power for most of that time, there's a long history of research into techniques for efficiently calculating various functions like the square root, logarithms, and so on. The texts we've mentioned so far in this chapter, such as Knuth [1981], Press [1986], and other texts on numerical methods, are good initial sources. For a practical, microprocessor-oriented approach, articles in such publications as *Embedded Systems Programming*, *EDN*, *Circuit Cellar Ink*, and several special interest publications of the IEEE are good sources.

However, you shouldn't feel compelled to code some complicated algorithm if you can use a table-based approach as we did earlier. Perhaps you won't have much symmetry to take advantage of, but as long as code space is not at a premium, the solution may be both computationally efficient and quick to write and debug. Also, don't fall into the trap of coding many generic functions (e.g., sine, cosine) if in fact your code uses some combination—say, the square root of the cosine of $4x$. If you don't need cosine as a separate function, perhaps you would benefit from looking at how to calculate the entire function most efficiently, not pieces. Again, your application will determine whether such a special function makes sense.

11.3.3 Arbitrary Waveforms

Complicated waveforms, such as those produced by a musical instrument, may be difficult to synthesize using sinusoids and filters; it may make more sense to sample one period of the waveform and just cycle through the wave table repeatedly to generate the signal. The problem with this time-domain method is how to change the pitch (fundamental period) of the waveform, as one would like to do to generate different notes, or with speech to create inflection. The interpolation method we used for the table-based sine routine (perhaps speeded up!) addresses the problem of sampling the table at noninteger points (otherwise, we'd be limited to speeding up or slowing down the playback by integer multiples). Human speech is trickier, as we'd like to change the pitch without changing the frequencies of resonances (formants) in the speech. However, some researchers have reported on algorithms that perform this pitch change acceptably using minimal computation.[13]

11.4 Summary

While truly "random" numbers can't be generated by algorithms on digital processors, many algorithms exist for generating pseudo-random number sequences with useful characteristics.

Linear congruence algorithms are particularly compact and well studied algorithms for generating pseudo-random number sequences. The sequence length before repeating is limited to 2^M-1 (where M is the number of bits used in the computations).

---

13. Research on speech synthesis using stored speech segments (called *diphones*) was done at the Applied Science and Engineering Laboratories, University of Delaware/A.I. du Pont Institute, though like many folks, they seemed to have moved toward DSP solutions recently.

The R250 algorithm is an extension of the feedback shift register pseudo-random bit sequence generator, generating an M-bit result using only the exclusive-or operation and some pointer arithmetic. The period is $2^{250}-1$.

Pseudo-random number generators typically produce sequences with a uniform distribution of values; however, other distributions, such as the normal or Gaussian distribution, are often more useful. Several methods exist for producing sequences with other distributions from a sequence with uniform distribution, including the simple method of adding about 10 uniformly-distributed numbers together to produce a number with normal distribution.

Polynomials can be evaluated efficiently using Horner's method; other methods allow one to trade additions for multiplications after "adapting" the coefficients.

The sine and cosine functions can be evaluated using polynomial approximations. Use of the pirad unit (vs. degrees or radians) simplifies sine and cosine routines.

Functions, including trigonometric functions, can also be quickly evaluated using table-based approximations. Tables can be viewed as zero-order polynomial approximations; tables may also be augmented with additional parameters to allow first- and higher-order approximations that allow a reduction in table size and/or an increase in function accuracy.

IIR filters can be used to generate sinusoids; such systems are marginally stable only when the arithmetic is ideal.

Resources

Knuth, Donald E., *The Art of Computer Programming, Vol. 2: Seminumerical Algorithms, 2nd edition.*, Reading, MA: Addison-Wesley, 1981.
All three volumes are worth having in your programming library at some point. Volume 1 is "fundamental algorithms," and Volume 3 is "sorting and searching." By the way, Knuth offers $2 to the first finder of any technical, typographical, or historical error in Volume 2. We were thinking of doing the same thing—have Knuth pay $2 for any error found in *this* book—but decided against it.

Embedded Systems Programming, ISSN 1040-3272, a monthly magazine with content of particular interest to, well, folks who do embedded systems programming. Plenty of good articles and useful advertisements. In particular, see the articles by Jack Crenshaw:
- "Getting the Sines Right," February 1992, pp. 30–50
- "Putting It Together," August 1995, pp. 9–24
- "Look It Up!" January 1995, pp. 13–24
- "More on Table Lookups," February 1995, pp. 19–28

Abramowitz, Milton and Stegun, Irene, editors, *Handbook of Mathematical Functions (with Formula, Graphs, and Mathematical Tables)*, New York: Dover, 1972.
If you ever need to know what an "Incomplete Beta" function looks like, or every trigonometric relation known to mathematicians, here's your book. If you don't know Dover Publications, they print a lot of excellent books (often "classics") at excellent prices. At about $23 for this (large format) 1000+ page book, it's quite a bargain for the mathematically inclined. (It may also be used to adjust monitors for optimum viewing angle in between consultations.)

Hamming, R. W., *Numerical Methods for Scientists and Engineers*, New York: Dover, 1973 (reprinted in 1986).

Another Dover "classic," this book's motto is "the purpose of computing is insight, not numbers." Polynomial approximations are but the tip of the iceberg. Yes, this is the Hamming of "Hamming window" fame. Was $15 just a year or so ago, another bargain for your library.

Press, William, Flannery, Brian, Teukolsky, Saul, and Vetterling, William, *Numerical Recipes, the Art of Scientific Computing*, New York: Cambridge University Press, 1986. (Also available as *Numerical Recipes in C*.)
Covers random numbers, polynomial evaluation, the FFT, sorting, and a lot of things you probably hope you never have to do. Widely cited, and source code is available on disk.

The C User's Journal (ISSN 0898-9788) is a monthly magazine that covers C programming. Past issues have included numerical methods, random numbers, signal processing, and embedded systems issues.

Dr. Dobb's Journal (ISSN 1044-789X), another monthly programmer's magazine, covers a wide range of programming issues. In particular, see:
Maier, W. L., "A *Fast* Pseudo Random Number Generator," May 1991, pp. 152–157.

A search of the World-Wide Web produced a couple of papers on pseudo-random number generation, including code for the R250 algorithm in C (based on Maier's code, it appears).

Howard, Scott, *Using the MC68HC16Z1 for Audio Tone Generation*, Motorola applications note AN1254/D, 1996.
Contains information on generating tones using tables, including for DTMF applications.

CHAPTER 12

Parting Words

or, a few of the bigger lies we told you

We could write volumes on what we didn't cover. But we want to say only a few words. Then we'll leave you in peace. Really. We're exhausted, too.

The two goals we have for this book divide into the immediate—be able to design and implement basic DSP on microcontrollers—and the longer term—be able to comfortably read more advanced texts for additional information when appropriate. But having a framework and language aren't all you'll need if you want to take this subject further. Ideally, you need a general idea of the lay of the land—a map. You might never visit some places, but knowing they're there could help you. Below is our "Signal Processing on $15 a Day" tour.

12.1 Signals and Linear Systems

Our treatment of signals was light on some important issues like causality and phase, just to name two. We generally ignored distinctions such as those between power signals and energy signals.[1] These omissions haven't hurt much so far, but without this background, it is difficult to rigorously justify the operations we've discussed in this book. Without careful attention to such details, we often run into theoretical conundra[2] in deeper pursuits.

We've also skipped some interesting topics having to do with linear systems. For example, state-variable representations are useful and compact ways of representing linear systems when

---

1. An energy signal is a signal with finite, nonzero energy, and a power signal has finite, nonzero power. An energy signal has zero power and so is not a power signal, and a power signal has infinite energy, so is not an energy signal. Just thought you'd like to know.

2. Conundrums.

one is interested in the "internal workings" of the system. As another example, the stability of systems can be analyzed with some methods that avoid explicitly finding the system poles. And so on. Some of these topics are not just mathematically interesting, but are central to efficient work with linear systems on a day-to-day basis (similar to "shortcuts" like using the Laplace transform to convert difficult differential equations into easier algebraic problems).

The subject of discrete-time signals and systems likewise has some jewels we haven't shown you, many of which depend on a more mathematical presentation. With the mathematics comes connections that tie together some of the subjects that in this book appear separate and unrelated. The idea of convolution, for example, can be very useful in explaining the effects of sampling, but to go that route requires quite a bit more foundation than we've studied here.

Signals and linear systems, in both the continuous and discrete-time domains, usually are studied early in an electrical engineering curriculum as these concepts are used not only in DSP but also in communications, power, control, and other areas. (Most texts won't be at all hesitant about using calculus and will probably assume some basic electronic circuit knowledge.) We've listed a few of these books at the end of prior chapters; almost all books with "signals" and "linear systems" in their titles will cover the same ideas.

And, by the way, the Gibbs Phenomenon wasn't named after Barry Gibbs of the Bee-Gees—it was his brother Robin.

12.2 Math

12.2.1 Transforms

We purposefully avoided including lots of transform equations in this book, but there's some basic understanding to be gained by having at least some familiarity with common transforms and their inverses, even if this means evaluating the occasional integral.

The properties of transforms give us important clues about signal processing. For example, we've mentioned—and used!—an important property of the Fourier transform relating convolution in the time domain with multiplication in the frequency domain. A half-dozen other properties of the Fourier transform, along with similar properties for other transforms, not only provide a justification for much of what we've done in this book, but also lead to other operations we haven't mentioned at all, like modulation.

Though the number of transforms we've mentioned in this book might surprise you,[3] there are still other transforms. For example, the *chirp-z* transform, a variation on the "plain" z-transform, is, like its plain cousin, related to the discrete Fourier transform, but instead of being evaluated on the unit circle in the z-plane, it is evaluated along a spiral. The chirp-z transform is useful for certain signal detection problems.

3. The Fourier transform, the discrete-time Fourier transform, the discrete Fourier transform, the Laplace transform, and the z-transform—not to mention the fast Fourier transform, the discrete cosine transform, and the discrete Walsh transform. And of course, all the inverse transforms.

The presentation of transforms, whether common or exotic, may seem overly abstract in many texts. Without this rigor,[4] however, it's harder to see the interrelationships that are present. (For example, some guy named Deller wrote an article you might like to check out on the relationship between the DFT and FT.[5]) Our hand waving here has taken us a certain distance, but it's likely some study of transforms would be quite helpful if you want to take your understanding further.

12.2.2 Probability and Stochastic Processes

Correlation and random numbers brought us in only brief contact with *stochastic*[6] *processes*, yet the study of signals and systems described by probabilistic measures is central to optimal filter design and problems in signal detection and estimation. For many simple DSP applications, it may not be necessary to delve into the theory of random processes; like so many topics we've mentioned, though, there is always the risk that not only are connections never seen, but efficient design might not be possible without this added set of tools.

12.3 Processing Multidimensional signals

We know we said it before, but we'd like to remind you that not all signals are a function of time, and two-dimensional signals that are a function of space (read "images" or "pictures") can be filtered, analyzed, and/or compressed with many of the same tools (or close relatives) as we've discussed in this book. For example, two-dimensional FIR filters and FFTs are relatively straightforward extensions of the one-dimensional versions we've seen. A close relative of the FFT, the discrete cosine transform, is useful, as we've noted, in compressing images.

Applications for DSP in image processing range from the artistic to the medical, and even include the synthesis of holograms. Whether you download a compressed "movie" over the Internet, or have your dental x-rays stored digitally, the same "sum of products" and other DSP building blocks are at work.

The advantages of processing an image using digital techniques continue to increase as the cost, size, and power consumption of digital processors drop. Chemical-based photography may, similar to its audio cousin the LP record, find itself phased out as the ability to digitally capture and process images catches up to and surpasses the "analog" techniques. (A number of texts address digital image processing specifically, which also involves nonlinear operations and decidedly nonfilter-like algorithms in addition to what should be now familiar to you as FIR filters.)

---

4. At least one of us fights off the temptation to append *mortis* here.

5. Deller, J. R. Jr., "Tom, Dick, and Mary Discover the DFT," *IEEE Signal Processing Magazine*, April 1994, pp. 36–50.

6. "Stochastic" is a $50 word for "random," sure to make an impression at your next social engagement.

12.4 Processing Music

We can't think of too many outright lies we told you about processing music, but we do admit to omitting virtually any mention of the audio effects you can create with DSP. Adding in an attenuated and delayed version of your input signal can create an echo, for example. Other more complex effects, like reverberation, "flanging," "chorusing," and so on require a bit more work but are just simple filters at heart, with some time-varying coefficients and/or time-varying delays. Orfanidis [1996] describes how several digital audio effects can be implemented in DSP, and the topic of digital audio effects is a common thread in the *comp.dsp* newsgroup. For the adventurous, you might even consider how DSP can be applied to recreate the (subtle?) qualities of vacuum-tube audio amplifiers versus solid-state. (Note that the operations will not be linear.)[7]

The pre-DSP implementations of some effects might add to your appreciation for DSP. Dale recalls assembling an amplifier that used a mechanism based on a metal spring to provide reverberation—shades of early delay-line-based computer storage. "Flanging" used to be done by applying friction to the flange of a reel on a reel-to-reel tape recorder.

Not many of us have a "flanging" button on our stereo receivers, but the chances are greater every year that a new receiver will include a DSP processor. Even inexpensive CD players have digital filters in them. The introduction of digital FM radio, or the high-quality audio of digital satellite TV, ensure that whether it's the Academy of St. Martin in the Fields or The Doors, it will eventually become more likely than not that an FIR or IIR filter will have had its hands on your music.

One point you might recall is that getting a good overall SNR requires good input and output design; high-quality music is demanding this way. Low-noise design is an art unto itself and probably a bit ambitious if you don't have the fundamentals of electronics down pat. However, sound cards for PCs provide an inexpensive but often good-quality way to input and output sound; provided you know the format of the files created (or the programming interface to the card itself), you can perform signal processing operations using your choice of languages or environments.

12.5 Processing Speech

We hope you've picked up on the fact that speech processing is of some interest to the authors! What can we say but that the discrete-time processing of speech signals is growing only more important as a variety of products shift to using digital means for communication and storage of speech. The advantages are great and include improvements in use of bandwidth, better SNR, ability to digitally encrypt the signal, and ability to store and process speech on computers.

Although there's some overlap, there are some distinct fields within the speech processing area. Speech *synthesis* (the best field, in Dale's humble opinion), is the best field. Many different

---

7. Though you might want to pursue this topic cautiously and with tact! The topic of the qualities (and desirability) of vacuum-tube amplification is perhaps the audiophile equivalent of "what's the best text editor" argument among programmers. (By the way, it's TECO, followed by Brief.)

techniques exist for synthesizing speech, depending on whether you start from actual speech (the "concentrated orange juice" model) or a simplified mathematical model of human speech (the artificial orange-flavored drink model[8]). LPC, IIR and FIR filters, and signal synthesis techniques are used in various forms of speech synthesis, as well as high-level symbolic processing that maps English text to the basic sounds the synthesizer can produce ("phonemes").

By the way, another true story. The late Dr. Dennis Klatt of M.I.T. did a lot of work on speech processing, especially on speech synthesis using "formant synthesizers" (an "artificially flavored" type of synthesis). This is the type of synthesizer upon which some of the world's best speech synthesizers are based, including DECTalk™. (Klatt also was personally involved in developing DECTalk™.) Although the underlying model for a formant synthesizer is idealized, the data driving the model are based on the observed formant (pole!) positions and movements in human speech. And for some of Klatt's research, the most convenient source of speech to analyze was his own voice. A fellow researcher of Klatt's described attending a lecture by Professor Stephen Hawking (*A Brief History of Time*) some time after Klatt's death. Professor Hawking uses a speech synthesizer to communicate—a DECTalk™, as it happens. And what voice of the several DECTalk™ voices had Hawking chosen? The voice based, apparently, most closely on Klatt's own voice, a somewhat eerie situation for Klatt's friend![9]

The speech *coding* field concerns itself with finding compact descriptions of speech signals; digital cellular phones, digital answering machines, and many other devices that store or transmit speech, benefit from the compression offered by techniques like LPC. Each application has its own set of requirements. Speech coding methods for use on cellular phones need to preserve intelligibility and a fair degree of the speaker's unique speech qualities, but, for example, methods used for communication between submerged submarines might not have that much bandwidth and, instead, would encode the speech using methods that preserve fewer of the original nuances. (Getting below 2,400 bits per second for telephone-quality speech is challenging; recall that the phone system uses on the order of 64,000 bits per second.)

Another field, speech *recognition*, also uses some DSP techniques we've discussed in this book, like autocorrelation, spectral analysis, and LPC analysis. However, this nut is a lot harder to crack, and techniques like "hidden Markov" modeling, "dynamic time warping," and "throwing raw fish heads out the window"[10] are also necessary, if not pleasant. (Many of these additional techniques are concerned with matching patterns that have variations in timing.)

Speaker recognition, where we're concerned with *who* is saying something rather than *what* they're saying, is another field related to speech synthesis and recognition. So, too (with

---

8. This analogy is from Dr. John Eulenberg of Michigan State University's Artificial Language Laboratory. That lab uses both "100% orange juice (from concentrate)" and "artificial orange-flavored" speech synthesizers to help disabled people communicate.

9. We should add that from comments in Klatt's writings it appears he felt that technology, including speech synthesis, should be used to help people with disabilities; that he should give voice, in a literal sense, to one of the great physicists of our times is quite fitting.

10. It works for us!

overlap), with the field of *speaker identification*, where we wish to verify the speaker's claim of identity. We can even seek to determine what language (*language identification*) a speaker is using—think about the calls that come into a long-distance operator!

Speech processing is a fascinating field[11] in part because it spans such a wide range of processing "levels." At the "low" end are the cut-and-dried techniques (e.g., IIR filters), but in many cases we build levels on top of those, and then levels on top of those levels. Problems like determining the proper inflection for a phrase, or picking the intended homonym ("hear" or "here"?), although not DSP, are very interesting challenges. Of course, implementing even the low-level DSP routines for speech processing has its own difficulties, especially in processors with limited bandwidth.

12.6 Numerical Methods

We chose just a few areas of numerical methods to discuss—random number generating and evaluating functions like sine and cosine—as being some generally useful and interesting techniques. The field of numerical methods is much larger than this, and some techniques may be quite useful to you in applications that employ DSP techniques. For example, many techniques have been developed for the evaluation of differential equations. Likewise, we commented that even the lowly polynomial can be evaluated in a more efficient manner than even Horner's Rule suggests. The field of numerical methods also includes efficient methods for computing matrix operations (recall the least-square error and LPC discussions), finding optimum solutions to complicated functions of many variables, sorting algorithms, root-finding, and so on.

Whether you need to implement some function or take an existing function and increase its efficiency, a text or two on numerical methods may be a useful addition to your library.

12.7 Embedded Systems

Finally, a few words about embedded systems, and, in particular, the areas of program structure and operating systems.

Although we've offered a few suggestions as far as programming your microcontroller for DSP tasks, the proper design and coding of an embedded systems application involves a number of skills. We've discussed some examples of assembly language DSP code, but in each case for every design decision noted, there are a half dozen decisions that we've ignored—in part, because we don't even see them any more. Your own code, no doubt, embodies a slightly different set of "invisible" design decisions from our own, based on experience, bias, or habit (good or bad!). As a concrete example, consider the structure we used to illustrate interpolation—a single, fast interrupt was used for both the (slow) input sampling rate (via a decrementing counter) and the (fast) output sampling rate. Many other structures could have been used instead, each leading to a different cost in terms of code size, execution speed, variation in sampling period, and so on.

11. It is!

Perhaps two externally generated interrupts would make more sense in one application and a mixture of polling and interrupts in another.

A thoughtful programmer once wrote of the process of structuring programs as one of choosing which routine to make the "driver" or main routine. If you picture the program as a net of routines, each connected by string to the routines it calls or is called by, this is like picking up one of the routines and watching how the other routines hang down below it. Applying that question to our DSP tasks, for example, we might ask whether our filter routine runs continuously, processing data whenever it is available, or instead if the filter routine is called only when needed by some other routine. Which routine "drives" the others?

One stage above this "program-level" structure is the issue of the operating system, which we've completely ignored. With the real-time nature of many embedded systems and the growing complexity of the tasks they are expected to perform, formal (commercial) real-time operating systems (vs. "roll-your-own") are common in even small microcontroller systems. This is yet another complication for the DSP programmer if there are issues with variation in interrupt latency, the ability to interrupt repeating MAC instructions,[12] use of coprocessor registers, and so on.

The literature on microcontroller programming is not as extensive as that for other types of programming, but a few new books appear each year. As long as you can handle the inevitable "continued next month" gamble of magazine articles, electronic and programming magazines can also be a good source of often practical embedded systems knowledge. You can also find book reviews relevant to embedded systems programming in most of these magazines. (By the way, many programming and technical magazines are now offering CD-ROM collections that span several years of issues—including program listings.) We've mentioned some of our favorite magazines throughout this book, but you might also find DSP/embedded systems content in magazines specific to a field of interest—for example, amateur radio operators have been using DSP for some years, and several "ham" magazines have run build-it-yourself articles on DSP-based projects.

12.8 Summary

Four legs good, two legs bad.[13]

> and

Go do something good.

---

12. By the way, the 68HC16 does allow an interruption in the middle of a RMAC loop. Just thought you'd want to know.

13. Aha, so you didn't read the note about notation in Chapter 2!

APPENDIX 1

Useful Mathematics

A1.1 Trigonometry

A brief summary:

$$\sin\theta = \frac{y}{r} \qquad \cos\theta = \frac{x}{r} \qquad \tan\theta = \frac{y}{x}$$

$$\csc\theta = \frac{r}{y} \qquad \sec\theta = \frac{r}{x} \qquad \cot\theta = \frac{x}{y}$$

$360° = 2\pi$ radians
$1° = \pi/180$ radians ≈ 0.01745 radians
1 radian $= 180°/\pi \approx 57.296°$

A few trigonometric identities:

$$\tan\theta = \frac{\sin\theta}{\cos\theta} \qquad \sin^2\theta + \cos^2\theta = 1$$

$$\sin(-\theta) = -\sin\theta \qquad \cos(-\theta) = \cos\theta$$

$$\cos\left(\frac{\pi}{2} - \theta\right) = \sin\theta \qquad \sin\left(\frac{\pi}{2} - \theta\right) = \cos\theta$$

Sine and cosine in terms of e^{jx}:

$$\sin x = \frac{e^{jx} - e^{-jx}}{2j} \qquad \cos x = \frac{e^{jx} + e^{-jx}}{2}$$

$$e^{\pm jx} = \cos x \pm j \sin x \quad \text{(this is Euler's theorem[1])}$$

A1.2 Complex Numbers

A1.2.1 Rectangular and Polar Forms

We've used complex numbers in this book in several ways, namely:
- to express magnitude and phase, as when we evaluate a system function $H(s)$;
- to express complex frequency in the s-plane;
- to express complex frequency in the z-plane.

(Needless to say, complex numbers are used in many other situations, as well, throughout engineering and mathematics.)

Complex numbers take the real number line, Figure A1-1, and add an additional dimension or axis called the imaginary or j axis.[2] (See Figure A1-2.) A real number can be thought of as indicating a point along a single axis (the "real" axis); a complex number defines a point in a plane, one axis of which is the real axis, the other axis the imaginary axis.

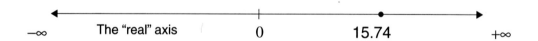

Figure A1-1 The real number line.

When a complex number is expressed as a real and imaginary pair, it is being expressed in rectangular form. For the example, we can write the number shown in the figure as either (15.74, 10.81), or the more usual form of $15.74 + 10.81j$. The rectangular (or *cartesian*) form is useful when we deal with the s-plane, because the real part of the complex number corresponds to exponential growth or decay in a signal, as the imaginary part corresponds to the frequency of that signal.

Just as we can express real numbers as a distance from 0 (the magnitude) and a direction (0°, the positive direction, or 180°, the negative direction), we can also express complex numbers

---

1. "Euler" is pronounced like "Houston Oiler(s)," but without the "Houston."

2. As we note elsewhere, electrical engineers use the letter j instead of i for $\sqrt{-1}$ because i is reserved for current.

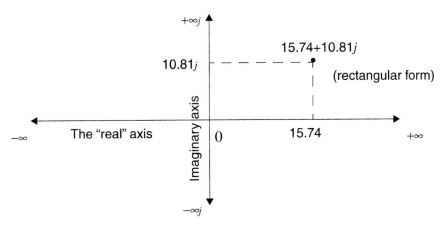

Figure A1-2 The complex number plane.

as a distance from the origin (magnitude) and a direction or angle. (See Figure A1-3.) Here, the distance is 19.09, and the angle is 34.48° or 0.60 radians. This is known as the *polar form*; in this book, we use the polar form when we interpret the magnitude/phase information from a system function and the complex frequency associated with a location in the z-plane.

Because the rectangular and polar forms are mathematically equivalent (they describe the same point in the plane), we are free to change between the forms as appropriate. As we'll see below, certain mathematical operations on complex numbers are easier in one form or the other.

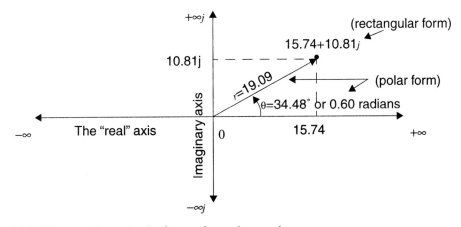

Figure A1-3 Rectangular and polar forms of complex numbers.

To convert between rectangular and polar forms of a complex number $p = a+bj = r\angle\theta$, use the equations below.[3]

Rectangular to polar form:

$$\theta = \angle p = \tan^{-1}\left(\frac{\text{Im}(p)}{\text{Re}(p)}\right) = \tan^{-1}\left(\frac{b}{a}\right) \quad \text{(warning--be careful using a calculator!)} \quad (A1.1)$$

$$r = |p| = \sqrt{(\text{Re}(p))^2 + (\text{Im}(p))^2} = \sqrt{a^2 + b^2}$$

where $Re(p)=a$ (i.e., the real part of p), and $Im(p)=b$ (i.e., the imaginary part of p). Note that the imaginary part of p does *not* include the "j."

Polar to rectangular form:

$$a = Re(p) = r\cos\theta \qquad b = Im(p) = r\sin\theta \qquad (A1.2)$$

Note that θ may be in either radians or degrees; further, you need to be sure you have the correct quadrant when you determine θ using the arctangent function ($\tan^{-1}$) on b/a, otherwise your angle could end up completely wrong. Calculators, as they take the arctangent of a single argument (b/a), have no way of knowing the proper quadrant.

Using Euler's identity, $e^{j\theta} = \cos\theta + j\sin\theta$, we can also express the polar form using a compact exponential form. Using Equation A1-2 and Euler's identity,

$$a + bj = r\cos\theta + jr\sin\theta = r(\cos\theta + j\sin\theta) = re^{j\theta} \qquad (A1.3)$$

which is admittedly a bit unfriendly if you're not comfortable with exponentials. However, it is very common notation in electrical engineering and digital signal processing. (Note that θ is usually in radians in this notation!)

A1.2.2 Arithmetic With Complex Numbers

Addition and subtraction with complex numbers is straightforward. Express both values in rectangular form, then separately add (subtract) the real and imaginary components. For example, we add $3-7j$ and $-10+1j$ to get $(3-10)+(-7+1)j$, or $7-6j$.

Multiplication and division are easiest with both values expressed in polar form. To multiply two complex numbers, multiply the two magnitudes together and add the angles. To divide, divide the magnitudes but subtract (not add) the angles. For example, $6e^{0.5j}$ multiplied by $2e^{0.2j}$ is $12e^{0.7j}$ (i.e., $(6\cdot2)e^{(0.5+0.2)j}$), and $6e^{0.5j}$ divided by $2e^{0.2j}$ is $3e^{0.3j}$ (i.e., $(6/2)e^{(0.5-0.2)j}$). If you

---

3. A bit of sloppy notation here—the "angle" symbol $\angle$ is being used to stand for both a function ("take the angle of the argument") and as a value ("what follows is an angle").

prefer, ($r=6$, $\theta=0.5$ radians) multiplied by ($r=2$, $\theta=0.2$ radians) is ($r=12$, $\theta=0.7$ radians), and dividing the two numbers produces ($r=3$, $\theta=0.3$ radians).

It's also possible to do multiplication of complex numbers using the rectangular form—just treat the j like a variable and multiply out like you would two binomials. For example, $(3+4j)\cdot(5+6j) = (15 + 20j + 18j + 24j^2)$. Just remember that j^2 is -1, and the answer is $-9+38j$.

The normal rules for exponentials apply to the compact exponential polar form for complex numbers. So, for example $(4e^{0.3j})^4$ is easily found as $4^4 e^{4(0.3)j}$ or $256 e^{1.2j}$.

A1.3 Transforms

Here is a summary of some of the transforms we've met in this book (and a few more to round things out):

The **Fourier series** (FS)[4]: The Fourier series is a revealing representation of a periodic (i.e., repeating), infinitely long, continuous signal. The signal is expanded on a set of sines and cosines at its harmonic frequencies by computing a set of amplitudes and phases at discrete frequencies (harmonics of the fundamental frequency of the signal). The FS gives us insight into the spectral content of periodic signals like squarewaves or other functions that can be represented as a mathematical equation, and we've used it in this book to plot the first few frequency components of the squarewave. The FS operates on a continuous signal whose period must be known, and produces a (potentially) infinite list of frequency components. (It is often the case that the harmonics of a signal die out quickly.)

The **Fourier transform** (FT): The Fourier transform takes continuous, infinitely long signals as input, just as the FS. However, there is no requirement that the input signal be periodic. (In fact, the input cannot be periodic, in theory.) However, the output is no longer magnitudes and phases of discrete frequencies (harmonics of the fundamental)—the output is a continuous function that describes an *amplitude density*. Kind of like a map that tells you that there are "4.5 people per square mile," the FT output tells you that if you take a range of frequencies and integrate the amplitude density over that range, you'll end up with the "average" amplitude over that frequency range. And, just like the map, if you take a small enough space (i.e., frequency range), you'll find fewer and fewer people (i.e., lower amplitude), until, at just a single point, even in the mostly densely populated areas, you won't find anyone. Again, because of the continuous input and also the continuous output, this isn't a transform we'll implement on a microprocessor, but it is important in signal processing. The FT, by the way, can be viewed as a restricted version of the Laplace Transform, which gave us the s-plane.

The **discrete-time Fourier transform** (DTFT): The first two transforms, the Fourier series and the Fourier transform, have taken continuous inputs. The discrete-time Fourier transform takes discrete-time samples as input, which is a little closer to what we want, but the output, like the FT, is a continuous amplitude density—we can't use the DTFT directly for

---

4. The label *transform* might not be appropriate here, but the FS is such a fundamental tool that we wanted to include it in this list.

microcontrollers. Because the DTFT uses a finite input length, we have to deal with the problem of discontinuities at the beginning and end of the sample sequence, but the solution to this—windowing—raises some issues like resolution and spectral leakage.

The **discrete Fourier transform** (DFT): Actually, we should call this the *discrete-time, discrete-frequency Fourier transform*. As suggested by our somewhat longer name for it, the DFT performs a transform between discrete-time and discrete-frequency representations. The DFT is a sampled version of the DTFT. However, the DFT acts as though the input signal represents a single period of an infinitely long periodic signal so, just like the FS, the outputs of the DFT are magnitudes of frequency components, *not* amplitude density. In using the DFT, we need to keep the resolution of the "sampling" in mind, as well as the resolution and spectral leakage issues we inherit from the DTFT.

The **fast Fourier transform** (FFT): This is a class of computationally efficient implementations of the DFT; they all, in principle, exactly compute the DFT; they just do it more efficiently than using the "basic definition" of the DFT.

Table A1-1 summarizes the key points of these different transforms. You don't have to memorize this; we just want to point out how each transform is specific to a certain type of input and output.

Appendix 1 • Useful Mathematics

		FS—Fourier series	FT—Fourier transform	DTFT—discrete-time Fourier transform	DFT—discrete Fourier transform
Input	Discrete- or continuous-time	Continuous	Continuous	Discrete	Discrete
	Input length	Finite (one period of a periodic [infinite] signal)	Infinite (need not be periodic)	Finite	Finite (transform "assumes" one period of periodic [infinite] sequence)
Output	Discrete- or continuous-frequency	Discrete	Continuous	Continuous	Discrete
	Periodicity	Nonperiodic	Nonperiodic	Periodic	Periodic
	Output length	Infinite	Infinite	One period	One period
Interpretation		Discrete magnitude and phase spectra	Continuous amplitude density, phase spectra	Continuous amplitude density, phase spectra	Discrete magnitude, phase spectra
Common uses		Harmonic analysis of waveforms; basic for DFT	"Paper and pencil" analysis of CT signals	"Paper and pencil" analysis of DT signals	Analysis/transformation of DT signals

Table A1-1

APPENDIX 2

Useful Electronics

Some familiarity with complex numbers will be useful for the AC electronics formulas. Other than that, simple algebra is the rule of the day.

A2.1 DC

This appendix contains a brief review of some of the important relationships in electronics. We begin with some basic relationships for DC circuits.

A2.1.1 Resistance, Ohm's Law

Ohm's Law describes the relationship between current (I), voltage (V), and resistance (R) in an electrical circuit. It is simply:

$$V = IR \qquad (A2.1)$$

Voltage is measured in volts (V), current in amps (A), and resistance in ohms (Ω).

For example, we might want to add an LED[1] to a +5 V power supply to indicate that it is on. We need to limit the amount of current flowing through the LED to about 10 mA (0.01 A) or the LED will burn out. The voltage drop across an LED is about 2 V (it actually depends on the

---

1. A light-emitting diode; for example, the small, usually red or green solid-state lights that nearly every appliance now sports to let us know they're on, off, or unhappy, are LEDs.

color and ranges from about 1.5 to 2.1 V); we need a resistor that will have 5-2, or 3 V across it when 10 mA flows through it. $R=V/I$, so $R=3/0.01$, or 300 ohms.

A2.1.2 Power

Power (P) is measured in watts (W); integrated over time, this gives us energy, measured in joules (J). Power is calculated from:

$$P=VI \qquad (A2.2)$$

or

$$P=I^2R \text{ or } V^2/R \qquad (A2.3)$$

(Either of the latter equations can be derived from the first, using Ohm's Law.)

A2.1.3 Simple Resistor, Capacitor, and Inductor Circuits

First, a quick review of just what resistors, capacitors, and inductors are. Resistors, first of all, are energy-dissipating devices. They are used to limit the flow of current or in conjunction with other resistors to divide down a voltage (more on this in a moment).

Capacitors do not dissipate energy—they store energy in an electric field and are happy to give it back whenever we like. Capacitors are used most often to
- provide a local reservoir of power for circuits that have "spiky" power needs (these are the famous *decoupling* capacitors that are sprinkled liberally on every well-designed circuit-board);
- pass only AC signals between circuits, blocking any DC signal;
- produce "RC" filters (in conjunction with resistors).

Physically, capacitors are made by manufacturing two conductive surfaces in very close proximity to each other, but separated by air or some other nonconductive material (usually a plastic).

Inductors also do not dissipate energy (ideally, that is)—they store energy in a magnetic field. Physically, they are made by coiling wire. (Transformers are just two inductors whose magnetic fields are coupled.) Inductors find much less use than resistors and capacitors at low frequencies, but it is very common to see cables or wires passing through ferrite beads, forming a simple inductor to get rid of high-frequency noise (emitted or received).

All three of these devices are—surprise—linear (in theory!); there are many other electronic devices that are nonlinear, including diodes, transistors of all types, and so on.

A2.1.3.1 Series and Parallel Connections

Resistors in series add their resistance together; the equivalent resistance of resistors in parallel is the reciprocal of the sum of the reciprocals of the resistances (got that?). (See Figure A2-1.) If we placed a 1-ohm, a 2-ohm, and a 3-ohm resistor in series, the total resistance would be 6 ohms (1+2+3). Connected in parallel, the resistance would be 1/(1/1+1/2+1/3), or 6/11 ohms (0.55 ohms).

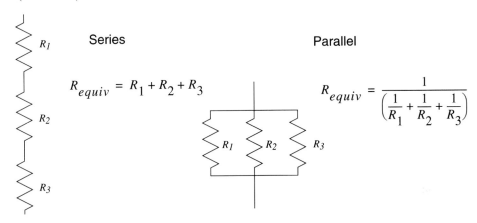

Figure A2-1 Series and parallel connections.

Inductors act just like resistors—in series, the inductances add, and in parallel, the inductance is the reciprocal of the sum of the reciprocals.

Capacitors, on the other hand, act the opposite. The capacitance of capacitors in parallel add, and the capacitance in series is the reciprocal of the sum of the reciprocals of the capacitances.

A2.1.3.2 Voltage Divider

A common circuit is the voltage divider, consisting of two resistors in series as shown in Figure A2-2. A voltage present across the divider will be attenuated based on the two resistor values. The equation is:

$$V_{out} = \frac{R_2}{R_1 + R_2} V_{in} \quad (A2.4)$$

In a moment, we'll replace a resistor with a capacitor to get a frequency-dependent voltage divider (an RC filter).

A2.1.4 Kirchhoff's Voltage/Current Laws

Two laws that are helpful in analyzing circuits are Kirchhoff's Voltage and Kirchhoff's Current

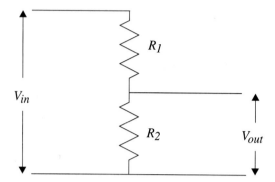

Figure A2-2 Voltage divider.

Law.[2] Kirchhoff's Voltage Law states that the sum of all the "voltage drops" around a closed circuit must be equal to zero. For example, if a 9-volt battery is connected to a resistor and LED in a closed loop, the voltage drop across the LED and resistor must be equal to 9 volts. (The voltage drop across the 9-volt battery is -9 volts, so the sum is zero.) In our earlier power LED example, we knew the resistor must be dropping 3 volts, as the LED drops a fixed 2 volts when properly operating, and we had to have a total voltage drop of 5 volts.

Kirchhoff's Current Law states that the sum of all the currents entering a point in a circuit must be zero. This region can be a wire, a connection, a group of components, or an entire system. In the case of simple regions with only one current flowing in and one out (the "flowing out" current is treated as a negative "flowing in" current), this is not a very interesting statement, but Kirchhoff's Current Law applies to arbitrarily complex circuits where it may not be at all obvious in what directions the currents are flowing.

Using Ohm's Law, Kirchhoff's Current and Voltage Laws, and derived relationships such as the voltage divider or parallel resistor equations, many simple circuits can be analyzed for static (i.e., unchanging) situations. However, when the voltages are changing, as is usually the case when signals are processed by circuits, we need some additional concepts.

A2.2 AC

AC signals are signals that change over time—in contrast to DC signals which are constant. AC circuit analysis is more complicated than DC analysis, as we can ask not only what the circuit does in the presence of a signal "after a long time" (steady-state), but also what the response of the circuit is when a new signal is received (the "transient" response). We're going to give you a brief overview of the steady-state case, as we really need only to justify some simple RC (resistor-capacitor) circuits here. The basic idea is to extend the idea of resistance to include a sort of

2. "Keer-cough."

Appendix 2 • Useful Electronics

frequency-dependent resistance called *impedance*.

A2.2.1 Impedance

Although it's possible to write out the differential equations describing the behavior of capacitors and inductors as a function of time, it's much easier to use the Laplace transform to find the complex impedance (Z) of devices. This complex impedance lets us treat capacitors and inductors just like we do resistors (i.e., algebraically), using the same formulas (e.g., for series, parallel, and voltage dividers), just substituting Z for R. Instead of calculus (as would be necessary were we to use the differential equations), we can use algebra to find the responses (transient and steady-state) of a circuit to AC signals.

A2.2.1.1 Resistors

Not much to be said here. The complex impedance of a resistor is just R, its resistance. The impedance is not frequency-dependent.

$$Z_R = R \qquad (A2.5)$$

A2.2.1.2 Capacitors

The complex impedance of capacitors is:

$$Z_C(s) = \frac{1}{sC} \qquad (A2.6)$$

where for steady-state analysis we can set $s = j\omega = j2\pi f$ (ω is in rad/sec, f is in Hz). The impedance of a 22-μF capacitor at 60 Hz is:

$$Z_C(60Hz) = \frac{1}{j2\pi 60 \cdot 22\times 10^{-6}} = -120.6j \qquad (A2.7)$$

Note the j here—when we calculate a voltage or current, the results will be complex. The magnitude (absolute value) of the results we calculate will be the amplitude of the voltage (or current) signal at the frequency we're examining (and not necessarily any other frequency!), while the angle of the result will be the phase of the signal.

A2.2.1.3 Inductors

Inductors have an impedance:

$$Z_L(s) = sL \qquad (A2.8)$$

where L is in Henry's (abbreviated as H). A 47-mH inductor at 60 Hz has the following impedance:

$$Z_L(60Hz) = j2\pi 60 \cdot 47 \times 10^{-3} = 17.7j \qquad (A2.9)$$

A2.2.2 RC Circuits

Using the impedances above, we can analyze a frequency-dependent voltage divider made up of a 10KΩ resistor and a 22-µF capacitor as in Figure A2-3. (We've just replaced the lower resistor of a voltage divider with a capacitor.)

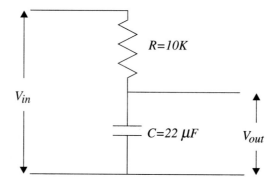

Figure A2-3 RC filter.

Replacing Rs with Zs, the voltage divider equation is:

$$V_{out}(s) = \frac{Z_C}{Z_R + Z_C} V_{in}(s) \qquad (A2.10)$$

Z_R is 10000, and we already calculated the impedance of the 22-µF capacitor at 60 Hz as $-120.6j$. The ratio of $Z_C/(Z_R+Z_C)$ is $0.0121\angle -1.56$ rad. (Or did you want that in compact exponential form: $0.0121e^{-1.56j}$?)

Note that the system function $H(s)$ is:

$$H(s) = \frac{V_{out}(s)}{V_{in}(s)} \qquad (A2.11)$$

If we have a 50-volt sine[3] wave at 60 Hz as our input, the output is a 0.60-volt sine wave with a phase shift of nearly -90° (-1.56 radians). This is a low-pass filter with a -3 dB point at 0.72 Hz—so we're quite a ways out into the stopband of this filter (the attenuation is 38.4 dB).

While there are plenty more issues involved in analyzing circuits, even complicated-looking circuits will give up many of their secrets using the basic laws we've given here and some additional simple rules governing "op-amps" and transistors (when used as simple switches). Texts such as Horowitz and Hill [1989] provide more details.

3. You can measure the sine wave either peak-to-peak or **RMS**; the answer will be in the respective form.

APPENDIX 3

68HC16 Sample FIR Program

The code that follows was written to be run on an M68HC16Z1EVB evaluation board; if your CPU or hardware is different, some register values may need to be changed.

The Motorola MASM assembler was used to assemble the code with the following command line:

```
masm -m 68hc16 -o c5fir1.o -l c5fir1.lst -a -y c5fir1.asm
```

A Motorola hex file was then created with:

```
hex c5fir1.o >c5fir1.s19
```

The background debug mode software we used (ICD16) also required a map file, which was generated using the following command:

```
mlst2map -o c5fir1.map c5fir1.lst.
```

File C5FIR1.ASM

```
*         Title:  C5FIR1.ASM
*         FIR filter example, uses on-chip ADC and an external
*         Burr-Brown PCM56P or Analog Devices AD766 DAC connected
*         via the Serial Peripheral Interface
*         (DAC installs in U12 on the 68HC16Z1EVB evaluation board).
*         11/18/96 rce
*
*         Uses RMAC instruction to evaluate FIR filter.
*         Accomodates filters other than powers of 2 without penalty
*         by using a buffer that is a power of 2, but only evaluating
*         the actual number of coefficients.
*         The input buffer, inbuff, starts at an appropriate address
```

```
*              to allow modulo addressing; the most recent sample is at
*              the address given by inptr.  Samples higher in memory are
*              older (i.e., inptr is decremented).
*
*              The file 'fircoef1.asm' should be present, and contains
*              the filter coefficients and other constants (see below).
*
*              Sample rate is based on the General Purpose Timer using
*              an interrupt from TOC2.
*
*              (Based on some code written by Steve Walker of
*              Hamilton/Avnet, from the Motorola BBS (firfilt))
*
**********************************************************************
****
* Memory map:  Assembled for the 68HC16Z1EVB (evaluation board)
*
*         $00 0000 - $00 ffff      Psuedo-ROM (64K RAM)
*           $00 0000 - $00 0007    Reset vectors
*           $00 0008 - $00 01ff    Interrupt vectors
*           $00 0200 - $00 ffff    Program memory (read only)
*         $02 0000 - $02 ffff     (Optional external RAM, not used, word-wide)
*         $0f 0000 - $0f 03ff     On-chip RAM (4K) (see INITRAM.ASM)
*         $0f f700 - $0f ffff     Registers
**********************************************************************
****

* Some constants:
TOCINCR   equ      419               ;value to give 10 KHz TOC rate

* Debugging/performance monitoring pins (using M68HC16Z1EVB)
* (Port F pins also double as modclk and IRQ1-7)
* Port F bit 0 (modclk) is available on connector P6, pin 4
* Port F bit 1 (*irq1) is available on connector P6, pin 5
* Port F bit 2 (*irq2) is available on connector P6, pin 6
* Port F bit 3 (*irq3) is available on connector P6, pin 7
* Port F bit 4 (*irq4) is available on connector P6, pin 8
* Port F bit 5 (*irq5) is available on connector P6, pin 9
* Port F bit 6 (*irq6) is available on connector P6, pin 10
* Port F bit 7 (*irq7) is available on connector P6, pin 11

* constants for debugging/monitoring pins [debug]
bit7      equ      %10000000
bit6      equ      %01000000
bit5      equ      %00100000
bit4      equ      %00010000
bit3      equ      %00001000
bit2      equ      %00000100
bit1      equ      %00000010
bit0      equ      %00000001

* This program:
* bit 1 of Port F is high while waiting and during irq
```

Appendix 3 • 68HC16 Sample FIR Program

```
        * bit 3 of Port F is high during interrupt processing
        * bit 5 of Port F is high during actual filtering operations
        * bit 7 of Port F is high during the RMAC instruction

                include 'equates.asm'   ;table of EQUates for register addresses

        *       Code space
                org     $f000           ;put coefs at top of program ROM
                                        ;(note--this is NOT bank f, it is bank 0)
                include 'fircoef1.asm'  ;filter coefficients (f_coefs),
                                        ;number of coefficients (numtaps),
                                        ;and bsize (next power of 2 up from numtaps)
                                        ;bsize must be <= 128 due to limits
                                        ;on modulo addressing

        bmask   equ     (bsize*2)-1     ;mask with n+1 bits set (bsize=2^n)

        *       Variables (in internal RAM)
                org     $0000           ;(in bank f, actually)
        inbuff  ds.w    bsize           ;input buffer storage
        topbuff equ     *-2             ;addr of last entry (starting point)
        * Note--you must start inbuff on an appropriate address boundary
        * (i.e. the address must have all 0's where bmask has 1's)
        * in order for modulo addressing to work properly
        inptr   ds.w    1               ;input buffer pointer

        *       Code space
        *       Reset and interrupt vectors
                org     $0000           ;put the following reset vector information
                                        ;at address $00000 of the memory map
                dc.w    $00f0           ;zk=0, sk=f, pk=0
                dc.w    $0200           ;pc=200 -- initial program counter
                dc.w    $03FE           ;sp=03fe -- initial stack pointer
                dc.w    $0000           ;iz=0 -- direct page pointer

        *       Interrupt vectors (vector number in decimal)
        *       Note that most (all but one!) vector points to BDM (end of this
        *       program), where there's a 'bgnd' instruction to put the
        *       system into background debug mode, since we don't expect
        *       to see any other interrupts running!
                dc.w    BDM             ;4      Breakpoint (BKPT)
                dc.w    BDM             ;5      Bus Error (BERR)
                dc.w    BDM             ;6      Software Interrupt (SWI)
                dc.w    BDM             ;7      Illegal Instruction
                dc.w    BDM             ;8      Divide by Zero
                dcb.w   6,BDM           ;9-14   Unassigned Reserved
                dc.w    BDM             ;15     Uninitialized Interrupt
                dc.w    BDM             ;16     (Unassigned Reserved)
                dc.w    BDM             ;17     Level 1 Interrupt Autovector
                dc.w    BDM             ;18     Level 2 Interrupt Autovector
                dc.w    BDM             ;19     Level 3 Interrupt Autovector
                dc.w    BDM             ;20     Level 4 Interrupt Autovector
                dc.w    BDM             ;21     Level 5 Interrupt Autovector
```

```
            dc.w    BDM             ;22     Level 6 Interrupt Autovector
            dc.w    BDM             ;23     Level 7 Interrupt Autovector
            dc.w    BDM             ;24     Spurious Interrupt
            dcb.w   31,BDM          ;25-55  Unassigned Reserved
            dcb.w   8,BDM           ;56-63 User Interrupt 1-8
*           Interrupts from the time (GPT) occur at User Interrupt 9..20
*           (as the GPT is configured by this program)
            dc.w    BDM             ;64     (prioritized GPT interrupt)
            dc.w    BDM             ;65     IC1
            dc.w    BDM             ;66     IC2
            dc.w    BDM             ;67     IC3
            dc.w    BDM             ;68     OC1
            dc.w    irq_sample      ;69     OC2 ** we use this one
            dc.w    BDM             ;70     OC3
            dc.w    BDM             ;71     OC4
            dc.w    BDM             ;72     IC4/OC5
            dc.w    BDM             ;73     TOF
            dc.w    BDM             ;74     PAOVF
            dc.w    BDM             ;75     PAIF
            dcb.w   180,BDM         ;76-255 User Interrupt 21-200

            org     $0200           ;start program after interrupt vectors

*           Initialization Routines
INITSYS:                            ;give initial values for extension
                                    registers
                                    ;and initialize system clock and COP
            orp     #$00e0          ; disable all interrupts
            ldab    #$0F            ; bank f, register/internal RAM access
            tbek                    ; point EK to bank F
            tbyk                    ; point YK to bank F
            tbzk                    ; point ZK to bank F
            ldab    #$00
            tbxk                    ; point XK to bank 0 (code space)

            ldd     #$0003          ; at reset, the CSBOOT block size is 512k.
            std     CSBARBT         ; this line sets the block size to 64k since
                                  ; that is what physically comes with the EVB16
            ldd     #$7830
            std     CSORBT          ;set CSBOOT (the chip select to the 'ROM')
                                    ;to be 0 wait state, word wide, user/super
                                    ;(the important one is 0 wait states--
                                    ;the reset default is 13!)
            LDAA    #$7F            ; w=0, x=1, y=111111
            STAA    SYNCR           ; set system clock to 16.78 Mhz
            CLR     SYPCR           ; turn COP (software watchdog) off,
                                    ; since COP is on after reset

INITRAM:                            ;initialize internal 1k SRAM and stack
            ldd     #$8000          ;make sure RAM is in stop mode
            std     RAMMCR
            ldd     #$00ff          ;(ff since high bits follow a19)
            std     RAMBAH          ; store high ram array, bank $0f
```

Appendix 3 • 68HC16 Sample FIR Program

```
        ldd     #$0000
        std     RAMBAL          ; store low ram array
        clr     RAMMCR          ; enable ram
        ldab    #$0f
        tbsk                    ; set SK to bank $0f for system stack
        lds     #$03FE          ; put SP at top of 1k internal SRAM

INITSCI:                        ;initialize the SCI
                                ;(Serial Communications Interface--
                                ;i.e. asynchronous serial port)
        ldd     #$0037
        std     SCCR0           ;set the SCI baud rate to 9600 baud

        ldd     #$000C
        std     SCCR1           ;enable the SCI receiver and transmitter

INITXRAM:                       ;Initialize external RAM (optional)
                                ;Initialize the Chip Selects.....
        ldd     #$0203
        std     CSBAR0          ;set U3 RAM base addr to $20000: 64k block
        std     CSBAR1          ;set U1 RAM base addr to $20000: 64k block
        ldd     #$5070          ; (was $5030)
        std     CSOR0           ;set Chip Select 0 (U3)
                                ;upper byte, emulate asynch, write only,
                                ;1 wait state, supervisor or user
        ldd     #$3070          ; (was $3030)
        std     CSOR1           ;set Chip Select 1 (U1)
                                ;lower byte, write only, 1 wait, super/user
        ldd     #$0203
        std     CSBAR2          ;set Chip Select 2 to fire at base addr $20000
                                ;$02 0000, in 64k blocks
        ldd     #$7870          ; (was $7830)
        std     CSOR2           ;set Chip Select 2, both bytes, read and write
                                ;both bytes, r/w, 1 wait, super/user
        ldd     #$3fff
        std     CSPAR0          ;set CSBOOT, CS0-CS5 to 16-bit ports

INITQSPI:                       ;Initialize the QSPI
                                ;(Queued Serial Peripheral Interface--
                                ;i.e. synchronous serial port, to DAC)
        ldaa    #$08
        staa    QPDR            ;set PSC0 high between serial transfers
        ldaa    #$0A
        staa    QPAR            ;assign Port D pins as PSC0/SS, MOSI
        ldaa    #$0E
        staa    QDDR            ;set data direction as output on MOSI, SCK pins
        ldd     #$8002
        std     SPCR0           ;master mode, 16 bits, clock low when inactive,
                                ;change data on falling edge of SCK
                                ;SCK is 4.19 MHz
        ldd     #$0201
        std     SPCR1           ;set delay between PCS0 and SCK=2 * SYS CLK
                                ;set delay between transfers=32 * SYS CLK
```

```
                                        ;enable =0 (for now)
            ldd     #$0000
            std     SPCR2           ;no irq, no wrap enable, no wrap to
                                    ; end queue ptr=0 (only 1 transfer)
                                    ; new queue ptr=0 (only 1 transfer)
*       Initialize QSPI Command RAM
            ldab    #$50            ;CONT=no (0), BITSE=8 bits (1), DT=none (0)
                                    ;DSCK=use DSCKL (1), PCS0= addr 0
            stab    CR0             ;QSPI command ram 0

INITGPT:                            ;Initialize the GPT (Timer)
            ldd     #$000f
            std     GPTMCR          ;ignore freeze, normal operation,
                                    ;user/supervisor access,
                                    ;irq arbitration at level f (highest)
            ldd     #$0000
            std     TMSK1           ;(/TMSK2) clr all interrupt mask
                                    ;and set CPR to /4 sys clk to TCNT
            ldd     #$0100
            std     TCTL1           ;(/TCTL2) toggle OC2 output on OC2 capture

            ldd     TFLG1           ;read to initialize flags

INITADC:                            ;Initialize ADC
            ldd     #$0083
            std     ADCTL0          ;10-bit conversion, 2 sample periods,
                                    ;2.1 MHz ADC clock
            ldd     #$0000
            std     ADCMCR          ;run, ignore freeze, user/supervisor access

INITPORT:                           ;Initialize I/O ports for [debug]
                                    ;Note--ports are 8-bit, not 16-bit!
                                    ;(Part of SIM module)
            ldab    #$ff
            stab    PORTF0          ;normally high outputs
            ldab    #$ff
            stab    DDRF            ;make output (vs input)
            ldab    #$01
            stab    PFPAR           ;enable above
                                    ;(0=user i/o; 1=system function irq1..7)

*       Done initializing modules

*       Clear on-chip RAM to 00's (for debug and clean startup)
*       (Note--cannot be done as a subroutine call--stack is here)
*       (YK pts to bank f)
            ldy     #$03fe          ;start at top
clrloop:
            clrw    $0000,y         ;($0000 is just an offset)
            aiy     #-2
            bne     clrloop

*       Set up the timer output compare register
```

Appendix 3 • 68HC16 Sample FIR Program

```
            ldd     TCNT            ;get current timer value
            addd    #TOCINCR        ;add to find next sample time
            std     TOC2            ;use TOC2
*       TOC2 will set a flag when TCNT (16-bit counter) matches TOC2
*       Enable interrupt from TOC2
            ldab    #$10            ;OC2I=enable, others disabled
            stab    TMSK1
            ldd     #$0540          ;set interrupt level, etc.
            std     ICR             ;interrupt level 5,
                                    ;interrupt vectors are $40-$4b,
                                    ;(TOC2 interrupt is at vector $45)

*       Initialize variables
            ldd     #topbuff        ;start at top of buffer
            std     inptr

*       set up "modulo" addressing registers (circular buffer)
            ldd     #bmask          ;modulo addressing mask
                                    ;XMSK (coefs) =$00 (modulo disabled)
                                    ;YMSK (data) = bmask (modulo enabled)
            tdmsk

*       Enable saturate mode, interrupts at level 5 and above
*       (the TOC2 interrupt is a level 5 interrupt)
            tpd                     ;get CCR
            ord     #$0010          ;set SM bit
            andd    #$ff1f          ;clear IP bits
            ord     #$0080          ;set to level 4 (enable 5 and up)
            tdp

*       Main loop
WAITLOOP:
            bset    PORTF0,#bit1    ;turn on bit 1 during wait [debug]
            wai                     ;sit here until interrupt
                                    ;(minimum and consistent interrupt latency)
            bclr    PORTF0,#bit1    ;turn off bit 1 [debug]
            bra     WAITLOOP

* Interrupt service routine, called at sample rate
irq_sample:
                                    ;(normally, stack registers here)
            bset    PORTF0,#bit3    ;turn on bit 3 during irq [debug]

*       clear the irq flags in the timer
            ldd     TFLG1           ;read the register first
            ldd     #$0000
            std     TFLG1           ;then write a negated flag back (i.e. 0)

            ldd     TOC2            ;set up TOC for next sample
            addd    #TOCINCR
            std     TOC2

*       read the ADC, put value in register D
```

```
                ldd     #$0000          ;single conversion sequence,
                                        ;one channel, 4 conversions
                                        ;channel 0
                std     ADCTL1          ;and start next conversion
                ldd     LJSRR0          ;grab last conversion

*       store sample in circular buffer
                ldy     inptr
                cpy     #inbuff         ;will decrement past start of buffer?
                bgt     nowrap
                ldy     #topbuff+2      ;start at top again (+2)
nowrap:
                aiy     #-2
                sty     inptr
                std     0,y             ;store actual data
                                        ;(this is the lowest valid location
                                        ;in buffer)

                bset    PORTF0,#bit5    ;turn on bit 5 during filtering [debug]

*       Evaluate filter
                clrm                    ;clear AM (the MAC accumulator)
                lde     #numtaps-1      ;number of times (-1) to do RMAC
                ldx     #f_coefs        ;start of coefs
                                        ;no need to load y--already set
                ldhi                    ;load H and I registers using ix,iy

                bset    PORTF0,#bit7    ;turn on bit 7 during RMAC [debug]

                rmac    $2,$2           ;repeating MAC instruction:
* RMAC performs the following operations:
* 1.    (AM)+(H)*(I) => AM      multiply & accumulate
* 2.    Qual(IX) => IX          increment (by 2) & adjust IX
* 3.    Qual(IY) => IY          increment (by 2) & adjust IY
* 4.    (M:M+1)X => H           load H using IX
* 5.    (M:M+1)Y => I           load I using IY
* 6.    (E)-1 => E              decrement count in E
* until E<0

                bclr    PORTF0,#bit7    ;turn off bit 7, done w/RMAC [debug]

                tmer                    ;transfer AM to E (saturated if necessary)

                bclr    PORTF0,#bit5    ;turn off bit 5, done w/filtering [debug]

*       done with filter, output data to DAC

                ste     TR0             ;store 16-bit data in QSPI output queue

                ldaa    #$80
                oraa    SPCR1           ;read SPCR1
                staa    SPCR1           ;just enable the QSPI
```

Appendix 3 • 68HC16 Sample FIR Program

```
                bclr    PORTF0,#bit3    ;turn off bit 3, done w/sample [debug]

                rti                     ;end isr


*       Other interrupts and exceptions go into background debug mode
BDM:
                bgnd                    ;exception vectors point here
                                        ;and put the user in background mode


*       end of file c5fir1.ASM *
```

File FIRCOEF1.ASM

```
        *       Coefficient file fircoef1.asm
        *       Generated 18-Nov-96
        *       "Fpb=2700 Hz, Fsb=3000 Hz, Fs=10000 Hz, Ap= 0.5 dB, As=40 dB"
        *
        *
        numtaps equ     51              ;number of coefficients
        bsize   equ     64              ;next power of 2 higher
        f_coefs:                        ;filter coefficients
                dc.w        -348        ;h(  0)=-0.010628
                dc.w        -618        ;h(  1)=-0.018856
                dc.w         -23        ;h(  2)=-0.000714
                dc.w         288        ;h(  3)= 0.008786
                dc.w         -92        ;h(  4)=-0.002800
                dc.w        -288        ;h(  5)=-0.008782
                dc.w         270        ;h(  6)= 0.008229
                dc.w         198        ;h(  7)= 0.006038
                dc.w        -406        ;h(  8)=-0.012383
                dc.w         -63        ;h(  9)=-0.001920
                dc.w         530        ;h( 10)= 0.016172
                dc.w        -165        ;h( 11)=-0.005050
                dc.w        -578        ;h( 12)=-0.017625
                dc.w         465        ;h( 13)= 0.014191
                dc.w         514        ;h( 14)= 0.015682
                dc.w        -824        ;h( 15)=-0.025147
                dc.w        -280        ;h( 16)=-0.008532
                dc.w        1205        ;h( 17)= 0.036783
                dc.w        -195        ;h( 18)=-0.005939
                dc.w       -1568        ;h( 19)=-0.047864
                dc.w        1054        ;h( 20)= 0.032174
                dc.w        1868        ;h( 21)= 0.057019
                dc.w       -2811        ;h( 22)=-0.085798
                dc.w       -2066        ;h( 23)=-0.063057
                dc.w       10200        ;h( 24)= 0.311294
                dc.w       18519        ;h( 25)= 0.565166
                dc.w       10200        ;h( 26)= 0.311294
                dc.w       -2066        ;h( 27)=-0.063057
                dc.w       -2811        ;h( 28)=-0.085798
                dc.w        1868        ;h( 29)= 0.057019
```

```
            dc.w       1054          ;h( 30)= 0.032174
            dc.w      -1568          ;h( 31)=-0.047864
            dc.w       -195          ;h( 32)=-0.005939
            dc.w       1205          ;h( 33)= 0.036783
            dc.w       -280          ;h( 34)=-0.008532
            dc.w       -824          ;h( 35)=-0.025147
            dc.w        514          ;h( 36)= 0.015682
            dc.w        465          ;h( 37)= 0.014191
            dc.w       -578          ;h( 38)=-0.017625
            dc.w       -165          ;h( 39)=-0.005050
            dc.w        530          ;h( 40)= 0.016172
            dc.w        -63          ;h( 41)=-0.001920
            dc.w       -406          ;h( 42)=-0.012383
            dc.w        198          ;h( 43)= 0.006038
            dc.w        270          ;h( 44)= 0.008229
            dc.w       -288          ;h( 45)=-0.008782
            dc.w        -92          ;h( 46)=-0.002800
            dc.w        288          ;h( 47)= 0.008786
            dc.w        -23          ;h( 48)=-0.000714
            dc.w       -618          ;h( 49)=-0.018856
            dc.w       -348          ;h( 50)=-0.010628


* End of file *
```

File EQUATES.ASM

```
*
*         MOTOROLA, INC.
*         Advanced MCU Division
*         Austin, Texas
*
*         Title : EQUATES
*         Description : This is a table of EQUates for all of the
*                       registers in the HC16.
*
*         Note : This program is written for the M68HC16Z1EVB
***************************************************************************
**


*****   SIM Module Registers   *****

SIMCR     EQU      $FA00          ;SIM Module Configuration Register
SIMTR     EQU      $FA02          ;System Integration Test Register
SYNCR     EQU      $FA04          ;Clock Synthesizer Control Register
RSR       EQU      $FA07          ;Reset Status Register
SIMTRE    EQU      $FA08          ;System Integration Test Register  (E Clock)
PORTE0    EQU      $FA11          ;Port E Data Register (same data as PORTE1)
PORTE1    EQU      $FA13          ;Port E Data Register (same data  as PORTE0)
DDRE      EQU      $FA15          ;Port E Data Direction Register
PEPAR     EQU      $FA17          ;Port E Pin Assignment Register
```

Appendix 3 • 68HC16 Sample FIR Program

```
        PORTF0   EQU    $FA19           ;Port F Data Register (same data as PORTF1)
        PORTF1   EQU    $FA1B           ;Port F Data Register (same data as PORTF0)
        DDRF     EQU    $FA1D           ;Port F Data Direction Register
        PFPAR    EQU    $FA1F           ;Port F Pin Assignment Register
        SYPCR    EQU    $FA21           ;System Protection Control Register
        PICR     EQU    $FA22           ;Periodic Interrupt Control Register
        PITR     EQU    $FA24           ;Periodic Interrupt Timing Register
        SWSR     EQU    $FA27           ;Software Service Register
        TSTMSRA  EQU    $FA30           ;Master Shift Register A
        TSTMSRB  EQU    $FA32           ;Master Shift Register B
        TSTSC    EQU    $FA34           ;Test Module Shift Count
        TSTRC    EQU    $FA36           ;Test Module Repetition Count
        CREG     EQU    $FA38           ;Test Submodule Control Register
        DREG     EQU    $FA3A           ;Distributed Register
        CSPDR    EQU    $FA41           ;Port C Data Register
        CSPAR0   EQU    $FA44           ;Chip-Select Pin Assignment Register 0
        CSPAR1   EQU    $FA46           ;Chip-Select Pin Assignment Register 1
        CSBARBT  EQU    $FA48           ;Chip-Select Boot Base Address Register
        CSORBT   EQU    $FA4A           ;Chip-Select Boot Option Register
        CSBAR0   EQU    $FA4C           ;Chip-Select 0 Base Address Register
        CSOR0    EQU    $FA4E           ;Chip Select 0 Option Register
        CSBAR1   EQU    $FA50           ;Chip-Select 1 Base Address Register
        CSOR1    EQU    $FA52           ;Chip-Select 1 Option Register
        CSBAR2   EQU    $FA54           ;Chip-Select 2 Base Address Register
        CSOR2    EQU    $FA56           ;Chip-Select 2 Option Register
        CSBAR3   EQU    $FA58           ;Chip-Select 3 Base Address Register
        CSOR3    EQU    $FA5A           ;Chip-Select 3 Option Register
        CSBAR4   EQU    $FA5C           ;Chip-Select 4 Base Address Register
        CSOR4    EQU    $FA5E           ;Chip-Select 4 Option Register
        CSBAR5   EQU    $FA60           ;Chip-Select 5 Base Address Register
        CSOR5    EQU    $FA62           ;Chip-Select 5 Option Register
        CSBAR6   EQU    $FA64           ;Chip-Select 6 Base Address Register
        CSOR6    EQU    $FA66           ;Chip-Select 6 Option Register
        CSBAR7   EQU    $FA68           ;Chip-Select 7 Base Address Register
        CSOR7    EQU    $FA6A           ;Chip-Select 7 Option Register
        CSBAR8   EQU    $FA6C           ;Chip-Select 8 Base Address Register
        CSOR8    EQU    $FA6E           ;Chip-Select 8 Option Register
        CSBAR9   EQU    $FA70           ;Chip-Select 9 Base Address Register
        CSOR9    EQU    $FA72           ;Chip-Select 9 Option Register
        CSBAR10  EQU    $FA74           ;Chip-Select 10 Base Address Register
        CSOR10   EQU    $FA76           ;Chip-Select 10 Option Register


        *****  SRAM Module Registers  *****

        RAMMCR   EQU    $FB00           ;RAM Module Configuration Register
        RAMTST   EQU    $FB02           ;RAM Test Register
        RAMBAH   EQU    $FB04           ;RAM Base Address High Register
        RAMBAL   EQU    $FB06           ;RAM Base Address Low Register


        *****  QSM Address Map  *****
```

```
QMCR    EQU    $FC00          ;QSM Module Configuration Register
QTEST   EQU    $FC02          ;QSM Test Register
QUILR   EQU    $FC04          ;QSM Interrupt Levels Register
QIVR    EQU    $FC05          ;QSM Interrupt Vector Register
SCCR0   EQU    $FC08          ;SCI Control Register 0
SCCR1   EQU    $FC0A          ;SCI Control Register 1
SCSR    EQU    $FC0C          ;SCI Status Register
SCDR    EQU    $FC0E          ;SCI Data Register
QPDR    EQU    $FC15          ;QSM Port Data Register
QPAR    EQU    $FC16          ;QSM Pin Assignment Register
QDDR    EQU    $FC17          ;QSM Data Direction Register
SPCR0   EQU    $FC18          ;QSPI Control Register 0
SPCR1   EQU    $FC1A          ;QSPI Control Register 1
SPCR2   EQU    $FC1C          ;QSPI Control Register 2
SPCR3   EQU    $FC1E          ;QSPI Control Register 3
SPSR    EQU    $FC1F          ;QSPI Status Register
RR0     EQU    $FD00          ;spi rec.ram 0
RR1     EQU    $FD02          ;spi rec.ram 1
RR2     EQU    $FD04          ;spi rec.ram 2
RR3     EQU    $FD06          ;spi rec.ram 3
RR4     EQU    $FD08          ;spi rec.ram 4
RR5     EQU    $FD0A          ;spi rec.ram 5
RR6     EQU    $FD0C          ;spi rec.ram 6
RR7     EQU    $FD0E          ;spi rec.ram 7
RR8     EQU    $FD00          ;spi rec.ram 8
RR9     EQU    $FD02          ;spi rec.ram 9
RRA     EQU    $FD04          ;spi rec.ram A
RRB     EQU    $FD06          ;spi rec.ram B
RRC     EQU    $FD08          ;spi rec.ram C
RRD     EQU    $FD0A          ;spi rec.ram D
RRE     EQU    $FD0C          ;spi rec.ram E
RRF     EQU    $FD0E          ;spi rec.ram F
TR0     EQU    $FD20          ;spi txd.ram 0
TR1     EQU    $FD22          ;spi txd.ram 1
TR2     EQU    $FD24          ;spi txd.ram 2
TR3     EQU    $FD26          ;spi txd.ram 3
TR4     EQU    $FD28          ;spi txd.ram 4
TR5     EQU    $FD2A          ;spi txd.ram 5
TR6     EQU    $FD2C          ;spi txd.ram 6
TR7     EQU    $FD2E          ;spi txd.ram 7
TR8     EQU    $FD30          ;spi txd.ram 8
TR9     EQU    $FD32          ;spi txd.ram 9
TRA     EQU    $FD34          ;spi txd.ram A
TRB     EQU    $FD36          ;spi txd.ram B
TRC     EQU    $FD38          ;spi txd.ram C
TRD     EQU    $FD3A          ;spi txd.ram D
TRE     EQU    $FD3C          ;spi txd.ram E
TRF     EQU    $FD3E          ;spi txd.ram F
CR0     EQU    $FD40          ;spi cmd.ram 0
CR1     EQU    $FD41          ;spi cmd.ram 1
CR2     EQU    $FD42          ;spi cmd.ram 2
CR3     EQU    $FD43          ;spi cmd.ram 3
CR4     EQU    $FD44          ;spi cmd.ram 4
```

Appendix 3 • 68HC16 Sample FIR Program

```
        CR5      EQU     $FD45           ;spi cmd.ram 5
        CR6      EQU     $FD46           ;spi cmd.ram 6
        CR7      EQU     $FD47           ;spi cmd.ram 7
        CR8      EQU     $FD48           ;spi cmd.ram 8
        CR9      EQU     $FD49           ;spi cmd.ram 9
        CRA      EQU     $FD4A           ;spi cmd.ram A
        CRB      EQU     $FD4B           ;spi cmd.ram B
        CRC      EQU     $FD4C           ;spi cmd.ram C
        CRD      EQU     $FD4D           ;spi cmd.ram D
        CRE      EQU     $FD4E           ;spi cmd.ram E
        CRF      EQU     $FD4F           ;spi cmd.ram F


        *****   GPT Module Registers   *****

        GPTMCR   EQU     $F900           ;GPT Module Configuration Register
        GPTMTR   EQU     $F902           ;GPT Module Test Register (Reserved)
        ICR      EQU     $F904           ;GPT Interrupt Configuration Register
        PDDR     EQU     $F906           ;Parallel Data Direction Register
        GPTPDR   EQU     $F907           ;Parallel Data Register
        OC1M     EQU     $F908           ;OC1 Action Mask Register
        OC1D     EQU     $F909           ;OC1 Action Data Register
        TCNT     EQU     $F90A           ;Timer Counter Register
        PACTL    EQU     $F90C           ;Pulse Accumulator Control Register
        PACNT    EQU     $F90D           ;Pulse Accumulator Counter
        TIC1     EQU     $F90E           ;Input Capture Register 1
        TIC2     EQU     $F910           ;Input Capture Register 2
        TIC3     EQU     $F912           ;Input Capture Register 3
        TOC1     EQU     $F914           ;Output Compare Register 1
        TOC2     EQU     $F916           ;Output Compare Register 2
        TOC3     EQU     $F918           ;Output Compare Register 3
        TOC4     EQU     $F91A           ;Output Compare Register 4
        TI4O5    EQU     $F91C           ;Input Capture 4 or Output Compare 5
        TCTL1    EQU     $F91E           ;Timer Control Register 1
        TCTL2    EQU     $F91F           ;Timer Control Register 2
        TMSK1    EQU     $F920           ;Timer Interrupt Mask Register 1
        TMSK2    EQU     $F921           ;Timer Interrupt Mask Register 2
        TFLG1    EQU     $F922           ;Timer Interrupt Flag Register 1
        TFLG2    EQU     $F923           ;Timer Interrupt Flag Register 2
        CFORC    EQU     $F924           ;Compare Force Register
        PWMC     EQU     $F924           ;PWM Control Register
        PWMA     EQU     $F926           ;PWM Register A
        PWMB     EQU     $F927           ;PWM Register B
        PWMCNT   EQU     $F928           ;PWM Counter Register
        PWMBUFA  EQU     $F92A           ;PWM Buffer Register A
        PWMBUFB  EQU     $F92B           ;PWM Buffer Register B
        PRESCL   EQU     $F92C           ;GPT Prescaler
```

```
*****   ADC Module Registers   *****

        ADCMCR  EQU     $F700           ;ADC Module Configuration Register
        ADTEST  EQU     $F702           ;ADC Test Register
        ADCPDR  EQU     $F706           ;ADC Port Data Register
        ADCTL0  EQU     $F70A           ;A/D Control Register 0
        ADCTL1  EQU     $F70C           ;A/D Control Register 1
        ADSTAT  EQU     $F70E           ;ADC Status Register
        RJURR0  EQU     $F710           ;Right Justified Unsigned Result Register 0
        RJURR1  EQU     $F712           ;Right Justified Unsigned Result Register 1
        RJURR2  EQU     $F714           ;Right Justified Unsigned Result Register 2
        RJURR3  EQU     $F716           ;Right Justified Unsigned Result Register 3
        RJURR4  EQU     $F718           ;Right Justified Unsigned Result Register 4
        RJURR5  EQU     $F71A           ;Right Justified Unsigned Result Register 5
        RJURR6  EQU     $F71C           ;Right Justified Unsigned Result Register 6
        RJURR7  EQU     $F71E           ;Right Justified Unsigned Result Register 7
        LJSRR0  EQU     $F720            ;Left Justified Signed Result Register 0
        LJSRR1  EQU     $F722            ;Left Justified Signed Result Register 1
        LJSRR2  EQU     $F724            ;Left Justified Signed Result Register 2
        LJSRR3  EQU     $F726            ;Left Justified Signed Result Register 3
        LJSRR4  EQU     $F728            ;Left Justified Signed Result Register 4
        LJSRR5  EQU     $F72A            ;Left Justified Signed Result Register 5
        LJSRR6  EQU     $F72C            ;Left Justified Signed Result Register 6
        LJSRR7  EQU     $F72E            ;Left Justified Signed Result Register 7
        LJURR0  EQU     $F730            ;Left Justified Unsigned Result Register 0
        LJURR1  EQU     $F732            ;Left Justified Unsigned Result Register 1
        LJURR2  EQU     $F734            ;Left Justified Unsigned Result Register 2
        LJURR3  EQU     $F736            ;Left Justified Unsigned Result Register 3
        LJURR4  EQU     $F738            ;Left Justified Unsigned Result Register 4
        LJURR5  EQU     $F73A            ;Left Justified Unsigned Result Register 5
        LJURR6  EQU     $F73C            ;Left Justified Unsigned Result Register 6
        LJURR7  EQU     $F73E            ;Left Justified Unsigned Result Register 7
```

APPENDIX 4

68HC16 Sample IIR Program

The same comments apply to the sample IIR code as to the FIR code of Appendix 4. The "equates.asm" file is identical for the two programs and so is not repeated here.

File C6IIR1.ASM

```
*       Title:  C6IIR1.ASM
*       IIR filter example, uses on-chip ADC and an external
*       Burr-Brown PCM56P or Analog Devices AD766 DAC connected
*       via the Serial Peripheral Interface
*       (DAC installs in U12 on the 68HC16Z1EVB evaluation board).
*       11/18/96 rce
*
*       Evaluates IIR filter as cascade of second-order sections
*       (SOS's or "biquads").  An external file 'iircoef1.asm'
*       is required, which contains the actual coefficients as
*       well as EQU's for implementing the SOS's.  Up to 6 biquad
*       sections are supported, and up to 1 FOS.  (More are very
*       easily added.)
*
*       Some optimization has been done, but there may be special
*       cases that can be exploited, or tradeoffs made for
*       quicker execution at the cost of some decrease in
*       signal-to-noise ratio.
*
*       Each biquad section takes approximately 120 clock cycles;
*       at 16 MHz and a 10 KHz sample rate, this is 16 biquad
*       evaluations/sample (about a 32nd order filter).
*
*       Sample rate is based on the General Purpose Timer using
*       an interrupt from TOC2.
*
```

```
*************************************************************************
****
* Memory map:  Assembled for the 68HC16Z1EVB (evaluation board)
*
*       $00 0000 - $00 ffff     Pseudo-ROM (64K RAM)
*         $00 0000 - $00 0007     Reset vectors
*         $00 0008 - $00 01ff     Interrupt vectors
*         $00 0200 - $00 ffff     Program memory (read only)
*       $02 0000 - $02 ffff     (Optional external RAM, not used, word-wide)
*       $0f 0000 - $0f 03ff     On-chip RAM (4K) (see INITRAM.ASM)
*       $0f f700 - $0f ffff     Registers
*************************************************************************
****
        CLIST   OFF             ;turn off listing of non-assembled
                                ;conditional code
        MLIST   ON              ;include macro expansion in listing

* Some constants:
TOCINCR equ     419             ;value to give 10 KHz TOC rate

* Debugging/performance monitoring pins (using M68HC16Z1EVB)
* (Port F pins also double as modclk and IRQ1-7)
* Port F bit 0 (modclk) is available on connector P6, pin 4
* Port F bit 1 (*irq1) is available on connector P6, pin 5
* Port F bit 2 (*irq2) is available on connector P6, pin 6
* Port F bit 3 (*irq3) is available on connector P6, pin 7
* Port F bit 4 (*irq4) is available on connector P6, pin 8
* Port F bit 5 (*irq5) is available on connector P6, pin 9
* Port F bit 6 (*irq6) is available on connector P6, pin 10
* Port F bit 7 (*irq7) is available on connector P6, pin 11

* constants for debugging/monitoring pins [debug]
bit7    equ     %10000000
bit6    equ     %01000000
bit5    equ     %00100000
bit4    equ     %00010000
bit3    equ     %00001000
bit2    equ     %00000100
bit1    equ     %00000010
bit0    equ     %00000001

* This program:
* bit 1 of Port F is high while waiting and during irq
* bit 3 of Port F is high during interrupt processing
* bit 5 of Port F is high during actual filtering operations

        include 'equates.asm'   ;table of EQUates for register addresses
        include 'iirmac.asm'    ;macros for IIR

*       Code space
        org     $f000           ;put coefs at top of program ROM
                                ;(note--this is NOT bank f, it is bank 0)
        include 'iircoef1.asm'  ;filter coefficients:
```

Appendix 4 • 68HC16 Sample IIR Program

```
                        ; A1, A2, B0, B1, B2 (1.15 format) per SOS
                        ;  (see text for relation of these to
                        ;  a1, a2, b0, etc.)
                        ; or, for FOS:
                        ;  A1, B0, B1
                        ;
                        ; place SOS's in order to be evaluated
                        ;number of SOS's:  num_sos
                        ;number of FOS's:  num_fos (0 or 1)

*          Variables (in internal RAM)
           org     $0000           ;(actually in bank f, $f0000)
w_of_n     ds.w    (num_fos*2+num_sos*3)   ;space for w(n) storage
                        ;order is W0,W1, next W0 ... (for FOS),
                        ; W0,W1,W2, next W0,W1,W2 ... (for SOS's)
                        ;Clear these before using!

*          Code space
*          Reset and interrupt vectors
           org     $0000           ;put the following reset vector information
                                   ;at address $00000 of the memory map
           dc.w    $00f0           ;zk=0, sk=f, pk=0
           dc.w    $0200           ;pc=200 -- initial program counter
           dc.w    $03FE           ;sp=03fe -- initial stack pointer
           dc.w    $0000           ;iz=0 -- direct page pointer

*          Interrupt vectors (vector number in decimal)
*          Note that most (all but one!) vector points to BDM (end of this
*          program), where there's a 'bgnd' instruction to put the
*          system into background debug mode, since we don't expect
*          to see any other interrupts running!
           dc.w    BDM             ;4     Breakpoint (BKPT)
           dc.w    BDM             ;5     Bus Error (BERR)
           dc.w    BDM             ;6     Software Interrupt (SWI)
           dc.w    BDM             ;7     Illegal Instruction
           dc.w    BDM             ;8     Divide by Zero
           dcb.w   6,BDM           ;9-14  Unassigned Reserved
           dc.w    BDM             ;15    Uninitialized Interrupt
           dc.w    BDM             ;16    (Unassigned Reserved)
           dc.w    BDM             ;17    Level 1 Interrupt Autovector
           dc.w    BDM             ;18    Level 2 Interrupt Autovector
           dc.w    BDM             ;19    Level 3 Interrupt Autovector
           dc.w    BDM             ;20    Level 4 Interrupt Autovector
           dc.w    BDM             ;21    Level 5 Interrupt Autovector
           dc.w    BDM             ;22    Level 6 Interrupt Autovector
           dc.w    BDM             ;23    Level 7 Interrupt Autovector
           dc.w    BDM             ;24    Spurious Interrupt
           dcb.w   31,BDM          ;25-55 Unassigned Reserved
           dcb.w   8,BDM           ;56-63 User Interrupt 1-8
*          Interrupts from the time (GPT) occur at User Interrupt 9..20
*          (as the GPT is configured by this program)
           dc.w    BDM             ;64    (prioritized GPT interrupt)
           dc.w    BDM             ;65    IC1
```

```
        dc.w    BDM                 ;66     IC2
        dc.w    BDM                 ;67     IC3
        dc.w    BDM                 ;68     OC1
        dc.w    irq_sample          ;69     OC2 ** we use this one
        dc.w    BDM                 ;70     OC3
        dc.w    BDM                 ;71     OC4
        dc.w    BDM                 ;72     IC4/OC5
        dc.w    BDM                 ;73     TOF
        dc.w    BDM                 ;74     PAOVF
        dc.w    BDM                 ;75     PAIF
        dcb.w   180,BDM             ;76-255 User Interrupt 21-200

        org     $0200               ;start program after interrupt vectors

*       Initialization Routines
INITSYS:                            ;give initial values for extension registers
                                    ;and initialize system clock and COP
        orp     #$00e0              ; disable all interrupts
        ldab    #$0F                ; bank f, register/internal RAM access
        tbek                        ; point EK to bank F
        tbyk                        ; point YK to bank F
        tbzk                        ; point ZK to bank F
        ldab    #$00
        tbxk                        ; point XK to bank 0 (code space)

        ldd     #$0003              ; at reset, the CSBOOT block size is 512k.
        std     CSBARBT             ; this line sets the block size to 64k since
                                    ; that is what physically comes with the EVB16
        ldd     #$7830
        std     CSORBT              ;set CSBOOT (the chip select to the 'ROM')
                                    ;to be 0 wait state, word wide, user/super
                                    ;(the important one is 0 wait states--
                                    ;the reset default is 13!)

        LDAA    #$7F                ; w=0, x=1, y=111111
        STAA    SYNCR               ; set system clock to 16.78 Mhz
        CLR     SYPCR               ; turn COP (software watchdog) off,
                                    ; since COP is on after reset

INITRAM:                            ;initialize internal 1k SRAM and stack
        ldd     #$8000              ;make sure RAM is in stop mode
        std     RAMMCR
        ldd     #$00ff              ;(ff since high bits follow a19)
        std     RAMBAH              ; store high ram array, bank $0f
        ldd     #$0000
        std     RAMBAL              ; store low ram array
        clr     RAMMCR              ; enable ram
        ldab    #$0f
        tbsk                        ; set SK to bank $0f for system stack
        lds     #$03FE              ; put SP at top of 1k internal SRAM

INITSCI:                            ;initialize the SCI
                                    ;(Serial Communications Interface--
                                    ;i.e. asynchronous serial port)
```

Appendix 4 • 68HC16 Sample IIR Program

```
            ldd     #$0037
            std     SCCR0               ;set the SCI baud rate to 9600 baud

            ldd     #$000C
            std     SCCR1               ;enable the SCI receiver and transmitter

INITXRAM:                               ;Initialize external RAM (optional)
                                        ;Initialize the Chip Selects.....
            ldd     #$0203
            std     CSBAR0              ;set U3 RAM base addr to $20000: 64k block
            std     CSBAR1              ;set U1 RAM base addr to $20000: 64k block
            ldd     #$5070              ; (was $5030)
            std     CSOR0               ;set Chip Select 0 (U3)
                                        ;upper byte, emulate asynch, write only,
                                        ;1 wait state, supervisor or user
            ldd     #$3070              ; (was $3030)
            std     CSOR1               ;set Chip Select 1 (U1)
                                        ;lower byte, write only, 1 wait, super/user
            ldd     #$0203
            std     CSBAR2              ;set Chip Select 2 to fire at base addr $20000
                                        ;$02 0000, in 64k blocks
            ldd     #$7870              ; (was $7830)
            std     CSOR2               ;set Chip Select 2, both bytes, read and write
                                        ;both bytes, r/w, 1 wait, super/user
            ldd     #$3fff
            std     CSPAR0              ;set CSBOOT, CS0-CS5 to 16-bit ports

INITQSPI:                               ;Initialize the QSPI
                                        ;(Queued Serial Peripheral Interface--
                                        ;i.e. synchronous serial port, to DAC)
            ldaa    #$08
            staa    QPDR                ;set PSC0 high between serial transfers
            ldaa    #$0A
            staa    QPAR                ;assign Port D pins as PSC0/SS, MOSI
            ldaa    #$0E
            staa    QDDR                ;set data direction as output on MOSI, SCK pins
            ldd     #$8002
            std     SPCR0               ;master mode, 16 bits, clock low when inactive,
                                        ;change data on falling edge of SCK
                                        ;SCK is 4.19 MHz
            ldd     #$0201
            std     SPCR1               ;set delay between PCS0 and SCK-2 * SYS CLK
                                        ;set delay between transfers=32 * SYS CLK
                                        ;enable =0 (for now)
            ldd     #$0000
            std     SPCR2               ;no irq, no wrap enable, no wrap to
                                        ; end queue ptr=0 (only 1 transfer)
                                        ; new queue ptr=0 (only 1 transfer)
*           Initialize QSPI Command RAM
            ldab    #$50                ;CONT=no (0), BITSE=8 bits (1), DT=none (0)
                                        ;DSCK=use DSCKL (1), PCS0= addr 0
            stab    CR0                 ;QSPI command ram 0
```

```
INITGPT:                            ;Initialize the GPT (Timer)
        ldd     #$000f
        std     GPTMCR              ;ignore freeze, normal operation,
                                    ;user/supervisor access,
                                    ;irq arbitration at level f (highest)
        ldd     #$0000
        std     TMSK1               ;(/TMSK2) clr all interrupt mask
                                    ;and set CPR to /4 sys clk to TCNT
        ldd     #$0100
        std     TCTL1               ;(/TCTL2) toggle OC2 output on OC2 capture

        ldd     TFLG1               ;read to initialize flags

INITADC:                            ;Initialize ADC
        ldd     #$0083
        std     ADCTL0              ;10-bit conversion, 2 sample periods,
                                    ;2.1 MHz ADC clock
        ldd     #$0000
        std     ADCMCR              ;run, ignore freeze, user/supervisor access

INITPORT:                           ;Initialize I/O ports for [debug]
                                    ;Note--ports are 8-bit, not 16-bit!
                                    ;(Part of SIM module)
        ldab    #$ff
        stab    PORTF0              ;normally high outputs
        ldab    #$ff
        stab    DDRF                ;make output (vs input)
        ldab    #$01
        stab    PFPAR               ;enable above
                                    ;(0=user i/o; 1=system function irq1..7)

*       Done initializing modules

*       set up pointer extension registers (different from FIR!)
        ldab    #$00
        tbyk                        ;y points to code space
        ldab    #$0f
        tbxk                        ;x points to data space

*       Clear on-chip RAM to 00's (for debug and clean startup)
*       (Note--cannot be done as a subroutine call--stack is here)
*       (XK points to bank f)

        ldx     #$03fe              ;start at top
clrloop:
        clrw    $0000,x             ;($0000 is just an offset)
        aix     #-2
        bne     clrloop

*       Set up the timer output compare register
        ldd     TCNT                ;get current timer value
        addd    #TOCINCR            ;add to find next sample time
        std     TOC2                ;use TOC2
```

Appendix 4 • 68HC16 Sample IIR Program

```
*           TOC2 will set a flag when TCNT (16-bit counter) matches TOC2
*           Enable interrupt from TOC2
            ldab    #$10            ;OC2I=enable, others disabled
            stab    TMSK1
            ldd     #$0540          ;set interrupt level, etc.
            std     ICR             ;interrupt level 5,
                                    ;interrupt vectors are $40-$4b,
                                    ;(TOC2 interrupt is at vector $45)

*           set up modulo addressing register
            ldd     #$0000          ;modulo addressing mask
                                    ;XMSK (coefs) =$00 (modulo addr disabled)
                                    ;YMSK (data) = $00 (modulo addr disabled)
            tdmsk

*           Enable saturate mode, interrupts at level 5 and above
*           (the TOC2 interrupt is a level 5 interrupt)
            tpd                     ;get CCR
            ord     #$0010          ;set SM bit
            andd    #$ff1f          ;clear IP bits
            ord     #$0080          ;set to level 4 (enable 5 and up)
            tdp

*           Main loop
WAITLOOP:
            bset    PORTF0,#bit1    ;turn on bit 1 during wait [debug]
            wai                     ;sit here until interrupt
                                    ;(minimum and consistent interrupt latency)
            bclr    PORTF0,#bit1    ;turn off bit 1 [debug]
            bra     WAITLOOP


* Interrupt service routine, called at sample rate
irq_sample:
                                    ;(normally, stack registers here)
            bset    PORTF0,#bit3    ;turn on bit 3 during irq [debug]

*           clear the irq flags in the timer
            ldd     TFLG1           ;read the register first
            ldd     #$0000
            std     TFLG1           ;then write a negated flag back (i.e. 0)

            ldd     TOC2            ;set up TOC for next sample
            addd    #TOCINCR
            std     TOC2

*           read the ADC, put value in register E
            ldd     #$0000          ;single conversion sequence,
                                    ;one channel, 4 conversions
                                    ;channel 0
            std     ADCTL1          ;and start next conversion
            lde     LJSRR0          ;grab last conversion (reg E)
```

```
*       filter
        tem                         ;e -> am[31:16], am[35:32]=am[31],
                                    ;am[15:0]=$00
        ldx     #w_of_n+2           ;start ix pointing at W1 (not W0!)
        ldy     #iircoefs           ;start of coefficients (A1)
        ldhi                        ;hr=(ix)=W1, ir=(iy)=A1

        bset    PORTF0,#bit5        ;turn on bit 5 during filter [debug]

*       Error message if wrong number of FOS's (can only be 0 or 1)
        ifgt    num_fos-1
        fail    "Wrong number of FOS's!"
        endc

        ifeq    num_fos-1
        fos     FOS0G1              ;do one first-order section [macro]
        endc

*       Error message if wrong number of SOS's
        ifgt    num_sos-6
        fail    "Wrong number of SOS's!"
        endc

        ifge    num_sos-1
        sos     SOS1G1,SOS1G2       ;do second-order section [macro]
        endc
        ifge    num_sos-2
        sos     SOS2G1,SOS2G2
        endc
        ifge    num_sos-3
        sos     SOS3G1,SOS3G2
        endc
        ifge    num_sos-4
        sos     SOS4G1,SOS4G2
        endc
        ifge    num_sos-5
        sos     SOS5G1,SOS5G2
        endc
        ifge    num_sos-6
        sos     SOS6G1,SOS6G2
        endc
                                    ;result is in AM

        tmer                        ;transfer AM to E (saturated if necessary)

        bclr    PORTF0,#bit5        ;turn off bit 5 [debug]

*       done with filter, output data to DAC
        ste     TR0                 ;store 16-bit data in QSPI output queue

        ldaa    #$80
        oraa    SPCR1               ;read SPCR1
```

Appendix 4 • 68HC16 Sample IIR Program

```
                staa    SPCR1               ;just enable the QSPI

                bclr    PORTF0,#bit3        ;turn off bit 3, done w/sample [debug]

                rti                         ;end isr


*               Other interrupts and exceptions go into background debug mode
BDM:
                bgnd                        ;exception vectors point here
                                            ;and put the user in background mode


* end of file c5fir1.ASM *
```

File IIRCOEF1.ASM

```
        *       Coefficient file iircoef1.asm
        *       Generated 1-Nov-96 by i6x013.m
        *       Butterworth Example
        *       Lowpass, Fsamp=10000 Hz
        *       Fp=1000 Hz, Ap=1 dB
        *       Fs=2000 Hz, As=40 dB
        *       Order=7
        *
        *
        num_fos equ     1                   ;number of FOSs
        num_sos equ     3                   ;number of SOSs
        *
        *
        iircoefs:                           ;start of coefficient storage

        * FOS0:
        * zero at (-1.000000+0j)
        * pole at ( 0.472925+0j)
        * b0= 1.000000, b1= 1.000000
        * a1=-0.472925
        * G= 0.263538
                dc.w    $3c89               ;A1= 0.472925
                dc.w    $4377               ;B0= 0.527075
                dc.w    $4377               ;B1= 0.527075
        FOS0G1  equ     $1                  ;G1= 0.500000

        * SOS1:
        * zeros at (-1.000000+0j), (-1.000000+0j)
        * poles at ( 0.491831+ 0.175154j) and conjugate
        * b0= 1.000000, b1= 2.000000, b2= 1.000000
        * a1=-0.983662, a2= 0.272577
        * G= 0.072229
                dc.w    $7de9               ;A1= 0.983662
                dc.w    $dd1c               ;A2=-0.272577
                dc.w    $24fb               ;B0= 0.288915
                dc.w    $49f6               ;B1= 0.577829
```

```
                dc.w      $24fb           ;B2= 0.288915
        SOS1G1  equ       $2              ;G1= 0.250000
        SOS1G2  equ       $0              ;G2= 1

        * SOS2:
        * zeros at (-1.000000+0j), (-1.000000+0j)
        * poles at ( 0.553874+ 0.355431j) and conjugate
        * b0= 1.000000, b1= 2.000000, b2= 1.000000
        * a1=-1.107747, a2= 0.433107
        * G= 0.081340
                dc.w      $46e5           ;A1= 0.553874
                dc.w      $e448           ;A2=-0.216553
                dc.w      $29a5           ;B0= 0.325360
                dc.w      $534b           ;B1= 0.650720
                dc.w      $29a5           ;B2= 0.325360
        SOS2G1  equ       $3              ;G1= 0.125000
        SOS2G2  equ       $1              ;G2= 2

        * SOS3:
        * zeros at (-1.000000+0j), (-1.000000+0j)
        * poles at ( 0.677344+ 0.542017j) and conjugate
        * b0= 1.000000, b1= 2.000000, b2= 1.000000
        * a1=-1.354687, a2= 0.752576
        * G= 0.099472
                dc.w      $56b3           ;A1= 0.677344
                dc.w      $cfd6           ;A2=-0.376288
                dc.w      $32ee           ;B0= 0.397889
                dc.w      $65dc           ;B1= 0.795779
                dc.w      $32ee           ;B2= 0.397889
        SOS3G1  equ       $3              ;G1= 0.125000
        SOS3G2  equ       $1              ;G2= 2
        *
        *
        * end of file *
```

File IIRMAC.ASM[1]

```
        *         iirmac.asm
        *         Macros for IIR routines
        *         10/31/96 rce

        *         rasrm:   repeating asrm
        *         call with 1 argument that evaluates to 0,1,2,3 or 4
        *         generates 0 to 4 asrm's
        rasrm     macro
                  ifeq     \1-1
                  asrm
                  endc
                  ifeq     \1-2
                  asrm
```

1. Ugly code warning!

```
            asrm
            endc
            ifeq    \1-3
            asrm
            asrm
            asrm
            endc
            ifeq    \1-4
            asrm
            asrm
            asrm
            asrm
            endc
            ifgt    \1-4
            fail    "Too many shifts for rasrm"
            endc
            endm

*       raslm:  repeating aslm
*       call with 1 argument that evaluates to 0,1,2,3 or 4,
*       generates 0 to 4 aslm's
raslm   macro
            ifeq    \1-1
            aslm
            endc
            ifeq    \1-2
            aslm
            aslm
            endc
            ifeq    \1-3
            aslm
            aslm
            aslm
            endc
            ifeq    \1-4
            aslm
            aslm
            aslm
            aslm
            endc
            ifgt    \1-4
            fail    "Too many shifts for raslm"
            endc
            endm


*       sos:   Create a second-order section (SOS) with arguments
*       G1 shift count (e.g. SOS1G1), and G2 shift count (e.g. SOS1G2).
*
```

```
*                  G1           G2     W0       B0
*         x(n) >----->-----(+)----->------------>----(+)-----> y(n)
*                           |                |         |
*                           |              [z^-1]      |
*                           |       A1       |    B1   |
*                           +-----<-------------->-----+
*                           |               |W1        |
*                           |                |         |
*                           |              [z^-1]      |
*                           |       A2       |    B2   |
*                           +-----<-------------->-----+
*                                          W2
*
*
*        (Refer to text for value of G1, G2, and calculation of
*        A1, A2, B0, B1, and B2 from a1, a2, b0, b1, and b2)
*
*        Assumes input is in AM, ix points to W1, iy points to A1,
*        HR and IR are loaded with W1 and A1 respectively.
*        Enable saturation; clear x and y masks.
*
*        Order of variable & coefficient storage is:
*        A1, A2, B0, B1, B2 , next A1, next A2, ..., yk set to page
*        W0, W1, W2, next W0, next W1, ..., xk set to page
*        All variables, coefficients in 1.15 signed format
*
*        Output is in AM, with ix, iy, HR and IR set up for
*        a call to another SOS.
*
*        Cycles:  Depends on number of shifts for G1, G2;
*        100 cycles + 4*G1shifts + 4*G2shifts (about 120 cycles
*        typical)
*
*        Example usage:
*
*        lde      (input)            ;e is 1.15 format input
*        tem                         ;e -> am
*        ldx      #w_of_n+2          ;ix points to W1 of SOS #1
*        ldy      #iircoefs          ;iy points to A1 of SOS #1
*        ldhi                        ;load hr with W1, ir with A1
*        sos      SOS1G1,SOS1G2      ;implement SOS#1
*        sos      SOS2G1,SOS2G2      ;SOS #2
*        tmer                        ;output -> e (rounded & saturated)
*        ste      (output)
*
*        Cycles per instruction noted in first column of comments
sos      macro                       ;   G1 shift, G2 shift
         rasrm    \1                 ;4*G1  generate right shifts for G1
                                     ;      am=G1*x, hr=(ix)=W1, ir=(iy)=A1
         mac      2,2                ;12    hr=(ix)=W2,ir=(iy)=A2, am=G1*x+A1*W1
         mac      -4,2               ;12    hr=(ix)=W0,ir=(iy)=B0,
                                     ;      am=G1*x+A1*W1+A2*W2
         raslm    \2                 ;4*G2  generate left shifts for G2
```

Appendix 4 • 68HC16 Sample IIR Program

```
            tmer                    ;6   am -> e, rounded & saturated
            ste    $00,x            ;6   W0=new result
            ldhi                    ;8   hr=(ix)=W0 (new!), ir=(iy)=B0 (again)
            clrm                    ;2   am=0
            mac    2,2              ;12  iz=hr=W0, hr=(ix)=W1, ir=(iy)=B1
                                    ;    am=B0*W0
            stz    $00,x            ;4   W1=W0
                                    ;note hr=old W1, prior to this write!
            mac    2,2              ;12  iz=hr=W1, hr=(ix)=W2, ir=(iy)=B2
                                    ;    am=B0*W0+B1*W1
            stz    $00,x            ;4   W2=W1 (note again that hr=old W2)
            mac    0,2              ;12  hr=(ix)=W1, ir=(iy)=next A1
                                    ;    am=B0*W0+B1*W1+B2*W2
            aix    #8               ;2   ix=next W1
            ldhi                    ;8   hr=(ix)=next W1, ir=(iy)=next A1
                                    ; result is in am
            endm
*  Note--The last mac above would like to be 'mac 8,2' which
*  would set up hr and ix for the next section without need
*  for a separate 'aix #8' and 'ldhi'--half of which is redundant
*  since we already have ir/iy correct.  Alas, the mac instruction
*  has a 4-bit *signed* increment for x, with a range of -8 to 7.


*           fos:  Create a first-order section (FOS) with argument
*           G1 shift count (e.g. FOS0G1).
*
*                    G1                W0      B0
*           x(n) >----->-----(+)----------------->----(+)-----> y(n)
*                             |                |       |
*                             |              [z^-1]    |
*                             |      A1        |   B1  |
*                             +-----<------------>-----+
*                                             W1
*
*
*           (Refer to text for value of G1 and calculation of
*           A1, B0, and B1 from a1, b0, and b1)
*
*           Assumes input is in AM, ix points to W1, iy points to A1,
*           HR and IR are loaded with W1 and A1 respectively.
*           Enable saturation; clear x and y masks.
*
*           Order of variable & coefficient storage is:
*           A1, B0, B1, next A1, ..., yk set to page
*           W0, W1, next W0, ..., xk set to page
*           All variables, coefficients in 1.15 signed format
*
*           Output is in AM, with ix, iy, HR and IR set up for
*           a call to a subsequent SOS.
*
*           Cycles:  Depends on number of shifts for G1;
*           64 cycles + 4*G1shifts (about 68 cycles typical)
```

```
*
*       Example usage:
*
*       lde     (input)             ;e is 1.15 format input
*       tem                         ;e -> am
*       ldx     #w_of_n+2           ;ix points to W1 of FOS #0
*       ldy     #iircoefs           ;iy points to A1 of FOS #0
*       ldhi                        ;load hr with W1, ir with A1
*       fos     FOS0G1              ;implement FOS#0
*       sos     SOS1G1,SOS1G2       ;SOS #1
*       sos     SOS2G1,SOS2G2       ;SOS #2
*       tmer                        ;output -> e (rounded & saturated)
*       ste     (output)
*
fos     macro                       ;    G1 shift
        rasrm   \1                  ;4*G1   generate right shifts for G1
                                    ;     am=G1*x, hr=(ix)=W1, ir=(iy)=A1
        mac     -2,2                ;12  hr=(ix)=W0,ir=(iy)=B0, am=G1*x+A1*W1

        tmer                        ;6   am -> e, rounded & saturated
        ste     $00,x               ;6   W0=new result
        ldhi                        ;8   hr=(ix)=W0 (new!), ir=(iy)=B0 (again)
        clrm                        ;2   am=0
        mac     2,2                 ;12  iz=hr=W0, hr=(ix)=W1, ir=(iy)=B1
                                    ;    am=B0*W0
        stz     $00,x               ;4   W1=W0
                                    ;note hr=old W1, prior to this write!
        mac     4,2                 ;12  hr=(ix)=next W1, ir=(iy)=next A1
                                    ;    am=B0*W0+B1*W1
                                    ; result is in am
        endm

* end of file iirmac.asm *
```

APPENDIX 5

68HC16 Sample Interpolation Program

The following program illustrates interpolation on the 68HC16. The separate coefficient file follows the main program listing. The same comments with respect to assembly and target environment apply as that for the FIR code of Appendix 4. The "equates.asm" file is identical as used elsewhere and is not repeated here.

File C10INT1.ASM

```
*         Title:  C10INT1.ASM
*         Interpolation example, uses on-chip ADC and an external
*         Burr-Brown PCM56P or Analog Devices AD766 DAC connected
*         via the Serial Peripheral Interface
*         (DAC installs in U12 on the 68HC16Z1EVB evaluation board).
*         1/8/97 rce
*
*         Illustrates efficient interpolation filter, arbitrary
*         interpolation factor L, filter length N is multiple of L,
*         and N/L can be any integer (i.e., not restricted to power of 2).
*
*         Shares many features with the FIR filter program C5FIR1.ASM.
*
*         The input buffer, inbuff, starts at an appropriate address
*         to allow modulo addressing; the most recent sample is at
*         the address given by inptr.  Samples higher in memory are
*         older (i.e., inptr is decremented).
*
*         The file 'c10int1.h' is required, and must contain
*         equates for:  Fs, Ni, Li, NoverLi, Coefi, and BSizei.
*
*         Sample rate is based on the General Purpose Timer using
*         an interrupt from TOC2.
*
```

```
************************************************************************
* Memory map:  Assembled for the 68HC16Z1EVB (evaluation board)
*
*       $00 0000 - $00 ffff     Psuedo-ROM (64K RAM)
*       $00 0000 - $00 0007       Reset vectors
*       $00 0008 - $00 01ff       Interrupt vectors
*       $00 0200 - $00 ffff       Program memory (read only)
*       $02 0000 - $02 ffff     (Optional external RAM, not used, word-wide)
*       $0f 0000 - $0f 03ff     On-chip RAM (4K) (see INITRAM.ASM)
*       $0f f700 - $0f ffff     Registers
************************************************************************


* Debugging/performance monitoring pins (using M68HC16Z1EVB)
* (Port F pins also double as modclk and IRQ1-7)
* Port F bit 0 (modclk) is available on connector P6, pin 4
* Port F bit 1 (*irq1) is available on connector P6, pin 5
* Port F bit 2 (*irq2) is available on connector P6, pin 6
* Port F bit 3 (*irq3) is available on connector P6, pin 7
* Port F bit 4 (*irq4) is available on connector P6, pin 8
* Port F bit 5 (*irq5) is available on connector P6, pin 9
* Port F bit 6 (*irq6) is available on connector P6, pin 10
* Port F bit 7 (*irq7) is available on connector P6, pin 11

* constants for debugging/monitoring pins [debug]
bit7    equ     %10000000
bit6    equ     %01000000
bit5    equ     %00100000
bit4    equ     %00010000
bit3    equ     %00001000
bit2    equ     %00000100
bit1    equ     %00000010
bit0    equ     %00000001

* This program:
* bit 1 of Port F is high while waiting and during irq
* bit 3 of Port F is high during interrupt processing
* bit 5 of Port F is high when a new sample is taken
* bit 7 of Port F is high during filter calculation

        include 'equates.asm'   ;table of EQUates for register addresses

*       Code space
        org     $f000           ;put coefs at top of program ROM
                                ;(note--this is NOT bank f, it is bank 0)
        include 'c10int1.h'     ;FIR interpolation filter coefs & constants
; Fs            output sampling rate in Hz
; Ni            FIR filter length, zero-pad if necessary so Ni/Li is an integer
; Li            interpolation factor
; NoverLi       Ni/Li, equals length of each filter
```

Appendix 5 • 68HC16 Sample Interpolation Program

```
; BSizei         input buffer size, power of 2 >= NoverLi (<=128!)
; Coefi          start of coefficients, stored in order used

* Some constants:
TOCINCR equ     4194304/Fs       ;value to give Fs Hz TOC rate

bmask   equ     (BSizei*2)-1     ;mask with n+1 bits set (BSizei=2^n)

*       Variables (in internal RAM)
        org     $0000            ;(in bank f, actually)
inbuff  ds.w    BSizei           ;input buffer storage
topbuff equ     *-2              ;addr of last entry (starting point)
* Note--you must start inbuff on an appropriate address boundary
* (i.e. the address must have all 0's where bmask has 1's)
* in order for modulo addressing to work properly
inptr   ds.w    1                ;input buffer pointer
cptr    ds.w    1                ;coef pointer
lcnt    ds.w    1                ;counter (which set of coefs to use)
lastout ds.w    1                ;last output value calculated

*       Code space
*       Reset and interrupt vectors
        org     $0000            ;put the following reset vector information
                                 ;at address $00000 of the memory map
        dc.w    $00f0            ;zk=0, sk=f, pk=0
        dc.w    $0200            ;pc=200 -- initial program counter
        dc.w    $03FE            ;sp=03fe -- initial stack pointer
        dc.w    $0000            ;iz=0 -- direct page pointer

*       Interrupt vectors (vector number in decimal)
*       Note that most (all but one!) vector points to BDM (end of this
*       program), where there's a 'bgnd' instruction to put the
*       system into background debug mode, since we don't expect
*       to see any other interrupts running!
        dc.w    BDM              ;4     Breakpoint (BKPT)
        dc.w    BDM              ;5     Bus Error (BERR)
        dc.w    BDM              ;6     Software Interrupt (SWI)
        dc.w    BDM              ;7     Illegal Instruction
        dc.w    BDM              ;8     Divide by Zero
        dcb.w   6,BDM            ;9-14  Unassigned Reserved
        dc.w    BDM              ;15    Uninitialized Interrupt
        dc.w    BDM              ;16    (Unassigned Reserved)
        dc.w    BDM              ;17    Level 1 Interrupt Autovector
        dc.w    BDM              ;18    Level 2 Interrupt Autovector
        dc.w    BDM              ;19    Level 3 Interrupt Autovector
        dc.w    BDM              ;20    Level 4 Interrupt Autovector
        dc.w    BDM              ;21    Level 5 Interrupt Autovector
        dc.w    BDM              ;22    Level 6 Interrupt Autovector
        dc.w    BDM              ;23    Level 7 Interrupt Autovector
        dc.w    BDM              ;24    Spurious Interrupt
        dcb.w   31,BDM           ;25-55 Unassigned Reserved
        dcb.w   8,BDM            ;56-63 User Interrupt 1-8
*       Interrupts from the time (GPT) occur at User Interrupt 9..20
```

```
*               (as the GPT is configured by this program)
                dc.w    BDM             ;64    (prioritized GPT interrupt)
                dc.w    BDM             ;65    IC1
                dc.w    BDM             ;66    IC2
                dc.w    BDM             ;67    IC3
                dc.w    BDM             ;68    OC1
                dc.w    irq_sample      ;69    OC2 ** we use this one
                dc.w    BDM             ;70    OC3
                dc.w    BDM             ;71    OC4
                dc.w    BDM             ;72    IC4/OC5
                dc.w    BDM             ;73    TOF
                dc.w    BDM             ;74    PAOVF
                dc.w    BDM             ;75    PAIF
                dcb.w   180,BDM         ;76-255 User Interrupt 21-200

                org     $0200           ;start program after interrupt vectors

*               Initialization Routines
INITSYS:                                ;give initial values for extension registers
                                        ;and initialize system clock and COP
                orp     #$00e0          ; disable all interrupts
                ldab    #$0F            ; bank f, register/internal RAM access
                tbek                    ; point EK to bank F
                tbyk                    ; point YK to bank F
                tbzk                    ; point ZK to bank F
                ldab    #$00
                tbxk                    ; point XK to bank 0 (code space)

                ldd     #$0003          ; at reset, the CSBOOT block size is 512k.
                std     CSBARBT         ; this line sets the block size to 64k since
                                    ;   that is what physically comes with the EVB16
                ldd     #$7830
                std     CSORBT          ;set CSBOOT (the chip select to the 'ROM')
                                        ;to be 0 wait state, word wide, user/super
                                        ;(the important one is 0 wait states--
                                        ;the reset default is 13!)
                LDAA    #$7F            ; w=0, x=1, y=111111
                STAA    SYNCR           ; set system clock to 16.78 Mhz
                CLR     SYPCR           ; turn COP (software watchdog) off,
                                        ; since COP is on after reset

INITRAM:                                ;initialize internal 1k SRAM and stack
                ldd     #$8000          ;make sure RAM is in stop mode
                std     RAMMCR
                ldd     #$00ff          ;(ff since high bits follow a19)
                std     RAMBAH          ; store high ram array, bank $0f
                ldd     #$0000
                std     RAMBAL          ; store low ram array
                clr     RAMMCR          ; enable ram
                ldab    #$0f
                tbsk                    ; set SK to bank $0f for system stack
                lds     #$03FE          ; put SP at top of 1k internal SRAM
```

Appendix 5 • 68HC16 Sample Interpolation Program

```
       INITSCI:                     ;initialize the SCI
                                    ;(Serial Communications Interface--
                                    ;i.e. asynchronous serial port)
            ldd      #$0037
            std      SCCR0          ;set the SCI baud rate to 9600 baud

            ldd      #$000C
            std      SCCR1          ;enable the SCI receiver and transmitter

       INITXRAM:                    ;Initialize external RAM (optional)
                                    ;Initialize the Chip Selects.....
            ldd      #$0203
            std      CSBAR0         ;set U3 RAM base addr to $20000: 64k block
            std      CSBAR1         ;set U1 RAM base addr to $20000: 64k block
            ldd      #$5070         ; (was $5030)
            std      CSOR0          ;set Chip Select 0 (U3)
                                    ;upper byte, emulate asynch, write only,
                                    ;1 wait state, supervisor or user
            ldd      #$3070         ; (was $3030)
            std      CSOR1          ;set Chip Select 1 (U1)
                                    ;lower byte, write only, 1 wait, super/user
            ldd      #$0203
            std      CSBAR2         ;set Chip Select 2 to fire at base addr $20000
                                    ;$02 0000, in 64k blocks
            ldd      #$7870         ; (was $7830)
            std      CSOR2          ;set Chip Select 2, both bytes, read and write
                                    ;both bytes, r/w, 1 wait, super/user
            ldd      #$3fff
            std      CSPAR0         ;set CSBOOT, CS0-CS5 to 16-bit ports

       INITQSPI:                    ;Initialize the QSPI
                                    ;(Queued Serial Peripheral Interface--
                                    ;i.e. synchronous serial port, to DAC)
            ldaa     #$08
            staa     QPDR           ;set PSC0 high between serial transfers
            ldaa     #$0A
            staa     QPAR           ;assign Port D pins as PSC0/SS, MOSI
            ldaa     #$0E
            staa     QDDR           ;set data direction as output on MOSI, SCK pins
            ldd      #$8002
            std      SPCR0          ;master mode, 16 bits, clock low when inactive,
                                    ;change data on falling edge of SCK
                                    ;SCK is 4.19 MHz
            ldd      #$0201
            std      SPCR1          ;set delay between PCS0 and SCK=2 * SYS CLK
                                    ;set delay between transfers=32 * SYS CLK
                                    ;enable =0 (for now)
            ldd      #$0000
            std      SPCR2          ;no irq, no wrap enable, no wrap to
                                    ; end queue ptr=0 (only 1 transfer)
                                    ; new queue ptr=0 (only 1 transfer)
       *         Initialize QSPI Command RAM
            ldab     #$50           ;CONT=no (0), BITSE=8 bits (1), DT=none (0)
```

```
                                        ;DSCK=use DSCKL (1), PCS0= addr 0
                stab    CR0             ;QSPI command ram 0

        INITGPT:                        ;Initialize the GPT (Timer)
                ldd     #$000f
                std     GPTMCR          ;ignore freeze, normal operation,
                                        ;user/supervisor access,
                                        ;irq arbitration at level f (highest)
                ldd     #$0000
                std     TMSK1           ;(/TMSK2) clr all interrupt mask
                                        ;and set CPR to /4 sys clk to TCNT
                ldd     #$0100
                std     TCTL1           ;(/TCTL2) toggle OC2 output on OC2 capture

                ldd     TFLG1           ;read to initialize flags

        INITADC:                        ;Initialize ADC
                ldd     #$0083
                std     ADCTL0          ;10-bit conversion, 2 sample periods,
                                        ;2.1 MHz ADC clock
                ldd     #$0000
                std     ADCMCR          ;run, ignore freeze, user/supervisor access

        INITPORT:                       ;Initialize I/O ports for [debug]
                                        ;Note--ports are 8-bit, not 16-bit!
                                        ;(Part of SIM module)
                ldab    #$ff
                stab    PORTF0          ;normally high outputs
                ldab    #$ff
                stab    DDRF            ;make output (vs input)
                ldab    #$01
                stab    PFPAR           ;enable above
                                        ;(0=user i/o; 1=system function irq1..7)

        *       Done initializing modules

        *       Clear on-chip RAM to 00's (for debug and clean startup)
        *       (Note--cannot be done as a subroutine call--stack is here)
        *       (YK pts to bank f)
                ldy     #$03fe          ;start at top
        clrloop:
                clrw    $0000,y         ;($0000 is just an offset)
                aiy     #-2
                bne     clrloop

        *       Set up the timer output compare register
                ldd     TCNT            ;get current timer value
                addd    #TOCINCR        ;add to find next sample time
                std     TOC2            ;use TOC2
        *       TOC2 will set a flag when TCNT (16-bit counter) matches TOC2
        *       Enable interrupt from TOC2
                ldab    #$10            ;OC2I=enable, others disabled
                stab    TMSK1
```

Appendix 5 • 68HC16 Sample Interpolation Program

```
            ldd     #$0540              ;set interrupt level, etc.
            std     ICR                 ;interrupt level 5,
                                        ;interrupt vectors are $40-$4b,
                                        ;(TOC2 interrupt is at vector $45)

*           Initialize variables
            ldd     #topbuff            ;start at top of buffer
            std     inptr
            ldaa    #Li                 ;count down from Li (byte wide)
            staa    lcnt
            ldd     #Coefi              ;init coef pointer
            std     cptr

*           set up "modulo" addressing registers (circular buffer)
            ldd     #bmask              ;modulo addressing mask
                                        ;XMSK (coefs) =$00 (modulo disabled)
                                        ;YMSK (data) = bmask (modulo enabled)
            tdmsk

*           Enable saturate mode, interrupts at level 5 and above
*           (the TOC2 interrupt is a level 5 interrupt)
            tpd                         ;get CCR
            ord     #$0010              ;set SM bit
            andd    #$ff1f              ;clear IP bits
            ord     #$0080              ;set to level 4 (enable 5 and up)
            tdp

*           Main loop
WAITLOOP:
            bset    PORTF0,#bit1        ;turn on bit 1 during wait [debug]
            wai                         ;sit here until interrupt
                                        ;(minimum and consistent interrupt latency)
            bclr    PORTF0,#bit1        ;turn off bit 1 [debug]
            bra     WAITLOOP

* Interrupt service routine, called at sample rate
irq_sample:
                                        ;(normally, stack registers here)
            bset    PORTF0,#bit3        ;turn on bit 3 during irq [debug]

*           clear the irq flags in the timer
            ldd     TFLG1               ;read the register first
            ldd     #$0000
            std     TFLG1               ;then write a negated flag back (i.e. 0)

            ldd     TOC2                ;set up TOC for next sample
            addd    #TOCINCR
            std     TOC2

*           if switch pressed, just pass through the ADC data
            brset   GPTPDR,#$01,noswitch ;br if no switch
            jmp     passthru
noswitch:
```

```
*       output last calculated value to DAC
                                ;(deterministic timing)
        lde     lastout
        ste     TR0             ;store 16-bit data in QSPI output queue
        ldaa    #$80
        oraa    SPCR1           ;read SPCR1
        staa    SPCR1           ;just enable the QSPI

*       time for next input sample?
        dec     lcnt
        bne     noinput

        bset    PORTF0,#bit5    ;turn on bit 5 if sampling [debug]

*       read the ADC, put value in register D
        ldd     #$0000          ;single conversion sequence,
                                ;one channel, 4 conversions
                                ;channel 0
        std     ADCTL1          ;and start next conversion
        ldd     LJSRR0          ;grab last conversion

*       store sample in circular buffer
        ldy     inptr
        cpy     #inbuff         ;will decrement past start of buffer?
        bgt     nowrap
        ldy     #topbuff+2      ;start at top again (+2)
nowrap:
        aiy     #-2
        sty     inptr
        std     0,y             ;store actual data
                                ;(this is the lowest valid location
                                ;in buffer)
*       reset lcnt, cptr
        ldaa    #Li             ;count down from Li (byte wide)
        staa    lcnt
        ldd     #Coefi          ;start coef pointer over at beginning
        std     cptr

        bclr    PORTF0,#bit5    ;turn off bit 5, done w/sampling [debug]

noinput:

        bset    PORTF0,#bit7    ;turn on bit 7 during filter calc [debug]

*       Evaluate filter
        clrm                    ;clear AM (the MAC accumulator)
        lde     #NoverLi-1      ;number of times (-1) to do RMAC
        ldx     cptr            ;x=coef ptr
        ldy     inptr           ;y=input data ptr
        ldhi                    ;load H and I registers using ix,iy

        rmac    $2,$2           ;repeating MAC instruction:
```

Appendix 5 • 68HC16 Sample Interpolation Program

```
                tmer                    ;transfer AM to E (saturated if necessary)
                ldd     #Li             ;scale: E is 1.15, D is 16.0
                emuls                   ;E * D -> E:D (E:D is 17.15, D is 1.15)
                std     lastout         ;save to output next time

                ldx     cptr            ;update coef ptr
                aix     #(NoverLi*2)
                stx     cptr

                bclr    PORTF0,#bit7    ;turn off bit 7, done w/filter calc [debug]
endirq:
                bclr    PORTF0,#bit3    ;turn off bit 3, done w/sample [debug]

                rti                     ;end isr

*       If IC1 (pin 19 of connector P4 on EVB) is taken low
*       (e.g. a switch to ground, with pull-up resistor to +5V)
*       this routine replaces interrupt routine above.
*       Just passes through ADC results at the slower rate.
passthru:
*       time for next input sample?
                dec     lcnt
                bne     ptnoinput
*       read the ADC, put value in register D
                ldd     #$0000          ;single conversion sequence,
                                        ;one channel, 4 conversions
                                        ;channel 0
                std     ADCTL1          ;and start next conversion
                ldd     LJSRR0          ;grab last conversion
                std     TR0             ;store 16-bit data in QSPI output queue
                ldaa    #$80
                oraa    SPCR1           ;read SPCR1
                staa    SPCR1           ;just enable the QSPI

*       reset lcnt, cptr
                ldaa    #Li             ;count down from Li (byte wide)
                staa    lcnt
                ldd     #Coefi          ;start coef pointer over at beginning
                std     cptr
ptnoinput:
                jmp     endirq



*       Other interrupts and exceptions go into background debug mode
BDM:
                bgnd                    ;exception vectors point here
                                        ;and put the user in background mode



* end of file c10int1.ASM *
```

File C10INT1.H

```
*       Coefficient file "c10int1.h"
*       Generated 8-Jan-97
*       Interpolation FIR filter coefficients
*
*       Fpb=300 Hz, Fsb=700 Hz, Fs (orig)=1000 Hz
*       Ap=0.5 dB, As=40 dB
*
*
*
*
Fs       equ     6000            ;output sampling rate
Ni       equ     30              ;coefficients, including zeros
Li       equ     6               ;Interpolation factor
NoverLi  equ     5               ;N/L, length of filters
BSizei   equ     8               ;next power of 2 higher than N/L
Coefi:                           ;filter coefficients in 1.15 format
                                 ;filter bank 0
         dc.w    -266            ;h(0)=-0.008111
         dc.w    290             ;h(6)= 0.008851
         dc.w    5127            ;h(12)= 0.156457
         dc.w    290             ;h(18)= 0.008851
         dc.w    -266            ;h(24)=-0.008111
                                 ;filter bank 1
         dc.w    -403            ;h(1)=-0.012311
         dc.w    1157            ;h(7)= 0.035323
         dc.w    4895            ;h(13)= 0.149389
         dc.w    -310            ;h(19)=-0.009458
         dc.w    0               ;h(25)= 0.000000
                                 ;filter bank 2
         dc.w    -586            ;h(2)=-0.017894
         dc.w    2206            ;h(8)= 0.067326
         dc.w    4244            ;h(14)= 0.129528
         dc.w    -623            ;h(20)=-0.019015
         dc.w    0               ;h(26)= 0.000000
                                 ;filter bank 3
         dc.w    -689            ;h(3)=-0.021038
         dc.w    3293            ;h(9)= 0.100493
         dc.w    3293            ;h(15)= 0.100493
         dc.w    -689            ;h(21)=-0.021038
         dc.w    0               ;h(27)= 0.000000
                                 ;filter bank 4
         dc.w    -623            ;h(4)=-0.019015
         dc.w    4244            ;h(10)= 0.129528
         dc.w    2206            ;h(16)= 0.067326
         dc.w    -586            ;h(22)=-0.017894
         dc.w    0               ;h(28)= 0.000000
                                 ;filter bank 5
         dc.w    -310            ;h(5)=-0.009458
         dc.w    4895            ;h(11)= 0.149389
         dc.w    1157            ;h(17)= 0.035323
```

```
            dc.w    -403            ;h(23)=-0.012311
            dc.w    0               ;h(29)= 0.000000


    * End of file *
```

Glossary

Numbers in parentheses refer to pages in the text where more information on the term can be found.

AC (alternating current)   An AC signal has an amplitude that varies with time. Compare to DC. (37)

ACTIVE FILTERS   Analog filters that include components with power gain (amplification), usually OP-AMPs. (Passive filters have no power gain components.) (85)

ADAPTIVE FILTERS   Filters whose behavior changes dynamically in response to changes in the signal being processed. (9)

ALIASING   The effect by which signal components with frequencies higher than the NYQUIST FREQUENCY are, once sampled, indistinguishable from sampled components with frequencies below the NYQUIST FREQUENCY. (95)

ALL-PASS FILTER   A filter that passes all frequencies without affecting magnitude. Usually employed for its phase response. (72)

AMPLITUDE   The instantaneous value of a signal at a particular time. A TIME DOMAIN measure, compare to MAGNITUDE and PHASE in the FREQUENCY DOMAIN. (21)

ANALOG SIGNAL   A signal that is continuous in both value and in time (or other dimension such as space). Compare to DIGITAL and DISCRETE-TIME. (4)

ANALOG-TO-DIGITAL CONVERTER (ADC)   A device that samples and quantizes an analog signal, producing a digital signal. Compare to DIGITAL-TO-ANALOG CONVERTER (DAC). (128)

ANTIALIASING FILTER   A filter that removes undesired frequency components that would otherwise be aliased by sampling. (102)

ANTIIMAGING FILTER   (or reconstruction filter) A filter designed to remove high frequency artifacts in the output from a DIGITAL-TO-ANALOG CONVERTER. (144)

APERTURE JITTER   The variation in sampling period from sample to sample in a SAM-

PLE AND HOLD AMPLIFIER (SHA). (132)

APERTURE TIME   The time during which an analog signal is sampled in a SAMPLE AND HOLD AMPLIFIER (SHA). (132)

ATTENUATION   A reduction in the magnitude of a signal, the opposite of gain. Note X dB of attenuation is the same as $-X$ dB of gain. (42)

AUTOCORRELATION   An operation that quantifies the similarity between two (possibly identical) sections of the same signal. Compare to CROSSCORRELATION. (397)

BANDLIMITED   A signal with all its energy limited to a specific range of frequencies. (127)

BANDPASS   A signal or filter with a limited frequency range (not including 0 Hz). (37)

BANDPASS (BP) FILTER   A filter that passes a limited range of frequencies, attenuating frequencies higher and lower than the passband. Compare to LOWPASS, HIGHPASS, and BANDSTOP filters. (69)

BANDSTOP   A filter that passes all but a (narrow) range of frequencies. Compare to LOWPASS, HIGHPASS, and BANDPASS filters. (69)

BANDWIDTH (BW)   The range of frequencies a filter or signal covers. Often measured between -3 dB points. (37)

BARREL SHIFTER   A unit in digital processors (e.g., microprocessors and DSP chips) that performs efficient, "single-cycle" shifts of data.

BASEBAND   A signal whose frequency range starts at 0 Hz (or close to it). Compare to BANDPASS. (37)

BIAS   The average value of a signal. Also called the *DC-bias*. (37)

BILINEAR Z-TRANSFORM (BZT)   A mathematical tool that transforms descriptions in the continuous complex frequency domain (s-domain) to the sampled or discrete complex frequency domain (z-domain). (240)

BIPOLAR   A bipolar signal has amplitudes above and below zero. Compare to UNIPOLAR. (119)

BRICK-WALL FILTER   A filter with an infinitely sharp transition between the PASSBAND and STOPBAND. Such a filter is NONCAUSAL and, thus, cannot be realized in the real world. (67)

BUTTERFLY   The core operation of the FAST FOURIER TRANSFORM (FFT), involving a complex multiplication, negation, and two additions. (359)

CAUSAL   A system whose current output depends only on current and past inputs and the current state of the system, and not on future input values. Compare to NONCAUSAL. (67)

CENTER FREQUENCY   In a (narrow) bandpass filter, the point of maximum gain, generally at the geometric mean of the passband cutoff frequencies. (78)

CODEC (COder/DECoder)   ADC/DAC combinations that may or may not include nonlinear compression. (120)

COMB FILTER   A filter with repetitive, usually equally spaced stopbands. Useful for removing signals and their harmonics. (72)

COMPLEX FREQUENCY   A mathematical concept that extends the notion of frequency to include an exponential growth/decay factor. See S-PLANE, Z-PLANE. (52)

COMPLEX NUMBER   A number with both a real and an "imaginary" component. May be expressed in polar or rectangular forms. (29)

CONTROL   1. Digital processing of digital inputs to control digital devices or systems. 2. Analog or digital signal processing to control analog devices using control system theory. (13)

CONVOLUTION   A mathematical operation involving the multiplication and summation of elements from two continuous or discrete signals to produce a third signal. One of these sequences is reversed in time. (163)

CROSSCORRELATION   An operation that quantifies the similarity between two signals. Compare to AUTOCORRELATION. (384)

DC (direct current)   A DC signal has a constant value over time. Compare to AC. Also stands for a frequency of 0 Hz. (37)

DECADE   A factor of ten. For example, 20 kHz is one decade above 2 kHz. Compare to OCTAVE. (36)

DECIBEL (dB)   A logarithmic measure of the ratio of the power in two signals. Defined as $10\log(Pa/Pb)$. (42)

DECIMATION   Decreasing the sampling rate of a discrete-time signal. Compare to INTERPOLATION. (413)

DIFFERENCE EQUATION   An equation describing a system output in terms of discrete-time input, system state, and output values. The discrete-time equivalent to a DIFFERENTIAL EQUATION. (114)

DIFFERENTIAL EQUATION   An equation describing a system output in terms of continuous time input, system state, and output values. Compare to the discrete-time equivalent, DIFFERENCE EQUATION. (114)

DIGITAL SIGNAL   A signal that is discrete in both value and time (or other dimension, such as space). Compare to ANALOG and DISCRETE-TIME. (2)

DIGITAL SIGNAL PROCESSING (DSP)   Using a digital process (e.g., a program running on a microprocessor) to modify a digital representation of a signal. (2)

DISCRETE COSINE TRANSFORM (DCT)   A transform that is closely related to the DFT but uses real math instead of the complex math of the DFT/FFT. (379)

DISCRETE FOURIER TRANSFORM (DFT)   A mathematical tool that transforms descriptions of systems or signals in the discrete-time domain into descriptions in the discrete (sampled) frequency domain. See also FFT. (161)

DISCRETE-TIME (DT)   Signals and systems with values or outputs that exist only at discrete points in time. The values themselves are not necessarily quantized. Compare to DIGITAL. (94)

DISCRETE-TIME FOURIER TRANSFORM (DTFT)   A mathematical tool that transforms descriptions of signals or systems in the discrete-time domain to descriptions that are functions of continuous frequency. See Z-TRANSFORM. (114)

DISTORTION   Changes to a signal imposed by a system. May be desired (e.g., amplitude distortion) or undesired (e.g., nonlinear distortion). (67)

DTMF (Dual Tone Multiple Frequency)   Also known as Touch-Tone,® signals consisting of two simultaneous tones, one from a set of four row frequencies and one from a set of four column frequencies. (21)

DYNAMIC RANGE   The ratio (usually expressed in dB) of the step size of a quantizer to the FULL-SCALE RANGE. (119)

ENERGY   The total amount of work or power delivered over a period of time, measured in joules. See WORK.

FAST CONVOLUTION   Using the fact that convolution in one domain (e.g., time) is equivalent to multiplication in the other domain (e.g., frequency), convolution can be carried out on large blocks of data efficiently using the FAST FOURIER TRANSFORM. (197)

FAST FOURIER TRANSFORM (FFT)   A family of computationally efficient algorithms for evaluating the DISCRETE FOURIER TRANSFORM (DFT). (333)

FILTER   A system (often linear, time-invariant) with desirable magnitude and phase responses. Usually used to pass or attenuate specific ranges of frequencies. (63)

FINITE WORD-LENGTH (FWL) EFFECTS   Differences arising from the use of a finite number of bits to represent and calculate values in a digital processor. (244)

FOURIER TRANSFORM (FT)   A mathematical tool that transforms descriptions of signals or systems in the continuous time domain to descriptions as a function of (continuous) frequency. See LAPLACE TRANSFORM. (55)

FPGA (Field Programmable Gate Array)   A digital logic chip whose function can be defined after production (i.e., "in the field") and which contains many thousands of gates. (2)

FREQUENCY COMPONENT   One of one or more single-frequency signals that compose an arbitrary signal. (26)

FREQUENCY DOMAIN   A method of describing signals and systems as functions of frequency. Compare to TIME DOMAIN. (6)

FREQUENCY RESPONSE   The magnitude and phase response of a system as a function of frequency. Contrast with TIME DOMAIN responses like the IMPULSE RESPONSE. (49)

FULL-SCALE (FS)   The acceptable range of input values for UNIPOLAR signals converted by a quantizer. Compare to FULL-SCALE RANGE. (119)

FULL-SCALE RANGE (FSR)   The acceptable range of input values for BIPOLAR signals converted by a quantizer. Compare to FULL-SCALE. (119)

GAUSSIAN (or NORMAL) DISTRIBUTION   The familiar "bell-shaped" distribution of values seen in many random processes. Most PSEUDO-RANDOM NUMBER GENERATORS, however, produce sequences with UNIFORM DISTRIBUTION.

GIBB'S PHENOMENON   The "ringing" that results from transforming a signal that has been truncated. For example, the ringing in the frequency domain associated with a truncated (time domain) impulse response. (173)

GOERTZEL ALGORITHM   An algorithm that can efficiently compute the discrete-frequency spectrum at a limited number of frequencies. Related to the DISCRETE FOURIER TRANSFORM. (362)

HALF-POWER POINT   The -3 dB point, at which the ratio of the power of the two signals (or signal and reference) is 0.5. (43)

HIGHPASS (HP) FILTER   A filter that passes high frequencies while attenuating low frequencies. Compare to LOWPASS, BANDPASS, and BANDSTOP filters. (68)

IMPEDANCE   A complex-valued, often frequency-dependent property of an analog device or system. A generalization of resistance, it is a function of COMPLEX FREQUENCY. (84)

IMPULSE RESPONSE The time domain output of a system in response to an input consisting of an impulse. (47)

IN-BAND NOISE Noise (undesired signals) with frequency components occupying the same frequency range as the desired signal. Compare to OUT-OF-BAND NOISE. (127)

INTERPOLATION Increasing the sampling rate of a discrete-time signal. Compare to DECIMATION. (413)

INTERRUPT LATENCY The delay between the moment an interrupt request is generated by an external device and when execution of the interrupt service routine begins. (106)

LAPLACE TRANSFORM (LT) A mathematical tool that transforms descriptions of signals or systems in the continuous time domain to the continuous complex frequency domain (s-domain). See FOURIER TRANSFORM. (55)

LEAST-SIGNIFICANT BIT (LSB) The binary digit with the smallest weight in a binary word. The rightmost or zero bit. Compare to MOST-SIGNIFICANT BIT (MSB). (119)

LEFT-HALF PLANE (LHP) In the s-plane, the region to the left of the origin. Poles in the LHP are associated with stable systems. (89)

LINEAR PHASE The property of a system such that all frequency components experience the same time delay. When phase delay is plotted for a system with linear phase, a straight line is produced. (82)

LINEAR SYSTEM A system whose properties include SCALING and SUPERPOSITION. (58)

LOBES Peaks in the magnitude frequency spectrum. The main lobe (the passband) is often accompanied by side lobes. (158)

LOWPASS (LP) FILTER A filter whose passband includes 0 Hz. Compare to BANDPASS, HIGHPASS, and BANDSTOP filters. (68)

MAC (Multiply and ACcumulate) An operation consisting of a multiplication followed by an addition of the result to an accumulator. Usually heavily optimized, this operation is central to digital filtering. (270)

MAGNITUDE A measure of the intensity of a signal or component of a signal, usually in the frequency domain. Compare to AMPLITUDE. See also PHASE. (24)

MAGNITUDE RESPONSE (also MAGNITUDE SPECTRUM) A plot of the magnitude output of a system as a function of frequency. Often accompanied by the PHASE RESPONSE. (52)

MARGINAL STABILITY A property of systems with poles located on the imaginary axis in the S-PLANE or the unit circle in the Z-PLANE; such systems are associated with oscillation that neither grows nor decays. (89)

MATCHED FILTER Crosscorrelation of an expected waveform with an input signal that may contain that waveform. (389)

MICROCONTROLLER A microprocessor that also includes on-chip data and program memory, and often a selection of I/O, timing, and communication peripherals. The MC68HC16Z1 is an example. (2)

MINIMUM STOPBAND ATTENUATION See STOPBAND RIPPLE. (78)

MIXED-SIGNAL PROCESSING System design that includes both analog and digital signal processing or issues. (15)

MOST-SIGNIFICANT BIT (MSB)   The binary digit with the largest weight in a binary word. The leftmost bit. (Bit 7 in a byte, bit 15 in a 16-bit word.) Compare to LEAST-SIGNIFICANT BIT (LSB). (119)

MULTIPLEXER   A circuit that can route any of N inputs to one output; useful in routing multiple signals to a single ADC.

NONCAUSAL   A system whose current output is described (at least partially) in terms of future inputs. Such a system requires seeing into the future and, thus, cannot be realized. (67)

NONLINEAR SYSTEM   A system that lacks the SCALING or SUPERPOSITION properties. Compare to LINEAR SYSTEM. (58)

NORMALIZED FREQUENCY   If the sampling rate is set to 1 Hz, all discrete-time signals have spectra with a normalized frequency range from 0 to 1 (or 0 to π if measured in rad/sec). Compare to REAL-WORLD FREQUENCY. (106)

NYQUIST FREQUENCY (or folding frequency)   One-half of the sampling frequency; the frequency above which ALIASING will occur. Compare to NYQUIST RATE. (96)

NYQUIST RANGE   The frequency range from 0 to one-half the sampling rate. (144)

NYQUIST RATE   The sampling rate above which one must sample to avoid aliasing any components of a given signal. Compare to NYQUIST FREQUENCY. (99)

OCTAVE   A doubling in frequency. For example, 6 kHz is one octave above 3 kHz. Compare to DECADE. (36)

OP-AMP (operational amplifier)   A two-input, one-output analog device that can realize a number of operations, including voltage gain. They are the building blocks for active analog filters. (85)

ORTHOGONAL   The nonspecialization of register usage in a processor. An orthogonal instruction set allows any register to be used in an instruction that references a register. (267)

OUT-OF-BAND NOISE   Noise with frequency components outside the frequency range of the desired signal and, thus, noise that can be removed by filtering. Compare to IN-BAND NOISE. (127)

OVERSHOOT   A time domain characteristic of a system. The amount by which the output of a system temporarily exceeds the "proper" output value. (64)

PARTIAL FRACTION EXPANSION (PFE)   A mathematical technique for breaking up a lengthy rational expression into a sum of smaller (usually second-order) expressions. (249)

PASSBAND CUTOFF   The frequency that marks the edge of a passband. Frequencies on the passband side of the passband cutoff have a magnitude response within the PASSBAND RIPPLE tolerance. (77)

PASSBAND DEVIATION   One way of specifying PASSBAND RIPPLE, the amount of variation allowed in the passband magnitude response of a system (filter). (77)

PASSBAND EDGE FREQUENCY   See PASSBAND CUTOFF FREQUENCY. (77)

PASSBAND RIPPLE   The allowed variation in magnitude response in the passband of a system (filter). Defined differently for analog/IIR and FIR filters. See also PASSBAND DEVIATION. (77)

PHASE   The shift along the time axis of a periodic signal (or frequency component) relative to some reference, expressed as an angular measure where a complete revolution (i.e., 360 degrees) represents a shift of one period. (29)

PHASE RESPONSE (also PHASE SPECTRUM)   A plot of the phase shift of the output of a system as a function of frequency. See also MAGNITUDE RESPONSE. (52)

POLE   The (complex) frequency at which the system function has infinite magnitude response (i.e., a zero of the denominator of the system function). Often mapped as a point in the *S*-PLANE or *Z*-PLANE. Compare to ZERO. (85)

POWER   The rate of delivering energy or of performing work. Power is measured in Joules/sec or watts (W). See ENERGY.

PSEUDO-RANDOM NUMBER GENERATOR (PRNG)   An algorithm for producing number sequences with characteristics similar to random number sequences. (431)

PULSE-CODED MODULATION (PCM)   A form of signal encoding where an analog signal is transmitted by sampling and quantizing, and the resulting digital words transmitted as a serial bit stream. (121)

Q or QUALITY FACTOR   The ratio of the center frequency to the bandwidth in a bandpass filter. (79)

QUANTIZATION   Conversion from continuous amplitude values of a signal to discrete values. Part of the analog-to-digital conversion process. (118)

QUANTIZER   A device that converts continuous amplitude values into a finite number of discrete amplitude values. (133)

REAL-WORLD FREQUENCY   What we would normally call *frequency*; just a label to differentiate actual frequency from NORMALIZED FREQUENCY. (106)

RECIPROCAL PAIRS   In the *z*-domain, zeros that are reciprocals of each other. For example if a zero is at Z1, its reciprocal will be 1/Z1. (Z1 is complex-valued.) (195)

RISE TIME   A time domain characteristic of a system. The rate of change of a system in response to a step change in input. (64)

RMS (root-mean-square)   A measure of an AC signal that tells you the DC voltage that would deliver the same amount of power. The "117 VAC" of house wiring is measured in RMS. (37)

SAMPLE-AND-HOLD AMPLIFIER (SHA)   A device that samples an analog signal and holds this amplitude value steady; often used in an ADC prior to quantization. See TRACK AND HOLD AMPLIFIER. (132)

SAMPLING   The process of converting an ANALOG signal into a DISCRETE-TIME signal, usually by noting the value of an analog signal at a regular SAMPLING RATE. (95)

SAMPLING FREQUENCY   The frequency with which an ANALOG signal is sampled in order to convert it to a DISCRETE-TIME signal. (95)

SAMPLING PERIOD   The reciprocal of the SAMPLING FREQUENCY; the time between sampling of an analog signal or the time between consecutive samples in a DISCRETE-TIME signal. (95)

SAMPLING RATE   See SAMPLING FREQUENCY. (95)

SATURATION   An operation that prevents large positive or large negative values from exceeding the limits of a register. Results that would normally cause an overflow are replaced with the largest positive (or negative) numbers that are valid. (271)

SCALING   The property of a linear system such that if the input to a system is scaled, the output will be scaled by the same amount. (58)

SECOND-ORDER SECTION (SOS) Complex conjugate poles (and zeros) pair off to form SOSs with real (vs. complex) coefficients. IIR filters are often expressed as the product of SOSs (and up to one first-order section (FOS)). (221)

SETTLING TIME A time-domain characteristic of a system. The amount of time a system's output takes to settle down to a new value after a step change in input. (64)

SIGNAL A physical quantity of interest, often a function of time. An example is the voltage in a telephone line. (3)

SIGNAL-TO-NOISE RATIO (SNR) The ratio of the power of desired frequency components in a signal to the power of the noise in that signal, expressed in decibels. (123)

SIGNAL-TO-QUANTIZATION-NOISE RATIO (SQNR) The ratio of desired frequency components of a signal to the noise introduced by quantization. See SIGNAL-TO-NOISE RATIO (SNR). (123)

SMEARING The blending of closely spaced frequency components in a DFT magnitude spectrum caused by the width of the main lobe of the window function used. (342)

SPECTRAL LEAKAGE Inaccuracies in the DFT magnitude spectrum caused by the side lobes of the windowing function used. (341)

S-PLANE A graphical representation of the COMPLEX FREQUENCY domain (s) that can be used to plot the POLES and ZEROS of systems. (54)

STABILITY A property of systems. One measure is whether the system's theoretical output remains finite for any finite amplitude input. (64)

STATIONARY A signal is said to be stationary or time invariant if its frequency content is constant over time. (26)

STEADY-STATE The condition of a system once transient conditions have settled down and if the input is STATIONARY. (50)

STEP SIZE The difference between two adjacent quantization levels in a QUANTIZER. (119, 120)

STOPBAND CUTOFF FREQUENCY The start (or end) of the frequency range that meets the STOPBAND RIPPLE tolerance. (78)

STOPBAND EDGE FREQUENCY See STOPBAND CUT-OFF FREQUENCY. (78)

STOPBAND RIPPLE The maximum gain in the stopband of a system. (Analogous to PASSBAND RIPPLE.) (78)

SUPERPOSITION The property of LINEAR SYSTEMS such that the frequency components of signals are processed the same, regardless of the presence of other frequency components. (58)

SYSTEM A device that modifies an input signal to produce an output signal. See also FILTER. (20)

SYSTEM FUNCTION A function ($H(s)$ or $H(z)$) that describes the output of a system as a function of COMPLEX FREQUENCY (s- or z-domain). (52)

TIME DOMAIN A method of describing signals and systems as functions of time. Compare to FREQUENCY DOMAIN. (6)

TIME INVARIANCE A property of systems such that for a given input, the output is the same (aside from a shift in time), regardless of when that input is presented. (59)

TRACK-AND-HOLD AMPLIFIER (THA) A device that tracks the amplitude of an analog signal and, when directed, holds this amplitude value steady. See SAMPLE AND HOLD AMPLIFIER (SHA). (132)

TRANSDUCER   A device that translates a signal in one form into a signal in another form. A microphone is one example, where a sound pressure signal is translated into a voltage signal. (4)

TRANSIENT   Temporary conditions that eventually give way to STEADY-STATE conditions. (64)

TWIDDLE FACTORS   The constants used in evaluating the discrete Fourier transform (DFT) and the fast Fourier transform (FFT). (355)

UNDERSAMPLING   Sampling a signal at less than the NYQUIST RATE in order to alias a (bandpass) signal on purpose. (104)

UNIFORM DISTRIBUTION   The probability of any particular value in a sequence is the same as for any other value. Compare to GAUSSIAN (NORMAL) DISTRIBUTION, where certain values are far more common. (438)

UNIPOLAR   A unipolar signal has amplitudes greater than or equal to zero. Compare to BIPOLAR. (119)

UNIT CIRCLE   In the z-plane, the circle of radius 1 centered on the origin. Poles inside the unit circle are associated with stable systems. (112)

UNITY GAIN   A gain of 1 (equal to 0 dB). (43)

WAVELET TRANSFORM   A family of mathematical tools similar to the Fourier transforms but using waveforms ("basis functions") other than sine and cosine. (378)

WINDOW FUNCTION   A weighting that is applied to a finite-length signal to minimize the effects of truncation. (176)

ZERO   A (complex) frequency at which the system function has zero magnitude response (i.e., a zero of the system function numerator). Often mapped as a point on the S-PLANE or Z-PLANE. See also POLE. (85)

ZERO PADDING   The process of adding additional zero-valued samples to the input of a DFT in order to increase the "sampling rate" of the DFT with respect to the underlying DTFT spectrum. (352)

Z-PLANE   A graphical representation of the discrete COMPLEX FREQUENCY domain (z) that can be used to plot the POLES and ZEROS of discrete-time systems. (109)

Z-TRANSFORM (ZT)   A mathematical tool that transforms descriptions of systems or signals in the discrete-time domain into the discrete complex frequency (z) domain. Related to the DISCRETE-TIME FOURIER TRANSFORM (DTFT). (113)

Index

Numerics

0 dB 43
1.15 (format) 287
-3 dB 43
    Butterworth filter 219

A

AC 37
    circuits 472
adaptive filtering 158
addressing 273
    bit-reversed 267
aliasing 95
analog-to-digital conversion 128–148
    antialiasing 126
    encoding 135
    gain 125
    impedance matching 125
    offset 125
    sample and hold 131–133
        amplifier (SHA) 132
        aperture time 132
        droop 132
        jitter 132
    transmit 135
analog-to-digital converters (ADC)
    direct converter 140
    flash 139
    half-flash 139
    integration 140
    multiplexers 139
    multistage 140
    parallel encoding 140
    sigma-delta 141
    subranging 139
    successive approximation 140
    voltage-to-frequency 140
Animal Farm 35
antialiasing filter 102, 126
    Bessel 233
    minimum attenuation 127
    specifications 127
anti-anti-aliasing filter 103
antiimaging filter 144, 418
    Bessel 233
arbitrary waveforms, generating 448
architects, use of guns by 49
architecture

Harvard 265
Von Neumann 266
The Art of Computer Programming 403, 431
The Art of Electronics 62, 155
assembly code
    absolute 278
    relocatable 278
asynchronous serial interface 136
attenuation 42
audio frequency analyzer 294, 336
autocorrelation 397–401
Avis 34

B

background debug mode (BDM) 299
bandpass 37
Bandpass Sampling Theorem 102
bandwidth (BW) 37, 78–79
barrel shifter 272
baseband 37
bejabbers 164
Bel 41
Bessel function, zero-order modified of the first kind 181
bias 37
bilinear z-transform (BZT) 240
biquads 315
blood flow 100
Bode plot 51
BT43367F 303

C

capacitors 470
    "vampire" effect 60
    dielectric absorption 60
Cartesian coordinates 462
causal filters 67
center frequency 78
central limit theorem 438
Chebychev polynomials 225, 441
chest pain (sharp). See a doctor immediately
chip selects 136
chirp-z transform 453

Cincinnati, hump that ate 347
circular buffers 267, 273
clipping 303
CODEC 120
compact disk (CD) 414
compensators 13
complex conjugates 90
complex frequency 52
complex numbers
    exponential form 465
    magnitude and phase 29
    polar coordinates 463
    review 461–465
complex signals 339
component drift 9
control, digital vs. control engineering 13
convention, mime (regional) 101
convolution 163
Cooley and Tukey 356
correlation 383–410
    See also autocorrelation, crosscorrelation
cosine. *See* sine
Crenshaw, Jack 440, 444
crosscorrelation 383–396
    and convolution 388
    detection of signal in additive noise 388
    implementation 396
        fast correlation 396
    lag 384
    matched filter 389
    normalized crosscorrelation sequence 385

D

data movement 275
    unaligned 275
data representations 287
    1.15 format 287
    signed fractional 287
DC 37
decade 37
decibel 42
decimation 415–418

Index

efficient 415
    stages 417
    See also rational interpolation/decimation
Deller, John R. (Jack) 397, 454
development system
    hardware 298
    software 299
difference equations 109, 114
digital audio tape (DAT) 414
digital filters
    design process 149
    implementation 263, 329
        architectures 264–277
            dedicated DSP chips 266
            DSP with microcontroller extensions 268
            DSP/microcontroller 265
            Harvard 269
            microcontroller 269
            microcontroller w/DSP extensions 268
            Von Neumann 266
        operations
            MAC 270
    notation 152
    overview 148
    selecting FIR vs. IIR 152
    See also FIR, IIR filters
digital-to-analog conversion 141–148
digital-to-analog converter (DAC)
    anti-imaging filter 144
    differential current output 143
    gain 147
    glitch in output 145
    impedance matching 147
    multiplying 143
    offset 147
    $\sin(x)/x$ compensation 146
Dirac impulse function 108
direct digital waveform synthesizers 23
discontinuity. *See* Gibb's Phenomenon
discrete cosine transform (DCT) 379
discrete Fourier transform (DFT) 161, 184, 337–362, 466
    See also fast Fourier transform (FFT)
    and DTFT 350
    frequency resolution 341
        increasing 352
    implementing 353–362
        computation 353
        twiddle factors 353
    inverse 377
    periodogram 352
    power spectrum 454
    relationship to FT 339
    scaling 342
    smearing 342
    spectral leakage 341
    windows 340, 347
    zero padding 352
discrete Walsh transform (DWT) 380
discrete-time (DT)
    signals
        describing 106
        sources 94
    systems
        describing 107
discrete-time discrete-frequency Fourier transform. *See* discrete Fourier transform (DFT)
discrete-time Fourier transform (DTFT) 114, 465
    amplitude density output 350
discrete-time signal processing 5
distortion 67
    amplitude 67
    nonlinear 68
    phase 68
droop (in SHA) 132
DSP
    applications 7
    definition 2
    limitations 12
    on microcontrollers 11
    vs. analog electronics 6
DSP56800 268
DTMF (dual tone multiple frequency) 21, 336
dynamic time warping 456

E

electroencephalogram 21
electromyographic activity 21
electronics (review) 469–476
EPROM emulator 300
Euler's identity 111, 187, 462

F

fast convolution 379
fast Fourier transform (FFT) 356–362, 466
    bit-reversal addressing 359
    butterfly 359
    decimation-in-frequency 356
    decimation in time (DIT) 356, 357
    implementing 362–377
    in-place 361
    length 357
    scaling 360, 375
        block floating-point 360
        floating-point 361
        unconditional 360
        unconditional per stage 360
    time-frequency analysis with 377
feature detection by crosscorrelation 392
feedback shift register (FSR). *See* pseudo-noise (PN)
filters
    "Brick-wall" 67
    all-pass 72
    analog 63–92
        active 85
        implementation 82–85
        passive 83
    anti-aliasing 102, 128
    anti-antialiasing 103
    arbitrary 69
    bandpass 69
    band-reject. *See* bandstop
    bandstop 69
    casual 67
    comb 72, 186
    highpass 68
    ideal vs. real 67
    lowpass 68
    multipassband. *See* filters, arbitrary.
    noncausal 67
    notch. *See* bandstop
    order 91
    peaking 69
    RC 83
    resonator 69
    specification 68–82
finite impulse response (FIR) filters. *See* FIR filters
finite word-length (FWL) effects 244, 301
    coefficient quantization 301
    limit cycles 303
        large-scale 303
        small-scale 303
FIR filters 149, 157–204
    68HC16 example 491–504
    design
        process 164
        tools 165
        via crosscorrelation 394
    fast convolution 197
    generating coefficients 166–199
        ad-hoc 195–197
        frequency sampling 183–194
            simple coefficient 186–194
            warping 186
        lowpass-to-highpass conversion 199
        optimal equiripple 166
        Parks-McClellan 166–171
        Remez exchange. *See* Parks-McClellan
        windowing 172–183
    implementation 303–315
    operation 161
    structures 199–203
        cascade 201
        direct 199
        linear phase 199
first-order hold 143
first-order section (FOS) 244
fish heads, raw 456
fixed-point 267
floating-point 267

Index

floobydust 298
folding frequency. *See* Nyquist frequency
Fourier series (FS) 465
Fourier transform (FT) 465
    relationship to Laplace 55
frequency analysis 333–380
    individual filters 335
    See also discrete Fourier transform (DFT)
frequency components 26
frequency domain
    plotting 38
frequency response
    function, continuous time 49
    steady-state, discrete-time 108

G

Gibb's Phenomenon 173, 453
Goertzel algorithm 336, 362

H

half-power point 43
harmonic distortion 60
hidden Markov model 454
Horowitz and Hill 62, 475
hotline, psychic, use in filter design 67

I

IIR filters 148, 207–261
    68HC16 example, 491–504
    design process 209
    direct design methods 210–212
        ad-hoc 210
        frequency-domain 211
        time-domain 211
    filter prototypes 213–239
        Bessel 233
        Butterworth 215
        Cauer *See* Elliptical
        Chebychev 222
        Chebychev 227
        Elliptical 235
        Thomson. *See* Bessel
    implementation 315–329
        example 321
    indirect design methods
        overview 214
    lowpass to other types 249
    mapping s to z 240–250
        BZT 243–250
        impulse invariance 249
        matched-z transform 249
        prewarping 243
    norms
        broadband 318
        L_1 318
        L_2 319
        L-infinity 318
    structures 252–258
        cascade 253–257
            canonic 254
            direct form 253
            direct form 253
            ordering and pairing poles and zeros 257
        parallel 257
impedance 84, 473
    input/ouput 125
impulse response 47
    and architects 49
    determine via crosscorrelation 392
    discrete-time 106
in-circuit emulator (ICE) 282
inductor 5, 470
infinite impulse response (IIR) filter. See IIR filters 148
inline assembly 286
interpolation 418–424
    68HC16 example 420, 505–516
    efficient 418
        polyphase filtering 418
    See also rational interpolation/decimation
interpolators, hardware 427
interrupt driven 313
interrupt latency 106, 313, 458
inverse discrete Fourier transform (IDFT) 377
inverse discrete-time Fourier transform (IDTFT) 172

J

Joan
- driving speed 29
- on Jack's usefulness xii
- tax software 165

Jones, Professor "Tubes" 7

K

Kaiser, Dr. James 181
Kirchhoff's Current Law 472
Kirchhoff's Voltage Law 472
Klatt, Dr. Dennis 456
Knuth, Donald 403, 431, 440
Kotel'nikov, a gentleman named 103
Kronecker sequence 108

L

Laplace transform 55
least-significant bit (LSB) 119
least-square error 211, 394
left-half plane 89
linear congruence. *See* pseudo-random numbers
linear phase 82
linear prediction 399–401
linear systems 58
linker 278
lobes 158
logarithmic scale 39

M

M68HC16 268
- overview xiv
- register model 304

M68HC16 instruction set summary 477
M68HC16Z1
- resources 11

MAC instruction 270, 271
magnitude 24
- plotting 39

magnitude frequency response 52
- monotonic 215

Maier, W.L. 434

make (utility) 299
marginal stability 188
marginally stable 89, 112
matched filter 389
MC56002 267
MC68302 267
MC68356 267
minimum attenuation 78
music 455
- flanging 455

N

negative frequency 90, 98
noise
- in-band 127
- out-of-band 127
- pink 439
- white 439

nonlinear systems 58
normalized frequency 106
North Inlet Estuary Meteorological Station 23
not-a-number 272
numeric conventions 288
numerical methods 457
Nyquist frequency 96
Nyquist rate 99

O

octave 36
Ohm's law 469
operating systems 458
Oppenheim and Schafer 35
optimization
- code size 296
- data size 298
- speed 291
  - addressing modes 295
  - branches and jumps 291
  - common subexpression elimination 293
  - conditionals 292
  - constant folding 292
  - constant propagation 292
  - index registers 293

Index   533

      inline functions 291
      look-up tables 295
      loop unwinding 293
      pointers 295
      register usage 293
      self-modifying code 294
      strength reduction 291
      variable hoisting 294
      zero-overhead looping 293
orthogonality 267, 275
overflow 271
overlays 297
overshoot 64

P

Parks-McClellan program 306, 422
    See also FIR filters
partial fraction expansion 249, 258
passband
    cutoff 77
    deviation 77
    edge frequency 77
    ripple 77
peaking filter 69
people, tiny (in equipment) 23
periodicity determined via autocorrelation 397
phase 29
    constant 82
    linear 82
    plotting 46
    relationship to time delay 36
    zero 81
phase response 52, 81
ping-pong buffers 298
pizza, faxing a 108
poles 85–91
polynomials, evaluating 439
    Horner's method 439
power 470
    plotting 39
program structure 457
programming 277–301
    interfacing C and assembly 283

    languages 277–287
        assembly 277
        BASIC 279
        C 279
            proprietary extensions 280
        mixing C and assembly 281
    macro preprocessor 277
pseudo-noise (PN) 401–408
    characteristics 403
    feedback shift register (FSR) 403
        68HC16 example 404
        C example 406
        maximal-length shift register 403
        software implementation 404
pseudo-random bit sequences (PRBS). *See* pseudo-noise (PN)
pseudo-random numbers 430–439
    additive algorithm 436
    linear congruence 431–434
        68HC16 example 432
        C example 433
    normal distribution 437
    R250 434
        68HC16 example 435
    randomizing 436
    spectral characteristics 439
    zero mean 438
pulse-coded modulation (PCM) 121

Q

quality (Q) factor 79
quantization 118–124, 133
    dynamic range 119
    errors
        nonlinearity 119
        offset error 119
        scale-factor 119
    full scale (FS) 119
    full-scale range (FSR) 119
    input
        differential 135
        single-ended 135
    linear 119
    noise 122

nonlinear 120
    A-Law 121
    μ-Law 121
  references 135
  thermometer code 133
quantizer 133
Queued Serial Module (QSM) 136

R

radian/degree conversion 36
radians 34
  conversion to degrees 36
RANDU 431
rational interpolation/decimation 424–427
RC filter 424
real-world frequency 106
reciprocal pairs 195
reconstruction filter. *See* antiimaging filter
resistor 470
resonator 69
ringing. *See* Gibb's phenomenon
rise time 64
RMAC instruction 271, 312
root-mean-square (RMS) 37
RS-232 136

S

sample-and-hold amplifier (SHA) 104, 132
sampling 95–107
  and convolution 453
sampling frequency 95
sampling period 95
sampling rate 95
  changing 413–427
    See also decimation, interpolation
  generating 105
  upper limit 102
saturate 271
scaling 58
second-order sections (SOS) 221, 245
self-modifying code 294, 373
serial interfaces 136
settling time 64

Shannon, Claude 103
Shannon's Sampling Theorem 103
shift instructions 404
signal averaging 409–410
  stimulus/response 409
signals
  analog 4, 19–46
    describing 21
    noise 21
    sources 20
  bandlimited 127
  bipolar 119
  complex 107
  digital 95
  discrete-time 94
  energy 451
  multidimensional 5
  power 451
  time-limited 334
  unipolar 119
signal-to-noise ratio (SNR) 123
signal-to-quantization-noise ratio (SQNR) 123
simulators 281
$\sin(x)/x$ 143
sine, computing 440–448
  68HC16 example 441
  polynomial approximation 440
  table-based approximation 444
  using IIR filters 447
spectrum analyzers 26
spectrum, reversing 100
speech processing 455
s-plane 54
stability 64
stack, hardware 267
standard deviation 438
state-variable representation 451
stationary 26, 335, 342
steady-state 50
  characteristics of filters 64
step response 49
step size 119, 120
stochastic processes 454

Index

stopband
    cut-off frequency 78
    edge frequency 78
    ripple 78
strain gauge 21
superposition 58
synchronous serial interface 136
synthesizing signals 429–448
system function
    continuous, $H(s)$ 52
    determining via crosscorrelation 392
    discrete-time, $H(z)$ 108
system identification 158, 394
    crosscorrelation 394
systems
    analog 46–60
    discrete-time 93

T

tapped shift register. *See* feedback shift register, under pseudo-noise (PN)
time delay 36
    determination using crosscorrelation 391
    See also phase delay 36
time domain response 47
time invariant 26
time-frequency analysis 377
time-invariance 59
timing loops, software 105
tool chain 299
track-and-hold amplifier (THA) 132
transcendental functions 440
transducers 4, 21
transfer function 49
transflibber, fractally inverted kryptonite 6
transient characteristics of filters 64
transient response 82
trigonometry (review) 461
two's complement 135

U

undersampling 104

unit-circle 112
unity gain 43
unsigned binary 135

V

variance 438
voice synthesizer xiii
voltage divider 83
voltage vs. power 40
    table 45

W

wagon wheel 100
wait states 314
wavelets 378
Whittaker 103
window function 176
windows
    Blackman 179
    cosine bell. *See* Hanning
    Hamming 179
    Hanning 179
    Kaiser 181
    Kaiser-Bessel. *See* Kaiser
    raised cosine. *See* Hanning
    rectangular 179
    von Hann. *See* Hanning

Z

zero-order hold (ZOH) 143
zero-overhead looping 295
zero-page 295
zeros 85–91
z-plane 109–112
z-transform (ZT) 113, 114

Keep Up-to-Date with
PH PTR Online!

We strive to stay on the cutting-edge of what's happening in professional computer science and engineering. Here's a bit of what you'll find when you stop by **www.phptr.com**:

@ **Special interest areas** offering our latest books, book series, software, features of the month, related links and other useful information to help you get the job done.

$ **Deals, deals, deals!** Come to our promotions section for the latest bargains offered to you exclusively from our retailers.

Need to find a bookstore? Chances are, there's a bookseller near you that carries a broad selection of PTR titles. Locate a Magnet bookstore near you at www.phptr.com.

! **What's New at PH PTR?** We don't just publish books for the professional community, we're a part of it. Check out our convention schedule, join an author chat, get the latest reviews and press releases on topics of interest to you.

✉ **Subscribe Today!** **Join PH PTR's monthly email newsletter!**

Want to be kept up-to-date on your area of interest? Choose a targeted category on our website, and we'll keep you informed of the latest PH PTR products, author events, reviews and conferences in your interest area.

Visit our mailroom to subscribe today! **http://www.phptr.com/mail_lists**

LICENSE AGREEMENT AND LIMITED WARRANTY

READ THE FOLLOWING TERMS AND CONDITIONS CAREFULLY BEFORE OPENING THIS CD PACKAGE. THIS LEGAL DOCUMENT IS AN AGREEMENT BETWEEN YOU AND PRENTICE-HALL, INC. (THE "COMPANY"). BY OPENING THIS SEALED CD PACKAGE, YOU ARE AGREEING TO BE BOUND BY THESE TERMS AND CONDITIONS. IF YOU DO NOT AGREE WITH THESE TERMS AND CONDITIONS, DO NOT OPEN THE CD PACKAGE. PROMPTLY RETURN THE UNOPENED CD PACKAGE AND ALL ACCOMPANYING ITEMS TO THE PLACE YOU OBTAINED THEM FOR A FULL REFUND OF ANY SUMS YOU HAVE PAID.

1. **GRANT OF LICENSE:** In consideration of your purchase of this book, and your agreement to abide by the terms and conditions of this Agreement, the Company grants to you a nonexclusive right to use and display the copy of the enclosed software program (hereinafter the "SOFTWARE") on a single computer (i.e., with a single CPU) at a single location so long as you comply with the terms of this Agreement. The Company reserves all rights not expressly granted to you under this Agreement.

2. **OWNERSHIP OF SOFTWARE:** You own only the magnetic or physical media (the enclosed CD) on which the SOFTWARE is recorded or fixed, but the Company and the software developers retain all the rights, title, and ownership to the SOFTWARE recorded on the original CD copy(ies) and all subsequent copies of the SOFTWARE, regardless of the form or media on which the original or other copies may exist. This license is not a sale of the original SOFTWARE or any copy to you.

3. **COPY RESTRICTIONS:** This SOFTWARE and the accompanying printed materials and user manual (the "Documentation") are the subject of copyright. The individual programs on the CD are copyrighted by the authors of each program. You may not copy the Documentation or the SOFTWARE, except that you may make a single copy of the SOFTWARE for backup or archival purposes only. You may be held legally responsible for any copying or copyright infringement which is caused or encouraged by your failure to abide by the terms of this restriction.

4. **USE RESTRICTIONS:** You may not network the SOFTWARE or otherwise use it on more than one computer or computer terminal at the same time. You may physically transfer the SOFTWARE from one computer to another provided that the SOFTWARE is used on only one computer at a time. You may not distribute copies of the SOFTWARE or Documentation to others. You may not reverse engineer, disassemble, decompile, modify, adapt, translate, or create derivative works based on the SOFTWARE or the Documentation without the prior written consent of the Company.

5. **TRANSFER RESTRICTIONS:** The enclosed SOFTWARE is licensed only to you and may not be transferred to any one else without the prior written consent of the Company. Any unauthorized transfer of the SOFTWARE shall result in the immediate termination of this Agreement.

6. **TERMINATION:** This license is effective until terminated. This license will terminate automatically without notice from the Company and become null and void if you fail to comply with any provisions or limitations of this license. Upon termination, you shall destroy the Documentation and all copies of the SOFTWARE. All provisions of this Agreement as to warranties, limitation of liability, remedies or damages, and our ownership rights shall survive termination.

7. **MISCELLANEOUS:** This Agreement shall be construed in accordance with the laws of the United States of America and the State of New York and shall benefit the Company, its affiliates, and assignees.

8. **LIMITED WARRANTY AND DISCLAIMER OF WARRANTY:** The Company warrants that the SOFTWARE, when properly used in accordance with the Documentation, will operate in substantial conformity with the description of the SOFTWARE set forth in the Documentation. The Company does not warrant that the SOFTWARE will meet your requirements or that the operation of the SOFTWARE will be uninterrupted or error-free. The Company warrants that the media on which the SOFTWARE is delivered shall be free from defects in materials and workmanship under

normal use for a period of thirty (30) days from the date of your purchase. Your only remedy and the Company's only obligation under these limited warranties is, at the Company's option, return of the warranted item for a refund of any amounts paid by you or replacement of the item. Any replacement of SOFTWARE or media under the warranties shall not extend the original warranty period. The limited warranty set forth above shall not apply to any SOFTWARE which the Company determines in good faith has been subject to misuse, neglect, improper installation, repair, alteration, or damage by you. EXCEPT FOR THE EXPRESSED WARRANTIES SET FORTH ABOVE, THE COMPANY DISCLAIMS ALL WARRANTIES, EXPRESS OR IMPLIED, INCLUDING WITHOUT LIMITATION, THE IMPLIED WARRANTIES OF MERCHANTABILITY AND FITNESS FOR A PARTICULAR PURPOSE. EXCEPT FOR THE EXPRESS WARRANTY SET FORTH ABOVE, THE COMPANY DOES NOT WARRANT, GUARANTEE, OR MAKE ANY REPRESENTATION REGARDING THE USE OR THE RESULTS OF THE USE OF THE SOFTWARE IN TERMS OF ITS CORRECTNESS, ACCURACY, RELIABILITY, CURRENTNESS, OR OTHERWISE.

IN NO EVENT, SHALL THE COMPANY OR ITS EMPLOYEES, AGENTS, SUPPLIERS, OR CONTRACTORS BE LIABLE FOR ANY INCIDENTAL, INDIRECT, SPECIAL, OR CONSEQUENTIAL DAMAGES ARISING OUT OF OR IN CONNECTION WITH THE LICENSE GRANTED UNDER THIS AGREEMENT, OR FOR LOSS OF USE, LOSS OF DATA, LOSS OF INCOME OR PROFIT, OR OTHER LOSSES, SUSTAINED AS A RESULT OF INJURY TO ANY PERSON, OR LOSS OF OR DAMAGE TO PROPERTY, OR CLAIMS OF THIRD PARTIES, EVEN IF THE COMPANY OR AN AUTHORIZED REPRESENTATIVE OF THE COMPANY HAS BEEN ADVISED OF THE POSSIBILITY OF SUCH DAMAGES. IN NO EVENT SHALL LIABILITY OF THE COMPANY FOR DAMAGES WITH RESPECT TO THE SOFTWARE EXCEED THE AMOUNTS ACTUALLY PAID BY YOU, IF ANY, FOR THE SOFTWARE.

SOME JURISDICTIONS DO NOT ALLOW THE LIMITATION OF IMPLIED WARRANTIES OR LIABILITY FOR INCIDENTAL, INDIRECT, SPECIAL, OR CONSEQUENTIAL DAMAGES, SO THE ABOVE LIMITATIONS MAY NOT ALWAYS APPLY. THE WARRANTIES IN THIS AGREEMENT GIVE YOU SPECIFIC LEGAL RIGHTS AND YOU MAY ALSO HAVE OTHER RIGHTS WHICH VARY IN ACCORDANCE WITH LOCAL LAW.

ACKNOWLEDGMENT

YOU ACKNOWLEDGE THAT YOU HAVE READ THIS AGREEMENT, UNDERSTAND IT, AND AGREE TO BE BOUND BY ITS TERMS AND CONDITIONS. YOU ALSO AGREE THAT THIS AGREEMENT IS THE COMPLETE AND EXCLUSIVE STATEMENT OF THE AGREEMENT BETWEEN YOU AND THE COMPANY AND SUPERSEDES ALL PROPOSALS OR PRIOR AGREEMENTS, ORAL, OR WRITTEN, AND ANY OTHER COMMUNICATIONS BETWEEN YOU AND THE COMPANY OR ANY REPRESENTATIVE OF THE COMPANY RELATING TO THE SUBJECT MATTER OF THIS AGREEMENT.

Should you have any questions concerning this Agreement or if you wish to contact the Company for any reason, please contact in writing at the address below.

Robin Short
Prentice Hall PTR
One Lake Street
Upper Saddle River, New Jersey 07458

About the CD-ROM

The CD-ROM contains assembly language source, data, reference manuals, and programs to accompany the book, *Digital Signal Processing and the Microcontroller* by Dale Grover and John R. Deller.

The CD-ROM is organized into 6 separate subdirectories: Assembly, Manual, Mfiles, Jojo, Mathview, and Acrobat. For a full description of the contents of these subdirectories, open the readme.txt file on the CD after installing it.

CD Installation:

1. Insert the CD-ROM into your CD-ROM drive.

2. Double click on "My Computer."

3. Double-click on the CD-ROM icon.

System/Platform Requirements:

Software on this CD-ROM requires Windows 3.1 or higher. Adobe Acrobat® Reader, Version 3.01 requires Windows® 95 or Windows® NT.

Prentice Hall does not offer technical support for this software. However, if there is a problem with the media, you may obtain a free replacement copy of e-mailing us with your problem at: ptr_techsupport@phptr.com.